T0212073

Mathematical Methods in Robust Control of Linear Stochastic Systems

Vasile Dragan • Toader Morozan
Adrian-Mihail Stoica

Mathematical Methods
in Robust Control of Linear
Stochastic Systems

Second Edition

 Springer

Vasile Dragan
Institute of Mathematics
    of the Romanian Academy
Bucharest, Romania

Toader Morozan
Institute of Mathematics
    of the Romanian Academy
Bucharest, Romania

Adrian-Mihail Stoica
University Politechnica of Bucharest
Bucharest, Romania

ISBN 978-1-4939-3870-4   ISBN 978-1-4614-8663-3 (eBook)
DOI 10.1007/978-1-4614-8663-3
Springer New York Heidelberg Dordrecht London

Mathematics Subject Classification (2010): 93E03, 93E11, 93E15, 93E20, 93E25, 60J22

Printed on acid-free paper

Springer is part of Springer Science+Business Media (www.springer.com)

To our wives, Viorica, Elena and Dana
for their love, patience and support.

Preface to the First Edition

This monograph presents a thorough description of the mathematical theory of robust linear stochastic control systems. The interest in this topic is motivated by the variety of random phenomena arising in physical, engineering, biological, and social processes. The study of stochastic systems has a long history, but two distinct classes of such systems drew much attention in the control literature, namely stochastic systems subjected to white noise perturbations and systems with Markovian jumping. At the same time, the remarkable progress in recent decades in the control theory of deterministic dynamic systems strongly influenced the research effort in the stochastic area. Thus, the modern treatments of stochastic systems include optimal control, robust stabilization, and H^2-and H^∞-type results for both stochastic systems corrupted with white noise and systems with jump Markov perturbations.

In this context, there are two main objectives of the present book. The first one is to develop a mathematical theory linear time-varying stochastic systems including both white noise jump Markov perturbations. From the perspective of this generalized theory the stochastic systems subjected only to white noise perturbations or to jump Markov perturbations can be regarded as particular cases. The second objective is to develop analysis and design methods for advanced control problems of linear stochastic systems with white noise and Markovianjumping as linear-quadratic control, robust stabilization, and disturbance attenuation problems. Taking into account the maj or role played by the Riccati equations in these problems, the book presents this type of equation in a general framework. Particular attention is paid to the numerical aspects arising in the control problems of stochastic systems; new numerical algorithms to solve coupled matrix algebraic Riccati equations are also proposed and illustrated by numerical examples.

The book contains seven chapters. Chapter 1 includes some prerequisites concerning measure and probability theory that will be used in subsequent developments in the book. In the second part of this chapter, detailed proofs of some new results, such as the Itô-type formula in a general case covering the classes of stochastic systems with white noise perturbations and Markovian jumping, are given. The Itô-type formula plays a cmcial role in the proofs of the main results of the book.

Chapter 2 is mainly devoted to the exponential stability of linear stochastic systems. It is proved that the exponential stability in the mean square of the considered class of stochastic systems is equivalent with the exponential stability of an appropriate class of deterministic systems over a finite-dimensional Hilbert space. Necessary and sufficient conditions for exponential stability for such deterministic systems are derived in terms of some Lyapunov-type equations. Then necessary and sufficient conditions in terms of Lyapunov functions for mean square exponential stability are obtained. These results represent a generalization of the known conditions conceming the exponential stability of stochastic systems subjected to white noise and Markovian jumping, respectively.

Some stmctural properties such as controllability, stabilizability, observability, and detectability linear stochastic systems subjected to both white noise andjump Markov perturbations are considered in Chapter 3. These properties play a key role in the following chapters of the book.

In Chapter 4 differential and algebraic generalized Riccati-type equations arising in the control problems of stochastic systems are introduced. Our attention tums to the maximal, minimal, and stabilizing solutions of these equations for which necessary and sufficient existence conditions are derived. The final part of this chapter provides an iterative procedure for computing the maximal solution of such equations.

In the fifth chapter of the book, the linear-quadratic problem on the infinite horizon for stochastic systems with both white noise and jump Markov perturbations is considered. The problem refers to a general situation: The considered systems are subjected to both state and control multiplicative white noise and the optimization is performed under the class of nonanticipative stochastic controls. The optimal control is expressed in terms of the stabilizing solution of coupled generalized Riccati equations. As an application of the results deduced in this chapter, we consider the optimal tracking problem.

Chapter 6 contains corresponding versions of some known results from the deterministic case, such as the Bounded Real Lemma, the Small Gain Theorem, and the stability radius, for the considered class of stochastic systems. Such results have been obtained separately in the stochastic framework for systems subjected to white noise and Markov perturbations, respectively. In our book, these results appear as particular situations of a more general class of stochastic systems including both types of perturbations.

In Chapter 7 the γ-attenuation problem of stochastic systems with both white noise and Markovian jumping is considered. Necessary and sufficient conditions for the existence of a stabilizing γ-attenuating controller are obtained in terms of a system of coupled game-theoretic Riccati equations and inequalities. These results allow one to solve various robust stabilization problems of stochastic systems subjected to white noise and Markov perturbations, as illustrated by numerical examples.

The monograph is based entirely on original recent results of the authors; some of these results have been recently published in control joumals and conferences proceedings. There are also some other results that appear for the first time in this book.

This book is not intended to be a textbook or a guide for control designers. We had in mind a rather larger audience, including theoretical and applied mathematicians and research engineers, as well as graduate students in all these fields, and, for some parts of the book, even undergraduate students. Since our intention was to provide a self-contained text, only the first chapter reviews known results and prerequisites used in the rest of the book.

The authors are indebted to Professors Gerhard Freiling and Isaac Yaesh for fruitful discussions on some of the numerical methods and applications presented in the book.

Finally, the authors wish to thank the Springer publishing staff and the reviewer for carefully checking the manuscript and for valuable suggestions.

Preface

This new edition has nine chapters and it includes some new developments and results in robust control of linear stochastic systems.

In Chapter 1 properties of homogeneous Markov processes with countable infinite number of states are given together with Itô–type formula for stochastic systems with white noise perturbations and infinite Markov jumping. Lebesgue's Theorem and Fatou's Lemma for discrete measures are also presented.

Chapter 2 is new. The properties of the Minkovski norm are presented. The main purpose is to provide a characterization for the exponential stability of the linear differential equations with positive evolution on ordered Banach spaces. This characterization is given in terms of the existence of some global defined and bounded solutions of some suitable forward or backward affine differential equations and in terms of some forward or backward affine differential inequalities.

The problem of robustness of exponential stability with respect to some additive perturbations modeled by positive operator valued functions is analized in the case when the involved operators are periodic. The last part of the chapter is devoted to the investigation of the properties of linear evolution operators associated to Lyapunov type differential equations on the Banach spaces S_n^d, S_n^∞ and $\ell^1(\mathbf{Z}_+, S_n)$. Criteria for exponential stability of the Lyapunov type differential equations on S_n^∞ and S_n^d (the latest being also presented in the second chapter of the first edition) are finally derived as direct consequences of the criteria obtained in the general case.

The novelty of Chapter 3 is the characterization of exponential stability in mean square for stochastic linear differential equations perturbed both by multiplicative white noise and by an infinite Markov process. This is based on the representation theorem of the anticausal linear evolution operator defined by a Lyapunov type differential equation on the space S_n^∞ and on the criteria of exponential stability of the corresponding linear differential equation presented in Chapter 2.

Chapter 4 is the third chapter of the first edition.

Most of the fifth chapter is new. Bounded global solutions for a wide class of nonlinear differential equations (called GRDE-Generalized Riccati Differential Equations) on an ordered Banach space of symmetric matrices are analyzed. They include as particular cases the systems of Riccati–type equations arising in the stochastic linear control (SGRDE Stochastic Generalized Riccati Differential Equations). Comparison theorems and necessary and sufficient conditions for bounded maximal, minimal and stabilizing solutions of the GRDE and then by consequence, the corresponding results for the SGRDE which are also included in Chapter 4 of the first edition, are provided.

For the sake of clarity, the fifth chapter of the first edition has been split in two chapters, namely Chapter 6 and Chapter 7 of this edition. A new section treating a Kalman filtering problem for stochastic systems with state–dependent noise and Markovian jumping has been included in Chapter 7.

Chapters 8 and 9 are just the sixth and the seventh chapters of the first edition. In the final part of Chapter 9, a new section presenting a mixed H_2/H_∞ filtering problem has been introduced.

The authors wish to thank to Professors G. Freiling, T. Damm, I. Yaesh, O.L.V. Costa, M.D. Fragoso and V. Ungureanu for fruitful discussions on some general properties of differential equations with positive evolution on ordered Banach spaces, numerical methods and applications presented in the book.

We also should like express our gratitude to Mrs. Viorica Dragan and Mr. Catalin Dragan for their hard work in typing the manuscript.

Finally, the authors are indebted to the Springer publishing staff for the valuable support and suggestions.

Contents

Chapter 1
Preliminaries to Probability Theory and Stochastic Differential Equations

This first chapter collects for the readers convenience some definitions and fundamental results concerning the measure theory and the stochastic processes theory which are needed in the following developments of the book. Classical results concerning the measure theory, integration, stochastic processes, and stochastic integrals are presented without proofs. Appropriate references are given; thus for the measure theory we mention [33, 55, 71, 75, 118, 138]; for the probability theory we refer to [32, 71, 119, 130, 138], and for the theory of stochastic processes and stochastic differential equations we cite [6, 18, 32, 71, 72, 85, 102, 105, 120, 121, 141, 152, 153]. However several results which can be found in some references less accessible are proved.

In Sect. 1.10 we prove a general version of Itô-type formula which plays a key role in the developments of Chaps. 3–5. The results concerning mean square exponential stability in Chap. 3 may be derived using an Itô-type formula which refers to stochastic processes which are solutions to a class of stochastic differential equations. This version of the Itô-type formula can be found in Theorem 1.11.4. Theorem 1.10.1 used in the proof of Itô-type formula and also in Chap. 8 in order to estimate the stability radius, appears for the first time in this book.

1.1 Elements of Measure Theory

1.1.1 Measurable Spaces

Definition 1.1.1. A *measurable space* is a pair $(\Omega, \mathcal{F})$ where Ω is a set and $\mathcal{F}$ is a *σ-algebra* of subsets of Ω, that is, $\mathcal{F}$ is a family of subsets $A \subset \Omega$ with the properties

  (i) $\Omega \in \mathcal{F}$;
 (ii) If $A \in \mathcal{F}$, then $\Omega - A \in \mathcal{F}$;
(iii) If $A_n \in \mathcal{F}, n \geq 1$, then $\cup_{n=1}^{\infty} A_n \in \mathcal{F}$.

V. Dragan et al., *Mathematical Methods in Robust Control of Linear Stochastic Systems*, DOI 10.1007/978-1-4614-8663-3_1, © Springer Science+Business Media New York 2013

If $\mathcal{F}_1$ and $\mathcal{F}_2$ are two σ-algebras of subsets of Ω, by $\mathcal{F}_1 \vee \mathcal{F}_2$ we denote the smallest σ-algebra of subsets of Ω which contains the σ-algebras $\mathcal{F}_1$ and $\mathcal{F}_2$.

By $\mathcal{B}(\mathbf{R}^n)$ we denote the σ-algebra of Borel subsets of $\mathbf{R}^n$, that is, the smallest σ-algebra containing all open subsets of $\mathbf{R}^n$.

For a family $\mathcal{C}$ of subsets of Ω, $\sigma(\mathcal{C})$ will denote the smallest σ-algebra of subsets of Ω containing $\mathcal{C}$; $\sigma(\mathcal{C})$ will be termed the σ-algebra generated by $\mathcal{C}$.

If $(\Omega_1, \mathcal{G}_1)$ and $(\Omega_2, \mathcal{G}_2)$ are two measurable spaces, by $\mathcal{G}_1 \otimes \mathcal{G}_2$ we denote the smallest σ-algebra of subsets of $\Omega_1 \times \Omega_2$ which contains all sets $A \times B, A \in \mathcal{G}_1$, $B \in \mathcal{G}_2$.

Definition 1.1.2. A collection $\mathcal{C}$ of subsets of Ω is called to be a *π-system* if

(i) $\phi \in \mathcal{C}$, and
(ii) if $A, B \in \mathcal{C}$, then $A \cap B \in \mathcal{C}$.

The next result proved in [154] is frequently used in the probability theory:

Theorem 1.1.1. *If $\mathcal{C}$ is a π-system and $\mathcal{G}$ is the smallest family of subsets of Ω such that*

  (i) $\mathcal{C} \subset \mathcal{G}$;
  (ii) If $A \in \mathcal{G}$, then $\Omega - A \in \mathcal{G}$;
  (iii) $A_n \in \mathcal{G}, n \geq 1$ and $A_i \cap A_j = \phi$ for $i \neq j$ implies $\cup_{n=1}^{\infty} A_i \in \mathcal{G}$, then
$$\sigma(\mathcal{C}) = \mathcal{G}.$$

Proof. Since $\sigma(\mathcal{C})$ verifies (i), (ii), and (iii) in the statement, it follows that $\mathcal{G} \subset \sigma(\mathcal{C})$.

To prove the opposite inclusion we show first that $\mathcal{G}$ is a π-system.

Let $A \in \mathcal{G}$ and define $\mathcal{G}(A) = \{B; B \in \mathcal{G} \text{ and } A \cap B \in \mathcal{G}\}$.

Since $A - B = \Omega - [(A \cap B) \cup (\Omega - A)]$, it is easy to check that $\mathcal{G}(A)$ verifies the conditions (ii) and (iii), and if $A \in \mathcal{C}$, then (i) is also satisfied. Hence for $A \in \mathcal{C}$ we have $\mathcal{G}(A) = \mathcal{G}$; consequently, if $A \in \mathcal{C}$ and $B \in \mathcal{G}$, then $A \cap B \in \mathcal{G}$. But this implies $\mathcal{G}(B) \supset \mathcal{C}$ and therefore $\mathcal{G}(B) = \mathcal{G}$ for any $B \in \mathcal{G}$. Hence $\mathcal{G}$ is a π-system and now, since $\mathcal{G}$ verifies (ii) and (iii) it is easy to verify that $\mathcal{G}$ is a σ-algebra and the proof is complete. $\square$

1.1.2 Measures and Measurable Functions

Definition 1.1.3. (a) Given a measurable space $(\Omega, \mathcal{F})$, a function $\mu : \mathcal{F} \to [0, \infty]$ is called a *measure* if:

  (i) $\mu(\phi) = 0$
  (ii) if $A_n \in \mathcal{F}, n \geq 1$ and $A_i \cap A_j = \phi$ for $i \neq j$, then

$$\mu(\cup_{n=1}^{\infty} A_n) = \sum_{n=1}^{\infty} \mu(A_n).$$

(b) A triplet $(\Omega, \mathcal{F}, \mu)$ is said to be a *space with measure*.

(c) If $\mu(\Omega) = 1$ we say that μ is a *probability* on $\mathcal{F}$ and in this case the triplet $(\Omega, \mathcal{F}, \mu)$ is termed a *probability space*.

(d) A measure μ is called to be σ-*finite* if there exists a sequence $A_n, n \geq 1, A_n \in \mathcal{F}$ with $A_i \cap A_j = \phi$ for $i \neq j$ and $\Omega = \cup_{n=1}^{\infty} A_n$ and $\mu(A_n) < \infty$ for every n.

Definition 1.1.4. Given a measurable space $(\Omega, \mathcal{F})$, a function $f : \Omega \longmapsto \mathbf{R}$ is said to be a *measurable function* if for every $A \in \mathcal{B}(\mathbf{R})$ we have $f^{-1}(A) \in \mathcal{F}$ where $f^{-1}(A) = \{\omega \in \Omega; f(\omega) \in A\}$.

It is easy to prove that $f : \Omega \longmapsto \mathbf{R}$ is measurable if and only if $f^{-1}((-\infty, \alpha)) \in \mathcal{F}$ for every $\alpha \in \mathbf{R}$.

Remark 1.1.1. It is not difficult to verify that if $(\Omega_1, \mathcal{F}_1)$ and $(\Omega_2, \mathcal{F}_2)$ are two measurable spaces and if $f : \Omega_1 \times \Omega_2 \to \mathbf{R}$ is $\mathcal{F}_1 \otimes \mathcal{F}_2$ measurable, then for each $\omega_2 \in \Omega_2$ the function $\omega_1 \longmapsto f(\omega_1, \omega_2)$ is $\mathcal{F}_1$ measurable and for each $\omega_1 \in \Omega_1$ the function $\omega_2 \longmapsto f(\omega_1, \omega_2)$ is $\mathcal{F}_2$ measurable.

Definition 1.1.5. A measurable function $f : \Omega \longmapsto \mathbf{R}$ is said to be a *simple measurable function* if it takes only a finite number of values.

We shall write a.a. and a.e. for *almost all* and *almost everywhere*, respectively; $f = g$ a.e. means $\mu(f \neq g) = 0$.

Definition 1.1.6. Let $(\Omega, \mathcal{F}, \mu)$ be a space with measure, $f_n : \Omega \to \mathbf{R}, n \geq 1$ and $f : \Omega \to \mathbf{R}$ be measurable functions.

(i) We say that f_n *converges to* f *for a.a.* $\omega \in \Omega$ or equivalently $\lim_{n \to \infty} f_n = f$ *a.e.* $(f_n \overset{a.e.}{\to} f)$ if

$$\mu\{\omega; \lim_{n \to \infty} f_n(\omega) \neq f(\omega)\} = 0;$$

(ii) We say that the sequence f_n *converges in measure to* f $(f_n \overset{\mu}{\to} f)$ if for every $\delta > 0$, we have $\lim_{n \to \infty} \mu\{\omega; |f_n(\omega) - f(\omega)| > \delta\} = 0$.

Theorem 1.1.2. *Assume that* $\lim_{n \to \infty} f_n = f$ *a.e. and that* $\mu(\Omega) < \infty$. *Then* $f_n \overset{\mu}{\to} f$.

Theorem 1.1.3 (Riesz's Theorem). *If* $f_n \overset{\mu}{\to} f$, *then there exists a subsequence* f_{n_k} *of the sequence* f_n *such that* $\lim_{k \to \infty} f_{n_k} = f$ *a.e.*

Corollary 1.1.4. *Let* $(\Omega, \mathcal{F}, \mu)$ *be a space with measure such that* $\mu(\Omega) < \infty$. *Then the following assertions are equivalent:*

(i) $f_n \overset{\mu}{\to} f$;

(ii) *Any subsequence of* f_n *contains a subsequence converging a.e. to* f.

As usual, in the measure theory two measurable functions f and g are identified if $f = g$ a.e. Moreover if $f : \Omega \to \bar{\mathbf{R}} = [-\infty, \infty]$ is measurable, that is $f^{-1}([-\infty, \alpha)) \in \mathcal{F}$

for every $\alpha \in \mathbf{R}$ and if $\mu(|f| = \infty) = 0$, then f will be identified with a function $g : \Omega \to \mathbf{R}$ defined as follows:

$$g(\omega) = \begin{cases} f(\omega) \text{ if } |f(\omega)| < \infty, \text{ and} \\ 0 \text{ if } |f(\omega)| = \infty. \end{cases}$$

Theorem 1.1.5. *If $(\Omega_1, \mathcal{F}_1, \mu_1)$ and $(\Omega_2, \mathcal{F}_2, \mu_2)$ are two spaces with σ-finite measures, then there exists a unique measure $\mu : \mathcal{F}_1 \otimes \mathcal{F}_2 \to [0, \infty]$ such that $\mu(A \times B) = \mu_1(A) \mu_2(B)$ for all $A \in \mathcal{F}_1$ and $B \in \mathcal{F}_2$. This measure μ will be denoted by $\mu_1 \times \mu_2$.*

1.1.3 Integration

Theorem 1.1.6. *Let $f \geq 0$ be a measurable function. Let us define*

$$f_n(\omega) = \sum_{i=1}^{2^n n + 1} \frac{i-1}{2^n} \chi_{A_{i,n}}(\omega),$$

where

$$A_{i,n} = \left\{ \omega; \frac{i-1}{2^n} \leq f(\omega) < \frac{i}{2^n} \right\}, i = 1, 2 \ldots, 2^n n,$$

$$A_{2^n n + 1, n} = \{ \omega; f(\omega) \geq n \},$$

and $\chi_A(\omega)$ is the indicator function *of the set A; that is $\chi_A(\omega) = 1$ if $\omega \in A$ and $\chi_A(\omega) = 0$ if $\omega \in \Omega - A$. Then we have:*

(i) $0 \leq f_n \leq f_{n+1}$ and $\lim_{n \to \infty} f_n(\omega) = f(\omega), \omega \in \Omega$;

(ii) $0 \leq a_n \leq a_{n+1}$ where $a_n = \sum_{i=1}^{2^n n + 1} \frac{i-1}{2^n} \mu(A_{i,n})$ (with the convention that $0 \cdot \infty = 0$).

$\square$

Definition 1.1.7. (i) Let $f \geq 0$ be a measurable function on a space with measure $(\Omega, \mathcal{F}, \mu)$ and $f_n, a_n, n \geq 1$ be the sequences defined in statement of Theorem 1.1.6. By definition $a_n = \int_\Omega f_n d\mu$ and $\int_\Omega f d\mu = \lim_{n \to \infty} a_n$;

(ii) A measurable function $f : \Omega \to \mathbf{R}$ is called an *integrable function* if $\int_\Omega |f| d\mu < \infty$ and in this case,

$$\int_\Omega f d\mu = \int_\Omega f^+ d\mu - \int_\Omega f^- d\mu$$

where

$$f^+ = \frac{1}{2}(|f|+f); \; f^- = \frac{1}{2}(|f|-f);$$

(iii) We say that the integral of a measurable function f exists if at least one of the integrals $\int_\Omega f^+ d\mu$ or $\int_\Omega f^- d\mu$ is finite; if $\int_\Omega f^+ d\mu = \infty$ and $\int_\Omega f^- d\mu < \infty$, then by definition, $\int_\Omega f d\mu = \infty$, and if $\int_\Omega f^+ d\mu < \infty$ and $\int_\Omega f^- d\mu = \infty$, by definition $\int_\Omega f d\mu = -\infty$.

Remark 1.1.2. It can be proved that the definition of the integral $\int_\Omega f d\mu$ in (a) is not dependent upon the choice of the particular monotonic increasing sequence of simple measurable functions f_n converging to f. If f is a simple measurable function with values $c_1, c_2, \dots, c_n$, then by definition

$$\int_\Omega f d\mu = \sum_{i=1}^n c_i \mu \left(\{ \omega; f(\omega) = c_i \} \right).$$

It is known that:

(i) $|\int_\Omega f d\mu| \le \int_\Omega |f| d\mu$;
(ii) If $f = g$ a.e., then $\int_\Omega f d\mu = \int_\Omega g d\mu$;
(iii) If $A \in \mathcal{F}$, by definition $\int_A f d\mu = \int_\Omega \chi_A f d\mu$.

By $L^p(\Omega), p \ge 1$ we denote the space of all measurable functions $f : \Omega \to \mathbf{R}$ with $\int_\Omega |f|^p d\mu < \infty$.

Let us define

$$\|f\|_p = \left(\int_\Omega |f|^p d\mu \right)^{\frac{1}{p}} iff \in L^p.$$

Regarding the integrable functions we recall the following useful results.

Theorem 1.1.7 (Holder's Inequality). *If $f \in L^p(\Omega), p > 1$ and $g \in L^q(\Omega)$ with $\frac{1}{p} + \frac{1}{q} = 1$, then $fg \in L^1(\Omega)$ and*

$$\|fg\|_1 \le \|f\|_p \|g\|_q.$$

Taking, in the above theorem, $p = \frac{s}{r}, f = |h|^r, g = 1$, one obtains the following result.

Corollary 1.1.8. *If $\mu(\Omega) < \infty$ and $1 \le r < s$, then $h \in L^s(\Omega)$ implies $h \in L^r(\Omega)$ and if $\mu(\Omega) = 1$, we have $\|h\|_r \le \|h\|_s$.* □

Definition 1.1.8. Let $f_n, f \in L^p$. We say that $f_n \to f$ in L^p or $f_n \xrightarrow{L^p} f$ if

$$\lim_{n\to\infty} \int_\Omega |f_n - f|^p d\mu = 0.$$

Theorem 1.1.9. *If $f_n \xrightarrow{L^p} f$, then $f_n \xrightarrow{\mu} f$.*

1.2 Convergence Theorems for Integrals

Let $(\Omega, \mathcal{F}, \mu)$ be a space with measure. The following results are well known in the measure theory.

Theorem 1.2.1 (Fatou's Lemma). *Let $f_n \geq 0, n \geq 1$ be a sequence of measurable functions. Then*

$$\int_\Omega (\underline{\lim} f_n) d\mu \leq \underline{\lim} \int_\Omega f_n d\mu.$$

Theorem 1.2.2 (Lebesgue's Theorem). *Let f_n, f be measurable functions and $|f_n| \leq g, n \geq 1$, a.e. where g is an integrable function. If $\lim_{n\to\infty} f_n = f$ a.e., then $f_n \xrightarrow{L^1} f$, and therefore $\lim_{n\to\infty} \int_\Omega f_n d\mu = \int_\Omega f d\mu$.* $\square$

Theorem 1.2.3. *Let f_n, f be measurable functions. If $|f_n| \leq g, n \geq 1$ for some integrable function g and $f_n \xrightarrow{\mu} f$, then $f_n \xrightarrow{L^1} f$.*

Theorem 1.2.4 ([32, 71, 133]). *Let f_n, f be integrable functions. Suppose that $\mu(\Omega) < \infty$ and there exists $\alpha > 1$ such that*

$$\sup_n \int_\Omega |f_n|^\alpha d\mu < \infty.$$

If $f_n \xrightarrow{\mu} f$, then $f_n \xrightarrow{L^1} f$ and therefore $\lim_{n\to\infty} \int_\Omega f_n d\mu = \int_\Omega f d\mu$.

Theorem 1.2.5 ([55, 118]). *If $f : [a,b] \to \mathbf{R}$ is an integrable function, then*

$$\lim_{h\to 0+} \frac{1}{h} \int_{\max\{t-h, a\}}^t f(s) ds = f(t) \quad a.e., t \in [a,b].$$

$\square$

Definition 1.2.1. Let μ_1 and μ_2 be two measures on the measurable space $(\Omega, \mathcal{F})$; we say that μ_1 is *absolute continuous with respect to* μ_2 (and we write $\mu_1 \ll \mu_2$) if $\mu_2(A) = 0$ implies $\mu_1(A) = 0$.

Theorem 1.2.6 (Radon–Nicodym Theorem). *If $\lambda \ll \mu, \lambda(\Omega) < \infty, \mu(\Omega) < \infty$, then there exists a unique (mod μ) integrable function f such that $\lambda(A) = \int_A f d\mu$ for all $A \in \mathcal{F}$.* $\square$

Theorem 1.2.7 (Fubini's Theorem). *Let $(\Omega_1, \mathcal{F}_1, \mu_1), (\Omega_2, \mathcal{F}_2, \mu_2)$ be two spaces with σ-finite measures μ_1 and μ_2, respectively. Then we have:*

(a) *If $f : \Omega_1 \times \Omega_2 \to \mathbf{R}_+$ is a measurable function (with respect to $\mathcal{F}_1 \otimes \mathcal{F}_2$), then the function $\omega_2 \longmapsto \int_{\Omega_1} f(\omega_1, \omega_2) d\mu_1$ is $\mathcal{F}_2$ measurable, the function $\omega_1 \longmapsto \int_{\Omega_2} f(\omega_1, \omega_2) d\mu_2$ is $\mathcal{F}_1$ measurable and*

$$\int_{\Omega_1 \times \Omega_2} f d(\mu_1 \times \mu_2) = \int_{\Omega_1} \left(\int_{\Omega_2} f(\omega_1, \omega_2) d\mu_2 \right) d\mu_1$$

$$= \int_{\Omega_2} \left(\int_{\Omega_1} f(\omega_1, \omega_2) d\mu_1 \right) d\mu_2;$$

(b) A measurable function $f : \Omega_1 \times \Omega_2 \to \mathbf{R}$ is integrable (on the space $(\Omega_1 \times \Omega_2, \mathcal{F}_1 \otimes \mathcal{F}_2, \mu_1 \times \mu_2))$ if and only if

$$\int_{\Omega_1} \left(\int_{\Omega_2} |f(\omega_1, \omega_2)| d\mu_2 \right) d\mu_1 < \infty;$$

(c) If $f : \Omega_1 \times \Omega_2 \to \mathbf{R}$ is an integrable function, then:

(i) For a.a. $\omega_1 \in \Omega_1$ the function $\varphi_1(\omega_1) = \int_{\Omega_2} f(\omega_1, \omega_2) d\mu_2$ is well defined, finite and measurable and integrable on the space $\{\Omega_1, \mathcal{F}_1, \mu_1\}$.

(ii) For a.a. $\omega_2 \in \Omega_2$ the function $\varphi_2(\omega_2) = \int_{\Omega_1} f(\omega_1, \omega_2) d\mu_1$ is well defined, finite and measurable and integrable on the space $\{\Omega_2, \mathcal{F}_2, \mu_2\}$.

(iii) $\int_{\Omega_1 \times \Omega_2} f d(\mu_1 \times \mu_2) = \int_{\Omega_1} \varphi_1 d\mu_1 = \int_{\Omega_2} \varphi_2 d\mu_2$. $\square$

At the end of this section we provide some useful applications of Fatou's Lemma and Lebesque's Theorem to the study of the series of real numbers.

Let $(\mathbf{Z}_+, \mathfrak{P}(\mathbf{Z}_+), \mu)$ be the space with measure, where $\mathbf{Z}_+$ is the set of nonnegative integers, $\mathfrak{P}(\mathbf{Z}_+)$ is the family of all subsets of $\mathbf{Z}_+$ and $\mu : \mathfrak{P}(\mathbf{Z}_+) \to \overline{\mathbf{R}}_+$ is defined by $\mu(A)$ is the number of elements of A if A is a finite subset, $\mu(A) = +\infty$ if A is an infinite subset and $\mu(\emptyset) = 0$. It is obvious that $\mu(\{i\}) = 1$ if $i \in \mathbf{Z}_+$. A function $\mathbf{a} : \mathbf{Z}_+ \to \mathbf{R}$ is a sequence of real numbers $\mathbf{a} = \{\mathbf{a}(i)\}_{i \in \mathbf{Z}_+}$. It is easy to see that every function $\mathbf{a} : (\mathbf{Z}_+, \mathfrak{P}(\mathbf{Z}_+)) \to (\mathbf{R}, \mathcal{B}(\mathbf{R}))$ is a measurable function. Definition 1.1.7 (ii) specialized to this framework allows us to say that a function $\mathbf{a} = \{\mathbf{a}(i)\}_{i \in \mathbf{Z}_+}$ is integrable if and only if $\sum_{i=0}^{\infty} |\mathbf{a}(i)| < \infty$. We have

$$\int_{\mathbf{Z}_+} \mathbf{a} d\mu = \sum_{i=0}^{\infty} \mathbf{a}(i)$$

if the right-hand side is well defined.

Applying Theorem 1.2.1 one obtains.

Corollary 1.2.8. *Let $\mathbf{a}_k, k \geq 0$ be a sequence of functions $\mathbf{a}_k = \{\mathbf{a}_k(i)\}_{i \in \mathbf{Z}_+}$ with the properties:*

(i) $\mathbf{a}_k(i) \geq 0$ for all $k, i \in \mathbf{Z}_+$;
(ii) $\lim_{k \to \infty} \mathbf{a}_k(i) = \mathbf{x}(i)$ for all $i \in \mathbf{Z}_+$.

Then $\sum_{i=0}^{\infty} \mathbf{x}(i) \leq \lim_{k \to \infty} \sum_{i=0}^{\infty} \mathbf{a}_k(i)$. $\square$

Applying Theorem 1.2.2 one obtains.

Corollary 1.2.9. *Let* $\mathbf{a}_k, k \geq 0$ *be a sequence of functions* $\mathbf{a}_k = \{\mathbf{a}_k(i)\}_{i \in \mathbf{Z}_+}$ *with the properties:*

(a) $\lim_{k \to \infty} \mathbf{a}_k(i) = \mathbf{x}(i)$ *for all* $i \in \mathbf{Z}_+$;
(b) $|\mathbf{a}_k(i)| \leq m(i), k \geq 0, i \geq 0$ *where* $\sum_{i=0}^{\infty} m(i) < \infty$.

Under these conditions the following hold:

(i) The series $\sum_{i=0}^{\infty} |\mathbf{a}_k(i)|, k \geq 0, \sum_{i=0}^{\infty} |\mathbf{x}(i)|$ *are convergent.*
(ii) $\lim_{k \to \infty} \sum_{i=0}^{\infty} |\mathbf{a}_k(i) - \mathbf{x}(i)| = 0.$
(iii) $\lim_{k \to \infty} \sum_{i=0}^{\infty} \mathbf{a}_k(i) = \sum_{i=0}^{\infty} \mathbf{x}(i).$ □

1.3   Elements of Probability Theory

Throughout this section and throughout this monograph $\{\Omega, \mathcal{F}, P\}$ is a given probability space (see Definition 1.1.3 (c)).

In the probability theory a measurable function is called *random variable* and the integral of a random variable f is called the *expectation* of f and it is denoted by Ef or $E(f)$, that is $Ef = \int_\Omega f dP$.

A random vector is a vector whose components are random variables. All random vectors are considered column vectors. In the probability theory the words *almost surely* (a.s.) and *with probability one* are often used instead of *almost everywhere*.

As usual, two random variables (random vectors) x, y are identified if $x = y$ a.s.

With this convention the space $L^2(\Omega, P)$ of all random variables x with $E|x^2| < \infty$ is a real Hilbert space with respect to the *inner product* $< x, y >= E(xy)$.

If $x_\alpha, \alpha \in \Delta$ is a family of random variables by $\sigma(x_\alpha, \alpha \in \Delta)$, we denote the smallest σ-algebra $\mathcal{G} \subset \mathcal{F}$ with respect to which all functions $x_\alpha, \alpha \in \Delta$ are measurable.

1.3.1   Gaussian Random Vectors

Definition 1.3.1. An n-dimensional random vector x is said to be *Gaussian* if there exist $m \in \mathbf{R}^n$ and K-$n \times n$ symmetric positive semidefinite matrix such that

$$Ee^{iu^T x} = e^{iu^T m - \frac{1}{2} u^T Ku}$$

for all $u \in \mathbf{R}^n$, where u^T denotes the transpose of u and $i := \sqrt{-1}$.

Remark 1.3.1. The above equality implies

$$m = Ex \text{ and } K = E(x - m)(x - m)^T. \tag{1.1}$$

Definition 1.3.2. A Gaussian random vector x is said to be *nondegenerate* if K is a positive definite matrix. If x is a nondegenerate Gaussian random vector, then

$$P(x \in A) = \frac{1}{((2\pi)^n \det K)^{\frac{1}{2}}} \int_A e^{-\frac{1}{2}(y-m)^T K^{-1}(y-m)} dy$$

for every $A \in \mathcal{B}(\mathbf{R}^n)$.

1.4 Independence

Definition 1.4.1. (i) The σ-algebras $\mathcal{F}_1, \mathcal{F}_2, \ldots, \mathcal{F}_n$, $\mathcal{F}_i \subset \mathcal{F}$ are *independent* if

$$P(\cap_{j=1}^n A_j) = \Pi_{j=1}^n P(A_j)$$

for all $A_j \in \mathcal{F}_j$, $1 \le j \le n$.
(ii) The random variables (random vectors) $x_1, x_2, \ldots, x_n$ are *independent* if the σ-algebras $\sigma(x_1), \sigma(x_2), \ldots, \sigma(x_n)$ are independent.
(iii) The set $\{x_1, x_2, \ldots, x_n\}$ of random variables (random vectors) is independent of the σ-algebra $\mathcal{G}$, $\mathcal{G} \subset \mathcal{F}$ if the σ-algebra $\sigma(x_i, 1 \le i \le n)$ is independent of $\mathcal{G}$.

Theorem 1.4.1. *(i) If $x_1, x_2, \ldots, x_n$ are independent random variables and if x_i are integrable, $1 < i \le n$, then the product $x_1 x_2 \ldots x_n$ is integrable and $E(x_1 x_2 \ldots x_n) = \Pi_{i=1}^n E(x_i)$.*
(ii) If the random vectors $x_1, x_2, \ldots, x_n$, $n \ge 2$ are independent, then $\sigma(x_1, \ldots, x_k)$ is independent of $\sigma(x_{k+1}, \ldots, x_n)$ for every $1 \le k \le n-1$. □

1.5 Conditional Expectation

Let $\mathcal{G} \subset \mathcal{F}$ be a σ-algebra and x an integrable random variable. By Radon–Nicodym Theorem (Theorem 1.2.6) it follows that there exists a unique (mod P) random variable y with the following properties:

(a) y is measurable with respect to $\mathcal{G}$.
(b) $E|y| < \infty$, and
(c) $\int_A y dP = \int_A x dP$ for all $A \in \mathcal{G}$.

The random variable y with these properties is denoted by $E[x|\mathcal{G}]$ and is called the *conditional expectation of x with respect to the σ-algebra $\mathcal{G}$.*
By definition, for all $A \in \mathcal{F}$

$$P(A|\mathcal{G}) := E[\chi_A|\mathcal{G}] \text{ and}$$

$$E[x|y_1, \ldots, y_n] := E[x|\sigma(y_1, \ldots, y_n)]$$

where χ_A denotes the indicator function of A.

If x is an integrable random variable and $A \in \mathcal{F}$ with $P(A) > 0$, then by definition

$$E[x|A] := \int_{\Omega} x dP_A$$

where

$$P_A : \mathcal{F} \to [0, \infty) \text{ by } P_A(B) = \frac{P(A \cap B)}{P(A)} \text{ for all } B \in \mathcal{F}.$$

$E[x|A]$ is called the *conditional expectation of x with respect to the event A.*
  Since

$$P_A(B) = \frac{1}{P(A)} \int_B \chi_A dP,$$

we have

$$E[x|A] = \frac{1}{P(A)} \int_{\Omega} (x\chi_A) dP = \frac{1}{P(A)} \int_A x dP.$$

By definition

$$P(B|A) := P_A(B), A \in \mathcal{F}, B \in \mathcal{F}, P(A) > 0.$$

Obviously, $P[B|A] = E(\chi_B|A)$.

Theorem 1.5.1. *Let x, y be integrable random variables and $\mathcal{G}, \mathcal{H} \subset \mathcal{F}, \sigma$-algebras, then the following assertions hold:*

  (i) $E(E[x|\mathcal{G}]) = Ex$;
  (ii) $E[E[x|\mathcal{G}]|\mathcal{H}] = E[x|\mathcal{H}]$ a.s. if $\mathcal{G} \supset \mathcal{H}$;
  (iii) $E[(\alpha x + \beta y)|\mathcal{G}] = \alpha E[x|\mathcal{G}] + \beta E[x|\mathcal{G}]$ a.s. $\alpha, \beta \in \mathbf{R}$;
  (iv) $E[xy|\mathcal{G}] = yE[x|\mathcal{G}]$ a.s. if y is measurable with respect to $\mathcal{G}$ and xy is integrable;
  (v) If x is independent of $\mathcal{G}$, then $E[x|\mathcal{G}] = Ex$;
  (vi) $x \geq 0$ implies $E[x|\mathcal{G}] \geq 0$ a.s.   □

Remark 1.5.1. It is easy to verify that:

  (i) If x is an integrable random variable and y is a simple random variable with values $c_1, \ldots, c_n$, then

$$E[x|y] = \sum_{j \in M} \chi_{y=c_j} E[x|y = c_j],$$

   where $M = \{j \in \{1, 2, \ldots, n\}; P(y = c_j) > 0\}$;
  (ii) If $A \in \mathcal{F}, \mathcal{G}_A = \{\Phi, \Omega, A, \Omega - A\}$ and x is an integrable random variable, then

$$E[x|\mathcal{G}_A] = \begin{cases} \chi_A E[x|A] + \chi_{\Omega - A} E[x|\Omega - A] \text{ if } 0 < P(A) < 1 \\ Ex \quad \text{if } P(A) = 0 \text{ or } P(A) = 1. \end{cases}$$

Therefore $E[x|\mathcal{G}_A]$ takes at most two values.

1.6 Stochastic Processes

In this section $J \subseteq \mathbf{R}$ is an interval. Let us introduce first the following definition:

Definition 1.6.1. An *m-dimensional stochastic process* is a function $x : J \times \Omega \rightarrow \mathbf{R}^m$ with the property that $x(t, \cdot)$ is a random vector for each $t \in J$.

Usually we denote a stochastic process by $\{x(t), t \in J\}, x = \{x(t)\}_{t \in J}$ or $x(t), t \in J$, the dependence upon the second argument ω being omitted. The function $t \rightarrow x(t, \omega)$ (with ω fixed) are called the *sample paths of the process*.

If $m = 1$, we shall simply say that x is a stochastic process.

Definition 1.6.2. (i) We say that the process $x = \{x(t)\}_{t \in J}$ is *continuous* if for a.a. ω the functions $x(\cdot, \omega)$ are continuous on J;

(ii) x is called to be *right continuous* if for a.a. ω the functions $x(\cdot, \omega)$ are right continuous on J;

(iii) the process $x = \{x(t)\}_{t \in J}$ is *continuous in probability* if $t_n \rightarrow t_0$ with $t_n, t_0 \in J$ implies $x(t_n) \overset{P}{\rightarrow} x(t_0)$;

(iv) x is called to be a *measurable process* if it is measurable on the product space with respect to the σ algebra $\mathcal{B}(J) \otimes \mathcal{F}$, $\mathcal{B}(J)$ being the σ-algebra of Borel sets in J.

Remark 1.6.1. (i) Every right continuous stochastic process is a measurable process.

(ii) From the Fubini theorem it follows that if $x : J \times \Omega \rightarrow \mathbf{R}$ is a measurable process and $E \int_J |x(t)| dt < \infty$, then for a.a. ω, $\int_J x(t) dt$ is a random variable.

Definition 1.6.3. Two stochastic process $x_1 = \{x_1(t)\}_{t \in J}, x_2 = \{x_2(t)\}_{t \in J}$ are called *stochastically equivalent* if $P\{x_1(t) \neq x_2(t)\} = 0$ for all $t \in J$. We then say that x_2 is a *version* of x_1.

Now let us consider a family $\mathcal{M} = \{\mathcal{M}_t\}_{t \in J}$ of σ-algebras $\mathcal{M}_t \subset \mathcal{F}$ with the property that $t_1 < t_2$ implies $\mathcal{M}_{t_1} \subset \mathcal{M}_{t_2}$.

Definition 1.6.4. We say that the process $x = \{x(t)\}_{t \in J}$ is *nonanticipative with respect to the family* $\mathcal{M}$, if:

(i) x is a measurable process;

(ii) for each $t \in J, x(t, \cdot)$ is measurable with respect to the σ-algebra $\mathcal{M}_t$.

When (ii) holds we say that $x(t)$ is $\mathcal{M}_t$-*adapted*.

As usual, by $L^p (J \times \Omega, \mathbf{R}^m), p \geq 1$, we denote the space of all m-dimensional measurable and p-integrable stochastic processes $x : J \times \Omega \rightarrow \mathbf{R}^m$. By $L^p_{\mathcal{M}} (J)$ we denote the space of all $x \in L^p (J \times \Omega, \mathbf{R}^m)$ which are nonanticipative with respect to the family $\mathcal{M} = (\mathcal{M}_t), t \in J$.

Theorem 1.6.1. *If for every* $t \in J$, *the* σ-*algebra* $\mathcal{M}_t$ *contains all sets* $M \in \mathcal{F}$ *with* $P(M) = 0$, *then* $L^p_{\mathcal{M}} (J \times \Omega, \mathbf{R}^m)$ *is a closed subspace of* $L^p (J \times \Omega, \mathbf{R}^m)$.

Proof. Let $x_n \in L^p_{\mathcal{M}}(J \times \Omega, \mathbf{R}^m), n \geq 1$, be a sequence which converges to $x \in L^p(J \times \Omega, \mathbf{R}^m)$. We have to prove that there exists $\hat{x} \in L^p_{\mathcal{M}}(J \times \Omega, \mathbf{R}^m)$ such that x_n converges to $\hat{x}$ in the space $L^p(J \times \Omega, \mathbf{R}^m)$. Indeed, since

$$\lim_{n \to \infty} \int_J E|x_n(t) - x(t)|^p \, dt = 0,$$

by Theorem 1.1.9 the sequence of functions $E|x_n(t) - x(t)|^p$ converges in measure to zero. Hence by virtue of the Riesz's Theorem there exists a subsequence x_{n_k} and a set $N \subset J$ with $\mu(N) = 0$ (μ being the Lebesgue measure) such that

$$\lim_{n \to \infty} E|x_n(t) - x(t)|^p = 0$$

for all $t \in J - N$. Let $t \in J - N$ be fixed. Applying again Theorem 1.1.9 and Riesz's Theorem one concludes that the sequence $x_{n_k}(t), k \geq 1$ has a subsequence which converges a.e. to $x(t)$. But $x_{n_k}(t)$ are $\mathcal{M}_t$-adapted and $\mathcal{M}_t$ contains all sets $M \in \mathcal{F}$ with $P(M) = 0$. Therefore $x(t)$ is measurable with respect to $\mathcal{M}_t$ for each $t \in J - N$. Now, define $\hat{x} : J \times \Omega \to \mathbf{R}^m$ as follows:

$$\hat{x}(t, \omega) = \begin{cases} x(t, \omega) \text{ if } t \in J - N, \, \omega \in \Omega \\ 0 \text{ if } t \in N \text{ and } \omega \in \Omega \end{cases}$$

Obviously $\hat{x} \in L^p_{\mathcal{M}}(J \times \Omega, \mathbf{R}^m)$ and $\lim_{n \to \infty} \int_J E|x_n(t) - \hat{x}(t)|^p \, dt = 0$. The proof is complete. $\square$

The next result is proved in [102, Chap. 4, Sect. 2].

Theorem 1.6.2. *Let $\mathcal{M} = \{\mathcal{M}_t\}_{t \in [a,b]}$ be an increasing family of σ-algebras with the property that for each t, $\mathcal{M}_t$ contains all sets $M \in \mathcal{F}$ with $P(M) = 0$. If $x = \{x(t)\}_{t \in [a,b]}$ is a nonanticipative process with respect to the family $\mathcal{M}$ and if $E \int_a^b |x(t)| dt < \infty$, then the process*

$$y = \{y(t)\}_{t \in [a,b]}, y(t) = \int_a^t x(s) ds$$

is nonanticipative with respect to the family $\mathcal{M}$. $\square$

1.7   Stochastic Processes with Independent Increments

Definition 1.7.1. An r-dimensional stochastic process $x(t), t \in [0, \infty)$ is said to be a stochastic process with *independent increments* if for all $0 \leq t_0 < t_1 < \ldots < t_k$, the random vectors $x(t_0), x(t_1) - x(t_0), \ldots, x(t_k) - x(t_{k-1})$ are independent.

Theorem 1.7.1. *If $x(t), t \geq 0$ is an r-dimensional stochastic process with independent increments, then $\sigma(x(t) - x(a), t \in [a,b])$ is independent of $\sigma(x(b+h) - x(b), h > 0)$ for all $0 \leq a < b$.*

Proof. Let $\mathcal{M}$ be the family of all sets of the form $\cap_{i=1}^p (x(t_i) - x(a))^{-1} (A_i)$ where $a < t_i \leq b$ and $A_i \in \mathcal{B}(\mathbf{R}^r)$, $1 \leq i \leq p$, and $\mathcal{N}$ be the family of all sets of the form $\cap_{i=1}^m (x(b+h_i) - x(b))^{-1} (B_i)$ where $0 < h_i$, $B_i \in \mathcal{B}(\mathbf{R}^r)$, $1 \leq i \leq m$. Obviously $\mathcal{M}$ and $\mathcal{N}$ are π-systems and

$$\sigma(\mathcal{M}) = \sigma(x(t) - x(a), t \in [a,b]), \ \sigma(\mathcal{N}) = \sigma(x(b+h) - x(b), h > 0).$$

First we prove that $P(M \cap N) = P(M) \cdot P(N)$ if $M \in \mathcal{M}$ and $N \in \mathcal{N}$. Indeed, let $M = \cap_{i=1}^p (x(t_i) - x(a))^{-1} (A_i)$, $N = \cap_{i=1}^m (x(b+h_i) - x(b))^{-1} (B_i)$ with

$$a < t_1 < \ldots < t_p \leq b, 0 < h_1 < \ldots < h_m, A_i \in \mathcal{B}(\mathbf{R}^r), B_i \in \mathcal{B}(\mathbf{R}^r).$$

Since

$$\sigma(x(t_i) - x(a), 1 \leq i \leq p)$$
$$= \sigma(x(t_1) - x(a), x(t_2) - x(t_1), \ldots, x(t_p) - x(t_{p-1}))$$

and

$$\sigma(x(b+h_i) - x(b), 1 \leq i \leq m)$$
$$= \sigma(x(b+h_1) - x(b), x(b+h_2) - x(b+h_1), \ldots,$$
$$x(b+h_m) - x(b+h_{m-1}))$$

from Theorem 1.4.1 (ii) it follows that $P(M \cap N) = P(M) \cdot P(N)$. Further by using Theorem 1.1.1 and the equality $A - B = A - (A \cap B)$ one can prove that $P(M \cap B) = P(M) \cdot P(B)$ if $M \in \mathcal{M}$ and $B \in \sigma(x(b+h) - x(b), h > 0)$ and then applying again Theorem 1.1.1, we prove that $P(A \cap B) = P(A) \cdot P(B)$ if $A \in \sigma(\mathcal{M})$ and $B \in \sigma(\mathcal{N})$. The proof is complete. $\square$

Theorem 1.7.2 ([133]). *If $x(t), t \geq 0$ is a continuous r-dimensional stochastic process with independent increments, then all increments $x(t_2) - x(t_1)$ are Gaussian random vectors.* $\square$

1.8   Wiener Processes and Markov process Processes

1.8.1   Standard Wiener Processes

In the next definitions, I is the interval $[0, \infty)$.

Definition 1.8.1. A continuous stochastic process $\beta = \{\beta(t)\}_{t \in I}$ is called a *standard Brownian motion* or a *standard Wiener process* if:

(i) $\beta(0) = 0$;
(ii) $\beta(t)$ is a stochastic process with independent increments;
(iii) $E\beta(t) = 0, t \in I, E|\beta(t) - \beta(s)|^2 = |t - s|$ with $t, s \in I$.

Definition 1.8.2. An r-dimensional stochastic process $w(t) = (w_1(t), \ldots, w_r(t))^T$, $t \in I$ is called an *r-dimensional standard Wiener process* if each process $w_i(t)$ is a standard Brownian motion and the σ-algebras $\sigma(w_i(t), t \in I), 1 \leq i \leq r$ are independent.

For each $t \geq 0$, by $\mathcal{F}_t$ we denote the smallest σ-algebra which contains all sets $M \in \mathcal{F}$ with $P(M) = 0$ and with respect to which all random vectors $\{w(s)\}_{s \leq t}$ are measurable.

For $t \geq 0, \mathcal{U}_t = \sigma(w(t + h) - w(t), h > 0)$.

From Theorem 1.7.1 it follows that for each $t \in I, \mathcal{F}_t$ is independent of $\mathcal{U}_t$.

Remark 1.8.1. (i) Since $w(t) - w(s)$ is independent of $\mathcal{F}_s$ if $t > s$ (see Theorem 1.7.1), from Theorem 1.5.1 (v) it follows that

$$E[(w(t) - w(s)) \mid \mathcal{F}_s] = 0, \tag{1.2}$$

$$E\left[(w(t) - w(s))(w(t) - w(s))^T \mid \mathcal{F}_s\right] = I_r(t - s), t > s, \text{ a.e.}$$

(ii) The increments $w(t) - w(s), t \neq s$ are nondegenerate Gaussian random vectors (see Theorem 1.7.2 and (1.1)).

The converse assertion in (i) is also valid.

Theorem 1.8.1 ([68, 102]). *Let $w(t), t \geq 0$ be a continuous r-dimensional stochastic process with $w(0) = 0$ and adapted to an increasing family of σ-algebras $\mathcal{F}_t, t \geq 0$ such that (1.2) holds. Then $w(t), t \geq 0$ is a standard r-dimensional Wiener process and all increments $w(t_2) - w(t_1), t_2 \neq t_1$ are nondegenerate Gaussian random vectors.* $\square$

Theorems 1.7.2 and 1.8.1 will not be used in this book but they are given because they are interesting by themselves and they give a more detailed view of the properties of these stochastic processes.

1.8.2 Standard Homogeneous Markov Processes

Throughout this book $\mathcal{D}$ is a finite or countable infinite set of positive integers. Without loss of generality we shall take $\mathcal{D} = \{1, 2, \ldots, d\}$ if $\mathcal{D}$ is a finite set and $\mathcal{D} = \mathbf{Z}_+ = \{0, 1, \ldots\}$, respectively, in the case when $\mathcal{D}$ is a countable infinite set.

Definition 1.8.3. A family $\{P(t)\}_{t>0}$ of matrices $P(t) = [p_{ij}(t)], i, j \in \mathcal{D} \times \mathcal{D}$, is said to be a *transition semigroup* if the following two conditions are satisfied:

(i) For each $t > 0$, $P(t)$ is a stochastic matrix, that is $0 \leq p_{ij}(t) \leq 1$ and $\sum_{j \in \mathcal{D}} p_{ij}(t) = 1, \forall i \in \mathcal{D}$;

(ii) $P(t+s) = P(t)P(s)$ for all $t > 0, s > 0$.

The equality (ii) is termed the *homogeneous Chapman–Kolmogorov relation*.

Definition 1.8.4. A stochastic process $\eta(t), t \in [0, \infty)$ is called a *standard homogeneous Markov process* with state-space the set $\mathcal{D}$ and the transition semigroup $P(t) = [p_{ij}(t)]_{(i,j) \in \mathcal{D} \times \mathcal{D}}, t > 0$,, if:

(i) $\eta(t, \omega) \in \mathcal{D}$ for all $t \geq 0$ and $\omega \in \Omega$;

(ii) $P[\eta(t+h) = j | \eta(s), s \leq t] = p_{\eta(t)j}(h)$ a.s. for all $t \geq 0, h > 0, j \in \mathcal{D}$;

(iii) $\lim_{h \to 0_+} p_{ij}(h) = \delta_{ij}$ where $\delta_{ij} = 1$ if $i = j$ and $\delta_{ij} = 0$ if $i \neq j$;

(iv) $\eta(t), t \geq 0$ is a right continuous stochastic process.

In fact, the above definition says that a standard homogeneous Markov process is a *triplet* $\{\eta(t), P(t), \mathcal{D}\}$ satisfying (i)–(iv), $P(t)$, $t > 0$ being a transition semigroup. If $\mathcal{D} = \{1, 2, \ldots, d\}$, then $\{\eta(t), P(t), \mathcal{D}\}$ is a standard homogeneous Markov process with a finite number of states.

The next result is proved in [18, 21, 32].

Theorem 1.8.2. *The standard homogeneous Markov process has the properties:*

(i) $P\{\eta(t+h) = j | \eta(t) = i\} = p_{ij}(h)$ *for all* $i, j \in \mathcal{D}, h > 0, t \geq 0$ *with* $P\{\eta(t) = i\} > 0$.

(ii) $P\{\eta(t+h) = j | \eta(s), s \leq t\} = P[\eta(t+h) = j | \eta(t)], t \geq 0, h > 0, j \in \mathcal{D}$, *a.s.*

(iii) *If x is a bounded random variable measurable with respect to the σ-algebra* $\sigma(\eta(s), s \geq t)$, *then* $E[x | \eta(u), u \leq t] = E[x | \eta(t)]$, *a.s.$t \geq 0$.*

(iv) $p_{ii}(t) > 0$ *for all* $i \in \mathcal{D}, t > 0$.

(v) *For each $i \in \mathcal{D}$ and $j \in \mathcal{D}$, $\lim_{t \to \infty} p_{ij}(t)$ exists.*

If $\{\eta(t), P(t), \mathcal{D}\}$ is a standard homogeneous Markov process with a finite number of states, then the following properties hold:

(vi) $\eta(t)$ *is continuous in probability.*

(vii) *There exists a constant matrix Q such that $P(t) = e^{Qt}, t > 0, Q = [q_{ij}]$ is a matrix with $q_{ij} \geq 0$ if $i \neq j$ and $\sum_{j=1}^{d} q_{ij} = 0$.* $\qquad\square$

In fact (ii) follows from (iii) since $\chi_{\eta(t+h)=j}$ is measurable with respect to the σ-algebra $\sigma(\eta(u), u \geq t)$.

The assertion (iii) in Theorem 1.8.2 is termed *the Markov property of the process* $\eta(t)$.

Under some additional assumptions, the assertions (vi)–(vii) are remaining valid in the case of a standard homogeneous Markov process with an infinite countable number of states. For more details we refer to Sect. 1.13.

If $\mathcal{D} = \{1, 2, \ldots, d\}$ the fact that a transition semigroup $P(t), t > 0$ with the property that $\lim_{t \to 0_+} p_{ij}(t) = \delta_{ij}$ admits an infinitesimal generator Q ($P(t) = e^{Qt}, t > 0$) follows from the general theory of semigroups in Banach algebras [79] but in the theory of Markov processes a probabilistic proof is given in [18, 21, 32, 71].

We assume in the following that $\pi_i := P\{\eta(0) = i\} > 0$ for all $i \in \mathcal{D}$.

Remark 1.8.2. From the above assumption and from the equality

$$P\{\eta(t) = i\} = \sum_{j \in \mathcal{D}} \pi_j P\{\eta(t) = i | \eta(0) = j\}$$

we deduce that $P\{\eta(t) = i\} \geq \pi_i p_{ii}(t) > 0, t \geq 0, i \in \mathcal{D}$.

In the following developments $\mathcal{G}_t, t \geq 0$ denotes the family of σ-algebras $\mathcal{G}_t = \sigma(\eta(s), 0 \leq s \leq t)$ and $\mathcal{V}_t, t \geq 0$ is the family of σ-algebras $\mathcal{V}_t = \sigma(\eta(s), s \geq t)$.

1.9   Stochastic Integral

Throughout this section and throughout the monograph we consider the pair $(w(t), \eta(t)), t \geq 0$ where $w(t)$ is an r-dimensional standard Wiener process and $\eta(t)$ is a standard homogeneous Markov process (see Definitions 1.8.2 and 1.8.4). Assume that the σ-algebra $\mathcal{F}_t$ is independent of $\mathcal{G}_t$ for every $t \geq 0$, where $\mathcal{F}_t$ and $\mathcal{G}_t$ have been defined in the preceding section.

Denote by $\mathcal{H}_t := \mathcal{F}_t \vee \mathcal{G}_t, t \geq 0$.

Let $\widetilde{\mathcal{G}} = \sigma(\eta(t), t \geq 0)$.

Theorem 1.9.1. *For every $t \geq 0$, $\mathcal{F}_t$ is independent of $\widetilde{\mathcal{G}}$ and $\mathcal{U}_t$ is independent of $\mathcal{F}_t \vee \widetilde{\mathcal{G}}$. Therefore $\mathcal{U}_t$ and $\mathcal{H}_t$ are independent σ-algebras for every $t \geq 0$.*

Proof. First one proves that $\mathcal{F}_t$ is independent of $\mathcal{G}_s$ for all $t \geq 0$, $s \geq 0$. Indeed if $t < s$ we have $\mathcal{F}_t \subset \mathcal{F}_s$. Since $\mathcal{F}_s$ is independent of $\mathcal{G}_s$ it follows that $\mathcal{F}_t$ and $\mathcal{G}_s$ are independent σ-algebras. Similarly one proves in case $t > s$.

Now let $\mathcal{M}_0$ be the family of all sets of the form $\cap_{k=1}^m \{\eta(t_k) = i_k\}$, with $t_k \geq 0, t_k \neq t_\ell$ if $k \neq \ell$ and $i_k \in \mathcal{D}$, $1 \leq k \leq m$,

$$\mathcal{M} = \{A; A \in \mathcal{M}_0 \text{ or } A = \emptyset\}, \mathcal{N}_t = \left\{ G \cap F; G \in \widetilde{\mathcal{G}}, F \in \mathcal{F}_t \right\},$$

and $\mathcal{S}_t$ be the family of all sets of the form $\cap_{i=1}^p (w(t+h_i) - w(t))^{-1}(B_i)$ with $h_i > 0, B_i \in \mathcal{B}(\mathbf{R}^r), 1 \leq i \leq p$. Obviously $\mathcal{M}, \mathcal{N}_t$, and $\mathcal{S}_t$ are π-systems and $\sigma(\mathcal{M}) = \widetilde{\mathcal{G}}, \sigma(\mathcal{N}_t) = \mathcal{F}_t \vee \widetilde{\mathcal{G}}$ and $\sigma(\mathcal{S}_t) = \mathcal{U}_t$.

Define $\mathcal{G}(F) = \left\{ G \in \widetilde{\mathcal{G}}; P(G \cap F) = P(G) P(F) \right\}$ for each $F \in \mathcal{F}_t$. Since $\mathcal{F}_t$ is independent of $\mathcal{G}_s$ for all $s \geq 0$ it follows that $\mathcal{M} \subset \mathcal{G}(F)$. By using the equality $F - G = F - (F \cap G)$ one verifies easily that $\mathcal{G}(F)$ satisfies conditions (ii) and (iii) in Theorem 1.1.1. Thus, by virtue of Theorem 1.1.1, $\mathcal{G}(F) = \widetilde{\mathcal{G}}$ for all $F \in \mathcal{F}_t$ and thus the first assertion in the theorem is proved.

Further, if $S \in \mathcal{S}_t, H \in \mathcal{N}_t, H = G \cap F, G \in \widetilde{\mathcal{G}}, F \in \mathcal{F}_t$, since $\mathcal{F}_u$ is independent of $\widetilde{\mathcal{G}}$ for every $u \geq 0$ and $\mathcal{U}_t$ is independent of $\mathcal{F}_t$ (see Theorem 1.7.1), we have

$$P(S \cap H) = P(S \cap G \cap F) = P(G) P(S \cap F)$$

$$= P(G) P(S) P(F) = P(S) P(H).$$

Therefore, by using Theorem 1.1.1, one gets $P(U \cap H) = P(U)P(H)$ for all $U \in \mathcal{U}_t, H \in N_t$ and applying Theorem 1.1.1 again, one concludes that $P(U \cap V) = P(U)P(V)$ if $U \in \mathcal{U}_t, V \in \mathcal{F}_t \vee \widetilde{\mathcal{G}}$. The proof is complete. $\qquad\square$

If $[a,b] \subset [0,\infty)$ we denote by $L^{2p}_{\eta,w}[a,b]$ the space of all nonanticipative processes $f(t), t \in [a,b]$ with respect to the family $\mathcal{H} = (\mathcal{H}_t), t \in [a,b]$ with $E \int_a^b f^{2p}(t)dt < \infty$.

Let $k \in \{1,\dots,r\}$ be fixed and let $\beta(t) = w_k(t), t \geq 0$.

Since the family of σ-algebras $\mathcal{H}_t, t \in [a,b]$ has the properties used in the theory of the *Itô stochastic integral*, namely:

(a) $\mathcal{H}_{t_1} \subset \mathcal{H}_{t_2}$ if $t_1 < t_2$;
(b) $\sigma(\beta(t+h) - \beta(t), h > 0)$ is independent of $\mathcal{H}_t$ (see Theorem 1.9.1);
(c) $\beta(t)$ is measurable with respect to $\mathcal{H}_t$;
(d) $\mathcal{H}_t$ contains all sets $M \in \mathcal{F}$ with $P(M) = 0$ for every $t \geq 0$,

we can define the Itô stochastic integral $\int_a^b f(t)d\beta(t)$ (see [18,68,71,102,120,121]) with $f \in L^2_{\eta,w}[a,b]$.

Definition 1.9.1. A stochastic process $f(t), t \in [a,b]$ is called a *step function* if there exists a partition $a = t_0 < t_1 \dots < t_m = b$ of $[a,b]$ such that $f(t) = f(t_i)$ if $t \in [t_i, t_{i+1}), 1 \leq i \leq m-1$.

If f is a nonanticipative step function, by definition

$$\int_a^b f(t)\,d\beta(t) = \sum_{i=1}^{m-1} f(t_i)\left(\beta(t_{i+1}) - \beta(t_i)\right).$$

Further, let us remind some properties of the integral $\int_a^b f(t)d\beta(t)$ which are proved in [68].

Theorem 1.9.2. *If $f \in L^2_{\eta,w}[a,b]$ we have:*

(i) *There exists a sequence f_n of step functions in $L^2_{\eta,w}[a,b]$ such that $E \int_a^b |f_n(t) - f(t)|^2 dt \to 0$ and the sequence $\int_a^b f_n(t)d\beta(t)$ is convergent in probability; its limit is by definition $\int_a^b f(t)d\beta(t)$.*

(ii) *$E[\int_a^b f(t)d\beta(t)|\mathcal{H}_a] = 0$ and therefore $E[\int_a^b f(t)d\beta(t)|\eta(a) = i] = 0, i \in \mathcal{D}$.*

(iii) *$E[(\int_a^b f(t)d\beta(t))^2|\mathcal{H}_a] = E[\int_a^b f^2(t)dt|\mathcal{H}_a]$ and therefore*

$$E[(\int_a^b f(t)d\beta(t))^2|\eta(a) = i] = E[\int_a^b f^2(t)dt|\eta(a) = i], i \in \mathcal{D}.$$

(iv) *If ξ is a bounded random variable measurable with respect to $\mathcal{H}_a$, then*

$$\int_a^b \xi f(t)d\beta(t) = \xi \int_a^b f(t)d\beta(t).$$

(v) *The process $x(t) = \int_a^t f(s)d\beta(s), t \in [a,b]$ admits a continuous version and $x(t)$ is $\mathcal{H}_t$ adapted.* $\qquad\square$

Theorem 1.9.3. *Let $f \in L^{2p}_{\eta,w}[a,b]$ where p is a positive integer. Then:*

$$E\left(\int_a^b f(t)d\beta(t)\right)^{2p} \leq [p(2p-1)]^p (b-a)^{p-1} E\left(\int_a^b f^{2p}(t)dt\right).$$

$\square$

Remark 1.9.1. (i) Since almost all sample paths of a Brownian motion have infinite variation on any finite interval (see [68]) the stochastic Itô integral cannot be defined in the usual Lebesgue–Stieljes sense, with ω fixed; therefore, the assertion (iv) in Theorem 1.9.2 is not trivial and it must be proved.

(ii) The stochastic Itô integral can be defined for nonanticipative functions f with the property $\int_a^b |f(t)|\,dt < \infty$ a.s., but the equalities in Theorem 1.9.2 (ii) and (iii) hold if $E\int_a^b |f(t)|^2\,dt < \infty$.

Remark 1.9.2. The proof of assertion (i) in Theorem 1.9.2 shows (see also Lemma 6.2 Chap. 4 in [68]) that if $f \in L^{2p}_{\mathcal{M}}([a,b])$ where the increasing family $\mathcal{M}$ of σ-algebras has the property in Theorem 1.6.1 then there exists a sequence f_n of step functions, $f_n \in L^{2p}_{\mathcal{M}}([a,b])$ such that $\lim_{n\to\infty} E\int_a^b |f_n - f|^{2p}\,dt = 0$.

The next result has been proved in [101].

Theorem 1.9.4. *If $f \in L^2_{\eta,w}[a,b]$ we have $E[\chi_{\eta(b)=i}\int_a^b f(t)d\beta(t)|\mathcal{H}_a] = 0$ for every $i \in \mathcal{D}$.*

Proof. We prove first that if $f \in L^2_{\eta,w}[a,b]$ is a step function, then

$$E\left(\chi_{\eta(b)=i}\int_a^b f(t)d\beta(t)\right) = 0.$$

Indeed let $f(t) = \sum_{k=0}^{m-1} f(t_k)\chi_{[t_k,t_{k+1}]}$, $f(t_k)$ being measurable with respect to $\mathcal{H}_{t_k}$. Since $\mathcal{H}_{t_k} \vee \sigma(\eta(b)) \subset \mathcal{F}_{t_k} \vee \widetilde{\mathcal{G}}$ by Theorem 1.9.1 it follows that $\beta(t_{k+1}) - \beta(t_k)$ is independent of the σ-algebra $\mathcal{H}_{t_k} \vee \sigma(\eta(b))$ and thus by Theorem 1.5.1 (v) one gets

$$E\left[(\beta(t_{k+1}) - \beta(t_k))|\mathcal{H}_{t_k} \vee \sigma(\eta(b))\right] = E(\beta(t_{k+1}) - \beta(t_k)) = 0.$$

Hence by using the properties of the conditional expectation (see Theorem 1.5.1) one can write

$$E\chi_{\eta(b)=i}\int_a^b f(t)d\beta(t) = \sum_{k=1}^{m-1} E\chi_{\eta(b)=i}f(t_k)(\beta(t_{k+1}) - \beta(t_k))$$

$$= \sum_{k=1}^{m-1} E\left(E\left[\chi_{\eta(b)=i}f(t_k)(\beta(t_{k+1}) - \beta(t_k))\right.\right.$$

$$\left.\left.|\mathcal{H}_{t_k} \vee \sigma(\eta(b))\right]\right)$$

$$= \sum_{k=1}^{m-1} E\left(\chi_{\eta(b)=i}f(t_k)E\left[(\beta(t_{k+1}) - \beta(t_k))\right.\right.$$

$$\left.\left.|\mathcal{H}_k \vee \sigma(\eta(b))\right]\right)$$

$$= 0$$

Further, by Theorem 1.9.2 let f_n be a sequence of step functions in $L^2_{\eta,w}[a,b]$ with $E\int_a^b |f_n(t) - f(t)|^2 dt \to 0$. We have by virtue of Corollary 1.1.8 and Theorem 1.9.2

$$\left| E\left(\chi_{\eta(b)=i}\int_a^b f(t)d\beta(t)\right)\right| = \left| E\left[\chi_{\eta(b)=i}\left(\int_a^b f(t)d\beta(t) - \int_a^b f_n(t)d\beta(t)\right)\right]\right|$$

$$\leq E\left|\int_a^b (f_n(t) - f(t))d\beta(t)\right|$$

$$\leq \left[E\left(\int_a^b (f_n(t) - f(t))d\beta(t)\right)^2\right]^{1/2}$$

$$= \left(E\int_a^b (f_n(t) - f(t))^2 dt\right)^{1/2} \to 0 \text{ for } n \to \infty.$$

Hence

$$E\chi_{\eta(b)=i}\int_a^b f(t)d\beta(t) = 0. \tag{1.3}$$

Let ξ be a bounded random variable measurable with respect to $\mathcal{H}_a$. Then it follows that $\xi f \in L^2_{\eta,w}[a,b]$ and hence (1.3) gives

$$E\chi_{\eta(b)=i}\int_a^b \xi f(t)d\beta(t) = 0.$$

But, according to Theorem 1.9.2 (iv), we can write

$$E\chi_{\eta(b)=i}\xi\int_a^b f(t)d\beta(t) = E\chi_{\eta(b)=i}\int_a^b \xi f(t)d\beta(t) = 0.$$

Hence, by Theorem 1.5.1 we have

$$E\left(\xi E\left[\chi_{\eta(b)=i}\int_a^b f(t)d\beta(t)|\mathcal{H}_a\right]\right) = E\left(E\left[\xi\chi_{\eta(b)=i}\int_a^b f(t)d\beta(t)|\mathcal{H}_a\right]\right)$$

$$= E\left[\xi\chi_{\eta(b)=i}\int_a^b f(t)d\beta(t)\right] = 0.$$

Taking in the above equality $\xi = \chi_A, A \in \mathcal{H}_a$ we get that

$$E[\chi_{\eta(b)=i} \int_a^b f(t) d\beta(t) | \mathcal{H}_a] = 0 \ a.s.$$

and the proof is complete.   □

Further let $\sigma = (\sigma_{kl})$ be an $n \times r$ matrix whose elements are in $L^2_{\eta,w}[a,b]$. Then the stochastic integral $\int_a^b \sigma(t) dw(t)$ is an n-column vector whose k-th component is given by

$$\sum_{\ell=1}^r \int_a^b \sigma_{k\ell}(t) dw_\ell(t), 1 \le k \le n,$$

where the integral $\int_a^b \sigma_{k\ell}(t) dw_\ell(t)$ is the Itô integral for $\beta = w_\ell$ with respect to the family of σ-algebra $\mathcal{H}_t$.

Here $w(t) = (w_1(t), \ldots, w_r(t))^T$.

Remark 1.9.3. From Theorem 1.9.2 it follows directly that if ξ is a bounded random variable measurable with respect to $\mathcal{H}_a$, then

$$\xi \int_a^b \sigma(t) dw(t) = \int_a^b \xi \sigma(t) dw(t) \ a.s.,$$

the elements of $\sigma(t)$ being in $L^2_{\eta,w}[a,b]$.

The next result follows from Theorem 1.9.2 and it can be found in all books containing the theory of the stochastic Itô integral.

Theorem 1.9.5. *If the elements of $\sigma(t)$ are in $L^2_{\eta,w}[a,b]$, then*

$$E \int_a^b \sigma(t) dw(t) = 0 \ \text{and} \ E \left| \int_a^b \sigma(t) dw(t) \right|^2 = E \int_a^b \|\sigma(t)\|^2 dt,$$

where

$$\|\sigma(t)\|^2 = \sum_{k,\ell} \sigma_{k,\ell}^2(t) = Tr\left(\sigma^T(t)\sigma(t)\right).$$

Theorem 1.9.3 implies the following result directly.

Theorem 1.9.6. *If all elements of the matrix $\sigma(t)$ are in $L^{2p}_{\eta,w}[a,b]$, p being a positive integer, then*

$$E \left| \int_a^b \sigma(t) dw(t) \right|^{2p} \le \left[nr^2 p(2p-1)\right]^p (b-a)^{p-1} \sum_{k,\ell} E \int_a^b \sigma_{k,\ell}^{2p}(t) dt.$$

Applying Theorems 1.9.5 and 1.9.6 for $\chi_{\eta(a)=i} \cdot \sigma$ and taking into account Remark 1.9.3 one gets the following results.

Theorem 1.9.7. *Under the assumption of Theorem 1.9.5 we have*

$$E\left[\int_a^b \sigma(t)\,dw(t)\mid \eta(a)=i\right]=0,$$

$$E\left[\left|\int_a^b \sigma(t)\,dw(t)\right|^2 \mid \eta(a)=i\right]=E\left[\int_a^b \|\sigma(t)\|^2\,dt \mid \eta(a)=i\right]$$

for all $i \in \mathcal{D}$. $\quad\square$

Theorem 1.9.8. *Under the assumption of Theorem 1.9.6 we have*

$$E\left[\left|\int_a^b \sigma(t)\,dw(t)\right|^{2p} \mid \eta(a)=i\right]$$

$$\leq [nr^2 p(2p-1)]^p (b-a)^{p-1}\sum_{k,\ell} E\left[\int_a^b \sigma_{k,\ell}^{2p}(t)\,dt \mid \eta(a)=i\right]$$

for all $i \in \mathcal{D}$. $\quad\square$

Definition 1.9.2. Let $x(t), t \in [t_0, T]$ be an n-dimensional stochastic process verifying

$$x(t)-x(t_0)=\int_{t_0}^t a(s)\,ds+\int_{t_0}^t \sigma(s)\,dw(s), a.s.[t_0,T],$$

where $a=(a_1,\ldots,a_n)^T, \sigma=(\sigma_{k\ell})$ with $1 \leq k \leq n, 1 \leq \ell \leq r$, and $a_k, \sigma_{k\ell}$ being in $L^2_{\eta,w}[t_0,T]$ for all k and ℓ. Then we say that $x(t)$ has a *stochastic differential* $dx(t)$ given by

$$dx(t)=a(t)dt+\sigma(t)dw(t), t \in [t_0,T]. \tag{1.4}$$

Obviously if $x(t_0)$ is measurable with respect to $\mathcal{H}_{t_0}$ and $E|x(t_0)|^2 < \infty$, the above stochastic process $x=(x(t)), t \in [t_0,T]$ is a continuous process and $x \in L^2_{\eta,w}[t_0,T]$.

Theorem 1.9.9 (Itô's Formula). *Let $v(t,x)$ be a continuous function in $(t,x) \in [0,T] \times \mathbf{R}^n$ together with its derivatives v_t, v_x, v_{xx}. If $x(t)$ verifies (1.4), then*

$$dv(t,x(t)) = \left[\frac{\partial v}{\partial t}(t,x(t))+\left(\frac{\partial v}{\partial x}(t,x(t))\right)^T a(t)\right.$$

$$\left.+\frac{1}{2}Tr\sigma^T(t)\frac{\partial^2 v}{\partial x\partial x}(t,x(t))\sigma(t)\right]dt$$

$$+\left(\frac{\partial v}{\partial x}(t,x(t))\right)^T \sigma(t)dw(t),$$

a.s., $t \in [t_0,T]$. $\quad\square$

1.10   An Itô's Type Formula

In this section and in the next two ones, we shall assume as in the most part of this monograph that $\mathcal{D} = \{1, 2, \ldots, d\}$. We are interested in the following to obtain an Itô's type formula for (1.4) with functions $v(t, x, i), i \in \mathcal{D}$, rather than $v(t, x)$, namely for functions depending upon the states i of the Markov process $\eta(t)$.

Since $\mathcal{H}_t$ incorporates properties of $\eta(t)$, we would like to exploit the properties of both $w(t)$ and $\eta(t)$. This fact will be more clear in the following developments when stochastic differential equations with Markovian jumping will be investigated.

A strong argument to consider functions $v(t, x, i)$ instead of $v(t, x)$ is that the Itô formula for the function $v(t, x)$ (Theorem 1.9.9) does not retain the fundamental elements of the process $\eta(t)$ as $p_{ij}(t)$ and q_{ij}.

We must emphasize the fact that by contrast with the Itô's formula given in Theorem 1.9.9 which is valid for a.a. $\omega \in \Omega$, when considering functions $v(t, x, i)$ we cannot expect to obtain a similar formula for $v(t, x(t), \eta(t))$ holding a.s. This is due to the fact that the coefficients q_{ij} are strongly related by considering the conditional expectation with respect to the events $\{\eta(t) = i\}$.

In order to prove an Itô-type formula for functions $v(t, x, i)$ we need the following result which is interesting by itself.

Let us denote by $\mathcal{R}_t = \mathcal{U}_t \vee \mathcal{V}_t, t \geq 0$ where the σ-algebras $\mathcal{U}_t$ and $\mathcal{V}_t$ have been defined in Sect. 1.8.

Theorem 1.10.1. *If ξ is an integrable random variable measurable with respect to $\mathcal{R}_t$ that is $\xi \in L^1(\Omega, \mathcal{R}_t, P)$, then $E[\xi|\mathcal{H}_t] = E[\xi|\eta(t)]$ a.s.*

Proof. The proof is made in two steps. In the first step we show the equality in the statement holds for $\xi = \chi_B$ for all $B \in \mathcal{R}_t$ and in the second step we consider the general situation when ξ is integrable.

Step 1 Define $z = E[\xi|\eta(t)]$. We have to prove that

$$E(z\chi_A) = E(\xi\chi_A)\, \text{for all}\, A \in \mathcal{H}_t. \tag{1.5}$$

First we shall prove that (1.5) holds in the particular case when $\xi = \chi_M \chi_N, M \in \mathcal{U}_t, N \in \mathcal{V}_t$.

Let $\mathcal{M}$ be the family of all sets $A \in \mathcal{F}$ verifying (1.5). It is obvious that $\mathcal{M}$ verifies (ii) and (iii) in Theorem 1.1.1.

Let $\mathcal{C} = \{F \cap G; F \in \mathcal{F}_t, G \in \mathcal{G}_t\}$; it is easy to check that $\mathcal{C}$ is a π system. We show now that $\mathcal{C} \subset \mathcal{M}$. Indeed, let $F \in \mathcal{F}_t, G \in \mathcal{G}_t$; we have to prove that $E(z\chi_F\chi_G) = E(\xi\chi_F\chi_G)$. But since χ_M is independent of $\{\chi_N, \eta(t)\}$ (see Theorem 1.9.1) we can write

$$\int_{\{\eta(t)=i\}} E(\chi_M) E[\chi_N|\eta(t)]\, dP$$

$$= E(\chi_M) \int_{\{\eta(t)=i\}} E[\chi_N|\eta(t)]\, dP$$

$$= E(\chi_M)E(\chi_N\chi_{\eta(t)=i}) = E(\chi_M\chi_N\chi_{\eta(t)=i})$$

$$= \int_{\{\eta(t)=i\}} \chi_M\chi_N dP$$

Hence $z = E(\chi_M)E[\chi_N|\eta(t)]$ (in our case $z = E[\chi_M\chi_N|\eta(t)]$).

From Theorem 1.8.2 (iii) we have $E[\chi_N|\eta(t)] = E[\chi_N|\mathcal{G}_t]$.

Further, since χ_M is independent of $\{\chi_F,\chi_G,\chi_N\}$ and χ_F is independent of $\{\chi_G, E[\chi_N|\eta(t)]\}$ (see Theorem 1.9.1), we can write, applying Theorems 1.4.1 and 1.5.1, that:

$$E(\xi\chi_F\chi_G) = E(\chi_M\chi_N\chi_F\chi_G) = E(\chi_M)E(\chi_N\chi_F\chi_G)$$

$$= E(\chi_M)E(\chi_F)E(\chi_N\chi_G),$$

$$E(z\chi_F\chi_G) = E(\chi_M)E(\chi_F\chi_G E[\chi_N|\eta(t)])$$

$$= E(\chi_M)E(\chi_F)E(\chi_G E[\chi_N|\eta(t)])$$

$$= E(\chi_M)E(\chi_F)E(\chi_G E[\chi_N|\mathcal{G}_t])$$

$$= (E\chi_M)(E\chi_F)E(E[\chi_G\chi_N|\mathcal{G}_t])$$

$$= E(\chi_M)E(\chi_F)E(\chi_N\chi_G)$$

Thus we proved that $\mathcal{C} \subset \mathcal{M}$. Hence by Theorem 1.1.1 $\sigma(\mathcal{C}) \subset \mathcal{M}$. But $\sigma(\mathcal{C}) = \mathcal{H}_t$, thus
$E[\chi_M\chi_N|\mathcal{H}_t] = E[\chi_M\chi_N|\eta(t)]$ for all $M \in \mathcal{U}_t, N \in \mathcal{V}_t$.

Now let $\mathcal{N}$ be the family of $B \in \mathcal{F}$ with $E[\chi_B|\mathcal{H}_t] = E[\chi_B|\eta(t)]$.

We know that $\mathcal{N}$ contains $\widehat{\mathcal{C}} = \{M \cap N, M \in \mathcal{U}_t, N \in \mathcal{V}_t\}$. $\widehat{\mathcal{C}}$ is a π system and since $\mathcal{N}$ verifies (ii) and (iii) in Theorem 1.1.1 it follows that $\mathcal{N} \supset \sigma\left(\widehat{\mathcal{C}}\right) = \mathcal{R}_t$.

Step 2 First assume that $\xi \geq 0$; by Theorem 1.1.6 there exists a sequence of simple random variables $\xi_n(\omega)$ with the properties $0 \leq \xi_n \leq \xi_{n+1}$, $\lim_{n\to\infty} \xi_n(\omega) = \xi(\omega)$ and ξ_n are measurable with respect to $\mathcal{R}_t$. For each $n \geq 1$ we have $E[\xi_n|\mathcal{H}_t] = E[\xi_n|\eta(t)]$.

Applying Theorem 1.2.2, the equality in the statement is valid in the case when ξ is nonnegative, integrable, and measurable with respect to $\mathcal{R}_t$.

In the general case we can write $\xi = \xi^+ - \xi^-$, where $\xi^+ = \frac{1}{2}(|\xi|+\xi)$ and $\xi^- = \frac{1}{2}(|\xi|-\xi), \xi^+ \geq 0, \xi^- \geq 0$ and thus the equality in the statement takes place for ξ^+ and ξ^- and therefore, according to Theorem 1.5.1 it results that the proof is complete. $\square$

Let us notice that the proof of the above theorem does not require the set $\mathcal{D}$ to be finite.

Theorem 1.10.2 (Itô-Type Formula). *Let us consider $a = (a_1,\ldots,a_n)^T$ with $a_k \in L^2_{\eta,w}([t_0,T]), 1 \leq k \leq n$, $\sigma = [\sigma_{ij}]_{1\leq i\leq n,1\leq j\leq r}$ with $\sigma_{ij} \in L^2_{\eta,w}([t_0,T])$ and ξ an n-dimensional random vector $\mathcal{H}_{t_0}$ measurable with $E|\xi|^2 < \infty$ and let the function*

$$v(t,x,i) = x^T K(t,i)x + 2k^T(t,i)x + k_0(t,i),$$

where $K : [t_0, T] \times \mathcal{D} \to \mathbf{R}^{n \times n}, K = K^T, k : [t_0, T] \times \mathcal{D} \to \mathbf{R}^n, k_0 : [t_0, T] \times \mathcal{D} \to \mathbf{R}$ *are* C^1*-functions with respect to t. Then the following equality is true:*

$$E\left[(v(t, x(t), \eta(t)) - v(t_0, \xi, i)) \mid \eta(t_0) = i\right]$$

$$= E\left[\int_{t_0}^{t} \left\{ \frac{\partial v}{\partial t}(s, x(s), \eta(s)) + a^T(s)\frac{\partial v}{\partial x}(s, x(s), \eta(s)) \right. \right. \tag{1.6}$$

$$\left. \left. + Tr(\sigma^T(s)K(s, \eta(s))\sigma(s)) + \sum_{j=1}^{d} v(s, x(s), j)q_{\eta(s), j} \right\} ds \mid \eta(t_0) = i\right]$$

for all $i \in \mathcal{D}$ *and for the stochastic process* $x(t), t \in [t_0, T]$ *verifying*

$$dx(t) = a(t)dt + \sigma(t)dw(t), t \in [t_0, T] \text{ and } x(t_0) = \xi.$$

Proof. The proof consists in three steps.

Step 1. Assume that ξ, a, σ satisfy the assumption in the statement and additionally ξ is a bounded random vector a, σ are bounded on $[t_0, T] \times \Omega$, and $a(t), \sigma(t)$ are, with probability one, right continuous functions on $[t_0, T]$.
Under these assumptions, applying Theorem 1.9.6, we deduce that

$$\sup_{t \in [t_0, T]} E|x(t)|^{2k} < \infty,$$

for all $k \in \mathbf{N}, k \geq 1$. We can write:

$$v(t + h, x(t + h), \eta(t + h)) - v(t, x(t), \eta(t))$$

$$= v(t + h, x(t + h), \eta(t + h)) - v(t, x(t), \eta(t + h))$$

$$+ v(t, x(t), \eta(t + h)) - v(t, x(t), \eta(t))$$

$$= \sum_{j=1}^{d} \chi_{\eta(t+h)=j}(v(t + h, x(t + h), j) - v(t, x(t), j))$$

$$+ v(t, x(t), \eta(t + h)) - v(t, x(t), \eta(t)),$$

where χ_M is the indicator function of the set M.
For each fixed $j \in \mathcal{D}$, we can apply the Itô formula (Theorem 1.9.9) and obtain

$$v(t + h, x(t + h), j) - v(t, x(t), j)$$
$$= \int_{t}^{t+h} m_j(s)ds + 2\int_{t}^{t+h}(x^T(s)K(s, j) + k^T(s, j))\sigma(s)dw(s)$$

where

$$m_j(s) = x^T(s)\dot{K}(s, j)x(s) + 2\dot{k}^T(s, j)x(s) + \dot{k}_0(s, j)$$

$$+ 2x^T(s)K(s, j)a(s) + 2k^T(s, j)a(s) + Tr(\sigma^T(s)K(s, j)\sigma(s)),$$

$j \in \mathcal{D}$. Using Theorem 1.9.4, we deduce that

$$E[\chi_{\eta(t+h)=j} \int_t^{t+h} [x^T(s)K(s,j) + k^T(s,j)]\sigma(s)dw(s)|\mathcal{H}_t] = 0.$$

Hence

$$E[\chi_{\eta(t+h)=j} \int_t^{t+h} (x^T(s)K(s,j) + k^T(s,j))\sigma(s)dw(s)|\eta(t_0) = i] = 0$$

and finally we deduce

$$E\left[(v(t+h,x(t+h),\eta(t+h)) - v(t,x(t),\eta(t+h)))|\eta(t_0) = i\right]$$

$$= \sum_{j=1}^d E\left[\chi_{\eta(t+h)=j} \int_t^{t+h} m_j(s)ds|\eta(t_0) = i\right]. \tag{1.7}$$

It is obvious that $m_j(s)$ is, with probability 1, right continuous, and hence we have

$$\lim_{h \searrow 0} \frac{1}{h} \int_t^{t+h} m_j(s)ds = m_j(t), \ t \in [t_0,T), j \in \mathcal{D}.$$

Since $\eta(t)$ is right continuous we can write:

$$\lim_{h \searrow 0} \frac{1}{h} \chi_{\eta(t+h)=j} \int_t^{t+h} m_j(s)ds = \chi_{\eta(t)=j} m_j(t). \tag{1.8}$$

On the other hand, since $\sup_{t \in [t_0,T]} E|x(t)|^4 < \infty$ we obtain that there exists $\beta > 0$ (not depending upon t, h) such that:

$$E\left|\frac{1}{h}\chi_{\eta(t+h)=j} \int_t^{t+h} m_j(s)ds\right|^2 \leq \beta.$$

Thus, from (1.7) and (1.8) and Theorem 1.2.4 it follows

$$\lim_{h \searrow 0} \frac{1}{h} E[(v(t+h,x(t+h),\eta(t+h)) - v(t,x(t),\eta(t+h)))|\eta(t_0) = i]$$

$$= \sum_{j=1}^d E[\chi_{\eta(t)=j} m_j(t)|\eta(t_0) = i] = E[\tilde{m}(t)|\eta(t_0) = i], \tag{1.9}$$

$t \in [t_0,T), i \in \mathcal{D}$, where

$$\tilde{m}(t) = x^T(t)\dot{K}(t,\eta(t))x(t) + 2\dot{K}(t,\eta(t))x(t) + \dot{k}_0(t,\eta(t))$$

$$+ 2\left[x^T(t)K(t,\eta(t)) + k^T(t,\eta(t))\right]a(t) + Tr(\sigma^T(t)K(t,\eta(t))\sigma(t))$$

where $\dot{K}(t,\eta(t)) = \frac{\partial}{\partial t}K(t,\eta(t))$. Further, by using Theorem 1.5.1 we can write

$$E\left[(v(t,x(t),\eta(t+h)) - v(t,x(t),\eta(t)))|\eta(t_0) = i\right]$$

$$= E\left[\left(\sum_{j=1}^{d}\chi_{\eta(t+h)=j}v(t,x(t),j) - v(t,x(t),\eta(t))\right)|\eta(t_0) = i\right] \qquad (1.10)$$

$$= \sum_{j=1}^{d}E\left[v(t,x(t),j)E[\chi_{\eta(t+h)=j}|\mathcal{H}_t]|\eta(t_0) = i\right]$$

$$-E\left[v(t,x(t),\eta(t))|\eta(t_0) = i\right].$$

By virtue of Theorem 1.10.1 we have

$$E[\chi_{\eta(t+h)=j}|\mathcal{H}_t] = E[\chi_{\eta(t+h)=j}|\eta(t)] = p_{\eta(t)j}(h). \qquad (1.11)$$

Hence from (1.10) and (1.11) we have

$$E[(v(t,x(t),\eta(t+h)) - v(t,x(t),\eta(t)))|\eta(t_0) = i]$$

$$= E\left[\sum_{j\neq\eta(t)}(v(t,x(t),j) - v(t,x(t),\eta(t)))p_{\eta(t)j}(h)|\eta(t_0) = i\right].$$

Recall that $P(h) = [p_{ij}(h)] = e^{Qh}, h > 0$ with $\sum_{j=1}^{d}q_{ij} = 0$. Applying Lebesque's Theorem we obtain that

$$\lim_{h\searrow 0}\frac{1}{h}E\left[(v(t,x(t),\eta(t+h)) - v(t,x(t),\eta(t))|\eta(t_0) = i\right] \qquad (1.12)$$

$$= \sum_{j=1}^{d}E\left[v(t,x(t),j)q_{\eta(t)j})|\eta(t_0) = i\right].$$

Combining (1.9) with (1.12) we conclude that

$$\lim_{h\searrow 0}\frac{1}{h}E\left[(v(t+h),x(t+h),\eta(t+h)) - v(t,x(t),\eta(t)))|\eta(t_0) = i\right]$$

$$= E\left[\left(\tilde{m}(t) + \sum_{j=1}^{d}v(t,x(t),j)q_{\eta(t)j}\right)|\eta(t_0) = i\right].$$

Denote

$$G_i(t) = E[v(t,x(t),\eta(t))|\eta(t_0) = i], i \in \mathcal{D}$$

and

$$h_i(t) = E\left[\left(\tilde{m}(t) + \sum_{j=1}^{d} v(t,x(t),j)q_{\eta(t)j}\right)|\eta(t_0) = i\right].$$

Since $\sup_{t\in[t_0,T]} E(\tilde{m}(t) + \sum_{j=1}^{d} v(t,x(t),j)q_{\eta(t)j})^2 < \infty$ it follows by Theorem 1.2.4 that $h_i(t)$ is right continuous and therefore

$$\lim_{h\searrow 0} \frac{1}{h} \int_t^{t+h} h_i(s)ds = h_i(t), t \in [t_0,T).$$

Hence

$$\lim_{h\searrow 0} \frac{1}{h}\left(G_i(t+h) - G_i(t) - \int_t^{t+h} h_i(s)ds\right) = 0, t \in [t_0,T), i \in \mathcal{D}. \qquad (1.13)$$

Since the process $\eta(t)$ is continuous in probability (see Theorem 1.8.2) it follows by using Corollary 1.1.4 that $v(t,x(t),\eta(t))$ is continuous in probability. Having $\sup_{t\in[t_0,T]} E|v(t,x(t),\eta(t))|^2 < \infty$ it follows from Theorem 1.2.4 that $G_i(t), i \in \mathcal{D}$ is a continuous function and thus from (1.13) we conclude that

$$G_i(t) - G_i(t_0) = \int_{t_0}^{t} h_i(s)ds, t \in [t_0,T], i \in \mathcal{D}$$

and so the equality (1.6) holds.

Step 2. Assume that ξ is $\mathcal{H}_{t_0}$-measurable and $E|\xi|^2 < \infty$, and a, σ are bounded on $[t_0,T] \times \Omega, a(t), \sigma(t)$ are $\mathcal{H}_t$-adapted. Let

$$\xi_k = \xi\chi_{|\xi|\le k},$$

$$a_k(t) = k\int_{max\{t-\frac{1}{k},t_0\}}^{t} a(s)ds,$$

$$\sigma_k(t) = \int_{max\{t-\frac{1}{k},t_0\}}^{t} \sigma(s)ds.$$

It is obvious that a_k and σ_k are continuous (with probability 1), bounded on $[t_0,T] \times \Omega$, and $\mathcal{H}_t$-adapted (see Theorem 1.6.2). From Theorem 1.2.5 and from Lebesgue's Theorem it follows that

$$\lim_{k\to\infty} \int_{t_0}^{T} \left(|a_k(t) - a(t)|^2 + \|\sigma_k(t) - \sigma(t)\|^2\right)dt = 0 \qquad (1.14)$$

and applying again the Lebesgue's Theorem we have

$$\lim_{k\to\infty} E\int_{t_0}^{T} \left(|a_k(t) - a(t)|^2 + \|\sigma_k(t) - \sigma(t)\|^2\right)dt = 0.$$

From Lebesgue's Theorem it follows that

$$\lim_{k \to \infty} E|\xi_k - \xi|^2 = 0.$$

It is easy to verify by using Theorem 1.9.5 that $\sup_{t \in [t_0,T]} E|x(t)|^2 < \infty$ and

$$\sup_{t \in [t_0,T]} E|x_k(t) - x(t)|^2 \le 3E\left[|\xi_k - \xi|^2 + (T - t_0) \int_{t_0}^{T} |a_k(t) - a(t)|^2 dt \right.$$

$$\left. + \int_{t_0}^{T} \|\sigma_k(t) - \sigma(t)\|^2 dt \right], k \ge 1,$$

where

$$x_k = \xi_k + \int_{t_0}^{t} a_k(s)ds + \int_{t_0}^{t} \sigma_k(s)dw(s).$$

Applying the result of Step 1 for each $k \ge 1$ we obtain

$$E\left[(v(t, x_k(t), \eta(t)) - v(t_0, \xi_k, i))|\eta(t_0) = i\right] \qquad (1.15)$$

$$= E\left\{ \int_{t_0}^{t} \left[x_k^T(s)K(s, \eta(s))x_k(s) + 2k^T(s, \eta(s))x_k(s) + \dot{k}_0(s, \eta(s)) \right.\right.$$

$$+ 2\left(x_k^T(s)K(s, \eta(s)) + k^T(s, \eta(s))\right) a_k(s) + Tr(\sigma_k^T(s)K(s, \eta(s))\sigma_k(s))$$

$$\left.\left. + \sum_{j=1}^{d} v(s, x_k(s), j)q_{\eta(s)j} \right] ds|\eta(t_0) = i \right\}.$$

Taking the limit for $k \to \infty$ we conclude that (1.6) holds.
Step 3. Now consider that ξ, a, σ verify the general assumptions in the statement.
Define

$$\bar{a}_k(t) = a(t)\chi_{|a(t)| \le k}$$

$$\bar{\sigma}_k(t) = \sigma(t)\chi_{|\sigma(t)| \le k}.$$

Applying Lebesgue's Theorem it follows that $\bar{a}_k$ and $\bar{\sigma}_k$ verify an equality of type (1.14). On the other hand it can be proved by using Theorem 1.9.5:

$$\sup_{t \in [t_0,T]} E|\bar{x}_k(t) - x(t)|^2 \le 2E\left[\int_{t_0}^{T} \left\{ (T - t_0)|\bar{a}_k(t) - a(t)|^2 + \|\bar{\sigma}_k(t) - \sigma(t)\|^2 \right\} dt \right]$$

where

$$\bar{x}_k(t) = \xi + \int_{t_0}^{t} \bar{a}_k(s)ds + \int_{t_0}^{t} \bar{\sigma}_k(s)dw(s).$$

Now, applying the results from Step 2 for $\xi, \bar{a}_k, \bar{\sigma}_k, \bar{x}_k$ we obtain an equality of type (1.15) with $\xi_k, a_k, \sigma_k, x_k$ replaced by $\xi, \bar{a}_k, \bar{\sigma}_k, \bar{x}_k$.

Taking again the limit for $k \to \infty$ we conclude that (1.6) holds and the proof is complete.   □

Remark 1.10.1.   (i) The proof of Theorem 1.10.2 has been performed in several steps since only poor information is available concerning a and σ, namely their elements are in $L^2_{\eta,w}([t_0, T])$.

(ii) The particular form for $v(t, x, i)$ is essentially used when making $k \to \infty$ in steps 2 and 3 of the proof.

(iii) The proof shows that the result is true for functions $v(t, x, i)$ in C^1 with respect to t and in C^2 with respect to x, the functions $v(t, x, i)$, $\frac{\partial v}{\partial t}(t, x, i)$ and $\frac{\partial v}{\partial x}(t, x, i)$ have increments with respect to x of the same type as the increments of the quadratic function used in the theorem. Moreover $\frac{\partial^2 v}{\partial x \partial x}(t, x, i)$ must be bounded on $[t_0, T] \times \mathbf{R}^n \times \mathcal{D}$.

1.11 Stochastic Differential Equations

Stochastic differential equations depending on the pair $(w(t), \eta(t))$ with the above properties are considered in [76, 101, 104], where stability and control problems are investigated.

In [151], Wonham emphasizes the importance of the differential equations subject to the white noise perturbations $w(t)$ and Markovian jumping $\eta(t)$ for control problems.

Consider the system of stochastic differential equations:

$$dx(t) = [f(t, x(t), \eta(t)) + a(t)]\,dt + [F(t, x(t), \eta(t)) + \sigma(t)]\,dw(t) \qquad (1.16)$$

where the processes $w(t) = (w_1(t), \ldots, w_r(t))^T$ and $\eta(t), t \geq 0$ have the properties in Sect. 1.9. Assume that a, σ, f, and F satisfy the following conditions:

(C1) $a : \mathbf{R}_+ \times \Omega \to \mathbf{R}^n, \sigma : \mathbf{R}_+ \times \Omega \to \mathbf{R}^{n \times r}$ and their elements are in $L^2_{\eta,w}[0, T]$, for all $T > 0$;

(C2) $f : \mathbf{R}_+ \times \mathbf{R}^n \times \mathcal{D} \to \mathbf{R}^n, F : \mathbf{R}_+ \times \mathbf{R}^n \times \mathcal{D} \to \mathbf{R}^{n \times r}$ and for each $i \in \mathcal{D}$, $f(\cdot, \cdot, i)$ and $F(\cdot, \cdot, i)$ are measurable with respect to $\mathcal{B}(\mathbf{R}_+ \times \mathbf{R}^n)$, where $\mathcal{B}(\mathbf{R}_+ \times \mathbf{R}^n)$ denotes the σ-algebra of Borel sets in $\mathbf{R}_+ \times \mathbf{R}^n$;

(C3) For each $T > 0$ there exists $\gamma(T) > 0$ such that

$$|f(t, x_1, i) - f(t, x_2, i)| + \|F(t, x_1, i) - F(t, x_2, i)\| \leq \gamma(T)|x_1 - x_2| \qquad (1.17)$$

for all $t \in [0, T], x_1, x_2 \in \mathbf{R}^n, i \in \mathcal{D}$, and

$$|f(t,x,i)| + \|F(t,x,i)\| \le \gamma(T)(1+|x|),\tag{1.18}$$

for all $t \in [0,T], x \in \mathbf{R}^n, i \in \mathcal{D}$.

Using the same technique as in the proof of Theorem 1.1 from [68], Chap. 5, one can prove the following result:

Theorem 1.11.1. *Assume that a, σ, f, and F satisfy the conditions (C1)–(C3). Then for all $t_0 \ge 0$ and ξ measurable with respect to $\mathcal{H}_{t_0}$ and $E|\xi|^2 < \infty$ there exists a unique continuous solution $x(t) = x(t,x_0,\xi), t \ge t_0$ of (1.16), verifying $x(t_0) = \xi$ and which components belong to $L^2_{\eta,w}[t_0,T]$ for all $T > t_0$. Moreover we have:*

$$\sup_{t_0 \le t \le T} E\left[|x(t)|^2 |\eta(t_0) = i\right]$$

$$\le K\left(1 + E\left[\left(|\xi|^2 + \int_{t_0}^T \left(|a(t)|^2 + \|\sigma(t)\|^2\right) dt\right) |\eta(t_0) = i\right]\right)$$

where K depends on T and $T - t_0$. The uniqueness must be understood in the sense that if $x_1(t)$ and $x_2(t)$ are two solutions of (1.16) satisfying $x_1(t_0) = x_2(t_0) = \xi$ and whose components are in $L^2_{\eta,w}[t_0,T]$, then $E|x_1(t) - x_2(t)| = 0, t \in [t_0,T]$. □

For the particular case when $a(t) = 0$ and $\sigma(t) = 0$ one obtains the following result.

Theorem 1.11.2. *Assume that f and F satisfy (C2), (C3) and $a(t) = 0$, $\sigma(t) = 0$, for all $t \ge 0$. Then for all $t_0 \ge 0$ and ξ measurable with respect to $\mathcal{H}_{t_0}$ with $E|\xi|^2 < \infty$, the system (1.16) has a unique continuous solution $x(t)$, $t \ge t_0$ verifying $x(t_0) = \xi$ whose elements are in $L^2_{\eta,w}[t_0,T]$ for all $T > t_0$. Moreover if $E|\xi|^{2p} < \infty$ then we have*

$$\sup_{t_0 \le t \le T} E\left[|x(t)|^{2p} |\eta(t_0) = i\right] \le K\left(1 + E\left[|\xi|^{2p} |\eta(t_0) = i\right]\right),\tag{1.19}$$

$i \in \mathcal{D}$, where K depends on T, $T - t_0$, and p.

Proof. Consider the sequence of successive approximations defined by

$$x_0(t) = \xi, t \in [t_0,T]$$

$$x_{m+1}(t) = \xi + \int_{t_0}^t f(s,x_m(s),\eta(s))ds + \int_{t_0}^t F(s,x_m(s),\eta(s))dw(s), m \ge 0.$$

Using (1.17), (1.18) and Theorem 1.9.8 it is easy to verify by induction that

$$E\left[|x_{m+1}(t)|^{2p} |\eta(t_0) = i\right] \le \left[c + c^2(t-t_0) + \ldots + c^{m+2}\frac{(t-t_0)^{m+1}}{(m+1)!}\right]$$

$$\times \left(1 + E\left[|\xi|^{2p} |\eta(t_0) = i\right]\right), t_0 \le t \le T, i \in \mathcal{D}, m \ge 0,$$

where $c > 0$ depends only on T, $T - t_0$, and p. Hence

$$E\left[|x_{m+1}(t)|^{2p}|\eta(t_0) = i\right] \le ce^{c(t-t_0)}\left(1 + E\left[|\xi|^{2p} \mid \eta(t_0) = i\right]\right).$$

Since $x_m(t) \to x(t)$ a.s. uniform on $[t_0, T]$ (see [68]) from Fatou's Lemma it follows that

$$E\left[|x(t)|^{2p}|\eta(t_0) = i\right] \le K\left(1 + E\left[|\xi|^{2p}|\eta(t_0) = i\right]\right), t \in [t_0, T], i \in \mathcal{D}$$

and the proof is complete.   □

With the same proof used for stochastic differential Itô systems (see [120, 141]) one can prove the following result.

Theorem 1.11.3. *Under the assumptions of Theorem 1.11.2, suppose that f and F are continuous functions for each $i \in \mathcal{D}$. Then the function*

$$(t, x) \in [t_0, \infty) \times \mathbf{R}^n \to x(t, t_0, x)$$

is a.s. continuous for each $t_0 \ge 0$, hence $x(t, t_0, \cdot)$ defined on $\mathbf{R}^n \times \Omega$ is measurable with respect to $\mathcal{B}(\mathbf{R}^n) \otimes \mathcal{H}_{t_0,t}, t > t_0$, where

$$\mathcal{H}_{t_0,t} = \sigma\left(w(s) - w(t_0), \eta(s); s \in [t_0, t]\right).$$

Based on the inequality (1.19) one can obtain an Itô-type formula for the solution of the system (1.16) in case $a = 0, \sigma = 0$ and in more general assumptions for the functions $v(t, x, i)$ than the ones used in Theorem 1.10.2.

The result giving such a formula has been proved in [101].

Theorem 1.11.4. *Assume that the hypothesis of Theorem 1.11.2 are fulfilled and additionally $f(\cdot, \cdot, i), F(\cdot, \cdot, i)$ are continuous on $\mathbf{R}_+ \times \mathbf{R}^n$, for all $i \in \mathcal{D}$. Let $v : \mathbf{R}_+ \times \mathbf{R}^n \times \mathcal{D}$ be a function which for each $i \in \mathcal{D}$ is continuous together with its derivatives $v_t, v_x,$ and v_{xx}.*

Assume also that there exists $\gamma > 0$ such that:

$$|v(t, x, i)| + \left|\frac{\partial v}{\partial t}(t, x, i)\right| + \left|\frac{\partial v}{\partial x}(t, x, i)\right| + \left\|\frac{\partial^2 v}{\partial x \partial x}(t, x, i)\right\|$$

$$\le K_T(1 + |x|^{\gamma}), t \in [0, T], x \in \mathbf{R}^n, i \in \mathcal{D}$$

where $K_T > 0$ depends on T. Then we have:

$$E\left[v(s, x(s), \eta(s))|\eta(t_0) = i\right] - v(t_0, x_0, i)$$

$$= E\left[\int_{t_0}^{s}\left\{\frac{\partial v}{\partial t}(t, x(t), \eta(t)) + \left(\frac{\partial v}{\partial x}(t, x(t), \eta(t))\right)^T f(t, x(t), \eta(t))\right.\right.$$

$$+\frac{1}{2}TrF^T(t,x(t),\eta(t))\frac{\partial^2 v}{\partial x \partial x}(t,x(t),\eta(t)) \tag{1.20}$$

$$\times F(t,x(t),\eta(t))+\sum_{j=1}^{d}v(t,x(t),\eta(t))q_{\eta(t)j}\Bigg\} dt\,\Big|\eta(t_0)=i\Bigg],$$

$$x(t)=x(t,t_0,x_0),\, x_0 \in \mathbf{R}^n,\, t \geq t_0 \geq 0,$$

for all $s \geq t_0, i \in \mathcal{D}$.

Proof. From Theorem 1.11.2 it follows that for all positive integer p we have

$$\sup_{t_0 \leq t \leq T} E[|x(t)|^{2p}|\eta(t_0)=i] \leq K(1+|x_0|^{2p})$$

Therefore using Theorem 1.2.4 for $\alpha = 2$ it follows that it is possible to take the limits in the integrals from the first step in the proof of Theorem 1.10.2, obtaining that

$$\lim_{h\to 0_+}\frac{1}{h}E\left[\{v((t+h),x(t+h),\eta(t+h))\right. \tag{1.21}$$

$$\left.-v(t,x(t),\eta(t))-\int_t^{t+h}m(s)ds\right\}\Big|\eta(t_0)=i\Bigg]$$

$$=0$$

where

$$m(t)=\frac{\partial v}{\partial t}(t,x(t),\eta(t))+\left(\frac{\partial v}{\partial x}(t,x(t),\eta(t))\right)^T f(t,x(t),\eta(t))$$

$$+\frac{1}{2}TrF^T(t,x(t),\eta(t))\frac{\partial^2 v}{\partial x \partial x}(t,x(t),\eta(t))$$

$$\times F(t,x(t),\eta(t))+\sum_{j=1}^{d}v(t,x(t),\eta(t))q_{\eta(t)j}.$$

Taking into account that $\eta(t)$ is continuous in probability and using again Theorems 1.11.2 and 1.2.4 for $\alpha = 2$ it follows immediately that

$$E\left[\left(v(t,x(t),\eta(t))-\int_{t_0}^t m(s)ds\right)\Big|\eta(t_0)=i\right]$$

is a continuous function and therefore from (1.21) it results that (1.20) holds and the proof is complete.   □

Remark 1.11.1. (i) The proof of the previous theorem shows that the result in the statement is also valid for random initial conditions ξ, $\mathcal{H}_{t_0}$-measurable, and $E\left[|\xi|^{2p}\right] < \infty$ for $p \geq \gamma + 2$.

(ii) From Theorems 1.11.1 and 1.11.2 it results that for the system (1.16) Theorem 1.11.4 is not applicable while in the case when $a(t) \equiv 0$ and $\sigma(t) \equiv 0$ we can use Theorem 1.11.4 due to the estimate (1.19).

(iii) In many cases, in the following developments we shall consider the system (1.16) with $a(t) \neq 0$ and $\sigma(t) \neq 0$, being thus obliged to use Theorem 1.10.2.

1.12 Stochastic Linear Differential Equations

Since the problems investigated in this book refer to stochastic linear controlled systems we recall here some facts concerning the solutions of stochastic linear differential equations.

Let us consider the system of linear differential equations:

$$dx(t) = A_0(t, \eta(t)) x(t) dt + \sum_{k=1}^{r} A_k(t, \eta(t)) x(t) dw_k(t) \qquad (1.22)$$

where $t \to A_k(t, i) : \mathbf{R}_+ \to \mathbf{R}^{n \times n}$, $i \in \mathcal{D}$, are bounded and continuous matrix valued functions.

The system (1.22) has two important particular forms:

(i) $A_k(t, i) = 0$, $k = 1, \ldots, r$, $t \geq 0$. In this case (1.22) becomes

$$\dot{x}(t) = A(t, \eta(t)) x(t), t \geq 0 \qquad (1.23)$$

where $A(t, \eta(t))$ stands for $A_0(t, \eta(t))$ and it corresponds to the case when the system is subject only to Markovian jumping;

(ii) $\mathcal{D} = \{1\}$; in this situation the system (1.22) becomes

$$dx(t) = A_0(t) x(t) dt + \sum_{k=1}^{r} A_k(t) x(t) dw_k(t), t \geq 0 \qquad (1.24)$$

where $A_k(t) := A_k(t, 1)$, $k = 0, \ldots, r$, $t \geq 0$, representing the case when the system is subject only to white noise-type perturbations.

Definition 1.12.1. We say that the system (1.22) is *time invariant* (or it is in the *stationary case*) if $A_k(t, i) = A_k(i)$, for all $k = 0, \ldots, r$, $t \in \mathbf{R}_+$ and $i \in \mathcal{D}$. In this case the system (1.22) becomes

$$dx(t) = A_0(\eta(t)) x(t) dt + \sum_{k=1}^{r} A_k(\eta(t)) x(t) dw_k(t). \qquad (1.25)$$

Applying Theorem 1.11.2, it follows that for each $t_0 \geq 0$ and each random vector $\xi, \mathcal{H}_{t_0}$-measurable and $E|\xi|^2 < +\infty$ the system (1.22) has a unique solution $x(t;t_0,\xi)$ which verifies $x(t_0,t_0,\xi) = \xi$. Moreover if $E|\xi|^{2p} < +\infty$, $p \geq 1$, then

$$\sup_{t \in [t_0,T]} E\left[|x(t,t_0,\xi)|^{2p} \mid \eta(t_0) = i\right] \leq cE\left[|\xi|^{2p} \mid \eta(t_0) = i\right],$$

$i \in \mathcal{D}$, $c > 0$ depending upon $T, T - t_0$ and p. For each $k \in \{1,2,\ldots,n\}$ we denote $\Phi_k(t,t_0) = x(t,t_0,e_k)$ where $e_k = (0,0,\ldots,1,0,\ldots,0)^T$ and set

$$\Phi(t,t_0) = (\Phi_1(t,t_0) \ \Phi_2(t,t_0) \ldots \Phi_n(t,t_0)).$$

$\Phi(t,t_0)$ is the matrix valued solution of the system (1.22) which verifies $\Phi(t_0,t_0) = I_n$. If ξ is a random vector $\mathcal{H}_{t_0}$-measurable with $E|\xi|^2 < \infty$, we denote $\tilde{x}(t) = \Phi(t,t_0)\xi$. By Remark 1.9.3 it is easy to verify that $\tilde{x}(t)$ is a solution of the system (1.22) verifying $\tilde{x}(t) = \xi$. Then, by uniqueness arguments we conclude that $\tilde{x}(t) = x(t,t_0,\xi)$ a.s., $t \geq t_0$. Hence we have the representation formula

$$x(t,t_0,\xi) = \Phi(t,t_0)\xi \text{ a.s.}$$

The matrix $\Phi(t,t_0), t \geq t_0 \geq 0$ will be termed the *fundamental matrix solution* of the system of stochastic linear differential equations (1.22). By uniqueness argument it can be proved that

$$\Phi(t,s)\Phi(s,t_0) = \Phi(t,t_0) \text{ a.s.} t \geq s \geq t_0 \geq 0.$$

Proposition 1.12.1. *The matrix $\Phi(t,t_0)$ is invertible and its inverse is given by:*

$$\Phi^{-1}(t,t_0) = \widetilde{\Phi}^T(t,t_0) \text{ a.s.} t \geq t_0 \geq 0,$$

where $\widetilde{\Phi}(t,t_0)$ is the fundamental matrix solution of the stochastic linear differential equation:

$$dy(t) = \left[-A_0^T(t,\eta(t)) + \sum_{k=1}^{r}\left(A_k^2(t,\eta(t))\right)^T\right]y(t)\,dt \qquad (1.26)$$
$$- \sum_{k=1}^{r} A_k^T(t,\eta(t))\,y(t)\,dw_k(t).$$

Proof. Applying the Itô's formula (Theorem 1.9.9) to the function

$$v(t,x,y) = y^T x, \ t \geq t_0, \ x,y \in \mathbf{R}^n$$

and to the systems (1.22) and (1.26) we obtain:

$$y^T \widetilde{\Phi}^T (t,t_0) \Phi(t,t_0) x - y^T x = 0 \text{ a.s.} t \geq t_0 \geq 0, \ x,y \in \mathbf{R}^n,$$

hence $\widetilde{\Phi}^T (t,t_0) \Phi(t,t_0) = I_n$ a.s. $t \geq t_0$, and the proof is complete.   □

Let us consider the affine system of stochastic differential equations:

$$dx(t) = [A_0(t, \eta(t))x(t) + f_0(t)] dt \qquad (1.27)$$

$$+ \sum_{k=1}^{r} [A_k(t, \eta(t))x(t) + f_k(t)] dw_k(t),$$

$t \geq 0$, where $f_k : \mathbf{R}_+ \times \Omega \to \mathbf{R}^n$ are stochastic processes with components in $L^2_{\eta,w}([0,T])$ for all $T > 0$. Using Theorem 1.11.1 we deduce that for all $t_0 \geq 0$ and for all random vector ξ, $\mathcal{H}_{t_0}$-measurable with $E|\xi|^2 < \infty$, the system (1.27) has a unique solution $x_f(t,t_0,\xi)$, $f = (f_0, f_1, \ldots, f_r)$. Additionally, for all $T > t_0$, there exists a positive constant c depending on T, $T - t_0$ such that

$$\sup_{t \in [t_0,T]} E\left[|x_f(t,t_0,\xi)|^2 \mid \eta(t_0) = i\right] \qquad (1.28)$$

$$\leq cE\left[\left(|\xi|^2 + \sum_{k=0}^{r} \int_{t_0}^{T} |f_k(s)|^2 ds\right) \mid \eta(t_0) = i\right].$$

Let $\Phi(t,t_0)$, $t \geq t_0 \geq 0$ be the fundamental matrix solution of the linear system obtained by taking $f_k = 0$ in (1.27) and set $z(t) = \Phi^{-1}(t,t_0)x_f(t,t_0,\xi)$. Applying the Itô's formula (Theorem 1.9.9) to the function $v(t,x,y) = y^T x$, $x,y \in \mathbf{R}^n$ and to the systems (1.26) and (1.27), we obtain

$$y^T z(t) = y^T z(t_0) + y^T \int_{t_0}^{t} \Phi^{-1}(s,t_0) \left[f_0(s) - \sum_{k=1}^{r} A_k(s, \eta(s)) f_k(s)\right] ds$$

$$+ \sum_{k=1}^{r} y^T \int_{t_0}^{t} \Phi^{-1}(s,t_0) f_k(s) dw_k(s) \text{ a.s.},$$

$t \geq t_0, y \in \mathbf{R}^n$. Since y is arbitrary in $\mathbf{R}^n$ we may conclude that

$$z(t) = \xi + \int_{t_0}^{t} \Phi^{-1}(s,t_0) \left[f_0(s) - \sum_{k=1}^{r} A_k(s, \eta(s)) f_k(s)\right] ds$$

$$+ \sum_{k=1}^{r} \int_{t_0}^{t} \Phi^{-1}(s,t_0) f_k(s) dw_k(s) \text{ a.s.},$$

$t \geq t_0$. Thus we obtained the following representation formula

$$x_f(t, t_0, \xi) = \Phi(t, t_0) \tag{1.29}$$

$$+\Phi(t, t_0) \int_{t_0}^t \Phi^{-1}(s, t_0) \left[f_0(s) - \sum_{k=1}^r A_k(s, \eta(s)) f_k(s) \right] ds$$

$$+ \sum_{k=1}^r \Phi(t, t_0) \int_{t_0}^t \Phi^{-1}(s, t_0) f_k(s) dw_k(s) \ a.s.,$$

$t \geq t_0$, which extends the well-known constants variation formula from the deterministic framework to the case of stochastic affine system (1.27).

1.13 Standard Homogeneous Markov Processes with a Countable Infinite Number of States

Several results derived in this book refer to stochastic controlled systems modeled by stochastic differential equations containing an r-dimensional standard Wiener process $\{w(t)\}_{t \geq 0}$ together with a standard homogeneous Markov process $\{\eta(t), P(t), \mathbf{Z}_+\}$. It is known (see [32,71]) that if $P(t), t > 0$ has the properties from Definition 1.8.3, together with (iii) from Definition 1.8.4, then there exist the limits:

$$\lim_{t \to 0_+} \frac{p_{ij}(t) - \delta_{ij}}{t} = q_{ij} \tag{1.30}$$

$i, j \in \mathbf{Z}_+$. We have $q_{ij} \in [0, \infty)$, if $i \neq j$ and $q_{ii} \leq 0 \ \forall i \in \mathbf{Z}_+$.

One knows (see [32,71]) that in the case $\mathcal{D} = \mathbf{Z}_+$ it is possible to have $q_{ii} = -\infty$ for some states $i \in \mathbf{Z}_+$. Even if $q_{ii} \in (-\infty, 0]$ for every $i \in \mathbf{Z}_+$ it is not sure that $\sum_{j=0}^\infty q_{ij} = 0, \ \forall i \in \mathbf{Z}_+$. *That is why, throughout this book, every time when we talk about a standard homogeneous Markov process $\{\eta(t), P(t), \mathbf{Z}_+\}$ we assume that the following conditions are fulfilled*:

$$-\infty < q_{ii} \leq 0$$
$$\sum_{j=0}^\infty q_{ij} = 0, i \in \mathbf{Z}_+, \tag{1.31}$$

$$q = \sup_{i \in \mathbf{Z}_+} |q_{ii}| < \infty \tag{1.32}$$

and

$$\pi_i = \mathcal{P}\{\eta(0) = i\} > 0, \forall i \in \mathbf{Z}_+. \tag{1.33}$$

One may show (see [71]) that the condition (1.31) is equivalent to the fact that $p_{ij}(t)$ satisfy the Kolmogorov's differential equations:

$$\frac{d}{dt}p_{ij}(t) = \sum_{k=0}^{\infty} q_{ik}p_{kj}(t), i,j \in \mathbf{Z}_+, t > 0. \tag{1.34}$$

Furthermore, if (1.31) and (1.32) hold, then we have $P(t) = e^{Qt}, t > 0$ where $e^{Qt} = \sum_{k=0}^{\infty} \frac{Q^k t^k}{k!}$.

This series of infinite matrices is convergent in the norm

$$\|Q\| = \|Q\|_{\infty} = \sup_{(i,j) \in \mathbf{Z}_+ \times \mathbf{Z}_+} |q_{ij}|.$$

Often the matrix $Q = (q_{ij})_{(i,j) \in \mathbf{Z}_+ \times \mathbf{Z}_+}$ will be named the generator matrix.

Based on Theorem 2.2 Chap. 6 in [32] one may derive the following important result:

Theorem 1.13.1. *(i) (1.32) is equivalent with $\lim_{t \to 0+} p_{ii}(t) = 1$ uniformly with respect to $i \in \mathbf{Z}_+$,*

(ii) $\lim_{t \to 0+} \frac{p_{ii}(t)-1}{t} = q_{ii}$ uniformly with respect to $i \in \mathbf{Z}_+$ if (1.32) is satisfied.

A classical example of a standard homogenous Markov process $\{\eta(t), P(t), \mathbf{Z}_+\}$ satisfying conditions (1.31) and (1.32) is represented by a homogeneous Poisson process with parameter $\lambda > 0$. This stochastic process is characterized by

$$p_{ij}(t) = \begin{cases} ((\lambda t)^{j-i}/(j-i)!)e^{-\lambda t} & \text{if } j \geq i, \\ 0 & \text{if } j < i \end{cases}$$

$t > 0, i, j \in \mathbf{Z}_+$.

In this case, the elements (q_{ij}) of the generator matrix Q are given by: $q_{ii} = -\lambda, q_{i,i+1} = \lambda$ and $q_{ij} = 0$ if $j \in \mathbf{Z}_+ - \{i, i+1\}, i \in \mathbf{Z}_+$.

A consequence of the properties (1.31) and (1.32) is.

Theorem 1.13.2. *If $\{\eta(t), P(t), \mathbf{Z}_+\}$ is a standard homogeneous Markov process with a countable number of states, then the stochastic process $\{\eta(t)\}_{t \geq 0}$ is continuous in probability.*

The proof is the same as in the case when $\mathcal{D}$ is finite (see [32]), using Theorem 1.13.1 (i). □

In the next chapter, beside the standard homogeneous Markov process $\{\eta(t), P(t), \mathbf{Z}_+\}$ an r-dimensional standard Wiener process $\{w(t)\}_{t \geq 0}$ is considered.

In this case, the meaning of the notations $\mathcal{F}_t, \mathcal{G}_t, \mathcal{H}_t, \mathcal{R}_t$ introduced in Sect. 1.9, will be preserved.

We will assume also that for any $t \geq 0$ the σ algebras $\mathcal{G}_t$ is independent of $\mathcal{F}_t$. In this case, Theorem 1.10.2 becomes.

Theorem 1.13.3. *Let* $a : [t_0, T] \times \Omega \to \mathbf{R}^n$, $\sigma : [t_0, T] \times \Omega \to \mathbf{R}^{n \times r}$ *be such that* $a = (a_1, a_2, \ldots, a_n)^T$, $\sigma = (\sigma_{ij})_{1 \le i \le n1 \le j \le r}$ *with* a_k, σ_{ij} *are in* $L^2_{\eta, w}([t_0, T])$. *Let* ξ *be an n-dimensional random vector, which is* $\mathcal{H}_{t_0}$-*measurable and* $E|\xi|^2 < \infty$.

Consider the functions $K(\cdot, i) : [t_0, T] \to \mathbf{R}^{n \times n}$, $k(\cdot, i) : [t_0, T] \to \mathbf{R}^n$ *and* $k_0(\cdot, i) : [t_0, T] \to \mathbf{R}$, $i \in \mathbf{Z}_+$ *which are assumed to be* C^1-*functions.*

Assume that $K(t, i) = K^T(t, i)$, $(t, i) \in [t_0, T] \times \mathbf{Z}_+$ *and*

$$\sup_{i \in \mathbf{Z}_+, t \in [t_0, T]} (|K(t, i)| + |\frac{d}{dt} K(t, i)| + |k(t, i)| + |\frac{d}{dt} k(t, i)| + |k_0(t, i)| + |\frac{d}{dt} k_0(t, i)|) < \infty.$$

If $v(t, x, i) = x^T K(t, i) x + 2k^T(t, i) x + k_0(t, i)$, $(t, x, i) \in [t_0, T] \times \mathbf{R}^n \times \mathbf{Z}_+$, *then the following equality holds:*

$$E\left[(v(t, x(t), \eta(t)) - v(t_0, \xi, i)) | \eta(t_0) = i\right]$$

$$= E\left[\int_{t_0}^t \left\{ \frac{\partial v}{\partial t}(s, x(s), \eta(s)) + a^T(s) \frac{\partial v}{\partial x}(s, x(s), \eta(s)) \right. \right. \tag{1.35}$$

$$\left. \left. + Tr(\sigma^T(s) K(s, \eta(s)) \sigma(s)) + \sum_{j=0}^\infty v(s, x(s), j) q_{\eta(s), j} \right\} ds | \eta(t_0) = i \right]$$

for all $i \in \mathbf{Z}_+$ *and for the stochastic process* $x(t), t \in [t_0, T]$ *verifying*

$$dx(t) = a(t)dt + \sigma(t)dw(t), \ t \in [t_0, T] \ and \ x(t_0) = \xi.$$

Proof. May be done following step by step the proof of Theorem 1.10.2 with an obvious change of $v(\cdot, \eta(t+h)) = \sum_{j=0}^\infty \chi_{\{\eta(t+h)=j\}} v(\cdot, j)$ instead of $v(\cdot, \eta(t+h)) = \sum_{j=1}^d \chi_{\{\eta(t+h)=j\}} v(\cdot, j)$.

The conditions (1.31) and (1.32) allow us (via Theorem 1.13.1) to apply Lebesque's Theorem to derive the analogous of (1.12). Also, the conditions (1.31) and (1.32) allow us to show that $h_i(t)$ are right continuous functions for each $i \in \mathbf{Z}_+$ and, by using Theorem 1.13.2 one concludes that $G_i.(t), i \in \mathbf{Z}$ is a continuous function.

Let us notice that all results derived in Sects. 1.11 and 1.12 are also valid, based on the same proof, in the case when $(\eta(t), P(t), \mathbf{Z}_+)$ is a standard homogeneous Markov process.

Chapter 2
Linear Differential Equations with Positive Evolution on Ordered Banach Spaces

In this chapter the problem of exponential stability of the zero state equilibrium of a class of linear differential equations on a real ordered Banach space is investigated.

The linear differential equations under consideration named *differential equations with positive evolution* are defined by a special class of strongly continuous operator valued functions. These differential equations are natural extensions to the time-varying case of linear differential equations with constant coefficients on an ordered Banach space defined by a linear and bounded operator with positive semigroup.

In the time-varying framework one distinguishes two kinds of positive evolutions: causal positive evolution and anticausal positive evolution. The linear differential equations with positive evolution studied in this chapter contain as special cases Lyapunov-type differential equations arising in a natural way in connection with the problem of exponential stability in mean square of a stochastic linear differential equation affected simultaneously by multiplicative white noise perturbations and Markovian jumping.

The main tool in the derivation of the criteria for exponential stability of linear differential equations with positive evolution is the Minkovski norm introduced based on a suitable convex subset. A list of useful properties of Minkovski norm together with the basic properties of linear and positive operators may be found in the first section of the chapter. This section contains also several examples of infinite dimensional ordered Banach spaces.

Properties of operator valued functions defining causal positive evolution or anticausal positive evolution are emphasized in Sect. 2.2. Both the case of the causal exponential stability and the case of the anticausal exponential stability are considered in the third section of the chapter. The criteria for the two kinds of exponential stability are derived in terms of the existence of some globally defined and bounded solutions of some suitable forward and backward affine differential equations and in terms of solvability of some forward (backward) differential inequations.

V. Dragan et al., *Mathematical Methods in Robust Control of Linear Stochastic Systems*,
DOI 10.1007/978-1-4614-8663-3_2, © Springer Science+Business Media New York 2013

The problem of robustness of exponential stability with respect to some additive perturbations modeled by positive operator valued functions is analyzed in the case when the involved operator valued functions are periodic.

The last part of the chapter is devoted to the investigation of the properties of linear evolution operators associated with Lyapunov-type differential equations on some suitable ordered Banach spaces of finite or infinite sequences of symmetric matrices. Criteria for exponential stability of a Lyapunov-type differential equation are finally derived as direct consequences of the criteria obtained in the general case in the first part of the chapter.

2.1 Convex Cones. Minkovski Norms

Throughout this chapter, $(\mathcal{X}, \|\cdot\|)$ is a real Banach space and $\mathcal{X}^*$ stands for its dual space. If $M \subset \mathcal{X}$ is a subset, then $intM$ or M stands for the set of interior elements of M with respect to the topology induces by the norm $\|\cdot\|$. By $\overline{M}$ we denote the smallest closed subset containing M, while ∂M denotes the border of the subset M. One can show that $\overline{M} = intM \cup \partial M$.

2.1.1 Some Basic Facts on Convex Cones

In this subsection we collect several basic definitions and results regarding the convex cones and ordered Banach spaces. For more details concerning the convex cones and ordered linear spaces we refer to [27,55,95,97] and references therein.

Definition 2.1.1. A subset $C \subset \mathcal{X}$ is called *convex cone* if:

(i) $C + C \subset C$
(ii) $\alpha C \subset C$ for all $\alpha \in \mathbf{R}, \alpha \geq 0$.

We recall that if A, B are two subsets of $\mathcal{X}$ and $\alpha \in \mathbf{R}$, then $A + B = \{x + y | x \in A, y \in B\}$ and $\alpha A = \{\alpha x | x \in A\}$.

It is easy to see that a cone C is a convex subset and thus we shall say *convex cone* when we refer to a cone.

A convex cone $C \subset \mathcal{X}$ induces an ordering "$\leq$" on $\mathcal{X}$, by $x \leq y$ (or equivalently $y \geq x$) if and only if $y - x \in C$. If C is a convex cone, then $x < y$ (or equivalently $y > x$) if and only if $y - x \in IntC$. Hence $C = \{x \in \mathcal{X} | x \geq 0\}$ and $IntC = \{x \in \mathcal{X} | x > 0\}$. That is why, in the rest of the chapter we shall use the notation $\mathcal{X}^+$ for the convex cone which induces the order relation on $\mathcal{X}$.

Definition 2.1.2. (i) A cone C is called a *pointed cone* if $C \cap (-C) = \{0\}$.
(ii) A cone C is called a *solid cone* if its interior $IntC$ is not empty.
(iii) A cone C is called *normal cone* if there exists a real number $\tilde{b} > 0$ such that $\|x\| \leq \tilde{b}\|y\|$ if $0 \leq x \leq y$.

Remark 2.1.1. (i) If in the definition of a normal cone we may take $\tilde{b} = 1$ we shall say that the norm $\| \cdot \|$ is *monotone* with respect to the convex cone C.

(ii) If C is a normal cone, then it is pointed cone. Indeed, if x is such that x and $-x$ are in C, then from $(1 + \frac{1}{n})x \in C$ we have $0 \leq -x \leq \frac{1}{n}x$. Hence $\|x\| \leq \frac{\tilde{b}}{n}\|x\|$. Taking the limit for $n \to \infty$ we deduce that $\|x\| = 0$ hence $x = 0$. Thus we obtained that C is pointed cone.

Definition 2.1.3. If $C \subset \mathcal{X}$ is a cone, then $C^* \subset \mathcal{X}^*$ is called the *dual cone* of C if C^* consists of all bounded and linear functionals $\phi \in \mathcal{X}^*$ with the property that $\phi(x) \geq 0$ for all $x \in C$.

Based on Ritz theorem for representation of a bounded linear functional on a Hilbert space one sees that if $\mathcal{X}$ is a real Hilbert space then the dual cone C^* of a convex cone C may be defined as $C^* = \{y \in \mathcal{X} \mid \,<y,x> \geq 0, \forall x \in C\}$ where $< \cdot, \cdot >$ is the inner product on $\mathcal{X}$.

If $\mathcal{X}$ is a real Hilbert space a cone C is called *selfdual* if $C^* = C$.

Remark 2.1.2. It is worth mentioning that if $\mathcal{X}$ is a real Hilbert space ordered by the order relation induced by a self dual convex cone $\mathcal{X}_+$, then the usual norm on $\mathcal{X}$ is monotone with respect to the cone $\mathcal{X}_+$. Indeed if $x, y \in \mathcal{X}_+$ are such that $x \leq y$, then both $y + x$ and $y - x$ lie in $\mathcal{X}_+$. Then we have $\langle y + x, y - x \rangle \geq 0$ because $\mathcal{X}_+^* = \mathcal{X}_+$. We obtain $\langle y, y \rangle - \langle x, x \rangle \geq 0$ or equivalently $\|x\|^2 \leq \|y\|^2$. Hence, $\| \cdot \|$ is monotone with respect to $\mathcal{X}_+$.

Remark 2.1.3. (i) If $C \subset \mathcal{X}$ is a convex cone, then the following equivalences hold:

$$C \neq \mathcal{X} \iff 0 \notin IntC \iff 0 \in \partial C.$$

(ii) If $\dim \mathcal{X} \geq 2$ and $C \subset \mathcal{X}$ is a convex cone such that $C \neq \mathcal{X}$ and $C \neq \{0\}$, then $\partial C \setminus \{0\}$ is an infinite subset.

The following two results will be used in the next section; they present also interest in themselves.

Lemma 2.1.1. *Let $C \subset \mathcal{X}$ be a solid convex cone, $C \neq \mathcal{X}$. Then, for every $\varphi \in C^* \setminus \{0\}$ we have $\varphi(x) > 0$ for all $x \in IntC$.*

Proof. Let $\varphi \in C^* \setminus \{0\}$. This means that there exists $x_0 \in \mathcal{X}$ such that $\varphi(x_0) \neq 0$ and $\varphi(x) \geq 0$ for all $x \in C$. Let us assume that there exists $x_1 \in IntC$ such that $\varphi(x_1) = 0$. Since $IntC$ is an open set we deduce that there exists $\varepsilon_0 > 0$ sufficiently small with the property that $x_t = x_1 + \frac{t}{\varphi(x_0)}x_0 \in IntC$, if $|t| < \varepsilon_0$. So, from $\varphi(x_t) \geq 0$ we deduce that $t \geq 0$ for arbitrary $t \in (-\varepsilon_0, \varepsilon_0)$ which is a contradiction. This completes the proof. $\qquad\square$

Theorem 2.1.2. *Let $C \subset \mathcal{X}$ be a solid convex cone $C \neq \mathcal{X}$. Then for each $x_0 \in \partial C$ there exists $\varphi \in C^* \setminus \{0\}$ such that $\varphi(x_0) = 0$.*

Proof. If $x_0 = 0 \in \partial C$, the conclusion is obvious. Let us take $x_0 \in \partial C \setminus \{0\}$. Let $\hat{C} = \{tx_0; t \geq 0\}$. One can see that $\hat{C}$ is a closed convex cone. If there exists $t > 0$

such that $tx \in IntC$, then $x \in IntC$. Hence $\hat{C} \cap IntC = \emptyset$. Since $IntC$ is an open convex set we may apply a separation theorem to obtain the existence of a linear bounded functional $\varphi \in \mathcal{X}^* \setminus \{0\}$ and a real number c, such that $\varphi(y) \geq c \geq \varphi(z)$ for arbitrary $y \in IntC$ and $z \in \hat{C}$ (for details see [55]). Since $C \subset \overline{IntC}$ (see [55]) we deduce $\overline{C} \subset \overline{IntC}$. This allows us to deduce that $\varphi(y) \geq c \geq \varphi(z)$ for any $y \in \overline{C}$ and $z \in \hat{C}$. Taking $y = z = 0$ we obtain that $c = 0$; this means that $\varphi \in C^* \setminus \{0\}$. On the other hand we have $\varphi(x_0) \geq 0 \geq \varphi(1 \cdot x_0)$ which leads to $\varphi(x_0) = 0$. The proof is complete.

$\square$

2.1.2 Several Examples of Ordered Banach Spaces

This subsection collects several examples of real ordered Banach spaces. Part of them will play an important role in the next chapters of the book.

As usual $|x|$ stands for the euclidian norm of a vector $x \in \mathbf{R}^n$, that is, $|x| = (x^T x)^{1/2}$. For a matrix $A \in \mathbf{R}^{m \times n}$, $|A|$ stands for the matrix norm induced by the euclidian norm $|\cdot|$, that is

$$|A| = \sup_{|x| \leq 1} |Ax|. \tag{2.1}$$

Also, we shall use the notation $|A|_2$ for the Frobenius norm of the matrix A, i.e.

$$|A|_2 = \left(Tr[A^T A] \right)^{1/2} \tag{2.2}$$

where $Tr[\cdot]$ stands for the trace operator. Beside the two norms introduced before, we shall use the norm

$$|A|_1 = Tr\left[(A^T A)^{1/2} \right] \tag{2.3}$$

where $(A^T A)^{1/2}$ is the unique positive semidefinite matrix X such that $X^2 = A^T A$.

Let $\mathcal{S}_n \subset \mathbf{R}^{n \times n}$ be the linear subspace of symmetric matrices of size $n \times n$, that is $S \in \mathcal{S}_n$ iff $S = S^T$.

The restrictions of the norm (2.1)–(2.3) to the subspace $\mathcal{S}_n$ take the equivalent form:

$$|S| = max\{|\lambda|; \lambda \in \Lambda(S)\} = \sup_{|x| \leq 1} \{|x^T S x|\} \tag{2.4}$$

$$|S|_2 = \left(\sum_{i=1}^n \lambda_i^2 \right)^{1/2} \tag{2.5}$$

$$|S|_1 = \sum_{i=1}^n |\lambda_i| \tag{2.6}$$

where $\lambda_1, \ldots, \lambda_n \in \Lambda(S)$ with $\Lambda(S)$ is the spectrum of the matrix S.

For a matrix $S \in \mathcal{S}_n$ the following hold:

$$|S| \leq |S|_2 \leq |S|_1 \leq n|S|. \tag{2.7}$$

As we established in the previous chapter $\mathcal{D}$ denotes either the set $\{1, 2, \ldots, d\}$ or the set $\mathbf{Z}_+$.

Example 2.1.1. Let $\mathcal{X} = \mathcal{S}_n^{\mathcal{D}} = \ell^{\infty}\{\mathcal{D}, \mathcal{S}_n\}$ be the linear space of the bounded sequences of symmetric matrices, that is

$$\ell^{\infty}\{\mathcal{D}, \mathcal{S}_n\} = \left\{ \mathbf{X} = \{X(i)\}_{i \in \mathcal{D}} | X(i) \in \mathcal{S}_n, i \in \mathcal{D}, \sup_{i \in \mathcal{D}} |X(i)| < +\infty \right\}.$$

The space $\mathcal{S}_n^{\mathcal{D}}$ equipped with the norm

$$\|\mathbf{X}\|_{\infty} = \sup_{i \in \mathcal{D}} |X(i)| \tag{2.8}$$

is a real Banach space. On $\mathcal{S}_n^{\mathcal{D}}$ we consider the ordering induced by the cone $\mathcal{X}_+ = \mathcal{S}_{n+}^{\mathcal{D}} = \ell^{\infty}\{\mathcal{D}, \mathcal{S}_{n+}\}$ where

$$\ell^{\infty}\{\mathcal{D}, \mathcal{S}_{n+}\} = \{\mathbf{X} = \{X(i)\}_{i \in \mathcal{D}}; X(i) \geq 0, i \in \mathcal{D}\}. \tag{2.9}$$

Here $X(i) \geq 0$ means that $X(i)$ is positive semidefinite. One verifies that $\mathcal{X}_+$ is a closed, solid, convex cone. Its interior $Int\mathcal{X}_+$ consists of the subset of the sequences $\mathbf{X} = \{X(i), i \in \mathcal{D}; X(i) \geq \delta I_n, \forall i \in \mathcal{D} \text{ for some } \delta > 0 \text{ independent of } i\}$.

Based on the monotonicity of the norm $|\cdot|$ on $\mathcal{S}_n$ one obtains that $\|\cdot\|_{\infty}$ introduced by (2.8) is monotone with respect to the cone $\mathcal{X}_+$. Hence $\mathcal{X}_+$ is a normal cone.

In the next chapters we shall use $\mathcal{S}_n^d$ instead of $\mathcal{S}_n^{\mathcal{D}}$ and $\mathcal{S}_{n+}^d$ for $\mathcal{S}_{n+}^{\mathcal{D}}$ when $\mathcal{D} = \{1, 2, .., d\}$, while $\mathcal{S}_n^{\infty}$ is used for $\mathcal{S}_n^{\mathcal{D}}$ when $\mathcal{D} = \mathbf{Z}_+$. In the last case, $\mathcal{S}_n^{\infty}$ stand for the convex cone $\ell^{\infty}(\mathbf{Z}_+, \mathcal{S}_{n+})$. It is obvious that $(\mathcal{S}_n^d, \|\cdot\|_{\infty})$ is a finite dimensional real ordered Banach space, while $(\mathcal{S}_n^{\infty}, \|\cdot\|_{\infty})$ is an infinite dimensional real ordered Banach space.

Example 2.1.2. Let $\mathcal{X} = \ell^1(\mathcal{D}, \mathcal{S}_n)$, where

$$\ell^1(\mathcal{D}, \mathcal{S}_n) = \{\mathbf{X} = \{X(i)\}_{i \in \mathcal{D}} \subset \mathcal{S}_n; \sum_{i \in \mathcal{D}} |X(i)|_1 < \infty\}.$$

Taking

$$\|\mathbf{X}\|_1 = \sum_{i \in \mathcal{D}} |X(i)|_1 \tag{2.10}$$

one obtains that $(\mathcal{X}, \|\cdot\|_1)$ is a real Banach space.

On the Banach space $(\mathcal{X}, \|\cdot\|_1)$ we consider the order relation induced by the convex cone $\mathcal{X}_+ = \ell^1(\mathcal{D}, \mathcal{S}_{n+}) = \{\mathbf{X} \in \ell^1(\mathcal{D}, \mathcal{S}_n); \mathbf{X} = \{X(i)\}_{i \in \mathcal{D}}, X(i) \geq 0\}$.

It is a closed convex cone. In the case $\mathcal{D} = \{1, 2, \ldots, d\}$, $\ell^1(\mathcal{D}, \mathcal{S}_n)$ coincides with $\mathcal{S}_n^d$ and $\ell^1(\mathcal{D}, \mathcal{S}_{n+})$ coincides with $\mathcal{S}_{n+}^d$. In the case $\mathcal{D} = \mathbf{Z}_+$, $\mathcal{X} = \ell^1(\mathbf{Z}_+, \mathcal{S}_n) \subset \mathcal{S}_n^\infty$. The convex cone $\ell^1(\mathbf{Z}_+, \mathcal{S}_{n+})$ has empty interior. Finally, let us remark that based on (2.7) we may introduce a new norm on $\mathcal{X}$, by

$$\|\mathbf{X}\|_1 = \sum_{i \in \mathcal{D}} |X(i)|. \tag{2.11}$$

Based on (2.7), (2.11) we deduce that the norms $\|\cdot\|_1$ and $\tilde{\|}\cdot\tilde{\|}_1$ are equivalent, more precisely we have:

$$\tilde{\|}\mathbf{X}\tilde{\|}_1 \le \|\mathbf{X}\|_1 \le n\tilde{\|}\mathbf{X}\tilde{\|}_1 \tag{2.12}$$

for all $\mathbf{X} = \{X(i)\}_{i \in \mathcal{D}} \in \ell^1(\mathcal{D}, \mathcal{S}_n)$.

Example 2.1.3. Let $\mathcal{X} = \ell^2(\mathcal{D}, \mathcal{S}_n) = \{\mathbf{X} = \{X(i)\}_{i \in \mathcal{D}} \subset \mathcal{S}_n; \sum_{i \in \mathcal{D}} (|X(i)|_2)^2 < \infty\}$.

On $\ell^2(\mathcal{D}, \mathcal{S}_n)$ we introduce the inner product:

$$\langle \mathbf{X}, \mathbf{Y} \rangle_2 = \sum_{i \in \mathcal{D}} Tr[X(i)Y(i)] \tag{2.13}$$

for all $\mathbf{X} = \{X(i)\}_{i \in \mathcal{D}}$, $\mathbf{Y} = \{Y(i)\}_{i \in \mathcal{D}}$ from $\ell^2(\mathcal{D}, \mathcal{S}_n)$.

To show that the sum from the right-hand side of (2.13) is convergent, let us remark that

$$\sum_{i \in \mathcal{D}} Tr[X(i)Y(i)] = \frac{1}{4} \sum_{i \in \mathcal{D}} \left\{ |X(i) + Y(i)|_2^2 - |X(i) - Y(i)|_2^2 \right\} =$$

$$\frac{1}{4} \left\{ \sum_{i \in \mathcal{D}} |X(i) + Y(i)|_2^2 - \sum_{i \in \mathcal{D}} |X(i) - Y(i)|_2^2 \right\} \in \mathbf{R}$$

because

$$\sum_{i \in \mathcal{D}} |X(i) + Y(i)|_2^2 < +\infty$$

$$\sum_{i \in \mathcal{D}} |X(i) - Y(i)|_2^2 < +\infty.$$

One may check that the inner product $\langle \cdot, \cdot \rangle_2$ induces a real Hilbert space structure on $\ell^2(\mathcal{D}, \mathcal{S}_n)$. Set

$$\|\mathbf{X}\|_2 = \langle \mathbf{X}, \mathbf{X} \rangle_2^{1/2}. \tag{2.14}$$

On the space $\ell^2(\mathcal{D}, \mathcal{S}_n)$ we consider the order relation induced by the convex cone $\mathcal{X}_+ = \ell^2(\mathcal{D}, \mathcal{S}_{n+}) = \{\mathbf{X} = \{X(i)\}_{i \in \mathcal{D}} \in \ell^2(\mathcal{D}, \mathcal{S}_n); X(i) \geq 0, i \in \mathcal{D}\}$.

The cone $\ell^2(\mathcal{D}, \mathcal{S}_{n+})$ is a closed cone. If $\mathcal{D} = \mathbf{Z}_+$, its interior is empty.

Let us show that it is a selfdual convex cone. First, let us show that $\mathcal{X}_+^* \subset \ell^2(\mathcal{D}, \mathcal{S}_{n+})$. Let $\mathbf{Y} = \{Y(i)\}_{i \in \mathcal{D}} \in \mathcal{X}_+^*$. This means that

$$\langle \mathbf{Y}, \mathbf{X} \rangle_2 \geq 0 \tag{2.15}$$

for all $\mathbf{X} = \{X(i)\}_{i \in \mathcal{D}} \in \ell^2(\mathcal{D}, \mathcal{S}_{n+})$. Choose $x \in \mathbf{R}^n$ and $i_0 \in \mathcal{D}$ be arbitrary but fixed. We define $\mathbf{X}_{i_0} = \{X_{i_0}(i)\}_{i \in \mathcal{D}}$ as follows: $X_{i_0}(i) = \begin{cases} 0, i \neq i_0 \\ xx^T, i = i_0 \end{cases}$. It is obvious that $\mathbf{X}_{i_0} \in \ell^2(\mathcal{D}, \mathcal{S}_{n+})$.

In this case (2.15) becomes $Tr[Y(i_0)xx^T] \geq 0$ or equivalently $x^T Y(i_0)x \geq 0$. Since $x \in \mathbf{R}^n$ is arbitrarily chosen we deduce that $Y(i_0) \geq 0$. Further, since i_0 is arbitrarily chosen in $\mathcal{D}$ we conclude that $\mathbf{Y} \in \ell^2(\mathcal{D}, \mathcal{S}_{n+})$. Now we prove that $\ell^2(\mathcal{D}, \mathcal{S}_{n+}) \subseteq \mathcal{X}_+^*$. Let $\mathbf{Y} = \{Y(i)\}_{i \in \mathcal{D}} \in \ell^2(\mathcal{D}, \mathcal{S}_{n+})$. We show that $\langle \mathbf{Y}, \mathbf{X} \rangle_2 \geq 0$ or equivalently $\sum_{i \in \mathcal{D}} Tr[X(i)Y(i)] \geq 0$ for all $\mathbf{X} = \{X(i)\}_{i \in \mathcal{D}} \in \mathcal{X}_+$. Let $\lambda_{i1}, \lambda_{i2}, \ldots, \lambda_{in}$ be real numbers and $e_{i1}, e_{i2}, \ldots, e_{in}$ be a system of orthogonal vectors such that $|e_{ij}| = 1$ and $Y(i) = \sum_{j=1}^n \lambda_{ij} e_{ij} e_{ij}^T$. Since $Y(i) \geq 0$ it follows that $\lambda_{ij} \geq 0, 1 \leq j \leq n$. We write

$$\sum_{i \in \mathcal{D}} Tr[X(i)Y(i)] = \sum_{i \in \mathcal{D}} \sum_{j=1}^n \lambda_{ij} Tr[X(i)e_{ij}e_{ij}^T] = \sum_{i \in \mathcal{D}} \sum_{j=1}^n \lambda_{ij} e_{ij}^T X(i)e_{ij} \geq 0$$

because $X(i) \geq 0, i \in \mathcal{D}$. Hence, the equality $\ell^2(\mathcal{D}, \mathcal{S}_{n+}) = (\ell^2(\mathcal{D}, \mathcal{S}_{n+}))^*$ is true.

Remark 2.1.4. (i) In the case $\mathcal{D} = \{1, 2, \ldots, d\}$ the linear spaces $\ell^\infty(\mathcal{D}, \mathcal{S}_n)$, $\ell^1(\mathcal{D}, \mathcal{S}_n)$, $\ell^2(\mathcal{D}, \mathcal{S}_n)$ coincide with $\mathcal{S}_n^d = \underbrace{\mathcal{S}_n \times \mathcal{S}_n \times \ldots \times \mathcal{S}_n}_{d}$.

On $\mathcal{S}_n^d$ we have three norms:

$\| \cdot \|_\infty$ introduced via (2.8),

$\| \cdot \|_1$ defined in (2.10) and

$\| \cdot \|_2$ introduced by (2.14) for $\mathcal{D} = \{1, 2, \ldots, d\}$. We have $\|S\|_\infty \leq \|S\|_2 \leq \|S\|_1 \leq nd\|S\|_\infty$ for all $S \in \mathcal{S}_n^d$. The convex cone $\ell^2(\mathcal{D}, \mathcal{S}_{n+})$ coincide with the convex cone $\mathcal{S}_{n+}^d = \underbrace{\mathcal{S}_{n+} \times \mathcal{S}_{n+} \ldots \mathcal{S}_{n+}}_{d}$ if $\mathcal{D} = \{1, 2, \ldots, d\}$. The cone $\mathcal{S}_{n+}^d$ is a closed, solid, selfdual convex cone. It is selfdual with respect to the inner product

$$\langle X, Y \rangle = \sum_{i=1}^d Tr[X(i)Y(i)] \tag{2.16}$$

for all $X = (X(1), X(2), \ldots, X(d)), Y = (Y(1), Y(2), \ldots, Y(d)) \in \mathcal{S}_n^d$ which is the special form of (2.13) for the case $\mathcal{D} = \{1, 2, \ldots, d\}$.

(ii) In some applications occur in a natural way linear spaces of the form $\mathcal{X} = \mathcal{S}_{n_1} \times \mathcal{S}_{n_2} \times \ldots \times \mathcal{S}_{n_d}$ where it is possible to have $n_j \neq n_k$ if $j \neq k$. This kind of linear spaces occur, for example, in the study of exponential stability in mean square as well as in some problems regarding the existence of the stabilizing solution of Riccati-type equations for linear stochastic systems modeled by singularly perturbed Itô differential equations (see, e.g., [40, 53, 54]). The space $\mathcal{X}$ introduced before is a finite dimensional ordered Banach space. The order relation is induced by the closed, solid, selfdual convex cone $\mathcal{X}_+ = \mathcal{S}_{n_1+} \times \mathcal{S}_{n_2+} \times \ldots \times \mathcal{S}_{n_d+}$.

In the case $\mathcal{D} = \mathbf{Z}_+$ we have the following result.

Proposition 2.1.3. *If $\ell^1(\mathbf{Z}_+, \mathcal{S}_n)$ and $\ell^2(\mathbf{Z}_+, \mathcal{S}_{n+})$ are the linear spaces introduced in Examples 2.1.2 and 2.1.3, respectively, for $\mathcal{D} = \mathbf{Z}_+$, then*

$$\ell^1(\mathbf{Z}_+, \mathcal{S}_n) \subset \ell^2(\mathbf{Z}_+, \mathcal{S}_{n+}).$$

Proof. Let $\mathbf{X} = \{X(i)\}_{i \in \mathbf{Z}_+} \in \ell^1(\mathbf{Z}_+, \mathcal{S}_n)$. This means that the series $\sum\limits_{i=0}^{\infty} |X(i)|_1$ is convergent. Since $\lim\limits_{i \to \infty} |X(i)|_1 = 0$, we deduce that there exists $i_0 \geq 1$ such that $|X(i)|_1 < 1$ for all $i \geq i_0$.

Invoking (2.7) we may write:

$$\sum_{i=0}^{\infty} |X(i)|_2^2 = \sum_{i=0}^{i_0-1} |X(i)|_2^2 + \sum_{i=i_0}^{\infty} |X(i)|_2^2 \leq \sum_{i=0}^{i_0-1} |X(i)|_2^2 + \sum_{i=i_0}^{\infty} |X(i)|_2 \leq$$

$$\leq \sum_{i=0}^{i_0-1} |X(i)|_2^2 + \sum_{i=i_0}^{\infty} |X(i)|_1.$$

So we have obtained $\sum\limits_{i=0}^{\infty} |X(i)|_2^2 \leq \sum\limits_{i=0}^{i_0-1} |X(i)|_2^2 + \|\mathbf{X}\|_1 < +\infty$. Hence $\mathbf{X} \in \ell^2(\mathbf{Z}_+, \mathcal{S}_n)$. Thus the proof ends.   □

At the end of this subsection we present two examples of infinite dimensional ordered linear spaces which are not involved in the developments of the next chapters, but they are interesting in themselves.

The first one is an example of an infinite dimensional real Hilbert space ordered by a closed, solid, selfdual, convex cone, while the second one is an example of infinite dimensional Banach space ordered by order relation induced by a solid, closed, convex cone.

Example 2.1.4. Let us consider $\mathcal{X} = \ell^2(\mathbf{Z}_+, \mathbf{R})$ where

$$\ell^2(\mathbf{Z}_+, \mathbf{R}) = \left\{ \mathbf{x} = (x_0, x_1, \ldots x_n, \ldots); x_i \in \mathbf{R}, \sum_{i=0}^{\infty} x_i^2 < \infty \right\}.$$

On $\mathcal{X}$ we consider the usual inner product $\langle \mathbf{x}, \mathbf{y} \rangle_{\ell^2} = \sum_{i=0}^{\infty} x_i y_i$ for all $\mathbf{x} = \{x_i\}_{i \geq 0}, \mathbf{y} = \{y_i\}_{i \geq 0}$. We set

$$\mathcal{X}_+ = \left\{ \mathbf{x} = \{x_i\}_{i \geq 0}; x_0 \geq 0, \sum_{i=1}^{\infty} x_i^2 \leq x_0^2 \right\}. \tag{2.17}$$

It is easy to see that $\mathcal{X}_+$ is a closed, pointed convex cone. In the finite dimensional case the analogous of this cone is known as a *circular cone*.

The interior $Int \mathcal{X}_+ = \{\mathbf{x} = \{x_i\}_{i \geq 0}; x_0 > 0, \sum_{i=1}^{\infty} x_i^2 < x_0^2 \}$. It remains to prove that $\mathcal{X}_+$ is selfdual. Let $\mathbf{y} \in \mathcal{X}_+^*$. Hence

$$\langle \mathbf{x}, \mathbf{y} \rangle_{\ell^2} \geq 0 \tag{2.18}$$

for all $\mathbf{x} = \{x_i\}_{i \geq 0} \in \mathcal{X}_+$. In particular, taking in (2.18) $\mathbf{x} = \{1, 0, 0, \ldots, 0\}$ one obtains $y_0 \geq 0$. It is easy to verify that if $y_0 = 0$ then $y_t = 0$ for all $t \geq 1$. Since $y_0 \geq 0$ it is obvious that if $y_t = 0$ for all $t \geq 1$ we have $\mathbf{y} \in \mathcal{X}_+$. Suppose now $\sum_{t=1}^{\infty} y_t^2 > 0$. We take $\tilde{\mathbf{x}} = \{\tilde{x}_i\}_{i \geq 0}$ defined by

$$\tilde{x}_0 = y_0, \tilde{x}_i = -\gamma y_i y_0 \tag{2.19}$$

with $\gamma = (\sum_{k=1}^{\infty} y_k^2)^{\frac{-1}{2}}$. Obviously $\tilde{\mathbf{x}} \in \mathcal{X}_+$. Replacing (2.19) in (2.18) one gets $\sum_{k=1}^{\infty} y_k^2 \leq y_0^2$ which shows that $\mathbf{y} \in \mathcal{X}_+$. Thus was shown that $\mathcal{X}_+^* \subset \mathcal{X}_+$. Further take $\mathbf{y} = \{y_i\}_{i \geq 0} \in \mathcal{X}_+$. We have to show that (2.18) holds for all $\mathbf{x} \in \mathcal{X}_+$. Indeed for $\mathbf{x} \in \mathcal{X}_+$ we have $|\sum_{k=1}^{\infty} x_k y_k|^2 \leq \sum_{k=1}^{\infty} x_k^2 \sum_{k=1}^{\infty} y_k^2 \leq x_0^2 y_0^2$ which leads to $|\sum_{k=1}^{\infty} x_k y_k| \leq x_0 y_0$. This is equivalent with $-x_0 y_0 \leq \sum_{k=1}^{\infty} x_k y_k \leq x_0 y_0$ which shows that (2.18) is fulfilled. Thus it was proved that $\mathcal{X}_+ \subset \mathcal{X}_+^*$. So, we may conclude that $\mathcal{X}_+$ is selfdual.

Example 2.1.5. Let $\mathcal{X} = \ell^1(\mathbf{Z}_+, \mathbf{R})$ be the space of sequences of real numbers $\mathbf{x} = \{x_i\}_{i \in \mathbf{Z}_+}$ satisfying the condition $\sum_{i=0}^{\infty} |x_i| < \infty$. On $\mathcal{X}$ we introduce the usual norm

$$\|\mathbf{x}\|_1 = \sum_{i=0}^{\infty} |x_i|. \tag{2.20}$$

It is known that $(\mathcal{X}, \|\cdot\|_1)$ is a Banach space. If $\mathbf{x} = (x_0, x_1, \ldots, x_n, \ldots) \in \mathcal{X}$, then we write $\mathbf{x} = (x_0, \hat{\mathbf{x}})$ where $\hat{\mathbf{x}} = (x_1, x_2, \ldots, x_n, \ldots)$. We have $\|\mathbf{x}\|_1 = |x_0| + \|\hat{\mathbf{x}}\|_1$.

Let $\mathcal{X}_+ \subset \mathcal{X}$ be defined by

$$\mathcal{X}_+ = \left\{ \mathbf{x} \in \ell^1(\mathbf{Z}_+, \mathbf{R}); \mathbf{x} = (x_0, \hat{\mathbf{x}}), x_0 \geq 0, \|\hat{\mathbf{x}}\|_1 \leq \alpha x_0 \right\} \tag{2.21}$$

where $\alpha \in (0, 1)$ is fixed. One obtains that $\mathcal{X}_+$ is a closed convex cone. Let $\xi = (1, 0, \ldots, 0, \ldots) \in \mathcal{X}_+$. Consider the ball $B(\xi, \alpha) = \{\mathbf{x} \in \mathcal{X}; \|\mathbf{x} - \xi\|_1 \leq \alpha\}$. We show that

$$\mathcal{X}_+ = \mathbf{R}_+ \cdot B(\xi, \alpha) = \{\mathbf{y} = t\mathbf{x}, t \in \mathbf{R}_+, \mathbf{x} \in B(\xi, \alpha)\}. \tag{2.22}$$

First we show that $B(\xi, \alpha) \subset \mathcal{X}_+$. Let $\mathbf{x} \in B(\xi, \alpha)$. This means that $\|\mathbf{x} - \xi\|_1 \leq \alpha$. Hence

$$|x_0 - 1| + \|\hat{\mathbf{x}}\|_1 \leq \alpha. \tag{2.23}$$

Thus $1 - x_0 \leq |x_0 - 1| \leq \alpha$ and $1 - \alpha \leq x_0 \leq 1 + \alpha$, hence $x_0 \geq 1 - \alpha > 0$. Also considering the cases $1 - \alpha \leq x_0 \leq 1$ or $1 \leq x_0 \leq 1 + \alpha$ we can prove that

$$\alpha - |x_0 - 1| \leq \alpha x_0. \tag{2.24}$$

Combining (2.23) and (2.24) we deduce that $\|\hat{\mathbf{x}}\|_1 \leq \alpha x_0$. This means that $\mathbf{x} \in \mathcal{X}_+$. Thus the inclusion

$$B(\xi, \alpha) \subset \mathcal{X}_+ \tag{2.25}$$

is true. As a consequence of this inclusion one obtains that $\xi \in Int\mathcal{X}_+$, i.e. $\mathcal{X}_+$ is a solid convex cone. From (2.25) one obtains

$$\mathbf{R}_+ \cdot B(\xi, \alpha) \subset \mathcal{X}_+. \tag{2.26}$$

To prove the converse inclusion of (2.26) we choose $\mathbf{x} = (x_0, \hat{\mathbf{x}}) \in \mathcal{X}_+$. Hence $x_0 \geq 0$. If $x_0 = 0$, then $\mathbf{x} = 0, \mathbf{x} = 0 \cdot \xi$ and obviously $\mathbf{x} \in \mathbf{R}_+ \cdot B(\xi, \alpha)$. Suppose now $x_0 > 0$. We have $\mathbf{x} = x_0 \mathbf{z}$ with $\mathbf{z} = (1, \frac{1}{x_0}\hat{\mathbf{x}})$. We see that $\|\mathbf{z} - \xi\|_1 = \frac{1}{x_0}\|\hat{\mathbf{x}}\|_1 \leq \alpha$. So we obtained $\mathbf{z} \in B(\xi, \alpha)$ which means that $\mathbf{x} \in \mathbf{R}_+ \cdot B(\xi, \alpha)$ and the equality (2.22) holds.

Further we prove that $\|\cdot\|_1$ is monotone with respect to the cone $\mathcal{X}_+$. Let $0 \leq \mathbf{x} \leq \mathbf{y}$. This means that $\mathbf{x} = (x_0, \hat{\mathbf{x}})$, $\mathbf{y} = (y_0, \hat{\mathbf{y}})$ with $x_0 \geq 0$, $\|\hat{\mathbf{x}}\|_1 \leq \alpha x_0$, $y_0 - x_0 \geq 0$, $\|\hat{\mathbf{y}} - \hat{\mathbf{x}}\|_1 \leq \alpha(y_0 - x_0)$. Based on these inequalities we may write successively: $\|\mathbf{x}\|_1 = x_0 + \|\hat{\mathbf{x}}\|_1 \leq x_0 + \|\hat{\mathbf{y}}\|_1 + \|\hat{\mathbf{y}} - \hat{\mathbf{x}}\|_1 \leq x_0 + \|\hat{\mathbf{y}}\|_1 + \alpha(y_0 - x_0)$. Thus we obtain $\|\mathbf{x}\|_1 \leq \|\mathbf{y}\|_1 - (1 - \alpha)(y_0 - x_0)$. Hence $\|\mathbf{x}\|_1 \leq \|\mathbf{y}\|_1$ for all $0 \leq \mathbf{x} \leq \mathbf{y}$. This means that $\|\cdot\|_1$ is monotone.

2.1.3 Minkovski Norms

In order to make clearer the role of each property of the cone in the characterization of the Minkovski norms, we assume for the beginning that $\mathcal{X}$ is a real Banach space ordered by an order relation induced by a solid convex cone $\mathcal{X}_+$ where $\mathcal{X}_+ \neq \mathcal{X}$. For a fixed $\xi \in Int\mathcal{X}_+$ we consider the open and convex subset

$$B_\xi = \{x \in \mathcal{X}; -\xi < x < \xi\}. \tag{2.27}$$

The Minkovski functional $|\cdot|_\xi : \mathcal{X} \to \mathbf{R}$ associated with the subset B_ξ is

$$|x|_\xi = inf\left\{t > 0; \frac{1}{t}x \in B_\xi\right\}. \tag{2.28}$$

The main properties of the Minkovski functional introduced by (2.28) are collected in the next theorem.

Theorem 2.1.4. *The Minkovski functional introduced in (2.28) has the properties:*

(i) $|x|_\xi \geq 0$ *and* $|0|_\xi = 0$.
(ii) $|\alpha x|_\xi = |\alpha| |x|_\xi$ *for all* $\alpha \in \mathbf{R}, x \in \mathcal{X}$.
(iii) $|x|_\xi < 1$ *if and only if* $x \in B_\xi$.
(iv) $|x+y|_\xi \leq |x|_\xi + |y|_\xi$ *for all* $x, y \in \mathcal{X}$.
(v) *There exists* $\beta(\xi) > 0$ *such that* $|x|_\xi \leq \beta(\xi) \|x\|, (\forall) x \in \mathcal{X}$.
(vi) $|x|_\xi = 1$ *if and only if* $x \in \partial B_\xi$.
(vii) $|x|_\xi \leq 1$ *if and only if* $x \in \bar{B}_\xi$.
(viii) *If* $\mathcal{X}_+$ *is closed, then* $\bar{B}_\xi = \{x \in \mathcal{X}; -\xi \leq x \leq \xi\}$.
(ix) $|\xi|_\xi = 1$.
(x) *The set* $\mathcal{T}(x) = \{t > 0; \frac{1}{t} x \in B_\xi\}$ *coincides with the interval* $(|x|_\xi, \infty)$.
(xi) *If* $x, y, z \in \mathcal{X}$ *are such that* $y \leq x \leq z$, *then* $|x|_\xi \leq \max\{|y|_\xi, |z|_\xi\}$.

Proof. Properties (i)–(iv), (vi) and (vii) can be proved in a more general setting of Minkovski functionals, associated with some open and convex subsets in linear topological spaces (see [55, 90, 125]). The other properties are based on the special form of the set B_ξ given in (2.27). For details see [51]. □

From (i) and (iv) in Theorem 2.1.4 one obtains that the Minkovski functional is a seminorm.

The next result provides a sufficient condition such that the Minkovski seminorm becomes a norm.

Proposition 2.1.5 ([51]). *If* B_ξ *is a bounded set, then the Minkovski seminorm* $|\cdot|_\xi$ *defined by (2.28) is a norm. Moreover there exists* $\alpha_\xi > 0$ *such that* $\|x\| \leq \alpha_\xi |x|_\xi$ *for all* $x \in \mathcal{X}$.

Proposition 2.1.6 ([51]). *If the cone* $\mathcal{X}_+$ *is normal, then for all* $\xi \in Int\mathcal{X}_+$ *the set* B_ξ *is bounded.*

Specializing the results from Theorem 2.1.4, Proposition 2.1.5, Proposition 2.1.6, to the Banach spaces given in Example 2.1.1, Example 2.1.4, and Example 2.1.5, we obtain:

Corollary 2.1.7. *In the case of the Banach space* $(\mathcal{S}_n^{\mathcal{D}}, \|\cdot\|_\infty)$ *introduced in the Example 2.1.1, the following hold:*

(i) *If* $\mathcal{D} = \{1, 2, \dots, d\}$ *and* $J^d = \underbrace{(I_n, I_n, \dots, I_n)}_{d} \in Int\mathcal{S}_{n+}^d$ *then the Minkovski norm defined by (2.28) for* $\xi = J^d$ *is:*

$$|X|_{J^d} = \|X\|_\infty \tag{2.29}$$

for all $X = (X(1), X(2), \dots, X(d)) \in \mathcal{S}_n^d$.
(ii) *If* $\mathcal{D} = \mathbf{Z}_+$ *and* $J^\infty = (I_n, I_n, \dots, I_n, \dots) \in Int\mathcal{S}_{n+}^\infty$ *then the Minkovski norm introduced by (2.28) for* $\xi = J^\infty$ *is given by*

$$|\mathbf{X}|_{J^\infty} = \|\mathbf{X}\|_\infty \tag{2.30}$$

for all $\mathbf{X} = \{X(i)\}_{i \in \mathbf{Z}_+}$.

Remark 2.1.5. For shake of brevity we shall use $|X|$ and $|\mathbf{X}|$ respectively, instead of $|X|_{J^d}$ and $|\mathbf{X}|_{J^\infty}$, respectively, for the Minkovski norms defined by (2.29) and (2.30), respectively, if no confusions are possible.

Corollary 2.1.8. *(i) In the case of the Banach space $\mathcal{X} = \ell^2\{\mathbf{Z}_+, \mathbf{R}\}$ ordered by the order relation induced by the cone (2.17) the Minkovski norm defined by the sequence $\xi = (1, 0, \ldots, 0, \ldots) \in Int\mathcal{X}_+$ is given by $|\mathbf{x}|_\xi = |x_0| + |\hat{\mathbf{x}}|_2, \forall \mathbf{x} = (x_0, \hat{\mathbf{x}}) \in \ell^2(\mathbf{Z}_+, \mathbf{R})$.*

(ii) In the case of the Banach space $\mathcal{X} = \ell^1\{\mathbf{Z}_+, \mathbf{R}\}$ ordered by the order relation induced by the cone (2.21) the Minkovski norm defined by the sequence $\xi = (1, 0, \ldots, 0, \ldots) \in Int\mathcal{X}_+$ is given by $|\mathbf{x}|_\xi = |x_0| + \frac{1}{\alpha}\|\hat{\mathbf{x}}\|_1, \forall \mathbf{x} = (x_0, \hat{\mathbf{x}}) \in \ell^1(\mathbf{Z}_+, \mathbf{R})$.

2.1.4 Linear Positive Operators on Ordered Banach Spaces

Let $(\mathcal{X}, \|\cdot\|)$ be a real Banach space ordered by the closed, solid, normal, convex cone $\mathcal{X}_+$.

If $(\mathcal{Y}, \|\cdot\|)$ is another Banach space, then $\mathbf{B}(\mathcal{X}, \mathcal{Y})$ stands for the space of linear and bounded operators defined on $\mathcal{X}$ and taking values in $\mathcal{Y}$.

When $\mathcal{X} = \mathcal{Y}$ we shall write $\mathbf{B}(\mathcal{X})$ instead of $\mathbf{B}(\mathcal{X}, \mathcal{X})$.

Under the considered assumptions, the Minkovski functional $|\cdot|_\xi$ is a norm equivalent with the norm $\|\cdot\|$ on $\mathcal{X}$.

If $T \in \mathbf{B}(\mathcal{X})$ then $\|T\|$ and $\|T\|_\xi$ are the operator norms of T, induced by $\|\cdot\|$ and $|\cdot|_\xi$, respectively. This means that

$$\|T\| = \sup_{\|x\| \leq 1} \|Tx\| \tag{2.31}$$

$$\|T\|_\xi = \sup_{|x|_\xi \leq 1} |Tx|_\xi. \tag{2.32}$$

Remark 2.1.6. Based on (2.31) and (2.32) and the equivalence of the norms $\|\cdot\|$ and $|\cdot|_\xi$ one deduces that there exists the positive constants c_1, c_2 such that

$$c_1\|T\| \leq \|T\|_\xi \leq c_2\|T\| \tag{2.33}$$

for all $T \in \mathbf{B}(\mathcal{X})$.

Definition 2.1.4. Let $(\mathcal{X}, \mathcal{X}_+)$, $(\mathcal{Y}, \mathcal{Y}_+)$ be two ordered linear spaces with the order relation induced by the convex cones $\mathcal{X}_+$ and $\mathcal{Y}_+$, respectively. An operator $T \in \mathbf{B}(\mathcal{X}, \mathcal{Y})$ is called positive operator if $T\mathcal{X}_+ \subset \mathcal{Y}_+$. In this case we shall write $T \geq 0$.

By definition, if $T_1, T_2 \in \mathbf{B}(\mathcal{X})$ then $T_1 \leq T_2$ or equivalently $T_2 \geq T_1$ if and only if $T_2 - T_1 \geq 0$.
 Also $T \leq 0$ if and only if $-T \geq 0$.

Remark 2.1.7. If $T : \mathcal{X} \to \mathcal{X}$ is a linear bounded and positive operator then T is a monotone operator. This means that $Tx \leq Ty$ if $x \leq y$. Indeed, $x \leq y$ iff $y - x \in \mathcal{X}_+$ hence, $T(y - x) \geq 0$ which is equivalent to $Tx \leq Ty$.

Proposition 2.1.9. *If $\mathcal{X}$ is a real Hilbert space ordered by the selfdual convex cone $\mathcal{X}_+$ and if $T \in \mathbf{B}(\mathcal{X})$ then $T \geq 0$ iff $T^* \geq 0$.*

Proof. "$\to$" Let us assume that $T \geq 0$. This means that $Tx \in \mathcal{X}_+$ if $x \in \mathcal{X}_+$. Since $\mathcal{X}_+ = \mathcal{X}_+^*$ we deduce that $\langle Tx, y \rangle \geq 0$ for arbitrary $y \in \mathcal{X}_+$. This is equivalent to $\langle x, T^*y \rangle \geq 0$ for all $x \in \mathcal{X}_+$. Hence $T^*y \in \mathcal{X}_+^*$ which leads to $T^*y \in \mathcal{X}_+$ because $\mathcal{X}_+ = \mathcal{X}_+^*$. Since $y \in \mathcal{X}_+$ is arbitrary, we may conclude that $T^* \geq 0$. The converse implication may be proved in a similar way using T^* instead of T and taking into account that $(T^*)^* = T$. Thus the proof in complete.   □

The next result will be repeatedly used in the developments of this chapter. It provides a simple formula of the operator norm of a bounded linear positive operator induced by the Minkovski norm.

Theorem 2.1.10. *Let $(\mathcal{X}, \|\cdot\|)$ be a real Banach space ordered by a solid, closed, normal, convex cone $\mathcal{X}_+$. Let $\xi \in Int\mathcal{X}_+$ be fixed. Then for every positive operator $T \in \mathbf{B}(\mathcal{X})$ we have $\|T\|_\xi = |T\xi|_\xi$.*

Proof. Based on (vii) and (viii) in Theorem 2.1.4 we deduce that $|x|_\xi \leq 1$ if and only if $-\xi \leq x \leq \xi$. Since $T \geq 0$ we deduce that it is a monotone operator; hence $-T\xi \leq Tx \leq T\xi$ for all $x \in \mathcal{X}$ with $|x|_\xi \leq 1$.
 Applying (xi) from Theorem 2.1.4 we infer that $|Tx|_\xi \leq |T\xi|_\xi$ for all $x \in \mathcal{X}$ with $|x|_\xi \leq 1$.
 Since $|\xi|_\xi = 1$ we get: $|T\xi|_\xi \leq \sup_{|x|_\xi \leq 1} |Tx|_\xi \leq |T\xi|_\xi$.
 Invoking (2.32) we may conclude that $\|T\|_\xi = |T\xi|_\xi$ which ends the proof.   □

Using the equality proved in the above theorem together with the monotonicity of the Minkovski norm, we obtain:

Corollary 2.1.11. *If $T_k \in \mathbf{B}(\mathcal{X})$, $k = 1, 2$ are such that $T_1 \leq T_2$ then $\|T_1\|_\xi \leq \|T_2\|_\xi$.*

2.2 Linear Differential Equations with Positive Evolution on Ordered Banach Spaces

2.2.1 Linear Evolution Operators on Ordered Banach Spaces

Let $(\mathcal{X}, \|\cdot\|)$ be a real Banach space. Let $\mathcal{I} \subset \mathbf{R}$ be an interval of real numbers. Let $\mathcal{L} : \mathcal{I} \to \mathbf{B}(\mathcal{X})$ be a strongly continuous operator valued function. This means that for each $x \in \mathcal{X}$ the vector valued function $t \to \mathcal{L}(t)x$ is continuous on $\mathcal{I}$.

We consider the linear differential equation on $\mathcal{X}$:

$$\frac{d}{dt}x(t) = \mathcal{L}(t)x(t). \tag{2.34}$$

Based on the developments in Chap. 3 of [31] we deduce that for each $(t_0, x_0) \in \mathcal{I} \times \mathcal{X}$ there exists a unique C^1-function $x(\cdot; t_0, x_0) : \mathcal{I} \to \mathcal{X}$ satisfying (2.34) and the initial condition $x(t_0; t_0, x_0) = x_0$.

In Chap. 3 of [31] it is shown that there exists an operator valued function $T_{\mathcal{L}} : \mathcal{I} \times \mathcal{I} \to \mathbf{B}(\mathcal{X})$ with the property that $x(t; t_0, x_0) = T_{\mathcal{L}}(t, t_0)x_0$ for all $t, t_0 \in \mathcal{I}$ and $x_0 \in \mathcal{X}$. The operator valued function $(t, \tau) \to T_{\mathcal{L}}(t, \tau)$ or $T_{\mathcal{L}}(t, \tau)$ for shortness is named *the linear evolution operator on $\mathcal{X}$* defined by the operator valued function $\mathcal{L}(\cdot)$ or equivalently the linear evolution operator defined on $\mathcal{X}$ by the linear differential equation (2.34).

Often, we shall write $T(t, \tau)$ instead of $T_{\mathcal{L}}(t, \tau)$ if confusions are not possible.

Remark 2.2.1. A linear evolution operator $T(t, \tau)$ on a Banach space $\mathcal{X}$ has the properties (see [31] Chap. 3 for details).

(i) $t \to T(t, \tau)$ is the unique solution of the problem with given initial value on $\mathcal{B}(\mathcal{X})$

$$\frac{d}{dt}X(t) = \mathcal{L}(t)X(t), \ X(\tau) = I_{\mathcal{X}}$$

where $I_{\mathcal{X}}$ is the identity operator on $\mathcal{X}$. More precisely,

$$\frac{d}{dt}T(t, \tau) = \mathcal{L}(t)T(t, \tau), t \in \mathcal{I} \tag{2.35}$$

$$T(\tau, \tau) = I_{\mathcal{X}}.$$

(ii) $\tau \to T(t, \tau) : \mathcal{I} \to \mathbf{B}(\mathcal{X})$ satisfies

$$\frac{d}{d\tau}T(t, \tau) = -T(t, \tau)\mathcal{L}(\tau) \forall \tau \in \mathcal{I}. \tag{2.36}$$

(iii)

$$T(t, \tau)T(\tau, s) = T(t, s), \ (\forall) \ t, \tau, s \in \mathcal{I}. \tag{2.37}$$

(iv) For each $(t,\tau) \in \mathcal{I} \times \mathcal{I}$, the operator $T(t,\tau)$ is invertible and $T^{-1}(t,\tau) \in \mathbf{B}(\mathcal{X})$. More precisely, we have $T^{-1}(t,\tau) = T(\tau,t)$.

(v) $\|T(t,\tau)\| \le e^{|\int_\tau^t \|\mathcal{L}(s)\| ds|}$, $(\forall)\ t, \tau \in \mathcal{I}$.

(vi) If $\mathcal{L}(t) = \mathcal{L} \in \mathbf{B}(\mathcal{X})$ then $T(t,\tau) = e^{\mathcal{L}(t-\tau)}$, where $e^{\mathcal{L}t} = \sum\limits_{k=0}^{\infty} \frac{t^k}{k!} \mathcal{L}^k$. This series is convergent in the topology induced by the operator norm uniformly on any compact subinterval of $\mathbf{R}$.

(vii) If $\mathcal{I} = \mathbf{R}$ and $\mathcal{L}(t+\theta) = \mathcal{L}(t)$ for all $t \in \mathbf{R}$ and some $\theta > 0$ then $T(t+k\theta, \tau + k\theta) = T(t,\tau)$ for any $t, \tau \in \mathbf{R}, k \in \mathbf{Z}$.

(viii) Based on the uniqueness of the linear evolution operator one shows that for any $\alpha \in \mathbf{R}$, $T_\alpha(t,\tau) := e^{\alpha(t-\tau)}T(t,\tau) = e^{\alpha I_\mathcal{X}(t-\tau)}T(t,\tau)$ is the linear evolution operator on $\mathcal{X}$ defined by

$$\frac{d}{dt}x(t) = (\alpha I_\mathcal{X} + \mathcal{L}(t))x(t).$$

The strongly continuous operator valued function $\mathcal{L}(\cdot)$ defines also the linear differential equation

$$\frac{d}{dt}y(t) + \mathcal{L}(t)y(t) = 0. \tag{2.38}$$

Applying the results from Chap. 3 of [31] to the operator valued function $t \to -\mathcal{L}(t)$ we deduce that for each $(t_0, y_0) \in \mathcal{I} \times \mathcal{X}$ the linear differential equation (2.38) has a unique solution $y(\cdot; t_0, y_0) : \mathcal{I} \to \mathcal{X}$ which satisfy the initial condition $y(t_0; t_0, y_0) = y_0$. One proves also that $y(t; t_0, y_0) = T_\mathcal{L}^a(t, t_0)y_0$ for all $(t; t_0, y_0) \in \mathcal{I} \times \mathcal{I} \times \mathcal{X}$ where $T_\mathcal{L}^a(t, t_0) : \mathcal{I} \times \mathcal{I} \to \mathbf{B}(\mathcal{X})$ is *the anticausal linear evolution operator on $\mathcal{X}$* generated by the operator valued function $\mathcal{L}(\cdot)$ or, equivalently, *the anticausal linear evolution operator on $\mathcal{X}$* generated by the linear differential equation (2.38).

In the sequel, we shall write $T^a(t, t_0)$ instead of $T_\mathcal{L}^a(t, t_0)$, if confusions are not possible.

Remark 2.2.2. Many of the assertions of Remark 2.2.1 remains valid if the causal linear evolution operator $T(t,\tau)$ is replaced by the anticausal linear evolution operator $T^a(t,\tau)$.

In the case of anticausal linear evolution operator, the statements (i), (ii), (vi) from Remark 2.2.1 become:

(i') for each $\tau \in \mathcal{I}$, $t \to T^a(t,\tau)$ is a C^1-function which satisfies the linear differential equation on $\mathbf{B}(\mathcal{X})$:

$$\frac{d}{dt}T^a(t,\tau) = -\mathcal{L}(t)T^a(t,\tau) \tag{2.39}$$

and the initial condition $T^a(\tau,\tau) = I_\mathcal{X}$.

(ii′) for each $t \in \mathcal{I}$, $\tau \to T^a(t, \tau)$ satisfies:

$$\frac{d}{d\tau}T^a(t, \tau) = T^a(t, \tau)\mathcal{L}(\tau).$$ (2.40)

(vi′) if $\mathcal{L}(t) = \mathcal{L} \in \mathbf{B}(\mathcal{X})$, $t \in \mathbf{R}$, then, $T^a_{\mathcal{L}}(t, \tau) = e^{\mathcal{L}(\tau-t)}$, $\forall\, t, \tau \in \mathbf{R}$.

Beside the linear differential equations (2.34) and (2.38) respectively, we associate the following affine differential equations:

$$\frac{d}{dt}x(t) = \mathcal{L}(t)x(t) + f(t)$$ (2.41)

and

$$\frac{d}{dt}y(t) + \mathcal{L}(t)y(t) + g(t) = 0$$ (2.42)

where $\mathcal{L}(\cdot)$ is an operator valued function as before, and $f : \mathcal{I} \to \mathcal{X}$, $g : \mathcal{I} \to \mathcal{X}$ are continuous vector valued functions.

The solutions of (2.41) and (2.42) have the following representation formulae:

$$x(t; t_0, x_0) = T(t, t_0)x_0 + \int_{t_0}^{t} T(t, s)f(s)ds$$ (2.43)

for all $t \in [t_0, \infty) \cap \mathcal{I}$, $x_0 \in \mathcal{X}$ and

$$y(t; t_0, y_0) = T^a(t, t_0)y_0 + \int_{t}^{t_0} T^a(t, s)g(s)ds$$ (2.44)

for all $t \in (-\infty; t_0] \cap \mathcal{I}$, $y_0 \in \mathcal{X}$.

In the development from this chapter, the affine differential equation of type (2.41) will be called *forward affine differential equation* while affine differential equations of type (2.42) will be named *backward affine differential equations*.

Remark 2.2.3. If $y(t)$ is a solution of the backward affine differential equation

$$\frac{d}{dt}y(t) + \mathcal{L}(t)y(t) + g(t) = 0$$

then $\hat{x}(t)$ defined by $\hat{x}(t) = y(-t)$ is a solution of the forward affine equation $\frac{d}{dt}x(t) = \hat{\mathcal{L}}(t)x(t) + \hat{f}(t)$ where $\hat{\mathcal{L}}(t) = \mathcal{L}(-t)$ and $\hat{f}(t) = g(-t)$.

Moreover if $T^a(t, t_0)$ is the anticausal evolution operator defined by the operator valued function $\mathcal{L}(\cdot)$, then $\hat{T}(t, t_0)$ defined by

$$\hat{T}(t, t_0) = T^a(-t, -t_0), \quad \forall t, t_0 \in \hat{\mathcal{I}} = \{t \in \mathbf{R}; -t \in \mathcal{I}\}$$ (2.45)

is the causal evolution operator defined by the operator valued function $\hat{\mathcal{L}} : \hat{\mathcal{I}} \to \mathbf{B}(\mathcal{X})$, $\hat{\mathcal{L}}(t) = \mathcal{L}(-t)$.

2.2.2 Linear Differential Equations with Positive Evolutions on Ordered Banach Spaces

Let $(\mathcal{X}, \|\cdot\|)$ be a real Banach space ordered by a solid, closed, normal, convex cone $\mathcal{X}_+$.

Let $\mathcal{L} : \mathcal{I} \to \mathbf{B}(\mathcal{X})$ be a strongly continuous operator valued function.

Definition 2.2.1. We say that the operator valued function $\mathcal{L}(\cdot)$ generates:

(i) *a causal positive evolution on* $\mathcal{X}$, or a positive evolution (for shortness) if $T_{\mathcal{L}}(t,t_0)\mathcal{X}_+ \subset \mathcal{X}_+$, for all $t \geq t_0, t, t_0 \in \mathcal{I}$.

(ii) *an anticausal positive evolution on* $\mathcal{X}$, if $T_{\mathcal{L}}^a(t,t_0)\mathcal{X}_+ \subset \mathcal{X}_+$, for all $t \leq t_0, t, t_0 \in \mathcal{I}$.

With other words, the operator valued function $\mathcal{L}(\cdot)$ generates a positive evolution on $\mathcal{X}$ if the solutions of the linear differential equation (2.34) have the property that $x(t;t_0,x_0) \in \mathcal{X}_+$ for all $t \geq t_0, t, t_0 \in \mathcal{I}$ if $x_0 \in \mathcal{X}_+$. In this case we shall say that the linear differential equation (2.34) defines a positive evolution on $\mathcal{X}$.

Similarly the operator valued function $\mathcal{L}(\cdot)$ generates an anticausal positive evolution on $\mathcal{X}$ if and only if the solutions of linear differential equation (2.38) have the property that $y(t;t_0,y_0) \in \mathcal{X}_+$, for all $t \leq t_0, t, t_0 \in \mathcal{I}$, if $y_0 \in \mathcal{X}_+$. In this case, we shall say that the linear differential equation (2.38) defines on anticausal positive evolution on $\mathcal{X}$.

Based on (2.45) we obtain.

Corollary 2.2.1. *Let* $\mathcal{L} : \mathcal{I} \to \mathbf{B}(\mathcal{X})$ *be a strongly continuous operator valued function and* $\hat{\mathcal{L}}(t) = \mathcal{L}(-t)$, $t \in \hat{\mathcal{I}}$. *Then the operator valued function* $\mathcal{L}(\cdot)$ *defines an anticausal positive evolution if and only if the operator valued function* $\hat{\mathcal{L}}(\cdot)$ *generates a causal positive evolution on* $\mathcal{X}$.

In the sequel we shall provide some useful properties of the operator valued functions which generate positive evolution on $\mathcal{X}$.

Definition 2.2.2. Let $\mathcal{L} \in \mathbf{B}(\mathcal{X})$. We say that the linear and bounded operator $\mathcal{L}$ is:

(i) *resolvent positive* if there exists $\lambda_0 \in \mathbf{R}$ such that $(\lambda I_{\mathcal{X}} - \mathcal{L})^{-1} \in \mathbf{B}(\mathcal{X})$ and $(\lambda I_{\mathcal{X}} - \mathcal{L})^{-1} \geq 0$, for all $\lambda > \lambda_0$.

(ii) quasimonotone if for all $x \in \partial\mathcal{X}_+$ there exists $\varphi \in \mathcal{X}_+^* \setminus \{0\}$ such that $\varphi(x) = 0$ and $\varphi(\mathcal{L}x) \geq 0$.

Theorem 2.2.2. *Let* $(\mathcal{X}, \|\cdot\|)$ *be a real Banach space ordered by a solid, closed, normal convex cone* $\mathcal{X}_+$. *If* $\mathcal{L} \in \mathbf{B}(\mathcal{X})$ *then the following are equivalent:*

(i) $e^{\mathcal{L}t} \geq 0$, $(\forall)t \geq 0$.

(ii) $\mathcal{L}$ *is a resolvent positive operator.*

(iii) *For all* $x \in \partial\mathcal{X}_+$ *and all* $\varphi \in \mathcal{X}_+^* \setminus \{0\}$ *with* $\varphi(x) = 0$ *it follows* $\varphi(\mathcal{L}x) \geq 0$.

(iv) $\mathcal{L}$ *is a quasimonotone operator.*

Proof. (i) → (ii) If (i) is true, then $e^{\mathcal{L}t}x_0 \geq 0$ for all $t \geq 0$ if $x_0 \in \mathcal{X}_+$. Obviously $e^{(\mathcal{L}-\lambda I_{\mathcal{X}})t}x_0 \in \mathcal{X}_+$ for all $t \geq 0$, $\lambda \in \mathbf{R}$, $x_0 \in \mathcal{X}_+$. On the other hand

$$\|e^{(\mathcal{L}-\lambda I_{\mathcal{X}})t}\| \leq e^{(\|\mathcal{L}\|-\lambda)t}, \; t \geq 0. \tag{2.46}$$

Chose $\lambda > \|\mathcal{L}\|$. Since $\|\int_0^\tau e^{(\mathcal{L}-\lambda I_{\mathcal{X}})t}x_0 dt\| \leq \int_0^\tau \|e^{(\mathcal{L}-\lambda I_{\mathcal{X}})t}x_0\| dt$ we obtain via (2.46) that $\mathcal{R}(\lambda, \mathcal{L})x_0 := \int_0^\infty e^{(\mathcal{L}-\lambda I_{\mathcal{X}})t}x_0 dt$ is well defined by $\int_0^\infty e^{(\mathcal{L}-\lambda I_{\mathcal{X}})t}x_0 dt = \lim_{\tau \to \infty} \int_0^\tau e^{(\mathcal{L}-\lambda I_{\mathcal{X}})t}x_0 dt$.

Moreover $\mathcal{R}(\lambda, \mathcal{L})$ is a linear operator. Using again (2.46) we deduce $\|\mathcal{R}(\lambda, \mathcal{L})x\| \leq \frac{1}{(\lambda-\|\mathcal{L}\|)}\|x\|$. This allows us to conclude that $\mathcal{R}(\lambda, \mathcal{L}) \in \mathbf{B}(\mathcal{X})$ for all $\lambda > \|\mathcal{L}\|$. On the other hand $\mathcal{R}(\lambda, \mathcal{L})\mathcal{X}_+ \subset \mathcal{X}_+$ because $\mathcal{X}_+$ is a closed convex cone. Further, by direct calculation one obtains that:

$$(\lambda I_{\mathcal{X}} - \mathcal{L})\mathcal{R}(\lambda, \mathcal{L})x = x$$

and

$$\mathcal{R}(\lambda, \mathcal{L})(\lambda I_{\mathcal{X}} - \mathcal{L})x = x,$$

for all $x \in \mathcal{X}$. Hence $\mathcal{R}(\lambda, \mathcal{L}) = (\lambda I_{\mathcal{X}} - \mathcal{L})^{-1}$, for all $\lambda > \|\mathcal{L}\|$. So we have shown that for arbitrary $\lambda > \|\mathcal{L}, (\lambda I_{\mathcal{X}} - \mathcal{L})^{-1}$ is well defined and it is a positive operator. Thus the proof of implication (i) → (ii) is complete.

We prove now that (ii) → (iii). Let $x \in \partial \mathcal{X}_+$ and $\varphi \in \mathcal{X}_+^* \setminus \{0\}$ be such that $\varphi(x) = 0$. If (ii) holds then there exists $\alpha_0 > 0$ such that $(I_{\mathcal{X}} - t\mathcal{L})^{-1} \geq 0$ for all $t \in [0, \alpha_0)$.

Let $\kappa : [0, \alpha_0) \to \mathbf{R}$ be defined by $\kappa(t) = \varphi((I_{\mathcal{X}} - t\mathcal{L})^{-1}x)$. It follows that $\kappa(t) \geq 0, \forall t \in [0, \alpha_0)$ because $\varphi \in \mathcal{X}_+^*$. Since $\kappa(0) = 0$ we obtain

$$\frac{d}{dt}\kappa(t)|_{t=0} = \varphi(\mathcal{L}x). \tag{2.47}$$

Thus we may conclude via (2.47) that $\varphi(\mathcal{L}x) \geq 0$ which confirms that (ii) → (iii) is true. Let us assume that (iii) holds and let us show that $\mathcal{L}$ is quasimonotone. Take $x_0 \in \partial \mathcal{X}_+$ be arbitrary. Based on Theorem 2.1.2 we deduce that there exists $\varphi \in \mathcal{X}_+^* \setminus \{0\}$ such that $\varphi(x_0) = 0$. If (iii) holds, it follows that $\varphi(\mathcal{L}x_0) \geq 0$. This confirms that $\mathcal{L}$ is quasimonotone.

Finally, let us show that (iv) → (i). Assume that (iv) holds. Choose $\xi \in Int \mathcal{X}_+$ and $\delta > 0$ a small parameter. Let $x_\delta(t), t \geq 0$ be the solution of the linear differential equation

$$\frac{d}{dt}x(t) = \mathcal{L}x(t) + \delta\xi \tag{2.48}$$

with the initial condition $x_\delta(0) = x_0 \in Int\mathcal{X}_+$. It is obvious that $x_\delta(t) \in Int\mathcal{X}_+$ if $t > 0$ is sufficiently small. We show that $x_\delta(t) \in Int\mathcal{X}_+$ for arbitrary $t > 0$.

Let us assume by contrary that there exists $\hat{t} > 0$ such that $x_\delta(\hat{t}) \notin Int\mathcal{X}_+$. Let $t_0 = \inf\{t > 0; x_\delta(t) \notin Int\mathcal{X}_+\}$. From the definition of t_0, one obtains that $x_\delta(t_0) \in \partial\mathcal{X}_+$ and $x_\delta(t) \in Int\mathcal{X}_+$ if $0 \le t < t_0$. From (iv) and Theorem 2.1.2 it follows that there exists $\varphi \in \mathcal{X}_+^* \setminus \{0\}$ such that $\varphi(x_\delta(t_0)) = 0$ and $\varphi(\mathcal{L}x_\delta(t_0)) \ge 0$. Let us define $g(t) = \varphi(x_\delta(t))$, $t \ge 0$. We have $g(t) > 0$ if $0 \le t < t_0$ and $g(t_0) = 0$. Thus $\frac{g(t)}{t-t_0} \le 0$ if $t < t_0$. This leads to $\frac{d}{dt}g(t)|_{t=t_0} \le 0$. On the other hand, from (2.48) one obtains

$$\frac{d}{dt}g(t_0) = \varphi(\mathcal{L}x_g(t_0)) + \delta\varphi(\xi).$$

Invoking Lemma 2.1.1 we may conclude that $0 \ge \frac{d}{dt}g(t_0) > 0$ which is a contradiction. Therefore we may conclude that $x_\delta(t) \in Int\mathcal{X}_+$ for all $t > 0$. On the other hand, from (2.43) we deduce that

$$x_\delta(t) = e^{\mathcal{L}t}x_0 + \delta \int_0^t e^{\mathcal{L}(t-s)}\xi ds.$$

Hence for each fixed $t > 0$ we have

$$\lim_{\delta \to 0_+} x_\delta(t) = e^{\mathcal{L}t}x_0. \tag{2.49}$$

This leads to $e^{\mathcal{L}t}x_0 \in \mathcal{X}_+$, $\forall t \ge 0$, and $x_0 \in Int\mathcal{X}_+$ because $\mathcal{X}_+$ is a closed convex cone.

Taking again into account that $\mathcal{X}_+$ is a closed convex cone and $\mathcal{X}_+ \subset \overline{Int\mathcal{X}_+}$ (see [55]), we deduce that (i) is true. Thus the proof is complete. $\square$

The result proved in the previous theorem may be extended to the time-varying case:

Theorem 2.2.3. *Let $(\mathcal{X}, \|\cdot\|)$ be a real Banach space, ordered by the closed, solid, normal convex cone $\mathcal{X}_+$.*

Let $\mathcal{L} : \mathcal{I} \to \mathbf{B}(\mathcal{X})$ be a strongly continuous operator valued function. Under these conditions the following are equivalent:

(i) $\mathcal{L}(\cdot)$ generates a positive evolution on $\mathcal{X}$;
(ii) for each $t \in \mathcal{I}$, $\mathcal{L}(t)$ is a resolvent positive operator.

Proof. (i) $\to$ (ii) Let us assume that $\mathcal{L}(\cdot)$ generates a positive evolution on $\mathcal{X}$ and we prove that $\mathcal{L}(t)$ is resolvent positive for all $t \in \mathcal{I}$. Let us assume by contrary that there exists $\tau \in \mathcal{I}$ such that $\mathcal{L}(\tau)$ is not resolvent positive. Based on the equivalence (ii) $\Leftrightarrow$ (iii) from Theorem 2.2.2 we deduce that there exists $x \in \partial\mathcal{X}_+ \setminus \{0\}$ and $\varphi \in \mathcal{X}_+^* \setminus \{0\}$ such that $\varphi(x) = 0$ and $\varphi(\mathcal{L}(\tau)x) < 0$. Let $x(t)$ be the solution of the problem with given initial values:

$$\frac{d}{dt}x(t) = \mathcal{L}(t)x(t)$$

$$x(\tau) = x.$$

Hence $x(t) \in \mathcal{X}_+$ for all $t \in [\tau, \infty) \cap \mathcal{I}$, because $\mathcal{L}(\cdot)$ generates a positive evolution on $\mathcal{X}$. On the other hand if $g(t) = \varphi(x(t))$ we have $g(\tau) = 0$ and $\frac{d}{dt} g(\tau) < 0$.

Therefore there exists $\delta > 0$ sufficiently small such that $g(\tau + t) < 0$ or equivalently $\varphi(x(\tau + \delta)) < 0$. Hence, $x(\tau + \delta) \notin \mathcal{X}_+$ which contradicts the fact that $\mathcal{L}(\cdot)$ generates a positive evolution. So we have shown that the implication (i) $\rightarrow$ (ii) holds.

The proof of the implication (ii) $\rightarrow$ (i) may be done in a similar way as the proof of implication (iv) $\rightarrow$ (i) in Theorem 2.2.2. Here we shall display only the parts which are specific to the time-varying case. Let $x_\delta(t)$ be the solution of the problem with given initial values

$$\frac{d}{dt} x_\delta(t) = \mathcal{L}(t) x_\delta(t) + \delta \xi \tag{2.50}$$

$$x_\delta(t_0) = x_0 \in Int\mathcal{X}_+,$$

where $\delta > 0$ is sufficiently small and $\xi \in Int\mathcal{X}_+$ is fixed. Following step by step the reasoning in the proof of implication (iv) $\rightarrow$ (i) in Theorem 2.2.2 we deduce that $x_\delta(t) \in Int\mathcal{X}_+$ for all $t \in [t_0, \infty) \cap \mathcal{I}$. Using (2.43) in the case of (2.50) we obtain $x_\delta(t) = T(t, t_0) x_0 + \delta \int_{t_0}^{t} T(t, s) \xi ds$. Letting $\delta \rightarrow 0_+$ and taking into account that $\mathcal{X}_+$ is a closed, convex cone, we deduce that

$$T(t, t_0) x_0 \geq 0, \forall t_0 \in \mathcal{I}, t \in [t_0, \infty) \cap \mathcal{I} \tag{2.51}$$

and $x_0 \in Int\mathcal{X}_+$. Using again the fact that $\mathcal{X}_+$ is a closed convex cone, we conclude that (2.51) holds for any $t \in [t_0, \infty) \cap \mathcal{I}$ and any $t_0 \in \mathcal{I}$, $x_0 \in \mathcal{X}_+$. Thus the proof is complete. □

Further we prove.

Lemma 2.2.4. *Let $(\mathcal{X}, \|\cdot\|)$ be a real Banach space ordered by the closed, solid, normal, convex cone $\mathcal{X}_+$. Let $\mathcal{L}_k \in \mathbf{B}(\mathcal{X})$, $k = 1, 2$, be such that $\mathcal{L}_1 \leq \mathcal{L}_2$. If $\mathcal{L}_1$ is resolvent positive operator, then $\mathcal{L}_2$ is resolvent positive operator too.*

Proof. Let $\lambda > \|\mathcal{L}_1\|$. As in the proof of implication (i) $\rightarrow$ (ii) of Theorem 2.2.2 one shows that $(\lambda I_\mathcal{X} - \mathcal{L}_1)^{-1} x = \int_{0}^{\infty} e^{(\mathcal{L}_1 - \lambda I_\mathcal{X})t} x dt, \forall x \in \mathcal{X}$. Moreover we have that $(\lambda I - \mathcal{L}_1)^{-1} x \in \mathcal{X}_+$ if $x \in \mathcal{X}_+$ and

$$\|(\lambda I_\mathcal{X} - \mathcal{L}_1)^{-1}\| \leq \frac{1}{\lambda - \|\mathcal{L}_1\|} \tag{2.52}$$

for all $\lambda > \|\mathcal{L}_1\|$. Further we write

$$(\lambda I_\mathcal{X} - \mathcal{L}_2) = (\lambda I_\mathcal{X} - \mathcal{L}_1)(I_\mathcal{X} - U(\lambda)) \tag{2.53}$$

where $U(\lambda) = (\lambda I_{\mathcal{X}} - \mathcal{L}_1)^{-1}(\mathcal{L}_2 - \mathcal{L}_1)$. We have $U(\lambda) \geq 0$ for all $\lambda > \|\mathcal{L}_1\|$ because $(\lambda I_{\mathcal{X}} - \mathcal{L}_1)^{-1} \geq 0$ and $(\mathcal{L}_2 - \mathcal{L}_1) \geq 0$.

Moreover if $\lambda > \lambda_0 := \|\mathcal{L}_1\| + \|\mathcal{L}_2 - \mathcal{L}_1\|$ we obtain that $\|U(\lambda)\| < 1$. Hence, $I_{\mathcal{X}} - U(\lambda)$ is invertible and $(I_{\mathcal{X}} - U(\lambda))^{-1} \in \mathbf{B}(\mathcal{X})$. Since $(I_{\mathcal{X}} - U(\lambda))^{-1} = \sum_{k=0}^{\infty} (U(\lambda))^k$ we deduce that $(I_{\mathcal{X}} - U(\lambda))^{-1} \geq 0$ for all $\lambda > \lambda_0$. Finally, from (2.53) we deduce that $(\lambda I_{\mathcal{X}} - \mathcal{L}_2)^{-1} = (I_{\mathcal{X}} - U(\lambda))^{-1}(I_{\mathcal{X}} - \mathcal{L}_1)^{-1}$, $\forall \lambda > \lambda_0 \geq 0$. So the proof is complete. $\square$

In the time-varying case the following result holds.

Theorem 2.2.5. *Let $(\mathcal{X}, \|\cdot\|)$ be a real Banach space ordered by the closed, solid, normal, convex cone $\mathcal{X}_+$. Let $\mathcal{L}_k : \mathcal{I} \to \mathbf{B}(\mathcal{X})$, $k = 1,2$, be two strongly continuous operator valued functions. Assume that $\mathcal{L}_1(t) \leq \mathcal{L}_2(t)$ for all $t \in \mathcal{I}$. Under these conditions, the following hold:*

(i) *If the operator valued function $\mathcal{L}_1(\cdot)$ generates a positive evolution on $\mathcal{X}$, then the operator valued function $\mathcal{L}_2(\cdot)$ generates a positive evolution on $\mathcal{X}$. Moreover, if $T_k(t,t_0)$ are the causal linear evolution operators on $\mathcal{X}$, defined by the linear differential equations*

$$\frac{d}{dt}x(t) = \mathcal{L}_k(t)x(t), \quad k = 1,2,$$

then $T_1(t,t_0) \leq T_2(t,t_0)$ for all $t \geq t_0$, $t,t_0 \in \mathcal{I}$.

(ii) *If the operator valued function $\mathcal{L}_1(\cdot)$ generates an anticausal evolution operator on $\mathcal{X}$, then the operator valued function $\mathcal{L}_2(\cdot)$ generates an anticausal positive evolution operator on $\mathcal{X}$. Furthermore, if $T_k^a(t,t_0)$, $k = 1,2$, are the anticausal linear evolution operators on $\mathcal{X}$ defined by the linear differential equations*

$$\frac{d}{dt}y(t) + \mathcal{L}_k(t)y(t) = 0, k = 1,2,$$

then $T_1^a(t,t_0) < T_2^a(t,t_0)$ $\forall t \leq t_0$, $t,t_0 \in \mathcal{I}$.

Proof. (i) The fact that $\mathcal{L}_2(\cdot)$ generates a causal positive evolution on $\mathcal{X}$ follows combining Theorem 2.2.3 and Lemma 2.2.4. Further, we write $\frac{d}{dt}x(t) = \mathcal{L}_2(t)x(t)$ as $\frac{d}{dt}x(t) = \mathcal{L}_1(t)x(t) + f(t)$ where $f(t) = (\mathcal{L}_2(t) - \mathcal{L}_1(t))x(t)$.

Applying (2.43) and using $x(t) = T_2(t,t_0)$ we obtain:

$$T_2(t,t_0)x_0 = T_1(t,t_0)x_0 + \int_{t_0}^{t} T_1(t,s)(\mathcal{L}_2(s) - \mathcal{L}_1(s))T_2(s,t_0)x_0 ds \quad (2.54)$$

$\forall t \geq t_0$, $t,t_0 \in \mathcal{I}$, $x_0 \in \mathcal{X}_+$.

Since $\mathcal{L}_k(\cdot)$ generate a positive evolution on $\mathcal{X}$ we have $T_1(t,s) \geq 0$, $T_2(s,t_0) \geq 0$ for all $t \geq s \geq t_0$, $t,t_0 \in \mathcal{I}$. Based on $\mathcal{L}_2(s) - \mathcal{L}_1(s) \geq 0$ we may conclude that $\int_{t_0}^{t} T_1(t,s)(\mathcal{L}_2(s) - \mathcal{L}_1(s))T_2(s,t_0)x_0 ds \geq 0$ for all $t \geq t_0$, $t,t_0 \in \mathcal{I}$, $x_0 \in \mathcal{X}_+$. So (2.54) leads to

$$(T_2(t,t_0) - T_1(t,t_0))x_0 \geq 0$$

for all $t \geq t_0$, $t,t_0 \in \mathcal{I}$, $\forall x_0 \in \mathcal{X}_+$. Hence $T_2(t,t_0) \geq T_1(t,t_0)$ and thus the proof of (i) is complete.

The proof of (ii) follows immediately from the first part, (i) and the Corollary 2.2.1.   □

Corollary 2.2.6. *Let $\mathcal{L}(\cdot), \Pi(\cdot)$ be two strongly continuous operator valued functions defined on $\mathcal{I}$, taking values in $\mathbf{B}(\mathcal{X})$. Assume that $\Pi(t) \geq 0$ for all $t \in \mathcal{I}$. Then the following are true:*

(i) *If $\mathcal{L}(\cdot)$ generates a positive evolution on $\mathcal{X}$, then $\mathcal{L}(\cdot) + \Pi(\cdot)$ generates a positive evolution on $\mathcal{X}$.*

(ii) *If $\mathcal{L}(\cdot)$ generates an anticausal positive evolution on $\mathcal{X}$, then $\mathcal{L}(\cdot) + \Pi(\cdot)$ generates an anticausal positive evolution on $\mathcal{X}$.*

(iii) *$\Pi(\cdot)$ generates both a causal positive evolution and anticausal positive evolution on $\mathcal{X}$.*

Proof. (i) and (ii) follows immediately from Theorem 2.2.5 taking $\mathcal{L}_1(t) = \mathcal{L}(t)$ and $\mathcal{L}_2(t) = \mathcal{L}(t) + \Pi(t)$. For (iii) one applies Theorem 2.2.5 for $\mathcal{L}_1(t) = 0$ and $\mathcal{L}_2(t) = \Pi(t)$.

2.3 Exponential Stability of Linear Differential Equations with Positive Evolution on Ordered Banach Spaces

In this section $(\mathcal{X}, \|\cdot\|)$ is a real Banach space ordered by a solid, closed, normal, convex cone $\mathcal{X}_+$.

Let $\mathcal{L} : \mathcal{I} \to \mathbf{B}(\mathcal{X})$ be a strongly continuous operator valued function. This function defines the forward linear differential equation (2.34) as well as the backward linear differential equation (2.38). In the developments of this section $T(t,\tau)$ stands for the causal linear evolution operator on $\mathcal{X}$ defined by the linear differential equation (2.34) while $T^a(t,\tau)$ denotes the anticausal linear evolution operator on $\mathcal{X}$ defined by the backward linear differential equation (2.38).

Definition 2.3.1. (i) We say that the zero state equilibrium of the linear differential equation (2.34) is exponentially stable, or equivalently, the operator valued function $\mathcal{L}(\cdot)$ generates an exponentially stable evolution if there exist the constants $\beta \geq 1$, $\alpha > 0$ such that

$$\|T(t,t_0)\| \leq \beta e^{-\alpha(t-t_0)} \tag{2.55}$$

for all $t \geq t_0$, $t,t_0 \in \mathcal{I}$.

(ii) We say that the zero state equilibrium of the linear differential equation (2.38) is anticausal exponentially stable, or equivalently, the operator valued function $\mathcal{L}(\cdot)$ generates an anticausal exponentially stable evolution on $\mathcal{X}$ if there exist the constants $\beta \geq 1$, $\alpha > 0$ such that

$$\|T^a(t,t_0)\| \leq \beta e^{\alpha(t-t_0)} \tag{2.56}$$

for all $t \leq t_0$, $t, t_0 \in \mathcal{I}$.

Since both (2.34) and (2.38) are linear differential equations we will often say that the linear differential equation (2.34) is exponentially stable and the linear differential equation (2.38) is anticausal exponentially stable, respectively, if (2.55) and (2.56), respectively, are fulfilled.

It is worth mentioning that under the considered assumptions for any $\xi \in Int\mathcal{X}_+$ the corresponding Minkovski functional $\|\cdot\|_\xi$ is a norm equivalent with the norm $\|\cdot\|$ (see (2.33)). Therefore, the above definition may be stated in terms of the operator norm $\|\cdot\|_\xi$ for some $\xi \in Int\mathcal{X}_+$.

The criteria for exponential stability derived in this section are different from the ones based on the method of Lyapunov functions.

In the developments of this section the following integrals $\int_{-\infty}^{t} T(t,s)f(s)ds$ and $\int_{t}^{\infty} T^a(t,s)g(s)ds$ will be involved. By definition, $\int_{-\infty}^{t} T(t,s)f(s)ds = \lim_{\tau \to -\infty} \int_{\tau}^{t} T(t,s)f(s)ds$ and $\int_{t}^{\infty} T^a(t,s)g(s)ds = \lim_{\tau \to \infty} \int_{t}^{\tau} T^a(t,s)g(s)ds$ if the limits from the right-hand side exist. In this case we shall say that $\int_{-\infty}^{t} T(t,s)f(s)ds$ and $\int_{t}^{\infty} T^a(t,s)g(s)ds$ are convergent. We shall say that these integrals are absolute convergent if $\int_{-\infty}^{t} \|T(t,s)f(s)\|ds < +\infty$ and $\int_{t}^{\infty} \|T^a(t,s)g(s)\|ds < +\infty$.

Remark 2.3.1. Based on the inequalities of the form:

$$\left\| \int_{\tau_1}^{t} T(t,s)f(s)ds - \int_{\tau_2}^{t} T(t,s)f(s)ds \right\| \leq \left| \int_{\tau_1}^{\tau_2} \|T(t,s)f(s)\|ds \right|, \quad \forall \tau_1, \tau_2 < t$$

or,

$$\left\| \int_{t}^{\tau_1'} T^a(t,s)g(s)ds - \int_{t}^{\tau_2'} T^a(t,s)g(s)ds \right\| \leq \left| \int_{\tau_1'}^{\tau_2'} \|T^a(t,s)g(s)\|ds \right|,$$

respectively, one obtains' via Cauchy criteria of the existence of the limit of a function that $\int\limits_{-\infty}^{t} T(t,s)f(s)ds$ and $\int\limits_{t}^{\infty} T^a(t,s)g(s)ds$ are convergent if they are absolute convergent.

The converse implication is not always true. However in the next subsections we shall see that they become equivalent under the exponential stability condition of linear differential equations (2.34) and (2.38), respectively.

The next result will be repeatedly used in the proofs of this section:

Lemma 2.3.1. *(i) Assume $\mathcal{I} = (-\infty; a)$ with $a \leq \infty$. Let $\mathcal{L} : \mathcal{I} \to \mathbf{B}(\mathcal{X})$ be a strongly continuous operator valued function and $f : \mathcal{I} \to \mathcal{X}$ be a continuous vector valued function. If for each $t \in \mathcal{I}$, $\int\limits_{-\infty}^{t} T(t,s)f(s)ds$ is convergent, then the function $\tilde{x}(t) = \int\limits_{-\infty}^{t} T(t,s)f(s)ds$ is differentiable and solves the affine differential equation $\frac{d}{dt}x(t) = \mathcal{L}(t)x(t) + f(t)$.*

(ii) Assume $\mathcal{I} = (a, \infty)$ with $a \geq -\infty$. Let $\mathcal{L} : \mathcal{I} \to \mathbf{B}(\mathcal{X})$ be strongly continuous operator valued function and $T^a(t,\tau)$ be the anticausal linear evolution operator on $\mathcal{X}$ generated by $\mathcal{L}(\cdot)$. Let $g : \mathcal{I} \to \mathcal{X}$ be a continuous vector valued function. If for each $t \in \mathcal{I}$, $\int\limits_{t}^{\infty} T^a(t,s)g(s)ds$ is convergent, then the function $\tilde{y}(t) = \int\limits_{t}^{\infty} T^a(t,s)g(s)ds$ is differentiable and solves the backward affine equation $\frac{d}{dt}y(t) + \mathcal{L}(t)y(t) + g(t) = 0, \forall t \in \mathcal{I}$.

Proof. (i) Let $\tau \in \mathcal{I}$ be fixed. Based on the properties of the linear evolution operator $T(t,s)$ (see Remark 2.2.1) we may write:

$$\tilde{x}(t) = T(t,\tau)x(\tau) + \int_{\tau}^{t} T(t,s)f(s)ds, \ \forall t \in \mathcal{I}.$$

Therefore (see [31]) $\tilde{x}(t)$ is differentiable and it solves the affine differential equation:

$$\frac{d}{dt}\tilde{x}(t) = \mathcal{L}(t)\tilde{x}(t) + f(t), \quad t \in \mathcal{I}.$$

(ii) May be proved similarly. The details are omitted.

□

2.3.1 Criteria for Causal Exponential Stability

Throughout this section $\xi \in Int\mathcal{X}_+$ is a fixed vector. We know that the Minkovski functional $|\cdot|_\xi$ is a norm equivalent with the norm $\|\cdot\|$ of the Banach space $\mathcal{X}$. Thus we obtain via (2.33) that the exponential stability of the linear differential equation (2.34) is equivalent to

$$\|T(t,t_0)\|_\xi \le \beta e^{-\alpha(t-t_0)} \tag{2.57}$$

for all $t \ge t_0$, $t,t_0 \in \mathcal{I}$, for some $\beta \ge 1$, $\alpha > 0$ not depending upon t,t_0.

The first result specific to the case of linear differential equations defined by operator valued functions which generates a positive evolution is the following.

Proposition 2.3.2. *Under the considered assumptions the following statements hold:*

(i) *If $\mathcal{L} : \mathcal{I} \to \mathbf{B}(\mathcal{X})$ is a bounded and strongly continuous operator valued function which generates a positive evolution, then the corresponding linear differential equation (2.34) is exponentially stable iff there exist $\beta \ge 1$, $\alpha > 0$ such that*

$$\|x(t;t_0,\xi)\| \le \beta e^{-\alpha(t-t_0)} \|\xi\|$$

for all $t \ge t_0$, $t,t_0 \in \mathcal{I}$.

(ii) *Let $\mathcal{L}_k : \mathcal{I} \to \mathbf{B}(\mathcal{X})$, $k = 1,2$ be two bounded and strongly continuous operator valued functions.*

Assume: (a) $\mathcal{L}_1(t) \le \mathcal{L}_2(t)$, $t \in \mathcal{I}$.
(b) $\mathcal{L}_1(\cdot)$ generates a positive evolution on $\mathcal{X}$.
(c) The linear differential equation $\frac{d}{dt}x(t) = \mathcal{L}_2(t)x(t)$ is exponentially stable.
Under these conditions the linear differential equation $\frac{d}{dt}x(t) = \mathcal{L}_1(t)x(t)$ is exponentially stable.

Proof. (i) Follows immediately from $x(t;t_0,\xi) = T(t,t_0)\xi$ together with (2.57) and Theorem 2.1.10

To prove (ii) we remark that from assumptions (a) and (b) together with Theorem 2.2.5 (i) we deduce that $\mathcal{L}_2(\cdot)$ generates also a positive evolution and $T_1(t,t_0) \le T_2(t,t_0)$. Further from Corollary 2.1.11 we deduce that $\|T_1(t,t_0)\|_\xi \le \|T_2(t,t_0)\|_\xi$, for all $t \ge t_0$, $t,t_0 \in \mathcal{I}$.

The conclusion follows from (2.57) written for $T_2(t,t_0)$. Thus the proof is complete.   □

Remark 2.3.2. The statement of Proposition 2.3.2 (i) emphasizes an interesting fact specific to linear differential equations with positive evolution. In the case of these kinds of linear differential equations, the exponentially stable behavior of a single solution $x(t;t_0,\xi)$ implies the exponentially stable behavior of all solutions, that is, the exponential stability of the considered differential equation.

We note that if $\mathcal{I} \subset \mathbf{R}$ is a right unbounded but left bounded interval, then without loss of generality, we may take $\mathcal{I} = \mathbf{R}_+$.

In this case we have:

Theorem 2.3.3. *Let $\mathcal{L} : \mathbf{R}_+ \to \mathbf{B}(\mathcal{X})$ be a bounded and strongly continuous operator valued function. Assume that $\mathcal{L}(\cdot)$ defines a positive evolution on $\mathcal{X}$. Then the following are equivalent:*

(i) $\mathcal{L}(\cdot)$ *defines an exponentially stable evolution on* $\mathcal{X}$.

(ii) *There exists* $\delta > 0$ *not depending upon* t *and* t_0 *such that* $\int_{t_0}^{t} \|T(t,s)\| ds \leq \delta$, *for all* $t \geq t_0 \geq 0$.

(iii) *There exists* $\delta > 0$ *not depending upon* t *and* t_0 *such that* $\int_{t_0}^{t} T(t,s)\xi ds \leq \delta\xi$, *for all* $t \geq t_0 \geq 0$.

(iv) *The solution with initial value* $x(0) = 0$ *of the affine differential equation*

$$\frac{d}{dt}x(t) = \mathcal{L}(t)x(t) + \xi \qquad (2.58)$$

is bounded.

(v) *For each bounded and continuous vector function* $f : \mathbf{R}_+ \to \mathcal{X}$ *the solution with initial value* $x(0) = 0$ *of the affine differential equation*

$$\frac{d}{dt}x(t) = \mathcal{L}(t)x(t) + f(t) \qquad (2.59)$$

is bounded.

Proof. The proof of the implications (i) $\to$ (ii) $\to$ (iii) $\to$ (iv) are obvious. We proof (iv) $\to$ (v). If (iv), holds then there exists $\mu > 0$ such that $| \int_0^t T(t,s)\xi ds |_\xi \leq \mu$ for all $t \in \mathbf{R}_+$. Let $f : \mathbf{R}_+ \to \mathcal{X}$ be a continuous and bounded vector function. This means that there exist the real numbers δ_1, δ_2 such that

$$\delta_1 \xi \leq f(s) \leq \delta_2 \xi$$

for all $s \geq 0$. Since $T(t,s) \geq 0$ we deduce that it is a monotone operator. Hence

$$\delta_1 T(t,s)\xi \leq T(t,s)f(s_\leq \delta_2 T(t,s)\xi$$

for all $t \geq s \geq 0$. This leads to

$$\delta_1 \int_0^t T(t,s)\xi ds \leq \int_0^t T(t,s)f(s)ds \leq \delta_2 \int_0^t T(t,s)\xi ds$$

for all $t \geq 0$. Applying Theorem 2.1.4 (xi) we deduce that

$$\left| \int_0^t T(t,s)f(s)ds \right|_\xi \leq \tilde{\delta} \left| \int_0^t T(t,s)\xi ds \right|_\xi$$

for all $t \geq 0$ where $\tilde{\delta} = max\{|\delta_1|, |\delta_2|\}$. Thus we have obtained

$$\left| \int_0^t T(t,s)f(s)ds \right|_\xi \leq \tilde{\delta}\mu$$

for all $t \geq 0$. This allows us to conclude that

$$\left\| \int_0^t T(t,s) f(s) ds \right\| \leq \tilde{\mu}$$

for all $t \geq 0$ which shows that (v) is true. The implication (v) $\rightarrow$ (i) is the infinite dimensional version of the Peron's theorem (see [74, 107]). $\square$

Remark 2.3.3. The equivalence (i) $\leftrightarrow$ (iv) of Theorem 2.3.3 reveals another interesting fact specific to linear differential equations with positive evolution. It is known from Peron's theorem that the exponential stability of a linear differential equation is characterized via the boundedness of the solutions with zero initial value of all affine differential equations of type (2.59) corresponding to all forcing terms $f(\cdot)$ which are bounded and continuous vector valued functions. In the case of linear differential equations with positive evolution, the exponential stability can be deduced from the boundedness of the solution with zero initial value of a single affine equation, namely (2.58).

Definition 2.3.2. We say that the vector valued function $f : \mathcal{I} \to \mathcal{X}_+$ is uniformly positive if there exists a constant $c = c(f) > 0$ such that $f(t) \geq c\xi$, for all $t \in \mathcal{I}$. In this case we shall write $f(t) \gg 0, t \in \mathcal{I}$. Also we shall write $f(t) \ll 0, t \in \mathcal{I}$ if and only if $-f(t) \gg 0, t \in \mathcal{I}$.

Remark 2.3.4. A more natural definition of uniform positivity of a vector valued function would be: $f : \mathcal{I} \to \mathcal{X}_+$ is a uniform positive vector valued function if there exists $\zeta = \zeta(f) \in Int\mathcal{X}_+$ such that $f(t) \geq \zeta$ for all $t \in \mathcal{I}$.

Let B_ζ be the open and convex set associated via (2.27) with the vector $\zeta \in Int\mathcal{X}_+$. Then, by (2.28), there exists $c > 0$ such that $c\xi \in B_\zeta$. That is $-\zeta < c\xi < \zeta$. Thus we obtain that $f(t) \geq c\xi$ for all $t \in \mathcal{I}$, and thus we proved that the two definitions of uniform positivity are equivalent.

The next auxiliary result will be repeatedly used in the developments in this section.

Lemma 2.3.4. *Let $\mathcal{L} : \mathcal{I} \to \mathbf{B}(\mathcal{X})$ be a bounded and strongly continuous operator valued function. Let $T(t,t_0)$ be the corresponding linear evolution operator. Then for each $\zeta \in Int\mathcal{X}_+$, there exists $\tau_0 = \tau_0(\zeta) > 0$ such that*

$$T(t,\tau)\zeta \geq \frac{1}{2}\zeta \tag{2.60}$$

for all $t, \tau \in \mathcal{I}, t \geq \tau, t - \tau \leq \tau_0$.

Proof. Let $\tau \in \mathcal{I}$ be arbitrary but fixed and $x_\tau(t) = T(t,\tau)\zeta$. We have

$$x_\tau(t) = \zeta + \int_\tau^t \mathcal{L}(s) x_\tau(s) ds. \tag{2.61}$$

Hence $\|x_\tau(t)\| \leq \|\zeta\| + \beta \int_\tau^t \|x_\tau(s)\| ds$ where $\beta = \sup_{s \in \mathcal{I}} \|\mathcal{L}(s)\|$. Applying Gronwall's
Lemma we deduce $\|x_\tau(t)\| \leq \|\zeta\| e^{\beta(t-\tau)}$, for all $t \geq \tau, t, \tau \in \mathcal{I}$.

Using the last inequality we obtain from (2.61) that:

$$\|x_\tau(t) - \zeta\| \leq \|\zeta\|(e^{\beta(t-\tau)} - 1).$$

Furthermore we deduce, via Theorem 2.1.4 (v) that

$$|x_\tau(t) - \zeta|_\zeta \leq \beta(\zeta)\|\zeta\|(e^{\beta(t-\tau)} - 1). \tag{2.62}$$

Choose $\tau_0 > 0$ such that $\beta(\zeta)\|\zeta\|(e^{\beta(t-\tau)} - 1) < \frac{1}{2}$ for all $t - \tau \leq \tau_0$. Hence (2.62)
becomes $|x_\tau(t) - \zeta|_\zeta < \frac{1}{2}$ or equivalently $|2(x_\tau(t) - \zeta)|_\zeta < 1$.

Applying Theorem 2.1.4 (iii) we get $2(x_\tau(t) - \zeta) \in B_\zeta$ or equivalently $-\zeta <$
$2(x_\tau(t) - \zeta) < \zeta$ which leads to $x_\tau(t) \geq \frac{1}{2}\zeta$, for all $t \geq \tau, t - \tau \leq \tau_0$ which completes
the proof. $\square$

Remark 2.3.5. The result proved in Lemma 2.3.4 remains valid in a more general
setting than the one adopted in this section. In the proof of this result we do not need
to assume $T(t, \tau) \geq 0$ or that the Minkovski functional $\| \cdot \|_\xi$ is a norm.

Corollary 2.3.5. *Assume $\mathcal{I} = (-\infty, a), a \leq +\infty$. Let $\mathcal{L} : \mathcal{I} \to \mathbf{B}(\mathcal{X})$ be a bounded
and strongly continuous operator valued function generating a positive evolution on
$\mathcal{X}$. Let $f : \mathcal{I} \to \mathcal{X}_+$ be a continuous vector valued function such that $f(t) \gg 0, t \in \mathcal{I}$.
If $x : \mathcal{I} \to \mathcal{X}_+$ is a solution of the affine differential equation $\frac{d}{dt}x(t) = \mathcal{L}(t)x(t) + f(t)$,
then $x(t) \gg 0, t \in \mathcal{I}$.*

Proof. Recall that $f(t) \gg 0, t \in \mathcal{I}$ means that there exists $c > 0$ such that $f(t) \geq c\xi$,
$t \in \mathcal{I}$. Since $T(t, s) \geq 0$ for $t \geq s$ we deduce that

$$T(t, s)f(s) \geq cT(t, s)\xi \tag{2.63}$$

for all $t \geq s, t, s \in \mathcal{I}$. Let $\tau_0 = \tau_0(\xi) > 0$ be provided by Lemma 2.3.4 for $\zeta = \xi$. We
write $x(t) = T(t, \tau)x(\tau) + \int_\tau^t T(t, s)f(s)ds$. We have $x(t) \geq \int_\tau^t T(t, s)f(s)ds$. Invoking
(2.63) we have:

$$x(t) \geq c \int_\tau^t T(t, s)\xi ds, t \geq \tau.$$

Taking $\tau = t - \tau_0$ we obtain via Lemma 2.3.4 that

$$x(t) \geq c(\frac{\tau_0}{2})\xi, \ (\forall) \ t \in \mathcal{I}.$$

Thus we obtain $x(t) \gg 0, t \in \mathcal{I}$ which completes the proof. $\square$

The next result provides a list of criteria for exponential stability for linear differential equations with positive evolution in the case when $\mathcal{I}$ is a left unbounded interval.

Theorem 2.3.6. *Assume* $\mathcal{I} = (-\infty, a)$ *with* $a \leq +\infty$. *Let* $\mathcal{L} : \mathcal{I} \to \mathbf{B}(\mathcal{X})$ *be a bounded and strongly continuous operator valued function. Assume that* $\mathcal{L}(\cdot)$ *generates a positive evolution on* $\mathcal{X}$. *Under these assumptions the following statements are equivalent:*

(i) $\mathcal{L}(\cdot)$ *generates an exponentially stable evolution.*

(ii) *For each* $t \in \mathcal{I}$ *the integral* $\int_{-\infty}^{t} T(t,s)\xi ds$ *is absolute convergent and there exists* $\delta > 0$ *not depending upon* t, *such that*

$$0 \leq \int_{-\infty}^{t} T(t,s)\xi ds \leq \delta\xi, \ \forall t \in \mathcal{I}. \tag{2.64}$$

(iii) *For each* $t \in \mathcal{I}$ *the integral* $\int_{-\infty}^{t} T(t,s)\xi ds$ *is convergent and there exists* $\delta > 0$ *not depending upon* t, *such that*

$$0 \leq \int_{-\infty}^{t} T(t,s)\xi ds \leq \delta\xi, \ \forall t \in \mathcal{I}. \tag{2.65}$$

(iv) *The affine differential equation*

$$\frac{d}{dt}x(t) = \mathcal{L}(t)x(t) + \xi \tag{2.66}$$

has a bounded and uniformly positive solution.

(v) *For each bounded, continuous, and uniformly positive vector valued function* $f : \mathcal{I} \to \mathcal{X}$ *the affine differential equation*

$$\frac{d}{dt}x(t) = \mathcal{L}(t)x(t) + f(t) \tag{2.67}$$

has a bounded and uniformly positive solution.

(vi) *There exists a bounded, continuous, and uniformly positive vector valued function* $\tilde{f} : \mathcal{I} \to \mathcal{X}$ *such that the corresponding affine differential equation (2.67) has a bounded solution* $\tilde{x} : \mathcal{I} \to \mathcal{X}_+$.

(vii) *There exists a* C^1 *vector valued function* $z : \mathcal{I} \to \mathcal{X}$ *bounded with bounded derivative* $z(t) \gg 0$, $t \in \mathcal{I}$ *which solves the following linear differential inequality:*

$$-\frac{d}{dt}z(t) + \mathcal{L}(t)z(t) \ll 0, \ t \in \mathcal{I}. \tag{2.68}$$

Proof. If (i) holds, then we obtain, via (2.57) that

$$\int_{-\infty}^{t} \|T(t,s)\|_{\xi} ds \le \delta, \ \forall t \in \mathcal{I} \tag{2.69}$$

with $\delta = \frac{\beta}{\alpha}$. Based on the equivalence of the operator norms, $\|\cdot\|$ and $\|\cdot\|_{\xi}$ we deduce that there exists a positive constant γ such that

$$\int_{-\infty}^{t} \|T(t,s)\| ds \le \gamma \int_{-\infty}^{t} \|T(t,s)\|_{\xi} ds \le \gamma\delta, \ \forall t \in \mathcal{I}.$$

This allows us to conclude that $\int_{-\infty}^{t} \|T(t,s)\xi\| ds \le \gamma\delta\|\xi\|$. Thus we proved that $\int_{-\infty}^{t} T(t,s)\xi ds$ is absolute convergent.

Applying Theorem 2.1.4 (vii) and (viii) for $x = \frac{T(t,s)\xi}{|T(t,s)\xi|_{\xi}}$ we deduce that

$$0 \le T(t,s)\xi \le |T(t,s)\xi|_{\xi}\xi = \|T(t,s)\|_{\xi}\xi, \ \forall t \ge 0, t,s \in \mathcal{I}. \tag{2.70}$$

Since $\mathcal{X}_{+}$ is a closed convex cone we deduce via (2.70) that $0 \le \int_{-\infty}^{t} T(t,s)\xi ds \le \delta\xi, \ \forall t \in \mathcal{I}$.

This shows that (2.64) is true and the implication (i) $\to$ (ii) is valid. The implication (ii) $\to$ (iii) is obvious. Now we prove (iii) $\to$ (iv). If (iii) is true, then we define

$$\tilde{x}(t) = \int_{-\infty}^{t} T(t,s)\xi ds, \ t \in \mathcal{I}. \tag{2.71}$$

We have

$$0 \le \tilde{x}(t) \le \delta\xi, \ t \in \mathcal{I}. \tag{2.72}$$

Applying Lemma 2.3.1 we deduce that $t \to \tilde{x}(t)$ is a differentiable function and it solves the affine equation (2.66). Moreover, from (2.72) it follows that $\tilde{x}(t)$ is a bounded and positive solution of (2.66). Finally applying Corollary 2.3.5 we conclude that $\tilde{x}(t)$ is a bounded and uniformly positive solution of affine differential equation (2.66) and thus (iii) $\to$ (iv) is confirmed.

The implication (iv) $\to$ (vii) is obvious since a bounded and uniformly positive solution of (2.66) verifies the linear differential inequality (2.68). Let us assume that $z: \mathcal{I} \to \mathcal{X}_{+}$ is a C^1 function bounded with bounded derivative, $z(t) \gg 0$ which solves (2.68).

Define

$$\tilde{f}(t) = \frac{d}{dt}z(t) - \mathcal{L}(t)z(t), \ t \in \mathcal{I}. \tag{2.73}$$

One sees that $\tilde{f} : \mathcal{I} \to \mathcal{X}_+$ is a bounded, continuous, and uniformly positive vector valued function. Moreover $z(t)$ is a bounded and positive solution of the corresponding affine equation of type (2.67) associated with $\tilde{f}(\cdot)$ defined by (2.73). Thus we proved that (vii) $\to$ (vi).

Now we prove (i) $\to$ (v).

Let $f : \mathcal{I} \to \mathcal{X}$ be a bounded and uniformly positive continuous vector valued function. This means that $f(t) \geq \mu_1 \xi$, $\forall t \in \mathcal{I}$ and $|f(t)|_\xi \leq \mu_2$, $t \in \mathcal{I}$ for some positive constants μ_1, μ_2. Applying Theorem 2.1.4 (vi) and (viii) for $x = \frac{f(t)}{|f(t)|_\xi}$ we deduce that

$$f(t) \leq |f(t)|_\xi \xi.$$

Thus we have

$$\mu_1 \xi \leq f(t) \leq \mu_2 \xi, \ \forall t \in \mathcal{I}. \tag{2.74}$$

Since $T(t,s) \geq 0$ for all $t \geq s, t, s \in \mathcal{I}$ we deduce, via (2.74) that

$$\mu_1 T(t,s)\xi \leq T(t,s)f(s) \leq \mu_2 T(t,s)\xi \tag{2.75}$$

$\forall t \geq s, \ t, s \in \mathcal{I}$.

Using the monotonicity of the Minkovski norm, we get

$$\mu_1 |T(t,s)\xi|_\xi \leq |T(t,s)f(s)|_\xi \leq \mu_2 |T(t,s)\xi|_\xi \tag{2.76}$$

$\forall t \geq s, \ t, s \in \mathcal{I}$.

Reasoning as in the proof of the implication (i) $\to$ (ii) we may prove that

$$\int_{-\infty}^{t} \|T(t,s)f(s)\| ds \leq \delta \tag{2.77}$$

for some $\delta > 0$ not depending upon t. This shows that $\int_{-\infty}^{t} T(t,s)f(s)ds$ is absolute convergent. Set $z(t) = \int_{-\infty}^{t} T(t,s)f(s)ds$. It follows that $z(t)$ is well defined for all $t \in \mathcal{I}$ and from (2.76) and (2.77) we deduce that $z(t) \in \mathcal{X}_+$ and $\|z(t)\| \leq \delta$. Applying Lemma 2.3.1 we obtain that $t \to z(t)$ is a differentiable function and it verifies the affine differential equation (2.67). Applying Corollary 2.3.5 we deduce that $z(t)$

is a bounded and uniformly positive solution of (2.67) and thus the proof of the implication (i) $\rightarrow$ (v) is complete. The implication (v) $\rightarrow$ (vi) is obvious.

It remains to prove the implication (vi) $\rightarrow$ (i). If (vi) is true, then there exists a bounded, continuous, and uniformly positive vector valued function $f : \mathcal{I} \rightarrow \mathcal{X}_+$ such that the corresponding affine differential equation (2.67) has a bounded solution $\tilde{x} : \mathcal{I} \rightarrow \mathcal{X}_+$. Further, we deduce via Corollary 2.3.5 that $\tilde{x}(t) \gg 0$, $t \in \mathcal{I}$. Hence, there exist positive constants μ_i, $1 \leq i \leq 4$ such that

$$\mu_1 \xi \leq f(t) \leq \mu_2 \xi, \qquad \mu_3 \xi \leq \tilde{x}(t) \leq \mu_4 \xi \tag{2.78}$$

$t \in \mathcal{I}$.

Since $T(t,s) \geq 0$ if $t \geq s$, we deduce that

$$\mu_1 T(t,s)\xi \leq T(t,s)f(s) \leq \mu_2 T(t,s)\xi, \ t \geq s, t,s \in \mathcal{I}. \tag{2.79}$$

Let $\tau \in \mathcal{I}$ be fixed. Let $z : (-\infty, \tau] \rightarrow \mathcal{X}$ be defined by

$$z(t) = T(\tau,t)\tilde{x}(t), \ t \in (-\infty; \tau]. \tag{2.80}$$

From the (2.78) one obtains

$$\mu_3 T(\tau,t)\xi \leq z(t) \leq \mu_4 T(\tau,t)\xi, \ \forall t \in (-\infty; \tau]. \tag{2.81}$$

Based on the properties of the linear evolution operators we obtain from (2.80) that $z(t)$ is differentiable and we have

$$\frac{d}{dt}z(t) = -T(\tau,t)\mathcal{L}(t)\tilde{x}(t) + T(\tau,t)\mathcal{L}(t)\tilde{x}(t) + T(\tau,t)f(t)$$

or

$$\frac{d}{dt}z(t) = T(\tau,t)f(t), \ \forall t \leq \tau.$$

Combining (2.79) and (2.81) we deduce that $\frac{d}{dt}z(t) \geq \alpha z(t)$, with $\alpha = \frac{\mu_1}{\mu_4} > 0$.
The last inequality is equivalent to

$$\frac{d}{dt}(e^{\alpha(\tau-t)}z(t)) \geq 0, \ t \leq \tau. \tag{2.82}$$

Let $g(t) = \frac{d}{dt}(e^{\alpha(\tau-t)}z(t))$, $t \leq \tau$. Since $\mathcal{X}_+$ is a closed convex cone, we deduce from (2.82) that

$$\int_{t_1}^{t_2} g(s)ds \geq 0, \ \forall t_1 \leq t_2 \leq \tau. \tag{2.83}$$

Integrating (2.82) we obtain via (2.83) that $e^{\alpha(\tau-t)}z(t) \leq z(\tau)$, $\forall t \leq \tau$. Using again (2.81) one gets

$$e^{\alpha(\tau-t)}T(\tau,t)\xi \leq \frac{\mu_4}{\mu_3}\xi.$$

Using the monotonicity of the Minkovski norm we conclude:

$$|T(\tau,t)\xi|_\xi \leq \frac{\mu_4}{\mu_3}e^{-\alpha(\tau-t)}, \ \forall t \leq \tau.$$

Applying Theorem 2.1.10 we conclude that the last inequality is equivalent to

$$\|T(\tau,t)\|_\xi \leq \frac{\mu_4}{\mu_3}e^{-\alpha(\tau-t)}, \ t \leq \tau. \tag{2.84}$$

Since τ was arbitrary chosen in $\mathcal{I}$, we deduce that (2.84) is just (2.57). Thus the proof is complete. $\square$

Concerning the bounded on $\mathcal{I}$ solutions of the affine differential equations of type (2.67) we have:

Theorem 2.3.7. *Assume that $\mathcal{I} = (-\infty; a)$ with $a \leq +\infty$. Let $\mathcal{L} : \mathcal{I} \to \mathbf{B}(\mathcal{X})$ be a bounded and strongly continuous operator valued function defining an exponentially stable evolution on $\mathcal{X}$. Under these conditions the following hold:*

(i) *For each bounded and continuous vector valued function $f : \mathcal{I} \to \mathcal{X}$ the affine differential equation*

$$\frac{d}{dt}x(t) = \mathcal{L}(t)x(t) + f(t) \tag{2.85}$$

has a unique bounded solution $\tilde{x} : \mathcal{I} \to \mathcal{X}$. Moreover this solution has the representation

$$\tilde{x}(t) = \int_{-\infty}^{t} T(t,s)f(s)ds, \ t \in \mathcal{I}. \tag{2.86}$$

(ii) *Assume $\mathcal{I} = \mathbf{R}$ and there exists $\theta > 0$ such that $\mathcal{L}(t+\theta) = \mathcal{L}(t)$, $f(t+\theta) = f(t)$, $(\forall) \ t \in \mathcal{I}$, then the unique bounded on $\mathcal{I}$ solution of (2.85) is a periodic function with the same period θ.*

(iii) *Assume $\mathcal{I} = \mathbf{R}$ and $\mathcal{L}(t) = \mathcal{L}$, $f(t) = f$ for all $t \in \mathbf{R}$. Then the unique bounded on $\mathbf{R}$ solution of (2.85) is constant. It is given by $\tilde{x} = -\mathcal{L}^{-1}f$ and it solves the linear equation $\mathcal{L}\tilde{x} + f = 0$.*

(iv) *If $\mathcal{L}(\cdot)$ generates a positive evolution on $\mathcal{X}$ and $f(t) \geq 0, t \in \mathcal{I}$, then the unique bounded solution $\tilde{x}$ of (2.85) satisfies $\tilde{x}(t) \geq 0, t \in \mathcal{I}$. Moreover, if $f(t) \gg 0$, $t \in \mathcal{I}$, then $\tilde{x}(t) \gg 0, t \in \mathcal{I}$.*

Proof. (i) Based on (2.55) we obtain

$$\int_{-\infty}^{t} \|T(t,s)f(s)\| ds \leq \frac{\beta}{\alpha} sup_{s \in \mathcal{I}} \|f(s)\| \tag{2.87}$$

for all $t \in \mathcal{I}$. This shows that for arbitrary $t \in \mathcal{I}$, the integral $\int_{-\infty}^{t} T(t,s)f(s)ds$ is absolute convergent; hence, it is convergent. So, $\tilde{x}(t)$ is well defined by (2.86). Applying Lemma 2.3.1 together with (2.87) we deduce that $t \to \tilde{x}(t)$ is differentiable and it is a bounded solution of the affine differential equation (2.85). Let $\hat{x}(t), t \in \mathcal{I}$ be another bounded solution of (2.85). Let $t \in \mathcal{I}$ be arbitrary but fixed. We may write

$$\hat{x}(t) = T(t,\tau)\hat{x}(\tau) + \int_{\tau}^{t} T(t,s)f(s)ds \tag{2.88}$$

for all $\tau < t$. Invoking again (2.55) and using the boundedness of $\hat{x}(\tau)$ we deduce that $\lim_{\tau \to -\infty} T(t,\tau)\hat{x}(\tau) = 0$. Letting $\tau \to -\infty$ in (2.88) one gets $\hat{x}(t) = \lim_{\tau \to -\infty} \int_{\tau}^{t} T(t,s)f(s)ds = \int_{-\infty}^{t} T(t,s)f(s)ds$ for all $t \in \mathcal{I}$. This allows us to conclude that $\hat{x}(t) = \tilde{x}(t)$ for all $t \in \mathcal{I}$. This confirms the uniqueness of the bounded on $\mathcal{I}$ solution of the differential equation (2.85).

(ii) Assume that both $\mathcal{L}(\cdot)$ and $f(\cdot)$ are periodic functions with period $\theta > 0$. We show that under these conditions the unique bounded solution of (2.85) is also a periodic function of period θ. By direct calculations one obtains from (2.86):

$$\tilde{x}(t+\theta) = \int_{-\infty}^{t+\theta} T(t+\theta,s)f(s)ds = \int_{-\infty}^{t} T(t+\theta,s+\theta)f(s+\theta)ds.$$

Further, the property (vii) from Remark 2.2.1 together with the periodicity property of $f(\cdot)$ allows us to deduce that $\tilde{x}(t+\theta) = \int_{-\infty}^{t} T(t,s)f(s)ds = \tilde{x}(t)$ for all $t \in \mathbf{R}$.

(iii) If $\mathcal{L}(t) = \mathcal{L}, f(t) = f$ for all $t \in \mathcal{I}$, then (2.86) becomes:

$$\tilde{x}(t) = \int_{-\infty}^{t} e^{\mathcal{L}(t-s)} f ds$$

for all $t \in \mathbf{R}$. A straightforward change of the variable of integration leads to

$$\tilde{x}(t) = \int_{-\infty}^{0} e^{\mathcal{L}s} f ds$$

for all $t \in \mathbf{R}$. This shows that under the considered assumptions the unique bounded on $\mathbf{R}$ solution of the differential equation (2.85) is constant. Therefore it solves the equation $\mathcal{L}\tilde{x} + f = 0$, because in this case we have $\frac{d}{dt}\tilde{x}(t) = 0$ for all $t \in \mathbf{R}$. On the other hand, the exponential stability of the differential equation

$$\frac{d}{dt}x(t) = \mathcal{L}x(t)$$

allows us to deduce that $\lambda = 0$ does not in the spectrum of the operator $\mathcal{L}$. Therefore the operator $\mathcal{L}$ is invertible and $\mathcal{L}^{-1} \in \mathbf{B}(\mathcal{X})$. Thus we obtained that $\tilde{x} = \mathcal{L}^{-1}f$.

(iv) If $\mathcal{L}(\cdot)$ generates a positive evolution on $\mathcal{X}$ and $f(s) \in \mathcal{X}_+$ for all $s \in \mathcal{I}$ then, $\int_{-\infty}^{t} T(t,s)f(s)ds \geq 0$ for all $t \in \mathcal{I}$ because $\mathcal{X}_+$ is a closed convex cone. Thus we obtained via (2.86) that under the considered assumptions the unique bounded solution of the differential equation (2.85) lies in $\mathcal{X}_+$. The second part follows from the Corollary 2.3.5. Thus the proof is complete.

$\square$

Remark 2.3.6. Based on (2.43) one sees that under the conditions of Theorem 2.3.7, if $\mathcal{I}$ is a left bounded interval, then all solutions of the affine differential equation (2.41) are bounded on $\mathcal{I}$ if $f(\cdot)$ is a bounded and continuous function. Therefore, in order to have a unique bounded on $\mathcal{I}$ solution of the differential equation (2.85) it is necessary that $\mathcal{I}$ be a left unbounded interval.

The next corollary summarizes a set of criteria for exponential stability of linear differential equations defined by periodic operator valued functions which generates a positive evolution.

Corollary 2.3.8. *Let $\mathcal{L} : \mathbf{R} \to \mathbf{B}(\mathcal{X})$ be a strongly continuous operator valued function which generates a positive evolution on $\mathcal{X}$. Assume that there exists $\theta > 0$ such that $\mathcal{L}(t + \theta) = \mathcal{L}(t), \ t \in \mathbf{R}$. Then the following are equivalent:*

(i) *The operator valued function $\mathcal{L}(\cdot)$ generates an exponentially stable evolution on $\mathcal{X}$.*

(ii) *The affine differential equation (2.66) has a θ-periodic solution $\tilde{x} : \mathbf{R} \to \mathcal{X}_+$ such that $\tilde{x}(t) \gg 0, t \in \mathbf{R}$.*

(iii) *For each continuous and θ-periodic vector valued function $f : \mathbf{R} \to \mathcal{X}_+$, $f(t) \gg 0, \ t \in \mathbf{R}$, the affine differential equation (2.67) has a θ-periodic solution, $\tilde{x} \gg 0, t \in \mathcal{R}$.*

(iv) *There exists a θ-periodic, uniformly positive, continuous vector valued function $\tilde{f} : \mathbf{R} \to \mathcal{X}_+$ such that the corresponding affine differential equation (2.67) has a θ-periodic solution $\tilde{x}(t) \geq 0, t \in \mathbf{R}$.*

(v) *There exists a C^1 function $y : \mathbf{R} \to \mathcal{X}_+$ periodic with period θ and uniformly positive which solves the linear differential inequality $\frac{d}{dt}y(t) - \mathcal{L}(t)y(t) \gg 0, \ t \in [0, \theta]$.*

(vi) *$\rho[T(\theta, 0)] < 1, \rho[\cdot]$ being the spectral radius.*

Proof. (i) $\leftrightarrow$ (ii) $\leftrightarrow$ (iii) $\leftrightarrow$ (iv) $\leftrightarrow$ (v) follow immediately combining Theorems 2.3.6 and 2.3.7 (i), (ii), (iv). The proof of (i) $\leftrightarrow$ (vi) follows in a standard way. Indeed, let $U = T(\theta, 0)$ and $\rho_0 = \rho(U)$. From [55] it follows that

$$\rho_0^n = \rho(U^n) \leq \|U^n\| = \|T(n\theta, 0)\|.$$

Therefore (i) implies that $\lim_{n \to \infty} \rho_0^n = 0$, hence $\rho_0 < 1$. Conversely, if $\rho_0 < 1$, then $\lim_{n \to \infty} (\|U^n\|)^{\frac{1}{n}} < 1$. We deduce that there exist $n_0 \geq 1$ and $\gamma \in (0,1)$ such that

$$\|T(kn_0\theta, 0)\| \leq \gamma^{n_0 k \theta}, k \geq 1.$$

Further, using Remark 2.2.1 (iii), (v), (vii) one can obtain (2.55) with $\alpha = \ln\gamma$. The proof is complete.   □

Let us consider the case when the function $t \to \mathcal{L}(t)$ is constant, then

$$T(t, t_0) = e^{\mathcal{L}(t - t_0)}, \ t \geq t_0, \ t, t_0 \in \mathbf{R}.$$

In this case we have:

Corollary 2.3.9. *Let $\mathcal{L} \in \mathbf{B}(\mathcal{X})$ be such that $e^{\mathcal{L}t} \geq 0$, if $t \geq 0$. Under this condition the following are equivalent:*

 (i) *The linear differential equation $\frac{d}{dt}x(t) = \mathcal{L}x(t)$ is exponentially stable.*
 (ii) *The linear equation $\mathcal{L}x + \xi = 0$ has a solution $\tilde{x} \in Int\mathcal{X}_+$.*
 (iii) *For each vector $f \in Int\mathcal{X}_+$ there exists $\tilde{x} \in Int\mathcal{X}_+$ which solves $\mathcal{L}\tilde{x} + f = 0$.*
 (iv) *There exists $y \in Int\mathcal{X}_+$ such that $\mathcal{L}y < 0$.*
 (v) *$Spec\mathcal{L} \subset \mathbf{C}^-$ where $\mathbf{C}^- = \{\lambda \in \mathbf{C}; Re\lambda < 0\}$ and $Spec\mathcal{L}$ is the spectrum of the operator $\mathcal{L}$.*

Proof. (i) $\leftrightarrow$ (ii) $\leftrightarrow$ (iii) $\leftrightarrow$ (iv) follow immediately combining Theorems 2.3.6 and 2.3.7, (i), (iii), (iv). (i) $\leftrightarrow$ (v) follows directly from Corollary 2.3.8 for $\theta = 1$. The proof is complete.   □

Based on the equivalence (i) $\leftrightarrow$ (ii) of Theorem 2.2.2 we infer that the criteria for exponential stability collected in Corollary 2.3.9 recover the criteria for exponential stability known in the case of resolvent positive operators (see [27,29]).

2.3.2   Criteria for Anticausal Exponential Stability

Based on the identity (2.45) together with the Definition 2.3.1 we obtain:

Corollary 2.3.10. *Let $\mathcal{L} : \mathcal{I} \to \mathbf{B}(\mathcal{X})$ be a strongly continuous operator valued function and $\hat{\mathcal{L}}(t) = \mathcal{L}(-t)$, $t \in \hat{\mathcal{I}} = \{t \in \mathbf{R}; -t \in \mathcal{I}\}$. Then the operator valued function $\mathcal{L}(\cdot)$ defines an anticausal exponentially stable evolution if and only if the operator valued function $\hat{\mathcal{L}}(\cdot)$ generates a causal exponentially stable evolution.*

The above corollary allows us to derive criteria for anticausal exponential stability of a linear differential equation defined by an operator valued function $\mathcal{L}(\cdot)$ directly from the criteria for causal exponential stability for the linear differential equation defined by the operator valued function $\hat{\mathcal{L}}(\cdot)$.

First we remark that due to the equivalence of the operator norms $\|\cdot\|$ and $\|\cdot\|_\xi$ we may say that the operator valued function $\mathcal{L}(\cdot)$ defines an anticausal exponentially stable evolution if there exist $\beta \geq 1$, $\alpha > 0$ such that $\|T^a(t,t_0)\|_\xi \leq \beta e^{\alpha(t-t_0)}$, $\forall t \leq t_0$, $t, t_0 \in \mathcal{I}$.

The analogous of Proposition 2.3.2 is the following.

Proposition 2.3.11. *Under the considered assumptions the following hold:*

(i) *If $\mathcal{L} : \mathcal{I} \to \mathbf{B}(\mathcal{X})$ is a bounded and strongly continuous operator valued function generating an anticausal positive evolution on $\mathcal{X}$, the following are equivalent:*

   (a) *$\mathcal{L}(\cdot)$ generates an anticausal exponentially stable evolution;*
   (b) *there exist $\alpha > 0$, $\beta \geq 1$ such that $|y(t;t_0,\xi)|_\xi \leq \beta e^{\alpha(t-t_0)}$, $\forall t \leq t_0$, $t, t_0 \in \mathcal{I}$, $y(t;t_0,\xi)$ being the solution of the differential equation (2.38) starting from ξ at the initial time $t = t_0$.*

(ii) *Let $\mathcal{L}_k : \mathcal{I} \to \mathbf{B}(\mathcal{X})$, $k = 1,2$ be two bounded and strongly continuous operator valued functions.*
   Assume (a) $\mathcal{L}_1(t) \leq \mathcal{L}_2(t)$, $t \in \mathcal{I}$.

   (b) *the operator valued function $\mathcal{L}_1(\cdot)$ defines an anticausal positive evolution.*
   (c) *the operator valued function $\mathcal{L}_2(\cdot)$ defines an anticausal exponentially stable evolution.*

Under these conditions, $\mathcal{L}_1(\cdot)$ generates an anticausal exponentially stable evolution, too.

Combining Corollary 2.3.10 and Theorem 2.3.6 we obtain the following result.

Theorem 2.3.12. *Assume $\mathcal{I} = (a,\infty)$ with $a \geq -\infty$. Let $\mathcal{L} : \mathcal{I} \to \mathbf{B}(\mathcal{X})$ be a bounded and strongly continuous operator valued function generating an anticausal positive evolution on $\mathcal{X}$.*

Under these conditions the following are equivalent:

(i) *The operator valued function $\mathcal{L}(\cdot)$ generates an anticausal exponentially stable evolution.*

(ii) *For each $t \in \mathcal{I}$ the integral $\int_t^\infty T^a(t,s)\xi ds$ is absolute convergent and there exists*

$$\delta > 0 \text{ not depending upon } t \text{ such that } 0 \leq \int_t^\infty T^a(t,s)\xi ds \leq \delta\xi, \ \forall t \in \mathcal{I}.$$

(iii) *For each $t \in \mathcal{I}$ the integral $\int_t^\infty T^a(t,s)\xi ds$ is convergent and there exists $\delta > 0$*

not depending upon t such that $0 \leq \int_t^\infty T^a(t,s)\xi ds \leq \delta\xi$, $\forall t \in \mathcal{I}$.

(iv) *The backward affine differential equations*

$$\frac{d}{dt}x(t) + \mathcal{L}(t)x(t) + \xi = 0 \tag{2.89}$$

has a bounded and uniformly positive solution.

(v) For each bounded, continuous, uniformly positive vector valued function f :
$\mathcal{I} \to \mathcal{X}$ the backward affine differential equation

$$\frac{d}{dt}x(t) + \mathcal{L}(t)x(t) + f(t) = 0 \tag{2.90}$$

has a bounded and uniformly positive solution.

(vi) There exists a bounded, continuous, and uniformly positive vector valued
function $\tilde{f} : \mathcal{I} \to \mathcal{X}$ with the property that the corresponding backward affine
differential equation of type (2.90) has a bounded solution $\tilde{x}(t) \geq 0, t \in \mathcal{I}$.

(vii) There exists a C^1 function $y : \mathcal{I} \to \mathcal{X}$ uniform positive, bounded with bounded
derivative which verifies the following linear differential inequality:

$$\frac{d}{dt}y(t) + \mathcal{L}(t)y(t) \ll 0, \ t \in \mathcal{I}.$$

The analogous of Theorem 2.3.7, in the case of backward affine differential
equations of type (2.90), is the following.

Theorem 2.3.13. *Assume* $\mathcal{I} = (a, \infty)$ *with* $a \geq -\infty$. *Let* $\mathcal{L} : \mathcal{I} \to \mathbf{B}(\mathcal{X})$ *be a bounded
and strongly continuous operator valued function defining an anticausal stable
evolution on* $\mathcal{X}$.

Then the following statements are true:

(i) *for each bounded and continuous vector valued function* $f : \mathcal{I} \to \mathcal{X}$ *the
backward affine differential equation*

$$\frac{d}{dt}x(t) + \mathcal{L}(t)x(t) + f(t) = 0, \ t \in \mathcal{I} \tag{2.91}$$

has a unique bounded on $\mathcal{I}$ *solution. Moreover, that solution has the following
representation*

$$\tilde{x}(t) = \int\limits_t^\infty T^a(t,s)f(s)ds, \ t \in \mathcal{I}. \tag{2.92}$$

(ii) *If there exists* $\theta > 0$ *such that* $\mathcal{L}(t+\theta) = \mathcal{L}(t), \ f(t+\theta) = f(t), \ t \in \mathcal{I}$, *then the
unique bounded solution* $\tilde{x}(t)$ *of the differential equation (2.91) is a periodic
function with the same period* θ.

(iii) *If* $\mathcal{L}(t) = \mathcal{L}, \ f(t) = f, \ \forall t \in \mathcal{I}$, *then the unique bounded solution of (2.91) is
constant. It solves the linear equation* $\mathcal{L}x + f = 0$.

(iv) *If the operator valued function* $\mathcal{L}(\cdot)$ *defines an anticausal positive evolution
and* $f : \mathcal{I} \to \mathcal{X}_+$ *is a bounded and continuous vector valued function, then the
unique bounded solution of (2.91) satisfies* $\tilde{x}(t) \geq 0$. *Moreover,* $\tilde{x}(t) \gg 0, t \in \mathcal{I}$,
if $f(t) \gg 0, \ t \in \mathcal{I}$.

Combining Theorems 2.3.12 and 2.3.13 (i), (ii), (iv) we obtain the following set of criteria for anticausal exponential stability of a linear differential equation defined by a periodic operator valued function.

Corollary 2.3.14. *Let* $\mathcal{L} : \mathbf{R} \to \mathbf{B}(\mathcal{X})$ *be a strongly continuous operator valued function which defines an anticausal positive evolution on* $\mathcal{X}$. *Assume that there exists* $\theta > 0$ *such that* $\mathcal{L}(t + \theta) = \mathcal{L}(t)$, $\forall t \in \mathbf{R}$.
Then the following are equivalent:

(i) *the operator valued function* $\mathcal{L}(\cdot)$ *generates an anticausal exponentially stable evolution;*
(ii) *the backward affine differential equation (2.89) has a* θ-*periodic and uniformly positive solution;*
(iii) *for each continuous,* θ-*periodic, and uniformly positive vector function* $f :$ $\mathbf{R} \to \mathcal{X}$ *the backward affine differential equation (2.90) has a* θ-*periodic and uniformly positive solution;*
(iv) *there exists a continuous, uniformly positive vector valued function* $\tilde{f} : \mathbf{R} \to \mathcal{X}$ *periodic, with period* θ *such that the corresponding affine differential equation (2.90) has a* θ-*periodic solution* $\tilde{x} : \mathbf{R} \to \mathcal{X}_+;$
(v) *there exists a* C^1 *vector valued function,* $y : \mathbf{R} \to \mathcal{X}$, *uniformly positive and periodic, with period* θ, *which solves*

$$\frac{d}{dt} y(t) + \mathcal{L}(t) y(t) \ll 0, \ t \in [0, \theta].$$

(vi) $\rho[T(0, \theta)] < 1$.

Remark 2.3.7. In the time invariant case (i.e., $\mathcal{L}(t) = \mathcal{L}$, $\forall t \in \mathbf{R}$), when combining Theorems 2.3.12 and 2.3.13 (i), (iii), (iv) one obtains the same list of criteria as in Corollary 2.3.9. This is not an unexpected fact, because, in the time invariant case, there exists no difference between the causal exponential stability and anticausal exponential stability.

2.4 The Case of Differential Equations with Positive Evolution on Ordered Hilbert Spaces

Throughout this section $(\mathcal{X}; \langle \cdot, \cdot \rangle)$ is a real Hilbert space, ordered by the ordering "$\leq$" induced by the closed, solid, selfdual, convex cone. Based on Remark 2.1.2 we deduce that the norm $\| \cdot \|$, induced by the inner product is monotone with respect to the cone $\mathcal{X}_+$. So, $\mathcal{X}_+$ is a normal cone with a constant $\tilde{b} = 1$.

An example of an infinite dimensional Hilbert space equipped with a closed, solid, self-dual, convex cone is provided by Example 2.1.4.

Throughout this section $\xi \in Int\mathcal{X}_+$ is fixed and $| \cdot |_\xi$ is the corresponding Minkovski norm. As we already seen, Propositions 2.1.5 and 2.1.6 guarantee the fact that $|\cdot|_\xi$ is equivalent with the norm $\|\cdot\|$ of the Hilbert space $\mathcal{X}$. It is known that, if $T \in \mathbf{B}(\mathcal{X})$ and T^* is its adjoint operator, then, $\|T^*\| = \|T\|$. The equality $\|T^*\|_\xi = \|T\|_\xi$ is not, in general, true. However, one can proof, via the equivalence of the operator norms $\|\cdot\|$ and $\|\cdot\|_\xi$, that there exist positive constants, $\tilde{c}_1, \tilde{c}_2$ such that

$$\tilde{c}_1\|T\|_\xi \leq \|T^*\|_\xi \leq \tilde{c}_2\|T\|_\xi, \ \forall \, T \in \mathbf{B}(\mathcal{X}). \tag{2.93}$$

Let $\mathcal{L} : \mathcal{I} \to \mathbf{B}(\mathcal{X})$ be a continuous operator valued function, $\mathcal{I} \subset \mathbf{R}$ being a right unbounded interval. In this case $t \to \mathcal{L}^*(t) : \mathcal{I} \to \mathbf{B}(\mathcal{X})$ is also a continuous operator valued function. It is known that if $T(t,\tau)$, $t,\tau \in \mathcal{I}$, is the linear evolution operator defined by the linear differential equation

$$\frac{d}{dt}x(t) = \mathcal{L}(t)x(t), \ t \in \mathcal{I} \tag{2.94}$$

then, $\tau \to T^*(t,\tau)$ verifies

$$\frac{\partial}{\partial \tau}T^*(t,\tau) = -\mathcal{L}^*(\tau)T^*(t,\tau) \tag{2.95}$$

$$T^*(t,t) = I_\mathcal{X}.$$

So we have:

$$T^*(t,\tau) = T^a_{\mathcal{L}^*}(\tau,t) \tag{2.96}$$

$\forall \ t,\tau \in \mathcal{I}$, where $T^a_{\mathcal{L}^*}(\tau,t)$ is the anticausal linear evolution operator defined by the operator valued function $\mathcal{L}^*(\cdot)$. This means that, $T^a_{\mathcal{L}^*}(\tau,t)$ is a linear evolution operator associated with the linear differential equation

$$\frac{d}{d\tau}y(\tau) = -\mathcal{L}^*(\tau)y(\tau). \tag{2.97}$$

Combining (2.93), (2.96) and the result stated in Proposition 2.1.9 one obtains the following result.

Proposition 2.4.1. *If $\mathcal{L} : \mathcal{I} \to \mathbf{B}(\mathcal{X})$ is a continuous operator valued function, then the following statements are true:*

(i) *The operator valued function $\mathcal{L}(\cdot)$ defines a causal positive evolution on $\mathcal{X}$, iff the operator valued function $\mathcal{L}^*(\cdot)$ defines an anticausal positive evolution on $\mathcal{X}$.*

(ii) *The operator valued function $\mathcal{L}(\cdot)$ defines an exponentially stable evolution on $\mathcal{X}$ iff the operator valued function $\mathcal{L}^*(\cdot)$ defines an anticausal exponentially stable evolution on $\mathcal{X}$.*

Applying Theorem 2.3.12 to the operator valued function $\mathcal{L}^*(\cdot)$ and taking into account the equality (2.96) and Proposition 2.4.1 we obtain the following set of criteria for causal exponential stability of linear differential equation (2.94). Such criteria are specific to the linear differential equations with positive evolution on ordered Hilbert spaces.

Theorem 2.4.2. *Assume that $\mathcal{I} = (a, \infty)$, $a \geq -\infty$. Let $\mathcal{L} : \mathcal{I} \rightarrow \mathbf{B}(\mathcal{X})$ be a bounded and continuous vector valued function which defines a positive evolution on the Hilbert space $\mathcal{X}$. Then the following statements are equivalent:*

  (i)   *the corresponding linear differential equation (2.94) is exponentially stable;*
  (ii) *there exist $\beta \geq 1$, $\alpha > 0$ such that $\|T^*(t, \tau)\| \leq \beta_1 e^{-\alpha(t-\tau)}$, $\forall t \geq \tau$, $t, \tau \in \mathcal{I}$;*
  (iii) *for each $t \in \mathcal{I}$, the integral $\int_t^{\infty} T^*(s,t)\xi ds$ is absolute convergent and there*

      exists $\delta > 0$, not depending upon t, such that $0 \leq \int_t^{\infty} T^*(s,t)\xi ds \leq \delta\xi$, $\forall t \in \mathcal{I}$;

  (iv) *for each $t \in \mathcal{I}$, the integral $\int_t^{\infty} T^*(s,t)\xi ds$ is convergent and there exists $\delta > 0$,*

      not depending upon t, such that $0 \leq \int_t^{\infty} T^*(s,t)\xi ds \leq \delta\xi$, $\forall t \in \mathcal{I}$;

  (v)   *the backward affine differential equation*

$$\frac{d}{dt}y(t) + \mathcal{L}^*(t)y(t) + \xi = 0 \tag{2.98}$$

      has a bounded and uniformly positive solution;
  (vi) *for each bounded and continuous vector valued function $f : \mathcal{I} \rightarrow \mathcal{X}_+$, $f(t) \gg 0$, $t \in \mathcal{I}$, the backward affine differential equation*

$$\frac{d}{dt}y(t) + \mathcal{L}^*(t)y(t) + f(t) = 0, \ t \in \mathcal{I} \tag{2.99}$$

      has a bounded and uniformly positive solution;
  (vii) *there exists a bounded, continuous, and uniformly positive vector valued function $\tilde{f} : \mathcal{I} \rightarrow \mathcal{X}_+$ with the property that the corresponding affine differential equation of type (2.99) has a bounded solution $\tilde{y} : \mathcal{I} \rightarrow \mathcal{X}_+$;*
  (viii) *there exists a C^1 vector valued function $y : \mathcal{I} \rightarrow \mathcal{X}_+$ bounded with bounded derivative and the scalars $\lambda_1 > 0$, $\lambda_2 > 0$ such that $\frac{d}{dt}y(t) + \mathcal{L}^*(t)y(t) \leq -\lambda_1\xi$, $y(t) \geq \lambda_2\xi$, $t \in \mathcal{I}$.*

The identity (2.96) and Proposition 2.4.1 allow us to obtain the following result.

Theorem 2.4.3. *Assume that $\mathcal{I} = (a, \infty)$, $a \geq -\infty$. Let $\mathcal{L} : \mathcal{I} \rightarrow \mathbf{B}(\mathcal{X})$ be a bounded and continuous vector valued function which defines an exponentially stable evolution on the Hilbert space $\mathcal{X}$. Then the following statements hold:*

  (i)   *for each bounded and continuous vector valued function $f : \mathcal{I} \rightarrow \mathcal{X}$ the backward affine differential equation*

$$\frac{d}{dt}x(t) + \mathcal{L}^*(t)x(t) + f(t) = 0, \ t \in \mathcal{I} \qquad (2.100)$$

has a unique bounded on $\mathcal{I}$ solution. Moreover, that solution has the following representation

$$\tilde{x}(t) = \int\limits_t^\infty T^*(s,t)f(s)ds, \ t \in \mathcal{I}. \qquad (2.101)$$

(ii) If there exists $\theta > 0$ such that $\mathcal{L}(t+\theta) = \mathcal{L}(t)$, $f(t+\theta) = f(t)$, $t \in \mathcal{I}$, then the unique bounded solution $\tilde{x}(t)$ of the differential equation (2.100) is a periodic function with the same period θ.

(iii) If $\mathcal{L}(t) = \mathcal{L}$, $f(t) = f$, $(\forall) t \in \mathcal{I}$, then the unique bounded solution of (2.100) is constant. It solves the linear equation $\mathcal{L}x + f = 0$.

(iv) If the operator valued function $\mathcal{L}(\cdot)$ defines a causal positive evolution and $f : \mathcal{I} \to \mathcal{X}_+$ is a bounded and continuous vector valued function, then the unique bounded solution of (2.100) satisfies $\tilde{x}(t) \geq 0$. Moreover, $\tilde{x}(t) \gg 0$, $t \in \mathcal{I}$, if $f(t) \gg 0$, $t \in \mathcal{I}$.

Combining Theorem 2.4.2 with Theorem 2.4.3 (i), (ii), (iv) we may obtain a list of criteria for exponential stability of a linear differential equation with periodic coefficients and positive evolution on an ordered Hilbert space.

Corollary 2.4.4. *Let $\mathcal{L} : \mathbf{R} \to \mathbf{B}(\mathcal{X})$ be a continuous operator valued function which defines a causal positive evolution on the ordered Hilbert space $\mathcal{X}$. Assume that there exists $\theta > 0$ such that $\mathcal{L}(t + \theta) = \mathcal{L}(t)$, $\forall t \in \mathbf{R}$.*
Then the following are equivalent:

(i) *the operator valued function $\mathcal{L}(\cdot)$ generates an exponentially stable evolution;*

(ii) *the backward affine differential equation (2.98) has a θ-periodic and uniformly positive solution;*

(iii) *for each continuous, θ-periodic and uniformly positive vector function $f : \mathbf{R} \to \mathcal{X}$ the backward affine differential equation (2.99) has a θ-periodic and uniformly positive solution;*

(iv) *there exists a continuous, uniformly positive vector valued function $\tilde{f} : \mathbf{R} \to \mathcal{X}$ periodic, with period θ such that the corresponding affine differential equation (2.99) has a θ-periodic solution $\tilde{x} : \mathbf{R} \to \mathcal{X}_+$;*

(v) *there exists a C^1 vector valued function $y : \mathbf{R} \to \mathcal{X}$, uniformly positive and periodic, with period θ, which solves*

$$\frac{d}{dt}y(t) + \mathcal{L}^*(t)y(t) \ll 0, \ t \in [0,\theta].$$

(vi) $\rho[T(\theta,0)] < 1$.

The Theorems 2.4.2 and 2.4.3 (i), (ii), (iv) yield:

Corollary 2.4.5. *Let $\mathcal{L} \in \mathbf{B}(\mathcal{X})$ be such that $e^{\mathcal{L}t} \geq 0$, if $t \geq 0$. Under this condition the following are equivalent:*

(i) *The linear differential equation $\frac{d}{dt}x(t) = \mathcal{L}x(t)$ is exponentially stable.*

(ii) *The linear equation $\mathcal{L}^*x + \xi = 0$ has a solution $\tilde{x} \in Int\mathcal{X}_+$.*

(iii) *For each vector $f \in Int\mathcal{X}_+$ there exists $\tilde{x} \in Int\mathcal{X}_+$ which solves $\mathcal{L}^*\tilde{x} + f = 0$.*

(iv) *There exists $y \in Int\mathcal{X}_+$ such that $\mathcal{L}^*y < 0$.*

2.5 Robustness of the Exponential Stability Under the Additive Perturbations

Let $(\mathcal{X}, \|\cdot\|)$ be a real Banach space equipped with a closed, solid, normal, convex cone $\mathcal{X}_+ \subset \mathcal{X}$. Let $\mathcal{L} : \mathcal{I} \to \mathbf{B}(\mathcal{X})$, $\Pi : \mathcal{I} \to \mathbf{B}(\mathcal{X})$ be two continuous operator valued functions. Assume that $\mathcal{L}(\cdot)$ generates a positive and exponentially stable evolution on $\mathcal{X}$, while $\Pi(t)$ is a positive operator on $\mathcal{X}$ for all $t \in \mathcal{I}$.

In this section we want to provide a set of necessary and sufficient conditions which guarantee that the zero state equilibrium of the perturbed linear differential equation

$$\frac{d}{dt}x(t) = \mathcal{L}(t)x(t) + \Pi(t)x(t) \tag{2.102}$$

is exponentially stable.

In the time invariant case (i.e., $\mathcal{L}(t) = \mathcal{L}$, $\Pi(t) = \Pi$, $t \in \mathcal{I}$) the answer to this problem may be obtained from the following theorem.

Theorem 2.5.1. *Let $\mathcal{X}$ be a real Banach space ordered by a closed, solid, normal convex cone $\mathcal{X}_+$. Suppose $\mathcal{L} \in \mathbf{B}(\mathcal{X})$ to be resolvent positive operator and $\Pi \in \mathbf{B}(\mathcal{X})$ to be positive operator and set $T = \mathcal{L} + \Pi$. Then the following are equivalent:*

(i) *T is stable, that is the spectrum of T is located in $\mathbf{C}^- = \{\lambda \in \mathbf{C}; Re\lambda < 0\}$;*

(ii) *T^{-1} is well defined and $-T^{-1} \geq 0$;*

(iii) *for all $f \in Int\mathcal{X}_+$ there exists $x \in Int\mathcal{X}_+$ such that $-Tx = f$;*

(iv) *there exists $x \in Int\mathcal{X}_+$ such that $-Tx \in Int\mathcal{X}_+$;*

(v) *there exists $x \in \mathcal{X}_+$ such that $-Tx \in Int\mathcal{X}_+$;*

(vi) *$\mathcal{L}$ is stable and $\rho[\mathcal{L}^{-1}\Pi] < 1$, $\rho(\cdot)$ being the spectral radius.*

Proof. The equivalences (i) $\leftrightarrow$ (iii) $\leftrightarrow$ (iv) $\leftrightarrow$ (v) follow from Corollary 2.3.9 and the equivalence (ii) $\leftrightarrow$ (iii) is obvious. The equivalence (i) $\leftrightarrow$ (vi) is proved in Theorem 2.11 in [29]. $\qquad\square$

In this section we provide the answer to this issue in the case of differential equations with periodic coefficients.

Let us assume that there exists $\theta > 0$ such that $\mathcal{L}(t + \theta) = \mathcal{L}(t)$, $\Pi(t + \theta) = \Pi(t)$ for all $t \in \mathcal{I}$. Without loss of generality, we may assume that $\mathcal{I} = \mathbf{R}$. Let $T(t, \tau)$, $t, \tau \in \mathbf{R}$ be the linear evolution operator defined on $\mathcal{X}$ by the linear differential equation

$$\frac{d}{dt}x(t) = \mathcal{L}(t)x(t). \tag{2.103}$$

If the zero state equilibrium of (2.103) is exponentially stable, then, $\rho[T(\theta,0)] < 1$, so, the linear and bounded operator $I_{\mathcal{X}} - T(\theta,0)$ is invertible and $(I_{\mathcal{X}} - T(\theta,0))^{-1} \in$ $\mathbf{B}(\mathcal{X})$. This allows us to associate the so-called Green operator defined by the differential equation (2.103):

$$G(t,s) = T(t,0)(I_{\mathcal{X}} - T(\theta,0))^{-1}T(\theta,s) + T(t,s)\chi_{[0,t]}(s)$$

for all $(t,s) \in [0,\theta] \times [0,\theta]$, where

$$\chi_{[0,t]}(s) = \begin{cases} 1, \text{ if } s \in [0,t]; \\ 0, \text{ in other case.} \end{cases}$$

is the indicator function on the interval $[0,t]$.

Let us define the operator $\mathfrak{P} : C_0\{[0,\theta],\mathcal{X}\} \to C_0\{[0,\theta],\mathcal{X}\}$ by

$$(\mathfrak{P}x)(t) = \int_0^\theta G(t,s)\Pi(s)x(s)ds \tag{2.104}$$

$0 \leq t \leq \theta$ where $C_0\{[0,\theta],\mathcal{X}\}$ is the space of continuous functions $x : [0,\theta] \to \mathcal{X}$ with the property that $x(0) = x(\theta)$.

The linear space $C_0\{[0,\theta],\mathcal{X}\}$ equipped with the norm $\|\cdot\|$ defined by $\|x\| = \sup\{\|x(t)\| \mid 0 \leq t \leq \theta\}$ becomes a real Banach space. Moreover $C_0\{[0,\theta],\mathcal{X}\}$ is an ordered Banach space ordered by the order relation induced by the closed, solid, normal convex cone, $C_0\{[0,\theta],\mathcal{X}_+\}$.

Lemma 2.5.2. *Assume: (a) $\mathcal{L}(\cdot),\Pi(\cdot)$ are continuous operator valued functions periodic with period θ;*

(b) $\mathcal{L}(\cdot)$ generates a positive and exponentially stable evolution and $\Pi(t)$ is a positive operator.

Then $\mathfrak{P}$ defined by (2.104) is a bounded linear operator. Moreover $\mathfrak{P}C_0\{[0,\theta],\mathcal{X}_+\} \subset C_0\{[0,\theta],\mathcal{X}_+\}$.

Proof. The fact that $\mathfrak{P}$ is a linear and bounded operator follows immediately from (2.104) and the formula of $G(t,s)$. Let $f \in C_0\{[0,\theta],\mathcal{X}_+\}$ and $\tilde{x} = \mathfrak{P}f$. Let $\hat{x} : \mathbf{R} \to \mathcal{X}$, $\hat{f} : \mathbf{R} \to \mathcal{X}$ be periodic functions such that $\hat{x}|_{[0,\theta]} = \tilde{x}$ and $\hat{f}|_{[0,\theta]} = f$. The periodicity of $\Pi(\cdot)$ implies that $\hat{x}(t)$ is a periodic solution with period θ of the affine equation

$$\frac{d}{dt}x(t) = \mathcal{L}(t)x(t) + g(t) \quad \text{with} \quad g(t) = \Pi(t)\hat{f}(t), t \in \mathbf{R}. \tag{2.105}$$

Since $\Pi(t) \geq 0$, $f(t) \geq 0$ it follows that $g(t) \geq 0$, $t \in \mathbf{R}$. Theorem 2.3.7 now yields $\hat{x}(t) \geq 0$. Thus we obtained that $(\mathfrak{P}f)(t) \geq 0, t \in [0, \theta]$ or equivalently $\mathfrak{P}f \in \mathbf{C}_0\{[0, \theta], \mathcal{X}_+\}$. The proof is complete. $\square$

Theorem 2.5.3. *Under the assumptions from Lemma 2.5.2 the following are equivalent:*

(i) *The zero state equilibrium of the perturbed linear differential equation (2.102) is exponentially stable.*

(ii) *The zero state equilibrium of the unperturbed linear differential equation (2.103) is exponentially stable and $\rho[\mathfrak{P}] < 1$.*

Proof. (i) $\Rightarrow$ (ii): If (i) holds, then based on the implication (i) $\Rightarrow$ (iv) in Theorem 2.3.6 one deduces that for an arbitrary $\xi \in \operatorname{Int}\mathcal{X}_+$ the forward equation

$$\frac{d}{dt}x(t) = [\mathcal{L}(t) + \Pi(t)]x(t) + \xi \tag{2.106}$$

has a bounded and uniformly positive solution $\tilde{x} : \mathbf{R} \to \operatorname{Int}\mathcal{X}_+$.

It can be seen that $\tilde{x}(\cdot)$ satisfies (2.67) with $f(t) = \Pi(t)\tilde{x}(t) + \xi$ and thus one concludes that $\mathcal{L}(\cdot)$ defines an E.S. evolution because $f(t) \gg 0, t \in \mathbf{R}$. Theorem 2.3.7 (ii) yields that $\tilde{x}(\cdot)$ is a periodic function with period θ. Hence from (2.106) we have that

$$\tilde{x}(t) = \int_0^\theta G(t, s)\Pi(s)\tilde{x}(s)\,ds + \int_0^\theta G(t, s)\xi\,ds \quad t \in [0, \theta]. \tag{2.107}$$

If $\hat{x}(t) = \tilde{x}|_{[0, \theta]}$ we obtain from (2.107) that $\hat{x}$ solves the equation

$$(-\mathcal{I}_{\mathbf{C}_0} + \mathfrak{P})\hat{x} + \tilde{g} = 0, \tag{2.108}$$

where $\mathcal{I}_{\mathbf{C}_0}$ is the identity operator on $\mathbf{C}_0\{[0, \theta], \mathcal{X}\}$, and $\tilde{g}(t) = \int_0^\theta G(t, s)\xi\,ds$ is the bounded solution on $\mathbf{R}$ of the affine equation

$$\frac{d}{dt}x(t) = \mathcal{L}(t)x(t) + \xi. \tag{2.109}$$

Applying Theorem 2.3.7 (iv) we deduce that $\tilde{g}(t) \gg 0$, $t \in \mathbf{R}$. That means that $\tilde{g} \in \operatorname{Int}\mathbf{C}_0\{[0, \theta], \mathcal{X}_+\}$. Using the implication (v) $\Rightarrow$ (vi) of Theorem 2.5.1 for $\mathcal{L} = -\mathcal{I}_{\mathbf{C}_0}$ and $P = \mathfrak{P}$ one obtains that $\rho(\mathfrak{P}) < 1$ and so (ii) is valid.

(ii) $\Rightarrow$ (i): Let $\xi \in \operatorname{Int}\mathcal{X}_+$. If (ii) holds, then $\tilde{g}$ is well defined by

$$\tilde{g}(t) = \int_0^\theta G(t, s)\xi\,ds\,, \quad t \in [0, \theta].$$

Let $\hat{g} : \mathbf{R} \to \mathcal{X}$ be a periodic function with period θ such that $\hat{g} = \tilde{g}$, $t \in [0, \theta]$. Then $\hat{g}(\cdot)$ is the unique bounded on $\mathbf{R}$ solution of (2.109). By Theorem 2.3.7 (iv) we deduce that $\hat{g}(t) \gg 0$, $t \in \mathbf{R}$. Hence $\tilde{g} \in \operatorname{Int}\mathbf{C}_0\{[0, \theta], \mathcal{X}_+\}$. Applying Theorem 2.5.1 we conclude that the equation

$$[-\mathcal{I}_{\mathbf{C}_0} + \mathfrak{P}]x + \tilde{g} = 0$$

has a solution $\hat{x} \in C_0\{[0,\theta], \mathcal{X}_+\}$. Let $\tilde{x} : \mathbf{R} \to \mathcal{X}$ be the periodic function with period θ such that $\tilde{x}(t) = \hat{x}(t)$ for all $t \in [0,\theta]$. It follows that $\tilde{x}(\cdot) \gg 0$ is a bounded solution of the equation

$$\frac{d}{dt}x(t) = [\mathcal{L}(t) + \Pi(t)]x(t) + \xi.$$

The implication (iv) $\Rightarrow$ (i) of Theorem 2.3.6 yields that $\mathcal{L} + \Pi$ generates an E.S. evolution. □

Let us consider the case when the operator valued function $\mathcal{L}(\cdot)$ defines an anticausal exponentially stable evolution on $\mathcal{X}$ and $\Pi(t) \geq 0$, $\forall t \in \mathbf{R}$. Our aim is to find necessary and sufficient conditions under which the zero solution of the perturbed equation

$$\frac{d}{dt}y(t) = -\mathcal{L}(t)y(t) - \Pi(t)y(t) \tag{2.110}$$

is anticausal exponentially stable.

The answer to this issue will be done also under the assumption that the operator valued functions are periodic with period θ.

Let $T^a(t,\tau)$ be the anticausal linear evolution operator defined by the linear differential equation

$$\frac{d}{dt}y(t) = -\mathcal{L}(t)y(t). \tag{2.111}$$

If the zero solution of the differential equation (2.111) is anticausal exponentially stable, then, $\rho[T^a(0,\theta)] < 1$. So, the linear operator $I_\mathcal{X} - T^a(0,\theta)$ is invertible and $(I_\mathcal{X} - T^a(0,\theta))^{-1} \in \mathbf{B}(\mathcal{X})$. Thus we may construct the Green operator associated with (2.111) as follows:

$$G^a(t,s) = T^a(t,\theta)(I_\mathcal{X} - T^a(0,\theta))^{-1}T^a(0,s) + T^a(t,s)\chi_{[t,\theta]}(s), \ \forall(t,s) \in [0,\theta] \times [0,\theta],$$

where $\chi_{[t,\theta]}(\cdot)$ is the indicator function of the set $[t,\theta]$.

Consider the operator $\mathfrak{P}^a : C_0\{[0,\theta], \mathcal{X}\} \to C_0\{[0,\theta], \mathcal{X}\}$ defined by

$$(\mathfrak{P}^a x)(t) = \int_0^\theta G^a(t,s)\Pi(s)x(s)ds. \tag{2.112}$$

One shows that $\mathfrak{P}^a$ is a bounded and positive linear operator. Combining Theorems 2.3.12 and 2.5.1 from above, one proves:

Theorem 2.5.4. *Let $\mathcal{L} : \mathbf{R} \to \mathbf{B}(\mathcal{X})$, $\Pi : \mathbf{R} \to \mathbf{B}(\mathcal{X})$ be two continuous operator valued functions which are periodic with period $\theta > 0$. Assume that $\mathcal{L}(\cdot)$ defines an anticausal positive evolution and $\Pi(t) \geq 0$, for all $t \in \mathbf{R}$. Under these conditions the following are equivalent:*

(i) the zero solution of the perturbed linear differential equation (2.110) is anticausal exponentially stable;

(ii) the zero solution of the unperturbed linear differential equation (2.111) is anticausal exponentially stable and $\rho[\mathfrak{P}^a] < 1$.

2.6 Lyapunov-Type Linear Differential Equations on the Space $\mathcal{S}_n^{\mathcal{D}}$

In this section we emphasize several properties of an important class of operator valued functions on the Banach spaces $\mathcal{S}_n^{\mathcal{D}}$ and $\ell^1(\mathbf{Z}_+, \mathcal{S}_n)$, respectively. These operators extend to this framework the well-known Lyapunov operators and they will play an important role in the characterization of the exponential stability in mean square of stochastic linear differential equation.

2.6.1 Extended Lyapunov Operators

Let $\mathcal{M}_{mn}^{\mathcal{D}} := \ell^{\infty}(\mathcal{D}, \mathbf{R}^{m \times n})$ be the space of the bounded sequences of matrices $A = \{A(i)\}_{i \in \mathcal{D}}$ where $A(i) \in \mathbf{R}^{m \times n}$. We introduce the norm $\|A\|_{\infty} = \sup_{i \in \mathcal{D}} |A(i)|$ where $|A(i)|$ is defined by (2.1). One obtains that $(\mathcal{M}_{mn}^{\mathcal{D}}, \|\cdot\|_{\infty})$ is a real Banach space. If $m = n$ we shall write $\mathcal{M}_n^{\mathcal{D}}$ instead of $\mathcal{M}_{nn}^{\mathcal{D}}$. In the special case $\mathcal{D} = \{1, 2, \ldots, d\}$ we often write $\mathcal{M}_{mn}^d$ and $\mathcal{M}_n^d$, respectively, instead of $\mathcal{M}_{mn}^{\mathcal{D}}$ and $\mathcal{M}_n^{\mathcal{D}}$. If $\mathcal{D} = \mathbf{Z}_+$, $\mathcal{M}_{mn}^{\infty}$ and $\mathcal{M}_n^{\infty}$, respectively, stand for $\mathcal{M}_{mn}^{\mathcal{D}}$ and $\mathcal{M}_n^{\mathcal{D}}$. It is obvious that $\mathcal{S}_n^{\mathcal{D}} \subset \mathcal{M}_n^{\mathcal{D}}$.

We make the following convention of notation:

(a) If $A = \{A(i)\}_{i \in \mathcal{D}} \in \mathcal{M}_{mn}^{\mathcal{D}}$, $X = \{X(i)\}_{i \in \mathcal{D}} \in \mathcal{M}_{np}^{\mathcal{D}}$, by $Y = AX$ we understand the sequence $Y = \{Y(i)\}_{i \in \mathcal{D}} \in \mathcal{M}_{mp}^{\mathcal{D}}$, $Y(i) = A(i)X(i)$, $i \in \mathcal{D}$.

(b) If $A = \{A(i)\}_{i \in \mathcal{D}} \in \mathcal{M}_{mn}^{\mathcal{D}}$, then $A^T = \{A^T(i)\}_{i \in \mathcal{D}} \in \mathcal{M}_{nm}^{\mathcal{D}}$.

We have

$$\|AX\|_{\infty} \leq \|A\|_{\infty} \|X\|_{\infty}, \forall A, X \in \mathcal{M}_n^{\mathcal{D}}, \tag{2.113}$$

$$\|A^T\|_{\infty} = \|A\|_{\infty}.$$

Let $A : \mathcal{I} \to \mathcal{M}_n^{\mathcal{D}}$ be a continuous function. This means that $A(t) = \{A(t, i)\}_{i \in \mathcal{D}}$, where $t \to A(t, i)$ are matrix valued functions which are continuous on $\mathcal{I}$ uniformly with respect to $i \in \mathcal{D}$.

The extended Lyapunov operators associated with $A(t)$:

$$\mathcal{L}_A(t) : \mathcal{S}_n^{\mathcal{D}} \to \mathcal{S}_n^{\mathcal{D}},$$

$$\mathfrak{L}_A(t) : \mathcal{S}_n^{\mathcal{D}} \to \mathcal{S}_n^{\mathcal{D}},$$

are defined as follows

$$\mathcal{L}_A(t)X = A(t)X + XA^T(t) \qquad (2.114)$$

$$\mathfrak{L}_A(t)X = A^T(t)X + XA(t) \qquad (2.115)$$

for all $X = \{X(i)\}_{i \in \mathcal{D}} \in \mathcal{S}_n^{\mathcal{D}}$.

According to the notation introduced at the beginning of this subsection the i-th component of (2.114) and (2.115), respectively, is:

$$[\mathcal{L}_A(t)X](i) = A(t,i)X(i) + X(i)A^T(t,i)$$

$$[\mathfrak{L}_A(t)X](i) = A^T(t,i)X(i) + X(i)A(t,i)$$

$i \in \mathcal{D}, t \in \mathcal{I}$.

Based on (2.113) we deduce that $\|\mathcal{L}_A(t)X\|_\infty \leq 2\|A(t)\|_\infty \|X\|_\infty$ and $\|\mathfrak{L}_A(t)X\|_\infty \leq 2\|A(t)\|_\infty \|X\|_\infty$. Hence, $\mathcal{L}_A(t), \mathfrak{L}_A(t) \in \mathbf{B}(\mathcal{S}_n^{\mathcal{D}})$. Moreover $t \to \mathcal{L}_A(t)$ and $t \to \mathfrak{L}_A(t)$ are continuous functions in the topology induced by the operator norm.

Remark 2.6.1. (i) From (2.114) and (2.115) one sees that both $\mathcal{L}_A(t)$ and $\mathfrak{L}_A(t)$ can be extended to $\mathcal{M}_n^{\mathcal{D}}$. This extension is not of interest for applications to the exponential stability in mean square of stochastic linear differential equations. That is why such extensions are not considered in this chapter.

(ii) To be sure that the linear differential equations (2.116), (2.121), respectively, defined by $\mathcal{L}_A(t)$ and $\mathfrak{L}_A(t)$ on $\mathcal{S}_n^{\mathcal{D}}$ have nice properties, would be sufficient to assume that $t \to \mathcal{L}_A(t)$ and $t \to \mathfrak{L}_A(t)$ are strongly continuous operator valued functions. This means that for each $X \in \mathcal{S}_n^{\mathcal{D}}$, $t \to \mathcal{L}_A(t)X$ and $t \to \mathfrak{L}_A(t)X$ are continuous vector valued functions. If we take $X = \{X(i)\}_{i \in \mathcal{D}}$ with $X(i) = I_n$, $\forall i \in \mathcal{D}$ one obtains that $t \to A^T(t) + A(t)$ must be continuous. This condition is not far from our assumption that $t \to A(t)$ is a continuous function.

Let us consider the extended Lyapunov equation

$$\frac{d}{dt}X(t) = \mathcal{L}_A(t)X(t), \ t \in \mathcal{I}. \qquad (2.116)$$

Let $T_A(t,t_0) \ t, t_0 \in \mathcal{I}$ be the linear operator defined by

$$(T_A(t,t_0)X)(i) = \Phi_i(t,t_0)X(i)\Phi_i^T(t,t_0) \qquad (2.117)$$

$\forall i \in \mathcal{D}$ and $X = \{X(i)\}_{i \in \mathcal{D}} \in \mathcal{S}_n^{\mathcal{D}}$, where $\Phi_i(t,t_0)$ is the fundamental matrix solution of the linear differential equation on $\mathbf{R}^n$:

$$\frac{d}{dt}x(t) = A(t,i)x(t).$$

This means that $t \to \Phi_i(t,t_0)$ verifies

$$\frac{d}{dt}\Phi_i(t,t_0) = A(t,i)\Phi_i(t,t_0) \tag{2.118}$$

$$\Phi_i(t_0,t_0) = I_n.$$

Based on the convention of notations introduced before we may write (2.117) in the compact form:

$$T_A(t,t_0)X = \Phi(t,t_0)X\Phi^T(t,t_0) \tag{2.119}$$

for all $t,t_0 \in \mathcal{I}$, where $\Phi(t,t_0) = \{\Phi_i(t,t_0)\}_{i\in\mathcal{D}}$. If $\mathcal{D} = \{1,2,\ldots,d\}$, one may check that $t \to \Phi(t,t_0)$ is differentiable map and it satisfies:

$$\frac{d}{dt}\Phi(t,t_0) = A(t)\Phi(t,t_0), \quad \Phi(t_0,t_0) = J^d = (I_n\ldots I_n).$$

By direct calculations one obtains from (2.119) that

$$\frac{d}{dt}T_A(t,t_0)X = \mathcal{L}_A(t)T_A(t,t_0)X \tag{2.120}$$

$$T_A(t_0,t_0)X = X$$

for all $t,t_0 \in \mathcal{I}, X \in \mathcal{S}_n^d$. Therefore $T_A(t,t_0)$ defined by (2.117), or equivalently by (2.119) is just the linear evolution operator on $\mathcal{S}_n^d$ defined by the linear differential equation (2.116).

It remains to show that (2.117) defines also the linear evolution operator generated by (2.116) on $\mathcal{S}_n^\infty$. To this end, let us remark that

$$|\Phi_i(t,s)| \le e^{\gamma(t-s)}$$

for all $i \in \mathbf{Z}_+, t,s \in \mathcal{I}$, where $\gamma = \sup_{t\in\mathcal{D}}\|A(t)\|_\infty$. Using also the fact that $t \to A(t,i)$ are continuous functions uniformly with respect to $i \in \mathbf{Z}_+$ we deduce that

$$\lim_{h\to 0}\frac{1}{|h|}|\Phi_i(t+h,t_0) - \Phi_i(t,t_0) - hA(t,i)\Phi_i(t,t_0)| = 0$$

uniformly with respect to $i \in \mathbf{Z}_+$.

This shows that $t \to \Phi(t,t_0) : \mathcal{I} \to \mathcal{M}_n^\infty$ is a differentiable map and it satisfies:

$$\frac{d}{dt}\Phi(t,t_0) = \mathcal{L}_A(t)\Phi(t,t_0), \quad \Phi(t_0,t_0) = J^\infty = (I_n\ldots I_n\ldots) \in \mathcal{S}_n^\infty.$$

Thus we may obtain that $T_A(t,t_0)$ defined by (2.119) for $\mathcal{D} = \mathbf{Z}_+$ is differentiable and satisfies (2.120).

Remark 2.6.2. From (2.117) one sees that $T_A(t,t_0)X \in \mathcal{S}_{n+}^{\mathcal{D}}$ if $X \in \mathcal{S}_{n+}^{\mathcal{D}}$. This shows that the operator valued function $\mathfrak{L}_A(\cdot)$ generates a positive evolution on the Banach space $\mathcal{S}_n^{\mathcal{D}}$.

Changing $A(t,i)$ with $A^T(t,i)$ in (2.117), (2.118) one obtains that the operator valued function $\mathfrak{L}_A(\cdot)$ generates also positive evolution on the Banach space $\mathcal{S}_n^{\mathcal{D}}$. However, concerning the operator valued function $\mathfrak{L}_A(\cdot)$ we are interested by the anticausal evolution operator $T_A^a(t,t_0)$ defined by the linear differential equation

$$\frac{d}{dt}Y(t) + \mathfrak{L}_A(t)Y(t) = 0. \tag{2.121}$$

Reasoning as in the case of (2.116) we may conclude that

$$(T_A^a(t,t_0)Y)(i) = \Phi_i^T(t_0,t)Y(i)\Phi_i(t_0,t) \tag{2.122}$$

for all $i \in \mathcal{D}$, $0 \le t \le t_0$, $Y = \{Y(i)\}_{i\in\mathcal{D}} \in \mathcal{S}_n^{\mathcal{D}}$.

From (2.122) one deduces that the operator valued function $\mathfrak{L}_A(\cdot)$ generates an anticausal positive evolution on the Banach space $\mathcal{S}_n^{\mathcal{D}}$.

2.6.2 Lyapunov-Type Differential Equations on the Space $\mathcal{S}_n^d$

Let $\mathcal{I} \subseteq \mathbf{R}$ be an interval and $A_k : \mathcal{I} \to \mathcal{M}_n^d$, $k = 0, \ldots, r$ be continuous functions

$$A_k(t) = (A_k(t,1), \ldots A_k(t,d)), \, k \in \{0, \ldots, r\}, t \in \mathcal{I}.$$

Denote by $Q \in \mathbf{R}^{d\times d}$ a matrix which elements q_{ij} verify the condition

$$q_{ij} \ge 0 \text{ if } i \ne j. \tag{2.123}$$

For each $t \in \mathcal{I}$ we define the linear operator $\mathcal{L}(t) : \mathcal{S}_n^d \to \mathcal{S}_n^d$ by

$$(\mathcal{L}(t)S)(i) = A_0(t,i)S(i) + S(i)A_0^T(t,i) \tag{2.124}$$

$$+ \sum_{k=1}^{r} A_k(t,i)S(i)A_k^T(t,i) + \sum_{j=1}^{d} q_{ji}S(j),$$

$i \in \mathcal{D}, S \in \mathcal{S}_n^d$. It is easy to see that $t \longmapsto \mathcal{L}(t)$ is a continuous operator valued function.

Definition 2.6.1. The operator $\mathcal{L}(t)$ defined by (2.124) is called the *Lyapunov operator associated with $A_0, \ldots, A_r$ and Q.*

The Lyapunov operator $\mathcal{L}(t)$ defines the following linear differential equation on $\mathcal{S}_n^d$:

$$\frac{d}{dt}S(t) = \mathcal{L}(t)S(t), \ t \in \mathcal{I}. \tag{2.125}$$

For each $t_0 \in \mathcal{I}$ and $H \in \mathcal{S}_n^d$, $S(t,t_0,H)$ stands for the solution of the differential equation (2.125) which verifies the initial condition $S(t_0,t_0,H) = H$.

Let us denote by $T(t,t_0)$ the linear evolution operator on $\mathcal{S}_n^d$ defined by the differential equation (2.125), that is

$$T(t,t_0)H = S(t,t_0,H); \ t,t_0 \in \mathcal{I}, H \in \mathcal{S}_n^d.$$

It is said that $T(t,t_0)$ is the *linear evolution operator associated with the system* $(A_0,\ldots,A_r;Q)$.

We have

$$\frac{d}{dt}T(t,t_0) = \mathcal{L}(t)T(t,t_0)$$

$$T(t_0,t_0) = \tilde{J}^d,$$

where $\tilde{J}^d : \mathcal{S}_n^d \to \mathcal{S}_n^d$ is the identity operator.

It is easy to check (see also Remark 2.2.1 for a more general case) that $T(t,s)T(s,\tau) = T(t,\tau)$ for all $t,s,\tau \in \mathcal{I}$. For all pairs $(t,\tau) \in \mathcal{I} \times \mathcal{I}$, the operator $T(t,\tau)$ is invertible and its inverse is $T^{-1}(t,\tau) = T(\tau,t)$.

If $T^*(t,\tau)$ denotes the adjoint operator of $T(t,\tau)$ with respect to the inner product (2.16), the following hold:

$$T^*(t,t_0) = T^*(s,t_0)T^*(t,s), \tag{2.126}$$

$$\frac{d}{dt}T^*(t,s) = T^*(t,s)\mathcal{L}^*(t), \tag{2.127}$$

$$\frac{d}{ds}T^*(s,t) = -\mathcal{L}^*(t)T^*(s,t). \tag{2.128}$$

It is not difficult to see that the adjoint operator $\mathcal{L}^*(t) : \mathcal{S}_n^d \to \mathcal{S}_n^d$ is given by

$$(\mathcal{L}^*(t)S)(i) = A_0^T(t,i)S(i) + S(i)A_0(t,i) \tag{2.129}$$

$$+ \sum_{k=1}^{r} A_k^T(t,i)S(i)A_k(t,i) + \sum_{j=1}^{d} q_{ij}S(j)$$

$i \in \mathcal{D}, S \in \mathcal{S}_n^d$.

Remark 2.6.3. (i) If $A_k(t,i)$, $k = 1,\ldots,r$ do not depend on t, then the operator $\mathcal{L}$ defined by (2.124) is independent of t. More precisely, if $A_k = (A_k(1),\ldots,A_k(d))$, then

$$(\mathcal{L}S)(i) = A_0(i)S(i) + S(i)A_0^T(i) + \Sigma_{k=1}^r A_k(i)S(i)A_k^T(i) \\ + \Sigma_{j=1}^d q_{ji}S(j), \tag{2.130}$$

$i \in \mathcal{D}$, $S \in \mathcal{S}_n^d$. In this situation the evolution operator defined by the differential equation

$$\frac{d}{dt}S(t) = \mathcal{L}S(t)$$

is given by

$$T(t,t_0) = e^{\mathcal{L}(t-t_0)} \tag{2.131}$$

where

$$e^{\mathcal{L}t} := \sum_{k=0}^{\infty} \frac{\mathcal{L}^k t^k}{k!}$$

(the above series being uniform convergent on every compact subset of the real axis). $\mathcal{L}^k$ stands for *the kth iteration of the operator* $\mathcal{L}$ and $\mathcal{L}^0 = \tilde{J}^d$;
(ii) If $A_k : \mathcal{I} \to \mathcal{M}_n^d$ are θ-periodic functions, then $T(t+\theta, t_0+\theta) = T(t,t_0)$ for all $t, t_0 \in \mathcal{I}$ such that $t+\theta, t_0+\theta \in \mathcal{I}$.

Theorem 2.6.1. *If $T(t,t_0)$ is a linear evolution operator on $\mathcal{S}_n^d$ defined by the linear differential equation (2.125), then the following hold:*

(i) $T(t,t_0) \geq 0, T^*(t,t_0) \geq 0$ *for all $t \geq t_0, t, t_0 \in \mathcal{I}$;*
(ii) *If $t \to A_k(t)$ are bounded functions, then there exist $\delta > 0, \gamma > 0$ such that:*

$$T(t,t_0)J^d \geq \delta e^{-\gamma(t-t_0)}J^d, \quad T^*(t,t_0)J^d \geq \delta e^{-\gamma(t-t_0)}J^d$$

for all $t \geq t_0, t, t_0 \in \mathcal{I}$.

Proof. To prove (i) we consider the linear operators $\mathcal{L}_1(t) : \mathcal{S}_n^d \to \mathcal{S}_n^d, \Pi(t) : \mathcal{S}_n^d \to \mathcal{S}_n^d$ defined by

$$(\mathcal{L}_1(t)H)(i) = \left(A_0(t,i) + \frac{1}{2}q_{ii}I_n\right)H(i) + H(i)\left(A_0(t,i) + \frac{1}{2}q_{ii}I_n\right)^T$$

$$(\Pi(t)H)(i) = \sum_{k=1}^{r} A_k(t,i)H(i)A_k^T(t,i) + \sum_{j=1, j\neq i}^{d} q_{ji}H(j), i \in \mathcal{D}$$

$H = (H(1), H(2),\ldots,H(d)) \in \mathcal{S}_n^d, t \in \mathcal{I}$.

It is easy to see that for each $t \in \mathcal{I}$, the operator $\Pi(t)$ is a positive operator on $\mathcal{S}_n^d$. On the other hand one sees that the operator $\mathcal{L}_1(t)$ coincides with the extended Lyapunov operator $\mathcal{L}_A(t) : \mathcal{S}_n^d \to \mathcal{S}_n^d$ associated via (2.114) to the matrices $A(t,i) = A_0(t,i) + \frac{1}{2} q_{ii} I_n$. Based on Remark 2.6.2 we infer that the operator valued function $\mathcal{L}_1(\cdot)$ defines a positive evolution on $\mathcal{S}_n^d$. Applying Corollary 2.2.6 (i) we obtain that $t \to \mathcal{L}(t) = \mathcal{L}_1(t) + \Pi(t)$ defines a positive evolution on $\mathcal{S}_n^d$. Therefore, $T(t,t_0) \geq 0$ for all $t \geq t_0, t, t_0 \in \mathcal{I}$. Applying Proposition 2.1.9 we conclude that $T^*(t,t_0) \geq 0$ for all $t \geq t_0, t, t_0 \in \mathcal{I}$.

(ii) Firstly, we show that there exist $\delta > 0, \gamma > 0$, such that

$$|T(t,t_0)H| \geq \delta e^{-\gamma(t-t_0)}|H| \qquad (2.132)$$

$$|T^*(t,t_0)H| \geq \delta e^{-\gamma(t-t_0)}|H|$$

for all $H \in \mathcal{S}_n^d, t \geq t_0, t, t_0 \in \mathcal{I}$.

Let us denote

$$v(t) = \frac{1}{2} |||T(t,t_0)H|||^2 = \frac{1}{2} \langle T(t,t_0)H, T(t,t_0)H \rangle,$$

where $||| \cdot |||$ denotes the norm induced by the inner product, that is $||| \cdot ||| := \langle \cdot, \cdot \rangle^{\frac{1}{2}}$. By direct calculation, we obtain

$$\frac{d}{dt} v(t) = \langle \mathcal{L}(t)T(t,t_0)H, T(t,t_0)H \rangle, t \geq t_0.$$

Under the considered assumptions there exists $\gamma > 0$ such that

$$\left| \frac{d}{dt} v(t) \right| \leq \gamma |||T(t,t_0)H|||^2,$$

$$\left| \frac{d}{dt} v(t) \right| \leq 2\gamma v(t), \quad t \geq t_0.$$

Further we have

$$\frac{d}{dt} v(t) \geq -2\gamma v(t), \quad t \geq t_0$$

or equivalently

$$\frac{d}{dt} \left[v(t) e^{2\gamma(t-t_0)} \right] \geq 0$$

for all $t \geq t_0$. Hence the function $t \to v(t)e^{2\gamma(t-t_0)}$ is not decreasing and $v(t) \geq e^{-2\gamma(t-t_0)}v(t_0)$. Considering the definition of $v(t)$ we conclude that there exists $\delta > 0$ such that

$$|T(t,t_0)H| \geq \delta e^{-\gamma(t-t_0)}|H|$$

which is the first inequality in (2.132).

To prove the second inequality (2.132), we consider the function

$$\hat{v}(s) = 1/2|||T^*(t,s)H|||^2, H \in \mathcal{S}_n^d, s \leq t, s, t \in \mathcal{I}.$$

By direct computation we obtain

$$\frac{d}{ds}\hat{v}(s) = -\langle \mathcal{L}^*(s)T^*(t,s)H, T^*(t,s)H \rangle.$$

Further we have

$$\left|\frac{d}{ds}\hat{v}(s)\right| \leq 2\gamma\hat{v}(s)$$

and

$$\frac{d}{ds}\left[\hat{v}(s)e^{2\gamma(t-s)}\right] \leq 0,$$

thus we obtain that the function $s \to \hat{v}(s)e^{2\gamma(t-s)}$ is not increasing and therefore $\hat{v}(s)e^{2\gamma(t-s)} \geq \hat{v}(t)$ for all $s \leq t$ hence

$$|||T^*(t,s)H||| \geq e^{-\gamma(t-s)}|||H|||.$$

Using the inequality $|S| \leq |||S||| \leq nd|S|$ for all $S \in \mathcal{S}_n^d$ we obtain the second inequality in (2.132).

Let $x \in \mathbf{R}^n, i \in \mathcal{D}$ be fixed; consider $\tilde{H} \in \mathcal{S}_n^d$ defined by

$$\tilde{H}(j) = \begin{cases} 0 \text{ if } j \neq i, \\ xx^T \text{ if } j = i \end{cases}.$$

We may write successively

$$x^T\left(T(t,t_0)J^d\right)(i)x = Tr\left[xx^T(T(t,t_0)J^d)(i)\right] = \left\langle \tilde{H}, T(t,t_0)J^d \right\rangle$$

$$= \left\langle T^*(t,t_0)\tilde{H}, J^d \right\rangle = \sum_{j=1}^d Tr\left[T^*(t,t_0)\tilde{H}\right](j)$$

$$\geq \sum_{j=1}^{d} \left| (T^*(t,t_0)\widetilde{H})(j) \right| \geq \max_{j \in \mathcal{D}} \left| (T^*(t,t_0)\widetilde{H})(j) \right|$$

$$= \left| T^*(t,t_0)\widetilde{H} \right| \geq \delta e^{-\gamma(t-t_0)} |x|^2$$

Since $x \in \mathbf{R}^n$ is arbitrary we get

$$\left(T(t,t_0)J^d \right)(i) \geq \delta e^{-\gamma(t-t_0)} I_n, (\forall) i \in \mathcal{D}, t \geq t_0, t, t_0 \in \mathcal{I}$$

or equivalently $T(t,t_0)J^d \geq \delta e^{-\gamma(t-t_0)} J^d, \forall t \geq t_0$. The second inequality in (ii) may be proved in the same way. $\qquad \square$

Having in mind the equality stated in the Corollary 2.1.7 (i) and the convention of notation made in Remark 2.1.5, we may deduce via Theorem 2.1.10 and Theorem 2.6.1 (i) some useful equalities.

Corollary 2.6.2. *If* $\|T(t,t_0)\|$ *and* $\|T^*(t,t_0)\|$ *are the norms induced by the usual norm* $\| \cdot \|_\infty$ *on the space* $\mathcal{S}_n^d$, *then we have:*

$$\|T(t,t_0)\| = |T(t,t_0)J^d|$$

$$\|T^*(t,t_0)\| = |T^*(t,t_0)J^d|.$$

Proof. The equalities from the statement are obtained applying Theorem 2.1.10 to the positive operators $T(t,t_0)$ and $T^*(t,t_0)$, respectively, and taking into account that the usual norm $\| \cdot \|_\infty$ on the space $\mathcal{S}_n^d$ coincides with the Minkovski norm associated with the element $\xi = J^d$.

Remark 2.6.4. (i) Combining the result in Theorem 2.1.4 (vii) for $\xi = J^d$ and $X = \frac{1}{|T(t,t_0)J^d|} T(t,t_0)J^d$ or $X = \frac{1}{|T^*(t,t_0)J^d|} T^*(t,t_0)J^d$, respectively, we obtain

$$T(t,t_0)J^d \leq \|T(t,t_0)\| J^d, \tag{2.133}$$

$$T^*(t,t_0)J^d \leq \|T^*(t,t_0)\| J^d$$

for all $t \geq t_0, t, t_0 \in \mathcal{I}$.

(ii) If the dependence $t \longmapsto \|\mathcal{L}(t)\|$ is a bounded function, we deduce easily that there exists $\hat{\gamma} > 0$ such that

$$\|T(t,t_0)\| \leq e^{\hat{\gamma}(t-t_0)},$$

$$\|T^*(t,t_0)\| \leq e^{\hat{\gamma}(t-t_0)}$$

for all $t \geq t_0, t, t_0 \in \mathcal{I}$.

Corollary 2.6.3. *Suppose that $A_k, 0 \le k \le r$ are continuous and bounded functions. Then there exist $\delta > 0$ and $\gamma > 0$ such that*

$$\delta e^{-\gamma(t-t_0)} J^d \le T(t,t_0) J^d \le e^{\gamma(t-t_0)} J^d,$$

$$\delta e^{-\gamma(t-t_0)} J^d \le T^*(t,t_0) J^d \le e^{\gamma(t-t_0)} J^d$$

for all $t \ge t_0, t, t_0 \in \mathcal{I}$.

At the end of this section let us remark two important particular cases:
Case (a) $A_k(t) = 0, k = 1, \ldots, r$; in this case the linear operator (2.124) becomes:

$$\left(\widehat{\mathcal{L}}(t)S\right)(i) = A_0(t,i) S(i) + S(i) A_0^T(t,i) \tag{2.134}$$

$$+ \sum_{j=1}^{d} q_{ji} S(j),$$

$i \in \mathcal{D}, S \in \mathcal{S}_n^d$. It is easy to check that the evolution operator $T(t,t_0)$ defined by (2.125) has the representation:

$$T(t,t_0) = \widehat{T}(t,t_0) + \int_{t_0}^{t} \widehat{T}(t,s) \mathcal{L}_2(s) T(s,t_0) ds,$$

$t \ge t_0, t, t_0 \in \mathcal{I}$, where $\widehat{T}(t,t_0)$ is the evolution operator on $\mathcal{S}_n^d$ defined by the differential equation

$$\frac{d}{dt} S(t) = \widehat{\mathcal{L}}(t) S(t)$$

and $\mathcal{L}_2(t) : \mathcal{S}_n^d \to \mathcal{S}_n^d$ is defined by:

$$(\mathcal{L}_2(t) H)(i) = \sum_{k=1}^{r} A_k(t,i) H(i) A_k^T(t,i),$$

$t \in \mathcal{I}, H \in \mathcal{S}_n^d, i \in \mathcal{D}$. Also, we have

$$\mathcal{L}(t) = \widehat{\mathcal{L}}(t) + \mathcal{L}_2(t). \tag{2.135}$$

Remark 2.6.5. (i) Since (2.134) is the special case of (2.124) for $A_k(t,i) = 0, 1 \le k \le r$ we deduce that the operator valued function $\widehat{\mathcal{L}}(\cdot)$ generates a positive evolution on $\mathcal{S}_n^d$. From (2.135) we deduce that $\widehat{\mathcal{L}}(t) \le \mathcal{L}(t)$ for all $t \in \mathcal{I}$. Applying Theorem 2.2.5

(i) we have that $T(t,t_0) \geq \widehat{T}(t,t_0)$ for all $t \geq t_0$, $t,t_0 \in \mathcal{I}$. Further, Corollaries 2.1.11 and 2.6.2 yield

$$\|T(t,t_0)\| \geq \left\|\widehat{T}(t,t_0)\right\|, t \geq t_0, t,t_0 \in \mathcal{I}. \tag{2.136}$$

The evolution operator $\widehat{T}(t,t_0)$ will be called the *evolution operator on the space* $\mathcal{S}_n^d$ *defined by the pair* (A_0,Q). In Sect. 3.1 we shall see that if additionally Q verifies

$$\sum_{j=1}^{d} q_{ij} = 0, \forall i \in \mathcal{D}$$

then (2.134) is the Lyapunov-type operator associated with the system (1.23); Case (b) $\mathcal{D} = \{1\}$ and $q_{11} = 0$. In this case $\mathcal{S}_n^d$ reduces to $\mathcal{S}_n$ and the operator $\mathcal{L}(t)$ is defined by

$$\mathcal{L}(t)S = A_0(t)S + SA_0^T(t) + \sum_{k=1}^{r} A_k(t)SA_k^T(t) \tag{2.137}$$

$t \in \mathcal{I}, S \in \mathcal{S}_n$ where we denoted $A_k(t) := A_k(t,1)$. The evolution operator $T(t,t_0)$ will be called the *evolution operator on* $\mathcal{S}_n$ *defined by the system* $(A_0,\ldots,A_r)$. In Sect. 3.1 we shall see that the operator (2.137) corresponds to the stochastic linear system (1.24).

2.6.3 Lyapunov-Type Differential Equations on the Space $\mathcal{S}_n^\infty$

Let $A_k : \mathcal{I} \to \mathcal{M}_n^\infty, 0 \leq k \leq r$ be continuous and bounded functions. This means that $A_k(t) = \{A_k(t,i)\}_{i \in \mathbf{Z}_+}$ are such that $t \to A_k(t,i)$ are continuous functions on $\mathcal{I}$ uniformly with respect to $i \in \mathbf{Z}_+$ and $\sup_{t \in \mathcal{I}} \|A_k(t)\|_\infty < \infty$. Let $Q = (q_{ij})_{i,j \in \mathbf{Z}_+}$ be an infinite real matrix whose elements satisfy the conditions:

$$q_{ij} \geq 0, \text{ if } i \neq j \tag{2.138}$$

and

$$\sup_{i \in \mathbf{Z}_+} \left(|q_{ii}| + \sum_{j=0, j \neq i}^{\infty} q_{ij}\right) = \nu < \infty. \tag{2.139}$$

It is worth mentioning that the conditions (2.138) and (2.139) are satisfied by the generator matrix of a standard homogeneous Markov process with an infinite countable number of states $(\eta(t), P, \mathbf{Z}_+)$ (see Sect. 1.13 for more details).

Based on the functions $t \to A_k(t,i)$ and the elements q_{ij} of the matrix Q, one constructs the operators $\mathcal{L}$ and $\mathfrak{L}$ by:

$$(\mathcal{L}(t)X)(i) = A_0(t,i)X(i) + X(i)A_0^T(t,i) + \sum_{k=1}^{r} A_k(t,i)X(i)A_k^T(t,i) + \sum_{j=0}^{\infty} q_{ji}X(j) \tag{2.140}$$

$$(\mathfrak{L}(t)X)(i) = A_0^T(t,i)X(i) + X(i)A_0(t,i) + \sum_{k=1}^{r} A_k^T(t,i)X(i)A_k(t,i) + \sum_{j=0}^{\infty} q_{ij}X(j) \tag{2.141}$$

for all sequences $X = \{X(i)\}_{i \in \mathbf{Z}_+}$.

Lemma 2.6.4. *If the real numbers q_{ij} satisfy conditions (2.138) and (2.139), then for each $t \in \mathcal{I}$, $\mathcal{L}(t) \in \mathbf{B}(\ell^1(\mathbf{Z}_+, \mathcal{S}_n))$ and $\mathfrak{L}(t) \in \mathbf{B}(\mathcal{S}_n^{\infty})$.*

Proof. If $X \in \ell^1(\mathbf{Z}_+, \mathcal{S}_n)$, then one obtains via (2.11), (2.138)–(2.140) that:

$$\|\mathcal{L}(t)X\|_1 = \sum_{i=0}^{\infty} |(\mathcal{L}(t)X)(i)| \leq \gamma(t)\|X\|_1$$

where

$$\gamma(t) = 2\|A_0(t)\|_{\infty} + \sum_{k=1}^{r} \|A_k(t)\|_{\infty}^2 + \nu. \tag{2.142}$$

Based on (2.12) we may write $\|\mathcal{L}(t)X\|_1 \leq n\|\mathcal{L}(t)X\|_1 \leq n\gamma(t)\|X\|_1$ which yields $\|\mathcal{L}(t)X\|_1 \leq n\gamma(t)\|X\|_1$. This shows that $\mathcal{L}(t)$ introduced by (2.140) defines a linear and bounded operator on $\ell^1(\mathbf{Z}_+, \mathcal{S}_n)$ and $\|\mathcal{L}(t)\|_1 \leq n\gamma(t)$, $t \geq 0$.

Similarly, if $X \in \mathcal{S}_n^{\infty}$ one obtains via (2.113), (2.138), (2.139), (2.141) that

$$\|\mathfrak{L}(t)X\|_{\infty} \leq \gamma(t)\|X\|_{\infty}$$

where $\gamma(t)$ is defined by (2.142). This completes the proof.   □

In the developments of this book the linear operator $\mathcal{L}(t)$ introduced via (2.140) will be named *the Lyapunov-type operator on the space $\ell^1(\mathbf{Z}_+, \mathcal{S}_n)$ defined by the system $(A_0, A_1, \ldots, A_r; Q)$* while $\mathfrak{L}(t)$ will be named *the Lyapunov-type operator on the space $\mathcal{S}_n^{\infty}$ defined by the system $(A_0, A_1, \ldots, A_r; Q)$*.

Proposition 2.6.5. *Under the considered assumptions, the operator valued function $\mathcal{L}(\cdot)$ introduced by (2.140) defines a positive evolution on $\ell^1(\mathbf{Z}_+, \mathcal{S}_n)$ while, the operator valued function $\mathfrak{L}(\cdot)$ introduced by (2.141) defines an anticausal positive evolution on the Banach space $\mathcal{S}_n^{\infty}$.*

Proof. From (2.140) and (2.141) one obtains the decomposition: $\mathcal{L}(t) = \mathcal{L}_A(t) + \Pi(t)$ and $\mathfrak{L}(t) = \mathfrak{L}_A(t) + \tilde{\Pi}(t)$ where $\mathcal{L}_A(t)$ and $\mathfrak{L}_A(t)$ are the extended Lyapunov operators associated via (2.114) and (2.115) with the sequence $A(t) = \{A(t,i)\}_{i \in \mathbf{Z}_+}$ with $A(t,i) = A_0(t,i) + \frac{1}{2}q_{ii}I_n$, $i \in \mathbf{Z}_+$, $t \in \mathcal{I}$

$$(\Pi(t)X)(i) = \sum_{k=1}^{r} A_k(t,i)X(i)A_k^T(t,i) + \sum_{j=0, j \neq i}^{\infty} q_{ji}X(j) \qquad (2.143)$$

and

$$(\tilde{\Pi}(t)X)(i) = \sum_{k=1}^{r} A_k^T(t,i)X(i)A_k(t,i) + \sum_{j=0, j \neq i}^{\infty} q_{ij}X(j). \qquad (2.144)$$

One obtains that for all $X \in \ell^1(\mathbf{Z}_+, \mathcal{S}_n)$ we have $\|\tilde{\Pi}(t)X\|_1 \leq \tilde{\gamma}(t)\|X\|_1$ where

$$\tilde{\gamma}(t) = \sum_{k=1}^{r} \|A_k(t)\|_\infty^2 + \nu. \qquad (2.145)$$

Also $\|\tilde{\Pi}(t)X\|_\infty \leq \tilde{\gamma}(t)\|X\|_\infty$ for all $X \in \mathcal{S}_n^\infty$. Hence $\Pi(t) \in \mathbf{B}(\ell^1(\mathbf{Z}_+, \mathcal{S}_n))$ and $\tilde{\Pi}(t) \in \mathbf{B}(\mathcal{S}_n^\infty)$. Based on (2.138), (2.139) and (2.143), (2.144) we deduce that $\Pi(t)X \in \ell^1(\mathbf{Z}_+, \mathcal{S}_{n+})$ if $X \in \ell^1(\mathbf{Z}_+, \mathcal{S}_{n+})$ and $\tilde{\Pi}(t)X \in \mathcal{S}_{n+}^\infty$ if $X \in \mathcal{S}_{n+}^\infty$).

The conclusion follows directly from Corollary 2.2.6 (i) and (ii). $\square$

Let $T(t,\tau)$, $(t,\tau) \in \mathcal{I} \times \mathcal{I}$ be the linear evolution operator on $\ell^1(\mathbf{Z}_+, \mathcal{S}_n)$ defined by the linear differential equation

$$\frac{d}{dt}X(t) = \mathcal{L}(t)X(t). \qquad (2.146)$$

This means that $\frac{d}{dt}T(t,\tau) = \mathcal{L}(t)T(t,\tau)$, $\quad T(\tau,\tau) = I_{\ell^1(\mathbf{Z}_+, \mathcal{S}_n)}$.

Consider, also $T^a(t,\tau)$ the anticausal linear evolution operator on $\mathcal{S}_n^\infty$ defined by the backward linear differential equation

$$\frac{d}{dt}X(t) + \mathfrak{L}(t)X(t) = 0. \qquad (2.147)$$

This means that

$$\begin{aligned} \frac{\partial}{\partial t}T^a(t,\tau) &= -\mathfrak{L}(t)T^a(t,\tau), \\ T^a(\tau,\tau) &= I_{\mathcal{S}_n^\infty}. \end{aligned} \qquad (2.148)$$

Remark 2.6.6. Under the considered assumptions the operator valued functions $t \to \mathcal{L}(t)$ and $t \to \mathfrak{L}(t)$ are continuous in the topology induced by the norms of Banach algebras $\mathbf{B}(\ell^1(\mathbf{Z}_+, \mathcal{S}_n))$ and $\mathbf{B}(\mathcal{S}_n^\infty)$, respectively.

In the previous subsection we saw that in the case $\mathcal{D} = \{1, 2, \ldots, d\}$ the analogous of the operator $\mathfrak{L}(t)$ coincides with the adjoint $\mathcal{L}^*(t)$ of the operator $\mathcal{L}(t)$.

In the case $\mathcal{D} = \mathbf{Z}_+$, such an equality is not possible because the operators $\mathcal{L}(t)$ and $\mathfrak{L}(t)$ act on different linear spaces.

In the next developments we shall see that under some additional assumptions the restriction of the operator $\mathfrak{L}(t)$ to the Hilbert space $(\ell_2(\mathbf{Z}_+, \mathcal{S}_n), \|\cdot\|_2)$ coincides with the adjoint operator of $\mathcal{L}(t)$.

First we prove an auxiliary result which could be also of interest in itself.

Lemma 2.6.6. *If $A, M \in \mathbf{R}^{n \times n}$ are given matrices, then $|AM|_2 \leq \min\{|A||M|_2,$ $|A|_2|M|\}$ where $|\cdot|$ and $|\cdot|_2$ are the norms introduced by (2.1) and (2.2).*

Proof. Let $0 \leq \lambda_i \in \mathbf{R}$, $1 \leq i \leq n$, and $e_i \in \mathbf{R}^n$, $1 \leq i \leq n$ be orthogonal vectors such that $|e_k| = 1$, $1 \leq k \leq n$ and $MM^T = \sum_{i=1}^{n} \lambda_i e_i e_i^T$. We have $|M|_2 = (\sum_{i=1}^{n} \lambda_i)^{\frac{1}{2}}$. We obtain

$|AM|_2^2 = Tr[(AM)^T AM] = Tr[A^T AMM^T] = \sum_{i=1}^{n} \lambda_i |Ae_i|^2$. Since $|Ae_i|^2 \leq |A|^2$ we infer

$|AM|_2^2 \leq |A|^2 \sum_{i=1}^{n} \lambda_i$ which leads to $|AM|_2 \leq |A||M|_2$. Changing the role of A with M we get also the inequality $|AM|_2 \leq |A|_2|M|$.

So the proof is complete. $\square$

Theorem 2.6.7. *Assume that beside the conditions (2.138) and (2.139) the real numbers q_{ij} satisfy the condition:*

$$\sup_{i \in \mathbf{Z}_+} \sum_{j=0}^{\infty} |q_{ji}| = \tilde{q} < +\infty. \tag{2.149}$$

Let $\tilde{\mathfrak{L}}(t) = \mathfrak{L}(t)|_{\ell^2(\mathbf{Z}_+, \mathcal{S}_n)}$ be the restriction of the operator $\mathfrak{L}(t)$ to $\ell^2(\mathbf{Z}_+, \mathcal{S}_n) \subset \mathcal{S}_n^{\infty}$. Under these conditions, for each $t \in \mathcal{I}$, the following hold:

(i) $\tilde{\mathfrak{L}}(t) \in \mathbf{B}(\ell^2(\mathbf{Z}_+, \mathcal{S}_n))$.
(ii) $\mathcal{L}(t) \in \mathbf{B}(\ell^2(\mathbf{Z}_+, \mathcal{S}_n))$.
(iii) $\tilde{\mathfrak{L}}(t) = \mathcal{L}^*(t)$.

Proof. (i) Let $X = \{X(i)\}_{i \in \mathbf{Z}_+} \in \ell^2(\mathbf{Z}_+, \mathcal{S}_n)$ be arbitrary but fixed. Based on (2.141) we obtain

$$|(\mathcal{L}(t)X)(i)|_2^2 \leq 4 \left[|A_0^T(t,i)X(i)|_2^2 + |X(i)A_0(t,i)|_2^2 + |\sum_{k=1}^{r} A_k^T(t,i)X(i)A_k(t,i)|_2^2 \right.$$

$$\left. + |\sum_{j=0}^{\infty} q_{ij}X(j)|_2^2 \right].$$

Based on Lemma 2.6.6 we deduce

$$|(\mathcal{L}(t)X)(i)|_2^2 \leq 4 \left[\gamma_1(t)|X(i)|_2^2 + \left(\sum_{j=0}^{\infty} |q_{ij}||X(j)|_2 \right)^2 \right] \tag{2.150}$$

where $\gamma_1(t) = 2\|A_0(t)\|_{\infty}^2 + r \sum_{k=1}^{r} \|A_k(t)\|_{\infty}^4$.

Let $N \in \mathbf{Z}_+, N \geq 1$ be arbitrary but fixed. We have

$$\left(\sum_{j=0}^{N} |q_{ij}| |X(j)|_2 \right)^2 \leq \sum_{j=0}^{N} |q_{ij}| \sum_{j=0}^{N} |q_{ij}| |X(j)|_2^2. \tag{2.151}$$

Using (2.139) we obtain:

$$\left(\sum_{j=0}^{N} |q_{ij}| |X(j)|_2 \right)^2 \leq \nu \sum_{j=0}^{N} |q_{ij}| |X(j)|_2^2.$$

Further we have $\sum_{i=0}^{N_1} \left(\sum_{j=0}^{N} |q_{ij}| |X(j)|_2 \right)^2 \leq \nu \sum_{j=0}^{N} \left(\sum_{i=0}^{N_1} |q_{ij}| |X(j)|_2^2 \right)$ for all $N_1 \in \mathbf{Z}_+, N_1 \geq 1$. Using (2.149) one gets:

$$\sum_{i=0}^{N_1} \left(\sum_{j=0}^{N} |q_{ij}| |X(i)|_2 \right)^2 \leq \nu \tilde{q} \|X\|_2^2$$

for all $N_1, N \in \mathbf{Z}_+$.

Taking the limit for $N \to \infty$, $N_1 \to \infty$ one obtains

$$\sum_{i=0}^{\infty} \left(\sum_{j=0}^{\infty} |q_{ij}| |X(j)|_2 \right)^2 \leq \sum_{i=0}^{\infty} \left(\sum_{j=0}^{\infty} |q_{ij}| |X(j)|_2 \right)^2 \leq \nu \tilde{q} \|X\|_2^2 \tag{2.152}$$

for all $i \in \mathbf{Z}_+$.

So, we have shown that the right-hand side of (2.150) is finite. Further, from (2.150) to (2.152) we deduce:

$$\sum_{i=0}^{\infty} |(\mathfrak{L}(t)X)(i)|_2^2 \leq 4(\gamma_1(t) + \nu \tilde{q}) \|X\|_2^2.$$

This shows that $(\mathfrak{L}(t)X) \in \ell^2(\mathbf{Z}_+, \mathcal{S}_n)$ if $X \in \ell^2(\mathbf{Z}_+, \mathcal{S}_n)$. Furthermore we have $\|\mathfrak{L}(t)X\|_2 \leq \gamma_2(t) \|X\|_2 \ \forall X \in \ell^2(\mathbf{Z}_+, \mathcal{S}_n)$, with

$$\gamma_2(t) = 2(\gamma_1(t) + \nu \tilde{q})^{\frac{1}{2}}. \tag{2.153}$$

Thus (i) is proved.

Further we show that (2.140) is well defined if $X = \{X(i)\}_{i \in \mathbf{Z}_+} \in \ell^2(\mathbf{Z}_+, \mathcal{S}_n)$. Proceeding as in the proof of (i), we show that

$$|(\mathcal{L}(t)X)(i)|_2^2 \le 4 \left(\gamma_1(t)|X(i)|_2^2 + \left(\sum_{j=0}^{\infty} |q_{ji}||X(i)|_2 \right)^2 \right) \qquad (2.154)$$

$i \in \mathbf{Z}_+$, $\gamma_1(t)$ being as in (2.150).

For each $N \ge 1$ we have $(\sum_{j=0}^{N} |q_{ji}||X(j)|_2)^2 \le \sum_{j=0}^{N} |q_{ji}| \sum_{j=0}^{N} |q_{ji}||X(j)|_2^2$ which yields

$(\sum_{j=0}^{N} |q_{ji}||X(j)|_2)^2 \le \tilde{q} \sum_{j=0}^{N} |q_{ji}||X(j)|_2^2$. Further we obtain $\sum_{i=0}^{N_1} \left(\sum_{j=0}^{N} |q_{ji}||X(j)|_2 \right)^2 \le$

$v\tilde{q}\|X\|_2^2$.

Taking the limits for $N \to \infty$ and $N_1 \to \infty$ we deduce $\sum_{i=0}^{\infty} \left(\sum_{j=0}^{\infty} |q_{ji}||X(j)|_2 \right)^2 \le$

$\sum_{i=0}^{\infty} (\sum_{j=0}^{\infty} |q_{ji}||X(j)|_2)^2 \le v\tilde{q}\|X\|_2^2$ for all $i \in \mathbf{Z}_+$, $X \in \ell^2(\mathbf{Z}_+, \mathcal{S}_n)$.

This shows that the right-hand side of (2.154) is finite for all $i \in \mathbf{Z}_+$. Furthermore we obtain that

$$\sum_{i=0}^{\infty} |(\mathcal{L}(t)X)(i)|_2^2 \le \gamma_2(t)\|X\|_2^2, \quad (\forall) \ X \in \ell^2(\mathbf{Z}_+, \mathcal{S}_n)$$

where $\gamma_2(t)$ is defined as in (2.153). Thus we have proved that $\mathcal{L}(t) \in \mathbf{B}(\ell^2(\mathbf{Z}_+, \mathcal{S}_n))$.

In order to prove (iii) one employs (2.13), (2.140), (2.141) to show that the equality $\langle \tilde{\mathcal{L}}(t)X, Y \rangle_2 = \langle X, \mathcal{L}(t)Y \rangle_2$ holds for all $X, Y \in \ell^2(\mathbf{Z}_+, \mathcal{L}_n)$. Thus the proof is complete.   □

Remark 2.6.7. The condition (2.149) is satisfied if there exist $h_1 \ge 0$, $h_2 \ge 0$ such that $q_{ij} = 0$ if $i < j - h_1$ or $i > j + h_2$. In this case (2.149) is satisfied with $\tilde{q} = (h_1 + h_2 + 1)v$ where v is the constant from (2.139).

By direct calculation one shows that $\tilde{\mathcal{L}} : \mathcal{I} \to \mathbf{B}(\ell^2(\mathbf{Z}_+, \mathcal{S}_n))$ is a strongly continuous operator valued function. This function defines the linear differential equation:

$$\frac{d}{dt}Y(t) + \tilde{\mathcal{L}}(t)Y(t) = 0 \qquad (2.155)$$

$t \in \mathcal{I}$ on the space $(\ell^2(\mathbf{Z}_+, \mathcal{S}_n), \|\cdot\|_2)$.

Let $T_{\tilde{\mathcal{L}}}^a(t, \tau)$, $t, \tau \in \mathcal{I}$, be the anticausal linear evolution operator on $\ell^2(\mathbf{Z}_+, \mathcal{S}_n)$ defined by the linear differential equation (2.155).

Corollary 2.6.8. *Under the assumptions of Theorem 2.6.7 we have:*

$$T_{\tilde{\mathcal{L}}}^a(\tau, t) = T^*(t, \tau), \quad \forall \ t, \tau \in \mathcal{I}.$$

$T(t, \tau)$ *being the linear evolution operator defined by* $\mathcal{L}(t) \in \mathbf{B}(\ell^2(\mathbf{Z}_+, \mathcal{S}_n))$.

Proof. follows from Theorem 2.6.7 (iii) and the equality (2.96).

2.7 Exponential Stability for Lyapunov-Type Differential Equations on $\mathcal{S}_n^d$

In this section $\mathcal{I} \subset \mathbf{R}$ denotes a right-unbounded interval. Consider the Lyapunov operator (2.124) on $\mathcal{S}_n^d$, where Q satisfies (2.123) and A_k are continuous and bounded functions. Let $T(t,t_0)$ be the linear evolution operator on $\mathcal{S}_n^d$ defined by (2.125).

Definition 2.7.1. We say that the Lyapunov-type operator $\mathcal{L}(t)$ generates an *exponentially stable evolution* or equivalently, the system $(A_0,\ldots,A_r;Q)$ *is stable*, if there exist the constants $\beta \geq 1$, $\alpha > 0$ such that

$$\|T(t,t_0)\| \leq \beta e^{-\alpha(t-t_0)}, t \geq t_0, t_0 \in \mathcal{I}. \tag{2.156}$$

Remark 2.7.1. (i) In (2.156) $\|T(t,t_0)\|$ is the norm of the linear evolution operator computed via (2.31) based on the usual norm of the Banach space $\mathcal{S}_n^d$. On the other hand from Corollary 2.1.7 (i) the usual norm of the space $\mathcal{S}_n^d$ coincides with the Minkovski norm $|\cdot|_\xi$ corresponding to $\xi = J^d$. Hence, if we have in mind the convention of notation made in Remark 2.1.5 and apply Corollary 2.6.2 we may rewrite (2.156) in the equivalent form:

$$|T(t,t_0)J^d| \leq \beta e^{-\alpha(t-t_0)}$$

for all $t \geq t_0, t, t_0 \in \mathcal{I}$.

(ii) From Remark 2.6.5 immediately follows that if $(A_0,\ldots,A_r;Q)$ is stable, then there exists $\beta \geq 1$ and $\alpha > 0$ such that

$$\left\|\widehat{T}(t,t_0)\right\| \leq \beta e^{-\alpha(t-t_0)}$$

for all $t \geq t_0, t, t_0 \in \mathcal{I}$, where $\widehat{T}(t,t_0)$ is the evolution operator on $\mathcal{S}_n^d$ defined by the pair (A_0, Q).

As usually we denote

$$\int_t^\infty T^*(s,t)H(s)\,ds := \lim_{\tau \to \infty} \int_t^\tau T^*(s,t)H(s)\,ds$$

each time when the limit in the right-hand side exists. In this case we say that the integral in the left-hand side is convergent.

Lemma 2.7.1. *Let $H : \mathcal{I} \to \mathcal{S}_n^d$ be a continuous function. Assume that the integral $\int_t^\infty T^*(s,t)H(s)\,ds$ is convergent for all $t \in \mathcal{I}$. Set*

$$K(t) := \int_t^\infty T^*(s,t)H(s)\,ds.$$

Then $K(t)$ is a solution of the affine differential equation

$$\frac{d}{dt}K(t) + \mathcal{L}^*(t)K(t) + H(t) = 0.$$

The proof is similar with the proof of Lemma 2.3.1. The details are omitted.

The next lemma shows that the integrals used in this section are absolute convergent.

Lemma 2.7.2. *Let $H : \mathcal{I} \to \mathcal{S}_n^d$ be a continuous function such that $H(t) \geq 0$ for all $t \in \mathcal{I}$. Then the following are equivalent:*

(i) The integral $\int_t^\infty |T^*(s,t)H(s)|\,ds$ is convergent for all $t \in \mathcal{I}$;
(ii) The integral $\int_t^\infty T^*(s,t)H(s)\,ds$ is convergent for all $t \in \mathcal{I}$.

Proof. (i) $\Rightarrow$ (ii) follows immediately.
   (ii) $\Rightarrow$ (i) Let

$$\gamma(t) = \left| \int_t^\infty T^*(s,t)H(s)\,ds \right|, t \in \mathcal{I}.$$

We have

$$\int_t^\infty T^*(s,t)H(s)\,ds \leq \gamma(t)J^d, t \in \mathcal{I},$$

which leads to

$$\int_t^\infty (T^*(s,t)H(s))(i)\,ds \leq \gamma(t)I_n, i \in \mathcal{D}, t \in \mathcal{I}.$$

Hence

$$\int_t^\infty Tr(T^*(s,t)H(s))(i)\,ds \leq n\gamma(t), i \in \mathcal{D}, t \in \mathcal{I}$$

from which we deduce that

$$\int_t^\tau Tr(T^*(s,t)H(s))(i)\,ds \leq n\gamma(t), \tau \geq t.$$

The above inequality gives

$$\int_t^\tau |(T^*(s,t)H(s))(i)|\,ds \leq n\gamma(t)$$

which leads to

$$\sum_{i=1}^{d} \int_{t}^{\tau} |(T^*(s,t)H(s))(i)|\, ds \le dn\gamma(t).$$

Since

$$|T^*(s,t)H(s)| \le \sum_{i=1}^{d} |(T^*(s,t)H(s))(i)|,$$

we get

$$\int_{t}^{\tau} |(T^*(s,t)H(s))|\, ds \le nd\gamma(t)$$

for all $\tau \ge t$ and the proof is complete.   □

The following result provides necessary and sufficient conditions ensuring exponential stability of the considered class of differential equations.

Theorem 2.7.3. *Let $\mathcal{I} = \mathbf{R}_+$. The following are equivalent:*

(i) *The system $(A_0, \ldots, A_r; Q)$ is stable;*
(ii) *There exists $\delta > 0$ such that*

$$\int_{t_0}^{t} \|T(t,s)\|\, ds < \delta$$

for all $t \ge t_0$, $t, t_0 \in \mathcal{I}$;
(iii) *There exists a constant $\delta > 0$ such that*

$$\int_{t_0}^{t} T(t,s) J^d\, ds < \delta J^d$$

for all $t \ge t_0$, $t, t_0 \in \mathcal{I}$.
(iv) *The solution of the initial value problem*

$$\frac{d}{dt}X(t) = \mathcal{L}(t)X(t) + J^d, \ t \in \mathbf{R}_+, X(0) = \mathbf{0}$$

is bounded.
(v) *For any bounded and continuous function $H : \mathbf{R}_+ \to \mathcal{S}_n^d$, the solution of initial value problem:*

$$\frac{d}{dt}X(t) = \mathcal{L}(t)X(t) + H(t), \ t \in \mathbf{R}_+, X(0) = \mathbf{0}$$

is bounded.

The proof follows directly from Theorem 2.3.3.

Definition 2.7.2. We say that the vector valued function $H : \mathcal{I} \to \mathcal{S}_{n+}^d$ is uniformly positive if there exists a constant $c = c(H) > 0$ such that $H(t,i) \geq cI_n$, for all $t \in \mathcal{I}, i \in \mathcal{D}$. In this case we shall write $H(t) \gg 0, t \in \mathcal{I}$. Also we shall write $H(t) \ll 0$, $t \in \mathcal{I}$ if and only if $-H(t) \gg 0, t \in \mathcal{I}$.

Since $\mathcal{S}_n^d$ is an ordered real Hilbert space we may apply the results from Sect. 2.4 to derive a list of necessary and sufficient conditions for exponential stability of Lyapunov-type equation (2.125).

Theorem 2.7.4. *The following are equivalent:*

(i) *The system $(A_0,\dots,A_r;Q)$ is stable.*
(ii) *There exist the constants $\beta_1 \geq 1$, $\alpha > 0$ such that*

$$\|T^*(t,t_0)\| \leq \beta_1 e^{-\alpha(t-t_0)},$$

for all $t \geq t_0$, $t,t_0 \in \mathcal{I}$;
(iii) *There exists a constant $\delta > 0$ such that*

$$\int_t^\infty \|T^*(s,t)\| \, ds \leq \delta$$

for all $t \in \mathcal{I}$.
(iv) *There exists $\delta > 0$ such that*

$$\int_t^\infty T^*(s,t) J^d \, ds \leq \delta J^d$$

for all $t \in \mathcal{I}$.
(v) *The affine differential equation:*

$$\frac{d}{dt} K(t) + \mathcal{L}^*(t) K(t) + J^d = 0 \qquad (2.157)$$

has a bounded and uniform positive solution on $\mathcal{I}$.
(vi) *For each $H : \mathcal{I} \to \mathcal{S}_n^d$ continuous, bounded and uniform positive function, the affine differential equation on $\mathcal{S}_n^d$:*

$$\frac{d}{dt} K(t) + \mathcal{L}^*(t) K(t) + H(t) = 0 \qquad (2.158)$$

has a bounded and uniform positive solution defined on $\mathcal{I}$.
(vii) *There exists a bounded uniform positive and continuous function $H : \mathcal{I} \to \mathcal{S}_n^d$, for which the affine Lyapunov-type equation (2.158) has a bounded solution $K_0(t) = (K_0(t,1),\dots,K_0(t,d))$ with $K_0(t,i) \geq 0, t \in \mathcal{I}$.*
(viii) *There exists a C^1-function $K : \mathcal{I} \to \mathcal{S}_n^d$, bounded with bounded derivative, $K \gg 0$ solving the differential inequality:*

$$\frac{d}{dt}K(t)+\mathcal{L}^*(t)K(t) \ll 0, t \in \mathcal{I}. \qquad (2.159)$$

The proof follows directly from Theorem 2.4.2.

The analogous of Theorem 2.4.3 is:

Theorem 2.7.5. *If the system $(A_0,\ldots,A_r;Q)$ is stable, then the following hold:*

(i) *For all bounded and continuous function $H : \mathcal{I} \to \mathcal{S}_n^d$, the corresponding Lyapunov-type equation (2.158) has a unique bounded solution given by*

$$\tilde{K}(t) = \int_t^\infty T^*(s,t)H(s)\,ds.$$

(ii) *If $t \longmapsto A_k(t,i)$, $k = 0,\ldots,r$, $t \longmapsto H(t,i)$, $i \in \mathcal{D}$ are θ-periodic functions of period θ, then the unique bounded solution of (2.158) is a θ-periodic function too.*

(iii) *If $A_k(t,i) = A_k(i)$, $k = 0,\ldots,r$ and $H(t,i) = H(i), t \in \mathcal{I}, i \in \mathcal{D}$, then the unique bounded solution of (2.158) is constant and it solves the algebraic equation*

$$\mathcal{L}^*K+H = 0.$$

(iv) *If $H(t,i) \geq 0$ for all $t \in \mathcal{I}$ and $i \in \mathcal{D}$, then the unique bounded solution $\tilde{K}(t)$ of (2.158) satisfies $\tilde{K}(t,i) \geq 0$ for all $t \in \mathcal{I}$ and $i \in \mathcal{D}$. Furthermore, if $H(t) \gg 0, t \in \mathcal{I}$, then $\tilde{K}(t) \gg 0, t \in \mathcal{I}$.*

If the matrix valued function $t \to A_k(t,i)$ are periodic of period θ without loss of generality we may take $\mathcal{I} = \mathbf{R}$. In this case, we may apply both Corollaries 2.3.8 and 2.4.4 in order to obtain a set of criteria for exponential stability of Lyapunov differential equation (2.125).

Theorem 2.7.6. *Assume that there exists $\theta > 0$ such that $A_k(t+\theta,i) = A_k(t,i)$ for all $t \in \mathbf{R}, i \in \mathcal{D}, 0 \leq k \leq r$. Then the following are equivalent:*

(i) *The system $(A_0,\ldots,A_r;Q)$ is stable.*

(ii) *The affine differential equation*

$$\frac{d}{dt}K(t) = \mathcal{L}(t)+J^d$$

has a θ-periodic solution $\tilde{K} : \mathbf{R} \to \mathcal{S}_{n+}^d$ such that $\tilde{K}(t) \gg 0, t \in \mathbf{R}$.

(iii) *For each continuous and θ-periodic vector valued function $H : \mathbf{R} \to \mathcal{S}_{n+}^d$, $H(t) \gg 0$, $t \in \mathbf{R}$, the affine differential equation*

$$\frac{d}{dt}K(t) = \mathcal{L}(t)+H(t) \qquad (2.160)$$

has a θ-periodic solution, $\tilde{K} \gg 0, t \in \mathcal{R}$.

(iv) *There exists a θ-periodic, uniformly positive, continuous vector valued function $\tilde{H} : \mathbf{R} \to \mathcal{S}_{n+}^d$ such that the corresponding affine differential equation (2.160) has a θ-periodic solution $K_0(t) \geq 0, t \in \mathbf{R}$.*

(v) *There exists a C^1 function $Y : \mathbf{R} \to \mathcal{S}_{n+}^d$ periodic with period θ and uniformly positive which solves the linear differential inequality $\frac{d}{dt}Y(t) - \mathcal{L}(t)Y(t) \gg 0, \ t \in [0, \theta]$.*

(vi) *The backward affine differential equation (2.157) has a θ periodic and uniformly positive solution.*

(vii) *For each continuous, θ-periodic and uniformly positive function $H : \mathbf{R} \to \mathcal{S}_{n+}^d$ the backward affine differential equation (2.158) has a θ-periodic and uniformly positive solution.*

(viii) *There exists a continuous, uniformly positive vector valued function $\tilde{H} : \mathbf{R} \to \mathcal{S}_{n+}^d$ periodic, with period θ such that the corresponding affine differential equation (2.158) has a θ-periodic solution $K_0 : \mathbf{R} \to \mathcal{S}_{n+}^d$.*

(ix) *There exists a C^1 function, $Y : \mathbf{R} \to \mathcal{S}_{n+}^d$, uniformly positive and periodic, with period θ, which solves*

$$\frac{d}{dt}Y(t) + \mathcal{L}^*(t)Y(t) \ll 0, \ t \in [0, \theta].$$

(x) *$\rho[T(\theta, 0)] < 1$.*

In the time invariant case we obtain directly from Corollaries 2.3.9 and 2.4.5 the following result.

Theorem 2.7.7. *Assume that $A_k(t, i) = A_k(i)$ for all $t \in \mathbf{R}, i \in \mathcal{D}, 0 \leq k \leq r$. Then the following are equivalent:*

(i) *The system $(A_0, \ldots, A_r; Q)$ is stable.*

(ii) *For all $H = (H(1), \ldots, H(d)) \in \mathcal{S}_n^d, H(i) > 0, i \in \mathcal{D}$ the algebraic linear equation on $\mathcal{S}_n^d$.*

$$\mathcal{L}^*K + H = 0 \qquad (2.161)$$

has a unique solution $K = (K(1), \ldots, K(d)) \in \mathcal{S}_n^d, K(i) > 0, i \in \mathcal{D}$.

(iii) *For each $H = (H(1), \ldots, H(d)) \in \mathcal{S}_n^d, H(i) > 0, i \in \mathcal{D}$ the linear inequality*

$$\mathcal{L}^*K + H < 0 \qquad (2.162)$$

has a solution $K = (K(1), \ldots, K(d)), K(i) > 0, i \in \mathcal{D}$.

(iv) *There exists $K \geq 0$ satisfying $\mathcal{L}^*K < 0$.*

(v) *For each $H \in \mathcal{S}_n^d, H > 0$, the linear equation on $\mathcal{S}_n^d$*

$$\mathcal{L}K + H = 0 \qquad (2.163)$$

has a unique positive solution $K = (K(1), \ldots, K(d))$.

(vi) For each $H \in \mathcal{S}_n^d$, $H > 0$ the linear inequality

$$\mathcal{L}K + H < 0 \qquad (2.164)$$

has a solution $K > 0$.

(vii) There exists $K \geq 0$ satisfying $\mathcal{L}K < 0$.

(viii) $Spec\mathcal{L} \subset \mathbf{C}^-$.

Remark 2.7.2. The affine differential equation (2.158) is the compact version of the following system of matrix linear differential equations:

$$\frac{d}{dt}K(t,i) + \left(A_0(t,i) + \frac{1}{2}q_{ii}I_n\right)^T K(t,i) + K(t,i)\left(A_0(t,i) + \frac{1}{2}q_{ii}I_n\right)$$

$$+ \sum_{k=1}^{r} A_k^T(t,i)K(t,i)A_k(t,i) + \sum_{j=1,j\neq i}^{d} q_{ij}K(t,j) + H(t,i) \qquad (2.165)$$

$$= 0, \ i \in \mathcal{D}.$$

In the time invariant case the algebraic equation $\mathcal{L}^*K + H = 0$ is the compact form of the following system of linear equations:

$$\left(A_0(i) + \frac{1}{2}q_{ii}I_n\right)^T K(i) + K(i)\left(A_0(i) + \frac{1}{2}q_{ii}I_n\right)$$

$$+ \sum_{k=1}^{r} A_k^T(i)K(i)A_k(i) + \sum_{j=1,j\neq i}^{d} q_{ij}K(j) + H(i) \qquad (2.166)$$

$$= 0$$

A consequence of Theorems 2.7.4 and 2.7.7 is the following corollary.

Corollary 2.7.8. If the system $(A_0, \ldots, A_r; Q)$ is stable, then for all $i \in \mathcal{D}$ the system of linear differential equations on $\mathbf{R}^n$

$$\frac{d}{dt}y_i(t) = \left(A_0(t,i) + \frac{1}{2}q_{ii}I_n\right)y_i(t), \ t \in \mathcal{I} \qquad (2.167)$$

defines an exponentially stable evolution.

In the invariant case, if the system $(A_0, \ldots, A_r; Q)$ is stable, then for all $i \in \mathcal{D}$, the eigenvalues of the matrices $A_0(i) + \frac{1}{2}q_{ii}I_n$ are located in the half plane $\mathbf{C}^- = \{z \in \mathbf{C} | Re(z) < 0\}$.

Proof. Since the system $(A_0, \ldots, A_r; Q)$ is stable, from Theorem 2.7.4 it follows that (2.165) has a uniform positive and bounded solution $\tilde{K}(t) = \left(\tilde{K}(t,1), \ldots, \tilde{K}(t,d)\right)$. For each $i \in \mathcal{D}$ we can write

$$\frac{d}{dt}\widetilde{K}(t,i) + \left(A_0(t,i) + \frac{1}{2}q_{ii}I_n\right)^T \widetilde{K}(t,i) +$$

$$\widetilde{K}(t,i)\left(A_0(t,i) + \frac{1}{2}q_{ii}I_n\right) + \widetilde{H}(t,i)$$

$$= 0$$

where

$$\widetilde{H}(t,i) := H(t,i) + \sum_{k=1}^{r} A_k^T(t,i)\widetilde{K}(t,i)A_k(t,i) + \sum_{j=1,j\neq i}^{d} q_{ij}\widetilde{K}(t,j).$$

It is obvious that $\widetilde{H}(t,i) \gg 0$ for all $t \in \mathcal{I}$. By standard Lyapunov function arguments we conclude that the system (2.167) is exponentially stable and the proof ends. $\square$

The next result shows that the bounded solution of (2.158) can be obtained as a limit of a sequence of bounded solutions of some Lyapunov equations.

Proposition 2.7.9. *Assume that the system* $(A_0,\ldots,A_r;Q)$ *is stable. Let* $H : \mathcal{I} \to \mathcal{S}_n^d$ *be a bounded and positive semidefinite continuous function,* $H(t) = (H(t,1),\ldots,H(t,d))$. *For each* $i \in \mathcal{D}$ *we define the sequence* $\left\{K_i^p(t)\right\}_{p\in\mathbf{Z}_+}$ *where* $t \longmapsto K_i^p(t)$ *is the unique bounded solution of the differential equation:*

$$\frac{d}{dt}K_i^p(t) + \left(A_0(t,i) + \frac{1}{2}q_{ii}I_n\right)^T K_i^p(t)$$

$$+K_i^p(t)\left(A_0(t,i) + \frac{1}{2}q_{ii}I_n\right) + H_i^p(t) \qquad (2.168)$$

$$= 0, \, i \in \mathcal{D}$$

with

$$H_i^p(t) \; := H(t,i) + \sum_{k=1}^{r} A_k^T(t,i)K_i^{p-1}(t)A_k(t,i) + \sum_{j=1,j\neq i}^{d} q_{ij}K_j^{p-1}(t),$$

$$p = 1,\ldots, \, t \in \mathcal{I} \text{ and } K_i^0(t) = 0.$$

The sequences $\left\{K_i^p(t)\right\}_{p\in N}$, $i \in \mathcal{D}$ *are increasing and bounded. If we denote*

$$K^{\infty}(t,i) = \lim_{p\to\infty} K_i^p(t), \; i \in \mathcal{D}, t \in \mathcal{I},$$

then $K^{\infty}(t) = (K^{\infty}(t,1),\ldots,K^{\infty}(t,d))$ *is the unique bounded solution of (2.158) or equivalently (2.165).*

Proof. Let $\widetilde{K}(t) = \left(\widetilde{K}(t,1),\ldots,\widetilde{K}(t,d)\right)$ be the unique bounded solution of (2.158). From Theorem 2.7.5 (iv) it follows that $\widetilde{K} \geq 0$; then we have

$$\frac{d}{dt}\widetilde{K}(t,i) + \left(A_0(t,i) + \frac{1}{2}q_{ii}I_n\right)^T \widetilde{K}(t,i) + \widetilde{K}(t,i)\left(A_0(t,i) + \frac{1}{2}q_{ii}I_n\right)$$
$$+ \sum_{k=1}^r A_k^T(t,i)\widetilde{K}(t,i)A_k(t,i) + \sum_{j=1,j\neq i}^d q_{ij}\widetilde{K}(t,j) + H(t,i) = 0, \; i \in \mathcal{D}, \; t \in \mathcal{I}.$$

By direct calculations we obtain

$$\frac{d}{dt}\left(\widetilde{K}(t,i) - K_i^p(t)\right) + \left(A_0(t,i) + \frac{1}{2}q_{ii}I_n\right)^T \left(\widetilde{K}(t,i) - K_i^p(t)\right)$$

$$+ \left(\widetilde{K}(t,i) - K_i^p(t)\right)\left(A_0(t,i) + \frac{1}{2}q_{ii}I_n\right) + \Delta_i^p(t) \tag{2.169}$$

$$= 0, \; i \in \mathcal{D},$$

where

$$\Delta_i^p(t) = \sum_{k=1}^r A_k^T(t,i)\left(\widetilde{K}(t,i) - K_i^{p-1}(t)\right)A_k(t,i)$$

$$+ \sum_{j=1,j\neq i}^d q_{ij}\left(\widetilde{K}(t,j) - K_j^{p-1}(t)\right),$$

$i \in \mathcal{D}$, $p \geq 2$ and for $p = 1$ we have

$$\Delta_i^1(t) = \sum_{k=1}^r A_k^T(t,i)\widetilde{K}(t,i)A_k(t,i) + \sum_{j=1,j\neq i}^d q_{ij}\widetilde{K}(t,j) \geq 0, \; i \in \mathcal{D}, \; t \in \mathcal{I}.$$

Since for each $i \in \mathcal{D}$, $A_0(t,i) + \frac{1}{2}q_{ii}I_n$ defines an exponentially stable evolution, from (2.169) for $p = 1$ we deduce that $\widetilde{K}(t,i) - K_i^1(t) \geq 0$, $i \in \mathcal{D}$, $t \in \mathcal{I}$. Further, by induction with respect to p we obtain that $\Delta_i^{p-1}(t) \geq 0$ which shows together with (2.169) that $\widetilde{K}(t,i) - K_i^p(t) \geq 0$ for all $p \geq 1$, $i \in \mathcal{D}$, $t \in \mathcal{I}$. Hence, the sequence $\left\{K_i^p(t)\right\}_{p\geq 0}$ is bounded. On the other hand, for each $p \geq 1$, (2.168) gives:

$$\frac{d}{dt}\left(K_i^{p+1}(t) - K_i^p(t)\right) + \left(A_0(t,i) + \frac{1}{2}q_{ii}I_n\right)^T \left(K_i^{p+1}(t) - K_i^p(t)\right)$$

$$+ \left(K_i^{p+1}(t) - K_i^p(t)\right)\left(A_0(t,i) + \frac{1}{2}q_{ii}I_n\right) + \widetilde{\Delta}_i^p(t) \tag{2.170}$$

$$= 0, \; i \in \mathcal{D},$$

where

$$\widetilde{\Delta}_i^p(t) = \sum_{k=1}^{r} A_k^T(t,i)\left(K_i^p(t) - K_i^{p-1}(t)\right)A_k(t,i) + \sum_{j=1,j\neq i}^{d} q_{ij}\left(K_i^p(t) - K_i^{p-1}(t)\right),$$

$i \in \mathcal{D}, p \geq 2$. For $p = 1$ we have

$$\widetilde{\Delta}_i^1(t) = \sum_{k=1}^{r} A_k^T(t,i)K_i^1(t)A_k(t,i) + \sum_{j=1,j\neq i}^{d} q_{ij}K_j^1(t) \geq 0.$$

By induction with respect to p, one can easily show that $\widetilde{\Delta}_i^p(t) \geq 0$ which implies that $K_i^{p+1}(t) - K_i^p(t) \geq 0$, $i \in \mathcal{D}$, $p \geq 0$. Therefore the sequence $\{K_i^p(t)\}_{p\geq 0}$ is increasing and hence the sequence is convergent. Let $K_i^\infty(t,i) = \lim_{p\to\infty} K_i^p(t)$. By standard arguments based on Lesbegue theorem (Chap. 1) we deduce that $t \longmapsto K^\infty(t,i)$, $i \in \mathcal{D}$, is a solution of the system (2.165). Since $K^\infty(t,i)$ is bounded with respect to t, it follows that $K^\infty(t,i) = \widetilde{K}(t,i)$ and the proof ends. $\square$

Remark 2.7.3. (i) In the time-invariant case the unique bounded solution of (2.168) is constant and it solves the standard Lyapunov equation

$$\left(A_0(i) + \frac{1}{2}q_{ii}I_n\right)^T K_i^p + K_i^p\left(A_0(i) + \frac{1}{2}q_{ii}I_n\right) + H_i^p = 0$$

where

$$H_i^p := \sum_{k=1}^{r} A_k^T(i)K_i^{p-1}A_k(i) + \sum_{j=1,j\neq i}^{d} q_{ij}K_j^{p-1} + H(i), \ i \in \mathcal{D};$$

(ii) If $t \longmapsto A_k(t)$, $t \longmapsto H(t)$ are θ-periodic functions, then for each p and $i \in \mathcal{D}$, the unique bounded solution on $\mathcal{I}$ of the Lyapunov differential equation (2.168) is a θ-periodic function. Therefore it is sufficient to compute only the values of $K_i^p(t)$ on the interval $[t_0, t_0 + \theta]$. We have

$$K_i^p(t) = \Phi_i^T(t_0 + \theta, t)K_i^p(t_0 + \theta)\Phi_i(t_0 + \theta, t)$$
$$+ \int_t^{t_0+\theta} \Phi_i^T(s,t)H_i^p(s)\Phi_i(s,t)ds, \ t \leq t_0 + \theta$$

$\Phi_i(s,t)$ denoting the fundamental matrix solution of (2.167). The periodicity condition $K_i^p(t) = K_i^p(t + \theta)$ shows that $K_i^p(t_0 + \theta)$ is a solution of the following algebraic discrete-time Lyapunov equation:

$$X_i = \Phi_i^T(t_0 + \theta, t_0)X_i\Phi_i(t_0 + \theta, t_0)$$
$$+ \int_{t_0}^{t_0+\theta} \Phi_i^T(s,t_0)H_i^p(s)\Phi_i(s,t_0)ds, \ i \in \mathcal{D}. \tag{2.171}$$

The eigenvalues of the matrices $\Phi_i(t_0 + \theta, t_0)$ which are the Floquet multipliers [74] of the system (2.167) are inside the unit disk $|\lambda| < 1$, $\lambda \in \mathbf{C}$. Therefore (2.171) has a unique positive semidefinite solution.

2.8 Exponential Stability for Lyapunov-Type Differential Equations on $\mathcal{S}_n^\infty$

In this section we shall provide a set of criteria for anticausal exponential stability and the exponential stability of linear differential equations (2.147) and (2.146), respectively. Throughout in this section $\mathcal{L}(t)$ and $\mathfrak{L}(t)$ are Lyapunov-type operators on $\ell^1(\mathbf{Z}_+, \mathcal{S}_n)$ and $\mathcal{S}_n^\infty$, respectively, defined by the system $(A_0, A_1, \ldots, A_r; Q)$. For each $0 \leq k \leq r$, $A_k : \mathcal{I} \to \mathcal{M}_n^\infty$ is a bounded and continuous function and Q is an infinite real matrix which elements satisfy the conditions (2.138), (2.139). Here $\mathcal{I} \subset \mathbf{R}$ is an right unbounded interval.

Definition 2.8.1. We say that the linear differential equation (2.147) is *anticausal exponentially stable* or equivalently, *the system* $(A_0, A_1, \ldots, A_r; Q)$ *defines an anticausal exponentially stable evolution on* $\mathcal{S}_n^\infty$ if there exist the constants $\beta \geq 1, \alpha > 0$ such that

$$\|T^a(t, t_0)\| \leq \beta e^{\alpha(t - t_0)} \qquad (2.172)$$

for all $t \leq t_0, t, t_0 \in \mathcal{I}$.

In (2.172) $\|T^a(t, t_0)\|$ is the norm of the anticausal linear evolution operator computed via (2.31) based on the usual norm $\| \cdot \|_\infty$ of the Banach space $\mathcal{S}_n^\infty$. On the other hand, according to Corollary 2.1.7 (ii), $\| \cdot \|_\infty$ coincides with the Minkovski norm induced by $\xi = J^\infty$. Hence we have $\|T^a(t, t_0)\| = \|T^a(t, t_0)\|_\xi$. Since the cone $\mathcal{S}_{n+}^\infty$ is a solid, closed, normal, convex cone we may apply the general results developed in Sect. 2.3.2 in order to obtain necessary and sufficient conditions of the anticausal exponentially stable evolution generated by the system $(A_0, A_1, \ldots, A_r; Q)$ on $\mathcal{S}_n^\infty$.

So, from Proposition 2.3.11 we obtain:

Corollary 2.8.1. *Under the considered assumptions the following are equivalent:*

(i) *The system* $(A_0, A_1, \ldots, A_r; Q)$ *generates an anticausal exponentially stable evolution on* $\mathcal{S}_n^\infty$.

(ii) *There exist* $\alpha > 0$, $\beta \geq 1$ *such that* $|Y(t; t_0, J^\infty)| \leq \beta e^{\alpha(t - t_0)}$, $\forall t \leq t_0, t, t_0 \in \mathcal{I}$, $Y(t; t_0, J^\infty)$ *being the solution of the differential equation (2.147) starting from* J^∞ *at the initial time* $t = t_0$.

We recall that according to the convention made in Remark 2.1.6 $|\cdot|$ stands for the Minkovski norm $|\cdot|_\xi$ when $\xi = J^\infty$.

From Theorem 2.3.12 we obtain the following result.

Theorem 2.8.2. *Assume* $\mathcal{I} = (a, \infty)$ *with* $a \geq -\infty$. *Let* $A_k : \mathcal{I} \to \mathcal{M}_n^\infty, 0 \leq k \leq r$ *be continuous and bounded functions and* Q *be an infinite real matrix whose elements verify (2.138) and (2.139).*

Under these conditions the following are equivalent:

(i) *The system $(A_0, A_1, \ldots, A_r; Q)$ generates an anticausal exponentially stable evolution.*

(ii) *For each $t \in \mathcal{I}$ the integral $\int_t^\infty T^a(t,s) J^\infty ds$ is absolute convergent and there exists $\delta > 0$ not depending upon t such that $0 \leq \int_t^\infty T^a(t,s) J^\infty ds \leq \delta J^\infty$, $\forall\, t \in \mathcal{I}$.*

(iii) *For each $t \in \mathcal{I}$ the integral $\int_t^\infty T^a(t,s) J^\infty ds$ is convergent and there exists $\delta > 0$ not depending upon t such that $0 \leq \int_t^\infty (T^a(t,s) J^\infty)(i) ds \leq \delta I_n$, $\forall\, t \in \mathcal{I}, i \in \mathbf{Z}_+$.*

(iv) *The backward affine differential equation*

$$\frac{d}{dt} Y(t) + \mathfrak{L}(t) Y(t) + J^\infty = 0 \qquad (2.173)$$

has a bounded and uniformly positive solution.

(v) *For each bounded, continuous, uniformly positive function $H : \mathcal{I} \to \mathcal{S}_n^\infty$ the backward affine differential equation*

$$\frac{d}{dt} Y(t) + \mathfrak{L}(t) Y(t) + H(t) = 0 \qquad (2.174)$$

has a bounded and uniformly positive solution.

(vi) *There exists a bounded, continuous, and uniformly positive function $\tilde{H} : \mathcal{I} \to \mathcal{S}_n^\infty$ with the property that the corresponding backward affine differential equation of type (2.174) has a bounded solution $\tilde{Y}(t)$ with the property that $\tilde{Y}(t,i) \geq 0$, for all $t \in \mathcal{I}, i \in \mathbf{Z}_+$.*

(vii) *There exists positive scalars μ_1, μ_2 and a C^1 function $Y : \mathcal{I} \to \mathcal{S}_n^\infty$, bounded with bounded derivative, which verifies the following linear differential inequality:*

$$\frac{d}{dt} Y(t) + \mathfrak{L}(t) Y(t) + \mu_1 J^\infty \leq 0, \ Y(t,i) \geq \mu_2 I_n$$

for all $t \in \mathcal{I}, i \in \mathbf{Z}_+$.

We recall that a function $H : \mathcal{I} \to \mathcal{S}_n^\infty$ is named uniform positive if there exists $\delta > 0$, such that $H(t,i) \geq \delta I_n$ for all $t \in \mathcal{I}, i \in \mathbf{Z}_+$.

In the case of periodic coefficients we obtain the following result.

Corollary 2.8.3. *Assume that $\mathcal{I} = \mathbf{R}$ and there exists $\theta > 0$ such that $A_k(t + \theta, i) = A_k(t,i)$, $\forall\, 0 \leq k \leq r, t \in \mathcal{I}, i \in \mathbf{Z}_+$.*

Then the following are equivalent:

(i) *The system $(A_0, A_1, \ldots, A_r; Q)$ generates an anticausal exponentially stable evolution on $\mathcal{S}_n^\infty$.*

(ii) *The backward affine differential equation (2.173) has a θ periodic and uniformly positive solution.*

(iii) *For each continuous, θ-periodic, and uniformly positive function $H : \mathbf{R} \to S_n^\infty$ the backward affine differential equation (2.174) has a θ-periodic and uniformly positive solution.*

(iv) *There exists a continuous, uniformly positive valued function $\tilde{H} : \mathbf{R} \to S_n^\infty$ periodic, with period θ such that the corresponding affine differential equation (2.174) has a θ-periodic solution $\tilde{Y} : \mathbf{R} \to S_{n+}^\infty$.*

(v) *There exist a scalar $\delta > 0$ and a C^1 valued function, $Y : \mathbf{R} \to S_n^\infty$, uniformly positive and periodic, with period θ, which solves*

$$\frac{d}{dt}Y(t) + \mathcal{L}(t)Y(t) + \delta J^\infty < 0, \ t \in [0,\theta].$$

Proof. It is a special case of Corollary 2.3.14.

In the time invariant case we have the following list of criteria for anticausal exponentially stable evolution generated by the Lyapunov-type operator $\mathcal{L}$.

Corollary 2.8.4. *Assume $A_k(t,i) = A_k(i)$ for all $t \in \mathcal{I}, i \in \mathbf{Z}_+, 0 \leq k \leq r$. Then the following are equivalent:*

(i) *The system $(A_0, A_1, \ldots, A_r; Q)$ defines an anticausal exponentially stable evolution on S_n^∞.*

(ii) *The linear equation $\mathcal{L}X + J^\infty = 0$ has a solution $\tilde{X} = \{X(i)\}_{i \in \mathbf{Z}_+}$ such that $\tilde{X}(i) \geq \delta I_n$ for all $i \in \mathbf{Z}_+$ where $\delta > 0$ does not depend upon i.*

(iii) *For each sequence $\mathbf{H} = \{H(i)\}_{i \in \mathbf{Z}_+} \in \mathit{Int}S_{n+}^\infty$ there exists $\tilde{X} = \{\tilde{X}(i)\}_{i \in \mathbf{Z}_+} \in \mathit{Int}S_{n+}^\infty$ which solves $\mathcal{L}\tilde{X} + \mathbf{H} = 0$.*

(iv) *There exists a scalar $\delta > 0$ and $\mathbf{Y} \in \mathit{Int}S_{n+}^\infty$ such that $\mathcal{L}\mathbf{Y} + \delta J^\infty \leq 0$.*

Definition 2.8.2. We say that the linear differential equation (2.146) is exponentially stable, or equivalently the system $(A_0, A_1, \ldots, A_r; Q)$ defines an exponentially stable evolution on the space $\ell^1(\mathbf{Z}_+, S_n)$ if there exist $\beta \geq 1$, $\alpha > 0$ such that

$$\|T(t,t_0)\|_1 \leq \beta e^{-\alpha(t-t_0)} \tag{2.175}$$

for all $t \geq t_0, t, t_0 \in \mathcal{I}$.

In (2.175), $\|T(t,t_0)\|_1$ is operator norm induced via (2.31) by the usual norm $\|\cdot\|_1$ on $\ell^1(\mathbf{Z}_+, S_n)$.

Since the cone $\ell^1(\mathbf{Z}_+, S_{n+})$ has empty interior we cannot apply the developments from Sect. 2.3 in order to derive criteria for exponential stability of the linear differential equation (2.146) on $\ell^1(\mathbf{Z}_+, S_n)$.

The result proved in Corollary 2.8.8 shows that the criteria for anticausal exponential stability of (2.147) could be used as necessary and sufficient conditions for the exponential stability of (2.146).

Let $\mathfrak{L}(t) = \mathfrak{L}_A(t) + \tilde{\Pi}(t)$ be the partition of the linear operator $\mathfrak{L}(t)$ considered in the proof of Proposition 2.6.5, $\tilde{\Pi}(t)$ being defined by (2.144).

We prove:

Lemma 2.8.5. *For any monotone and bounded sequence $\{X_k\}_{k \in \mathbf{Z}_+} \subset \mathcal{S}_n^\infty$ we have:*

(i) $\lim\limits_{k \to \infty} (\tilde{\Pi}(t)[X_k])(i) = (\tilde{\Pi}(t)[X])(i)$ *for all $i \in \mathbf{Z}_+$, $t \in \mathcal{I}$.*

(ii) $\lim\limits_{k \to \infty} (T^a(t,t_0)X_k)(i) = (T^a(t,t_0)X)(i)$ *for all $i \in \mathbf{Z}_+$, $t \leq t_0$, $t, t_0 \in \mathcal{I}$, where $X = \{X(i)\}_{i \in \mathbf{Z}_+} \in \mathcal{S}_n^\infty$ is defined by $X(i) = \lim\limits_{k \to \infty} X_k(i)$, $i \in \mathbf{Z}_+$.*

Proof. Without loss of generality we may assume that $\{X_k\}_{k \in \mathbf{Z}}$ is a increasing and bounded sequence. This means that there exist $\mu_j \in \mathbf{R}$, $j = 1, 2$ such that

$$\mu_1 I_n \leq X_k(i) \leq X_{k+1}(i) \leq \mu_2 I_n, \quad \forall (k,i) \in \mathbf{Z}_+ \times \mathbf{Z}_+ \qquad (2.176)$$

Therefore, for each $i \in \mathbf{Z}_+$, $X(i) \in \mathcal{S}_n$ is well defined by

$$X(i) = \lim_{k \to \infty} X_k(i). \qquad (2.177)$$

Based on (2.176) we infer that $X = \{X(i)\}_{i \in \mathbf{Z}_+} \in \mathcal{S}_n^\infty$. From (2.144) we obtain

$$(\tilde{\Pi}(t)X_k)(i) - (\tilde{\Pi}(t)X)(i) = \sum_{l=1}^{r} A_l^T(t,i)(X_k(i) - X(i))A_l(t,i) + \sum_{\substack{j=0 \\ j \neq i}}^{\infty} q_{ij}(X_k(j) - X(j)). \quad (2.178)$$

First, from (2.176) we obtain

$$\lim_{k \to \infty} \sum_{k=1}^{r} A_l^T(t,i)(X_k(i) - X(i))A_l(t,i) = 0 \qquad (2.179)$$

On the other hand applying Corollary 1.2.9 for $a_k(j) = q_{ij}|X_k(j) - X(j)|$ we deduce that $\lim_{k \to \infty} \sum_{j=0, j \neq i}^{\infty} q_{ij}|X_k(j) - X(j)| = 0$ which leads to

$$\lim_{k \to \infty} \sum_{j=0, j \neq i}^{\infty} q_{ij}(X_k(j) - X(j)) = 0. \qquad (2.180)$$

Combining (2.178)–(2.180) we obtain that (i) is true.

Let us now prove that (ii) holds. To this end, let us denote $Y_k(t) = T^a(t,t_0)X_k$, $t \in (-\infty, t_0] \cap \mathcal{I}$, $t_0 \in \mathcal{I}$ being fixed. Since $T^a(t,t_0)$ is a positive operator, if $t \leq t_0$ the inequalities (2.176) yield

$$\mu_1(T^a(t,t_0)J^\infty)(i) \leq Y_k(t,i) \leq Y_{k+1}(t,i) \leq \mu_2(T^a(t,t_0)J^\infty)(i) \qquad (2.181)$$

for all $i \in \mathbf{Z}_+, t \leq t_0, t \in \mathcal{I}$. From (2.181) we obtain that the matrices $Z(t,i)$ are well defined by

$$Z(t,i) = \lim_{k \to \infty} Y_k(t,i), i \in \mathbf{Z}_+, t \in (-\infty; t_0] \cap \mathcal{I}.$$

Furthermore (2.181) yields $|Z(t,i)| \leq \mu_3 |T^a(t,t_0)J^\infty| = \mu_3 \|T^a(t,t_0)\|$ for all $i \in \mathbf{Z}_+$. This leads to

$$|Z(t)| \leq \mu_3 \|T^a(t,t_0)\|, \quad \forall t \in \mathcal{I}, t \leq t_0,$$

where $Z(t) = \{Z(t,i)\}_{i \in \mathbf{Z}_+}$.

Since $t \to \|\mathfrak{L}(t)\|_\infty$ is a bounded function we deduce that $\|T^a(t,t_0)\| \leq e^{c(t_0-t)}$, for all $t \in \mathcal{I}, t \leq t_0$. This leads to

$$|Z(t)| \leq \mu_3 e^{c(t_0-t)}. \tag{2.182}$$

Reasoning in the same way we obtain from (2.181)

$$|Y_k(t)| \leq \mu_3 e^{c(t_0-t)} \tag{2.183}$$

for all $t \in \mathcal{I}, t \leq t_0, k \in \mathbf{Z}_+$.

Let $T_A^a(t,s)$ be the anticausal linear evolution operator on $\mathcal{S}_n^\infty$ defined by the extended Lyapunov operator $\mathfrak{L}_A(t)$. We have the representation formula

$$Y_k(t) = T_A^a(t,t_0)X_k + \int_t^{t_0} T_A^a(t,s)\tilde{\Pi}(s)Y_k(s)ds$$

for all $t \leq t_0, t \in \mathcal{I}$.

Based on (2.122) written for $A(t,i)$ replaced by $A_0(t,i) + \frac{1}{2}q_{ii}I_n$, we obtain the component wise representation formula

$$Y_k(t,i) = \Phi_i^T(t_0,t)X_k(i)\Phi_i(t_0,t) \int_t^{t_0} \Phi_i^T(s,t)(\tilde{\Pi}(s)Y_k(s))(i)\Phi_i(s,t_0)ds \tag{2.184}$$

for all $i \in \mathbf{Z}_+, t \leq t_0, t \in \mathcal{I}$, where $\Phi_i(s,t)$ is the fundamental matrix solution of the differential equation

$$\frac{d}{dt}x(t) = (A_0(t,i) + \frac{1}{2}q_{ii}I_n)x(t).$$

Using the result proved in the part (i) of the lemma, we obtain that

$$\lim_{k \to \infty} \Phi_i^T(s,t)(\tilde{\Pi}(s)Y_k(s))(i)\Phi_i(s,t) = \Phi_i^T(s,t)(\tilde{\Pi}(s)Z(s))(i)\Phi_i(s,t) \tag{2.185}$$

for all $i \in \mathbf{Z}_+, t \leq s \leq t_0$.

We recall that the boundedness of the function $s \to \|A_0(s)\|_\infty$ together with (2.139) allow us to deduce that

$$|\Phi_i(s,t)| \leq e^{c_1(s-t)}, \tag{2.186}$$

$\forall t \leq s \leq t_0, t \in \mathcal{I}$, where $c_1 > 0$ is a constant not depending upon s,t.

Further, from (2.184), (2.186) together with the boundedness of the functions $s \to \|A_l(s)\|_\infty, 0 \leq l \leq r$ yield

$$x^T \Phi_i^T(s,t)(\tilde{\Pi}(s)Y_k(s))(i)\Phi_i(s,t)x \leq \tilde{\beta}e^{\tilde{c}(s-t)}|x|^2 \tag{2.187}$$

for all $t \leq s \leq t_0$, where $\tilde{\beta}, \tilde{c}$ are positive constants. Applying Lebesque's Theorem we obtain via (2.185) and (2.187) that $\lim\limits_{k \to \infty} \int_t^{t_0} x^T \Phi_i^T(s,t)(\tilde{\Pi}(s)Y_k(s))(i)\Phi_i(s,t)x ds = \int_t^{t_0} x^T \Phi_i^T(s,t)(\tilde{\Pi}(s)Z(s))(i)\Phi_i(s,t)x ds$ for all $x \in \mathbf{R}^n$. By a standard procedure, one obtains finally that

$$\lim_{k \to \infty} \int_t^{t_0} \Phi_i^T(s,t)(\tilde{\Pi}(s)Y_k(s))(i)\Phi_i(s,t)ds = \int_t^{t_0} \Phi_i^T(s,t)(\tilde{\Pi}(s)Z(s))(i)\Phi_i(s,t)ds$$

for all $t \leq t_0, t \in \mathcal{I}$. Taking the limit for $k \to \infty$ in (2.184) we obtain that

$$Z(t,i) = \Phi_i^T(t_0,t)X(i)\Phi_i(t_0,t) + \int_t^{t_0} \Phi_i^T(s,t)(\tilde{\Pi}(s)Z(s))(i)\Phi_i(s,t)ds$$

for all $i \in \mathbf{Z}_+, t \leq t_0, t \in \mathcal{I}$.

The above equality may be rewritten in a compact form:

$$Z(t) = T_A^a(t,t_0)X + \int_t^{t_0} T_A^a(t,s)\tilde{\Pi}(s)Z(s)ds. \tag{2.188}$$

Under the considered assumptions the identity (2.188) allows us to deduce that $t \to Z(t)$ is differentiable and additionally it solves the problem with given terminal condition:

$$\frac{d}{dt}Z(t) + \mathcal{L}(t)Z(t) = 0, \ t \leq t_0 \tag{2.189}$$

$$Z(t_0) = X.$$

From the uniqueness of the solution of the problem (2.189) we conclude that

$$Z(t,i) = (T^a(t,t_0)X)(i) \tag{2.190}$$

for all $i \in \mathbf{Z}_+, t \leq t_0, t \in \mathcal{I}$.

The conclusion follows now from (2.190). So the proof is complete. $\qquad\square$

Lemma 2.8.6. *Assume that the assumptions of Theorem 2.6.7 are fulfilled. Let* $\mathbf{H}_i^x = \{H_i^x(j)\}_{j \in \mathbf{Z}_+}$ *be defined by*

$$H_i^x(j) = \begin{cases} 0, if\, j \neq i \\ xx^T, if\, j = i. \end{cases} \tag{2.191}$$

where $x \in \mathbf{R}^n$ *and* $i \in \mathbf{Z}_+$ *are arbitrary but fixed.*
   Under the considered assumptions we have:

$$\|T(t,\tau)\mathbf{H}_i^x\|_1 = x^T[(T^a(\tau,t)J^\infty)(i)]x \tag{2.192}$$

for all $t \geq \tau,\, t, \tau \in \mathcal{I}.$

Proof. First we notice that $\mathbf{H}_i^x \in \ell^1(\mathbf{Z}_+, \mathcal{S}_n)$ and $\|\mathbf{H}_i^x\|_1 = |x|^2$. Therefore $T(t,\tau)\mathbf{H}_i^x$ is well defined and we have

$$\|T(t,\tau)\mathbf{H}_i^x\|_1 = \sum_{j=0}^\infty Tr[(T(t,\tau)H_i^x)(j)] \tag{2.193}$$

for all $t, \tau \in \mathcal{I}.$
   For each $k \in \mathbf{Z}_+$ we consider $J_k^\infty = \{J_k^\infty(j)\}_{j \in \mathbf{Z}_+}$ where

$$J_k^\infty(j) = \begin{cases} I_n, if\, 0 \leq j \leq k \\ 0, if\, j > k. \end{cases}$$

It is obvious that $J_k^\infty \in \ell^2(\mathbf{Z}_+, \mathcal{S}_n) \subset \mathcal{S}_n^\infty$ and we have $\|J_k^\infty\|_2 = (k+1)\sqrt{n}$ and $\|J_k^\infty\|_\infty = 1$. Also we have $J_k^\infty \leq J_{k+1}^\infty \leq J^\infty$ for all $k \in \mathbf{Z}_+$. This yields: $T^a(\tau,t)J_k^\infty \leq T^a(\tau,t)J_{k+1}^\infty \leq T^a(\tau,t)J^\infty$, for all $k \in \mathbf{Z}_+, t \geq \tau, t, \tau \in \mathcal{I}$ because $T^a(\tau,t) \geq 0$ for all $t \geq \tau$
   This allows us to obtain

$$x^T[(T^a(\tau,t)J_k^\infty)(i)]x \leq x^T[(T^a(\tau,t)J_{k+1}^\infty)(i)]x \leq x^T[(T^a(\tau,t)J^\infty)(i)]x \tag{2.194}$$

for all $k \in \mathbf{Z}_+$. Moreover, applying Lemma 2.8.5 (ii) for $X_k = J_k^\infty$ we obtain that

$$\lim_{k \to \infty} x^T(T^a(\tau,t)J_k^\infty)(i)x = x^T(T^a(\tau,t)J^\infty)(i)x. \tag{2.195}$$

On the other hand, from Theorem 2.6.7 (i) we deduce that $T^a(\tau,t)J_k^\infty \in \ell^2(\mathbf{Z}_+, \mathcal{S}_n)$. Therefore we may write:

$$x^T[(T^a(\tau,t)J_k^\infty)(i)]x = Tr[(T^a(\tau,t)J_k^\infty)(i)xx^T] =$$

$$= \sum_{j=0}^\infty Tr[(T^a(\tau,t)J_k^\infty)(j)H_i^x(j)] = \langle T^a(\tau,t)J_k^\infty, \mathbf{H}_i^x \rangle_2.$$

Further, the equality proved in Theorem 2.6.7 (iii) together with (2.96) yield:

$$\langle T^a(\tau,t)J_k^\infty,\mathbf{H}_i^x\rangle_2 = \langle T^*(t,\tau)J_k^\infty,\mathbf{H}_i^x\rangle_2 = \langle J_k^\infty,T(t,\tau)\mathbf{H}_i^x\rangle_2 = \sum_{j=0}^{k} Tr[(T(t,\tau)\mathbf{H}_i^x)(j)].$$

Thus we obtain

$$x^T[(T^a(\tau,t)J_k^\infty)(i)]x = \sum_{j=0}^{k} Tr[(T(t,\tau)\mathbf{H}_i^x)(j)]. \tag{2.196}$$

Based on (2.196) we get

$$\|T(t,\tau)\mathbf{H}_i^x\|_1 = \lim_{k\to\infty}\sum_{j=0}^{k} Tr[(T(t,\tau)\mathbf{H}_i^x)(j)] = \lim_{k\to\infty} x^T[(T^a(\tau,t)J_k^\infty)(i)]x. \tag{2.197}$$

The conclusion follows from (2.197) and (2.195). Thus the proof is complete.   □

Theorem 2.8.7. *Assume that the assumptions of Theorem 2.6.7 are fulfilled. Then we have*

$$\|T(t,\tau)\|_1 \le \|T^a(\tau,t)\| \quad \forall\, t \ge \tau,\, t,\tau \in \mathcal{I}. \tag{2.198}$$

Proof. Let $i \in \mathbf{Z}_+$ be arbitrary but fixed and $\psi_i : \ell^1(\mathbf{Z}_+,\mathcal{S}_n) \to \ell^1(\mathbf{Z}_+,\mathcal{S}_n)$ be defined by

$$\psi_i(\mathbf{X})(j) = \begin{cases} 0, \text{if } j \ne i \\ X(i), \text{if } j = i \end{cases} \tag{2.199}$$

for any $\mathbf{X} = \{X(j)\}_{j\in\mathbf{Z}_+} \in \ell^1(\mathbf{Z}_+,\mathcal{S}_n)$. We have $\left\|\mathbf{X} - \sum\limits_{i=0}^{k}\psi_i(\mathbf{X})\right\|_1 = \sum\limits_{i=k+1}^{\infty}|X(i)|_1$

which leads to $\lim\limits_{k\to\infty}\left\|\mathbf{X} - \sum\limits_{i=0}^{k}\psi_i(\mathbf{X})\right\|_1 = 0.$

Hence $\mathbf{X} = \sum\limits_{i=0}^{\infty}\psi_i(\mathbf{X})$ for all $\mathbf{X} \in \ell^1(\mathbf{Z}_+,\mathcal{S}_n)$.

Further we have

$$T(t,\tau)\mathbf{X} = \sum_{i=0}^{\infty} T(t,\tau)\psi_i(\mathbf{X}) \tag{2.200}$$

because $T(t,\tau) \in \mathbf{B}(\ell^1(\mathbf{Z}_+,\mathcal{S}_n))$.

Let $\lambda_{i1},\lambda_{i2},\dots,\lambda_{in}$ be real numbers and $e_{i1},e_{i2},\dots,e_{in} \in \mathbf{R}^n$ be orthogonal vectors such that $|e_{ij}| = 1$, $1 \le j \le n$ and $X(i) = \sum\limits_{j=1}^{n} \lambda_{ij}e_{ij}e_{ij}^T$. Combining (2.191) and (2.199) we deduce:

$$\psi_i(\mathbf{X}) = \sum_{j=1}^{n} \lambda_{ij} \mathbf{H}_i^{e_{ij}} \tag{2.201}$$

where $\mathbf{H}_i^{e_{ij}}$ is defined as in (2.191) with e_{ij} instead of x.

For each $k \geq 1$ we write

$$\left\| \sum_{i=0}^{k} T(t,\tau)\psi_i(\mathbf{X}) \right\|_1 \leq \sum_{i=0}^{k} \|T(t,\tau)\psi_i(\mathbf{X})\|_1 \leq \sum_{i=0}^{k} \sum_{j=1}^{n} |\lambda_{ij}| \left\| T(t,\tau)\mathbf{H}_i^{e_{ij}} \right\|_1.$$

Applying Lemma 2.8.6 we obtain

$$\sum_{i=0}^{k} \|T(t,\tau)\Psi_i(\mathbf{X})\|_1 \leq \sum_{i=0}^{k} \sum_{j=1}^{n} |\lambda_{ij}| e_{ij}^T [(T^a(\tau,t)J^\infty)(i)] e_{ij} \leq \sum_{i=0}^{k} \sum_{j=1}^{h} |\lambda_{ij}| |(T^a(\tau,t)J^\infty)(i)|.$$

Invoking (2.6) we infer $\sum_{i=0}^{k} \|T(t,\tau)\psi_i(\mathbf{X})\|_1 \leq \|T^a(\tau,t)J^\infty\|_\infty \sum_{i=0}^{\infty} |X(i)|_1$ for all $k \geq 1$.

Hence we have shown that

$$\sum_{i=0}^{\infty} \|T(\tau,t)\psi_i(\mathbf{X})\|_1 \leq \|T^a(\tau,t)J^\infty\|_\infty \|\mathbf{X}\|_1. \tag{2.202}$$

Further, from (2.200) to (2.202) we get:

$$\|T(t,\tau)\mathbf{X}\|_1 \leq \|T^a(\tau,t)J^\infty\|_\infty \|\mathbf{X}\|_1, \quad (\forall)\ \mathbf{X} \in \ell^1(\mathbf{Z}_+,\mathcal{S}_n),\ t \geq \tau,\ t,\tau \in \mathcal{I}.$$

So we may conclude that $\|T(t,\tau)\|_1 \leq \|T^a(\tau,t)J^\infty\|_\infty$ for all $t \geq \tau, t, \tau \in \mathcal{I}$. To show that the last inequality coincides with (2.198) we apply Theorem 2.1.10 in the special case of the positive operator $T^a(\tau,t)$ together with Corollary 2.1.7 (ii) and obtain that $\|T^a(\tau,t)J^\infty\|_\infty = \|T^a(\tau,t)\|$. This ends the proof. □

Corollary 2.8.8. *Under the assumptions of Theorem 2.6.7 the following are equivalent:*

(i) the operator valued function $\mathfrak{L}(\cdot)$ defines an exponentially stable anticausal evolution on $\mathcal{S}_n^\infty$;

(ii) the operator valued function $\mathcal{L}(\cdot)$ defines an exponentially stable evolution on $\ell^1(\mathbf{Z}_+,\mathcal{S}_n)$.

Proof. (i) → (ii). If (i) holds, then there exist $\beta \geq 1$, $\alpha > 0$ such that $\|T^a(\tau,t)\| \leq \beta e^{-\alpha(t-\tau)}$ for all $\tau,t \in \mathcal{I}, t \geq \tau$. Then from Theorem 2.8.7 we get

$$\|T(t,\tau)\|_1 \leq \beta e^{-\alpha(t-\tau)} \tag{2.203}$$

for all $t \geq \tau, t, \tau \in \mathcal{I}$. This shows that (ii) is true.

Let us prove now that (ii) $\rightarrow$ (i). If (ii) holds, then there exist $\beta \geq 1$, $\alpha > 0$ such that (2.203) is true. From Lemma 2.8.6 we have $x^T(T^a(\tau,t)J^\infty)(i)x \leq \|T(t,\tau)\|_1\|\mathbf{H}_i^x\|_1$ which yields $x^T(T^a(\tau,t)J^\infty)(i)x \leq \beta e^{-\alpha(t-\tau)}|x|^2$ for all $t \geq \tau \in \mathcal{I}$, $x \in \mathbf{R}^n$, $i \in \mathbf{Z}_+$. Therefore $|(T^a(\tau,t)J^\infty)(i)| \leq \beta e^{\alpha(\tau-t)}$, $(\forall)$ $i \in \mathbf{Z}_+$, which leads to

$$\|T^a(\tau,t)J^\infty\|_\infty \leq \beta e^{\alpha(\tau-t)}. \tag{2.204}$$

Applying Theorem 2.1.10 to the positive operator $T^a(\tau,t)$ and using Corollary 2.1.7 (ii) we obtain from (2.204) that $\|T^a(\tau,t)\| \leq \beta e^{\alpha(\tau-t)}$, for all $t \geq \tau$, $t,\tau \in \mathcal{I}$. This confirms that the implication (ii) $\rightarrow$ (i) is true. So the proof is complete. $\square$

Notes and References

The term *resolvent positive* seems to have been coined by Arendt, e.g., [5]. A resolvent positive operator is also called *Metzler operator*, e.g., in [62,88]. Condition (i) from Theorem 2.2.2 is often called *exponential positivity* or *exponential nonnegativity*, e.g., [9,10,96]. The notion of *quasi-monotonic* is introduced in [60]. Operators satisfying (iii) in Theorem 2.2.2 are called *cross-positive* in [128]. In the finite dimensional case Theorem 2.2.2 is proved in [60], while Theorem 2.2.3 is proved in [50]. Lemma 2.2.4 and Theorem 2.5.1 can be found in [27,29]. The result stated in Theorem 2.1.2 was proved in [97]. Many results of this chapter may be found in [52].

Chapter 3
Exponential Stability in Mean Square

In this chapter the problem of mean square exponential stability of the zero solution to the stochastic differential equations of type (1.22) is studied. The stability of a steady-state is one of the main tasks which appears in many design problems of controllers with prescribed performances.

In the case of stochastic systems there are several possibilities to define the concept of stability of a steady-state. Among them, one of the most popular is the so-called *exponential stability in mean-square* (ESMS). The ESMS has the advantage that it may be characterized by some conditions easy to be checked. Moreover in some particular cases as the time invariant case or in the periodic case, the ESMS is equivalent with other types of stability in mean square. From the representation formula proved in Theorems 3.1.1 and 3.1.4 one obtains that the ESMS of the zero-solution of (1.22) is equivalent with the exponential stability of the zero-solution of a deterministic linear differential equation on a finite or infinite dimensional linear space adequately chosen. The deterministic differential equations are defined by the so-called *Lyapunov-type operators* acting on a space of symmetric matrices. Criteria for exponential stability of the zero solution of Lyapunov differential equations were derived in a more general setting in Chap. 2.

This chapter starts with several theorems, named *representation theorems*, which emphasize the relationship between the linear evolution operators defined by Lyapunov differential equations and the fundamental matrix solution of (1.22)

In the last section of the chapter some useful estimates of the solutions of affine equations are derived. Some aspects concerning the ESMS of the zero state equilibrium for nonlinear stochastic differential equations of type (1.16) will be discussed in Chap. 8.

3.1   Representation Theorems

3.1.1   The First Representation Theorem

A. The case $\mathcal{D} = \{1, 2, \ldots, d\}$.

Consider the system of linear stochastic differential equations (1.22) where $\eta(t)$ is a standard homogeneous Markov process with the finite set of states $\mathcal{D} = \{1, 2, \ldots, d\}$. Based on the coefficients $A_k(t, i), 0 \le k \le r$ of the system (1.22) and the elements q_{ij} of the generator matrix Q of the Markov process we define the Lyapunov-type operators $\mathcal{L}(t)$ via (2.124). We also recall $T(t, t_0)$ stands for the linear evolution operator on $\mathcal{S}_n^d$ defined by the linear differential equation (2.125).

In order to motivate the definition of the Lyapunov operator $\mathcal{L}(t)$ and its corresponding evolution operator $T(t, t_0)$, we shall prove the following result which establishes the relationship between the evolution operator $T(t, t_0)$ and the fundamental matrix solution of a system of stochastic linear differential equations of type (1.22).

Theorem 3.1.1. *Assume that* $\mathcal{I} = \mathbf{R}_+$ *and that the elements of* Q *satisfy (2.123) and the additional condition* $\sum_{j=1}^d q_{ij} = 0, i \in \mathcal{D}$. *Under these assumptions we have*

$$(T^*(t, t_0)H)(i) = E\left[\Phi^T(t, t_0)H(\eta(t))\Phi(t, t_0) | \eta(t_0) = i\right]$$

for all $t \ge t_0 \ge 0$, $H \in \mathcal{S}_n^d$, $i \in \mathcal{D}$, *where* $\Phi(t, t_0)$ *is the fundamental matrix solution of the system (1.22).*

Proof. Let $\mathcal{U}(t, t_0) : \mathcal{S}_n^d \to \mathcal{S}_n^d$ be defined by

$$(\mathcal{U}(t, t_0)(H))(i) = E\left[\Phi^T(t, t_0)H(\eta(t))\Phi(t, t_0) | \eta(t_0) = i\right],$$

$H \in \mathcal{S}_n^d$, $i \in \mathcal{D}$, $t \ge t_0$.

Take $H \in \mathcal{S}_n^d$, we define $v(t, x, i) = x^T H(i)x$, $x \in \mathbf{R}^n$, $i \in \mathcal{D}$, $t \ge 0$.

Applying Theorem 1.10.2 from Chap. 1 (the Itô-type formula) to the function $v(t, x, i)$ and to (1.22) we obtain

$$x^T\left(\mathcal{U}(t, t_0)(H)\right)(i)x - x^T H(i)x = x^T\left(\int_{t_0}^t \left(\mathcal{U}(s, t_0)(\mathcal{L}^*(s)H)\right)(i)ds\right)x,$$

and hence

$$\frac{d}{dt}\mathcal{U}(t, t_0) = \mathcal{U}(t, t_0)\mathcal{L}^*(t).$$

Since $\mathcal{U}(t_0, t_0) = T^*(t_0, t_0)$ it follows using (2.127) that

$$\mathcal{U}(t, s) = T^*(t, s)$$

$t \ge s$ and the proof is complete.   □

As we shall see in Sect. 3.2, the above result allows us to reduce the study of the exponential stability for the linear stochastic system (1.22) to the problem of the exponential stability for a deterministic system of type (2.125).

Remark 3.1.1. (i) If in the system (1.22) we have $A_k(t + \theta) = A_k(t), t \geq 0, i \in \mathcal{D}$, then from Theorem 3.1.1 and the Remark 2.6.3 (ii) we deduce that

$$E\left[|\Phi(t + \theta, t_0 + \theta)x_0|^2 \mid \eta(t_0 + \theta) = i\right]$$

$$= E\left[|\Phi(t, t_0)x_0|^2 \mid \eta(t_0) = i\right]$$

for all $t \geq t_0 \geq 0$, $i \in \mathcal{D}$, $x_0 \in \mathbf{R}^n$;

(ii) If the system (1.22) is time invariant, then according to Theorem 3.1.1 and the Remark 2.6.3 (i), we have

$$E\left[|\Phi(t, t_0)x_0|^2 \mid \eta(t_0) = i\right]$$

$$= E\left[|\Phi(t - t_0, 0)x_0|^2 \mid \eta(0) = i\right]$$

for all $t \geq t_0 \geq 0$, $i \in \mathcal{D}$, $x_0 \in \mathbf{R}^n$.

Let $\hat{\mathcal{L}}(t)$ be the Lyapunov-type operator defined by the pair $(A; Q)$ via (2.134) with $A_0(t, i)$ replaced by $A(t, i)$. The following result provides a relationship between the linear evolution operator $\hat{T}(t, t_0)$ defined by the linear differential equation $\frac{d}{dt}S(t) = \hat{\mathcal{L}}(t)S(t)$ and the fundamental matrix solution $\hat{\Phi}(t, t_0)$ of (1.23).

Proposition 3.1.2. *Under the assumptions of Theorem 3.1.1 we have*

$$\left(\hat{T}^*(t, t_0)H\right)(i) = E\left[\hat{\Phi}^T(t, t_0)H(\eta(t))\hat{\Phi}(t, t_0)|\eta(t_0) = i\right]$$

for all $t \geq t_0 \geq 0$, $H \in \mathcal{S}_n^d$, $i \in \mathcal{D}$, where $\hat{\Phi}(t, t_0)$ is the fundamental matrix solution of the system (1.23).

Let us consider now the case $\mathcal{D} = \{1\}$ and $q_{11} = 0$. In this case $\mathcal{S}_n^d$ reduces to $\mathcal{S}_n$. Let $\mathcal{L}(t)$ be the linear Lyapunov operator defined by the system $(A_0, \dots, A_r)$ by (2.137) where $A_k(t) = A_k(t, 1)$. The evolution operator $T(t, t_0)$ will be called the *evolution operator on $\mathcal{S}_n$ defined by the system* $(A_0, \dots, A_r)$. The operator (2.137) corresponds to the stochastic linear system (1.24).

Proposition 3.1.3. *If $\mathcal{I} = \mathbf{R}_+$ and $T(t, t_0)$ is the linear evolution operator on $\mathcal{S}_n$ defined by the Lyapunov operator (2.137), then we have the following representation formulae*

$$T(t, t_0)S = E\left[\Phi(t, t_0)S\Phi^T(t, t_0)\right],$$

$$T^*(t, t_0)S = E\left[\Phi^T(t, t_0)S\Phi(t, t_0)\right]$$

for all $t \geq t_0 \geq 0, S \in \mathcal{S}_n$, $\Phi(t, t_0)$ denoting the fundamental matrix solution of the system (1.24).

Proof. The second equality follows directly from Theorem 3.1.1 and the first follows from the second one and the definition of the adjoint operator.   □

B. The case $\mathcal{D} = \mathbf{Z}_+$

Let us assume that the system (1.22) is affected by a standard homogeneous Markov chain with an infinite countable set of states $(\eta(t), P(t), \mathbf{Z}_+)$. In the sequel $\mathfrak{L}(t)$ is the Lyapunov-type operator on $\mathcal{S}_n^\infty$ defined by (2.141) using the coefficients of the system (1.22) and the elements q_{ij} of the generator matrix Q. The following result establishes a relationship between the anticausal linear evolution operator $T^a(t, \tau)$ generated by $\mathfrak{L}(t)$ and the fundamental matrix solution $\Phi(t, t_0)$ of the system (1.22).

Theorem 3.1.4. *Assume that*

(a) *The functions $t \to A_k(t, i)$ are continuous on $\mathbf{R}_+$ uniformly with respect to $i \in \mathbf{Z}_+$ and $\sup_{t \geq 0} \|A_k(t)\|_\infty < \infty, 0 \leq k \leq r$.*
(b) *The elements of the generator matrix Q satisfy the conditions (1.31) and (1.32).*
(c) *The initial distributions of the Markov process satisfy (1.33).*

Under these conditions we have

$$(T^a(\tau,t)H)(i) = E[\Phi^T(t,\tau)H(\eta(t))\Phi(t,\tau)|\eta(\tau) = i] \ \forall i \in \mathbf{Z}_+, \qquad (3.1)$$

$$H = \{H(i)\}_{i \in \mathbf{Z}_+} \in \mathcal{S}_n^\infty, \ t \geq \tau \geq 0.$$

Proof. For $t \geq \tau \geq 0$ consider the linear bounded operators, $\mathcal{V}(t, \tau) : \mathcal{S}_n^\infty \to \mathcal{S}_n^\infty$, defined by

$$(\mathcal{V}(t,\tau)H)(i) = E[\Phi^T(t,\tau)H(\eta(t))\Phi(t,\tau)|\eta(\tau) = i], \ \forall i \in \mathbf{Z}_+, H \in \mathcal{S}_n^\infty. \ (3.2)$$

Let $\tau \geq 0$ be fixed. Applying Itô-type formula (1.35) for $x(t; t_0, x_0) = \Phi(t, t_0)x_0, \ x_0 \in \mathbf{R}^n, K(t) = H \in \mathcal{S}_n^\infty, k(t) = 0, k_0(t) = 0$, one obtains

$$x_0^T[(\mathcal{V}(t,\tau)H)(i)]x_0 = x_0^T H(i)x_0 + x_0^T \left(\int_\tau^t E[\Phi^T(s,\tau)(\mathfrak{L}(s)H)(\eta(s))\Phi(s,\tau)|\eta(\tau) = i]ds \right) x_0$$

for all $i \in \mathbf{Z}_+, \ t \geq \tau, \ x_0 \in \mathbf{R}^n$. Invoking (3.2) and taking into account that x_0, i are arbitrary, we deduce that

$$\mathcal{V}(t,\tau)H = H + \int_\tau^t \mathcal{V}(s,\tau)(\mathfrak{L}(s)H)ds.$$

This shows that $t \to \mathcal{V}(t, \tau)$ verifies the following integral equation on $\mathbf{B}(\mathcal{S}_n^\infty)$

$$\mathcal{V}(t,\tau) = I_{\mathcal{S}_n^\infty} + \int_\tau^t \mathcal{V}(s,\tau)\mathfrak{L}(s)ds \qquad (3.3)$$

for all $t \geq \tau$. On the other hand, from the properties of the linear evolution operators we obtain

$$\frac{\partial}{\partial s}T^a(\tau,s) = T^a(\tau,s)\mathcal{L}(s).$$

Integrating the last equation one obtains

$$T^a(\tau,t) = I_{\mathcal{S}_n^\infty} + \int_\tau^t T^a(\tau,s)\mathcal{L}(s)ds \qquad (3.4)$$

for all $t \geq \tau$. From (3.3) and (3.4) one sees that both $t \to \mathcal{V}(t,\tau)$ and $t \to T^a(\tau,t)$ are two solutions of the same integral equation on $\mathbf{B}(\mathcal{S}_n^\infty)$. From the uniqueness of the solution of the above integral equation we may conclude that

$$\mathcal{V}(t,\tau) = T^a(\tau,t) \qquad (3.5)$$

for all $t \geq \tau$. Since τ was arbitrary chosen we deduce that (3.5) is true for all $t \geq \tau \geq 0$. From (3.5) and (3.2) we conclude that (3.1) is fulfilled and this completes the proof. $\qquad\square$

Remark 3.1.2. (i) The properties of the fundamental matrix solution $\Phi(t,t_0)$ of (1.22) displayed in the Remark 3.1.1 remain true also in the case when the system (1.22) is affected by a standard homogeneous Markov process with an infinite countable number of states.

(ii) Taking $H(i) = I_n, i \in \mathbf{Z}_+$ in (3.1) we obtain $E[\Phi^T(t,t_0)\Phi(t,t_0)|\eta(t_0) = i] = [T^a(t_0,t)J^\infty](i)$, for all $i \in \mathbf{Z}_+$, $t \geq t_0 \geq 0$. Based on (2.8) we get $E[|\Phi(t,t_0)|^2|\eta(t_0) = i] \leq \|T^a(t_0,t)J^\infty\|_\infty$. Further Corollary 2.1.7 (ii), Theorem 2.1.10 and Proposition 2.6.5 yield

$$E[|\Phi(t,t_0)|^2|\eta(t_0) = i] \leq \|T^a(t_0,t)\|$$

for all $i \in \mathbf{Z}_+, t \geq t_0 \geq 0$. This allows us to deduce via Remark 2.2.1 (v) the estimates

$$E[|\Phi(t,t_0)|^2|\eta(t_0) = i] \leq e^{\gamma(t-t_0)} \qquad (3.6)$$

for all $i \in \mathbf{Z}_+$,

$$E[|\Phi(t,t_0)|^2] \leq e^{\gamma(t-t_0)} \qquad (3.7)$$

for all $t \geq t_0 \geq 0$, where $\gamma = sup_{t \geq 0}\|\mathcal{L}(t)\|$.

3.1.2 The Second Representation Theorem

In this subsection we provide some representation formulae of the causal evolution operators $T(t,t_0)$ defined on $\mathcal{S}_n^d$ and $\ell^1(\mathbf{Z}_+, \mathcal{S}_n)$, respectively.

For the beginning we consider the case $\mathcal{D} = \{1, 2, \ldots, d\}$.

Remark 3.1.3. Although in Theorem 3.1.1 we determined a representation formula for the adjoint operator $T^*(t,t_0)$, a representation formula for $T(t,t_0)$ can be also be given, namely

$$(T(t,t_0)H)(j) = \sum_{i=1}^{d} E\left[\Phi(t,t_0)H_i\Phi^T(t,t_0)\chi_{\eta(t)=j} \mid \eta(t_0) = i\right], \qquad (3.8)$$

$t \geq t_0 \geq 0$, $j \in \mathcal{D}$, $H \in \mathcal{S}_n^d$. Indeed, we have for $T = T(t,t_0)$,

$$\langle TH, G \rangle = \langle H, T^*G \rangle = \sum_{i=1}^{d} Tr\left[H_i E\left[\Phi^T(t,t_0)G(\eta(t))\Phi(t,t_0) \mid \eta(t_0) = i\right]\right]$$

$$= \sum_{i=1}^{d}\sum_{j=1}^{d} Tr\left[H_i E\left[\Phi^T(t,t_0)G(j)\Phi(t,t_0)\chi_{\eta(t)=j} \mid \eta(t_0) = i\right]\right]$$

$$= \sum_{i=1}^{d}\sum_{j=1}^{d} E\left[Tr[(H_i\Phi^T(t,t_0)G(j)\Phi(t,t_0))\chi_{\eta(t)=j} \mid \eta(t_0) = i]\right]$$

$$= \sum_{i=1}^{d}\sum_{j=1}^{d} E\left[Tr[(G(j)\Phi(t,t_0)H_i\Phi^T(t,t_0))\chi_{\eta(t)=j} \mid \eta(t_0) = i]\right]$$

$$= \sum_{j=1}^{d} Tr\left[G_j\left(\sum_{i=1}^{d} E\left[\Phi(t,t_0)H_i\Phi^T(t,t_0)\chi_{\eta(t)=j} \mid \eta(t_0) = i\right]\right)\right],$$

from which (3.8) directly follows.

Consider now the case $\mathcal{D} = \mathbf{Z}_+$.

For each $\mathbf{H} = \{H(i)\}_{i \in \mathbf{Z}_+} \in \ell^1(\mathbf{Z}_+, \mathcal{S}_n)$ we define:

$$(\Upsilon\mathbf{H})(i) = \sum_{j=0}^{\infty} E[\Phi(t,t_0)H(j)\Phi^T(t,t_0)\chi_{\{\eta(t)=i\}} \mid \eta(t_0) = j]. \qquad (3.9)$$

We have $|E[\Phi(t,t_0)H(j)\Phi^T(t,t_0)\chi_{\{\eta(t)=i\}} \mid \eta(t_0) = j]| \leq E[|\Phi(t,t_0)|^2\chi_{\{\eta(t)=i\}} \mid \eta(t_0) = j]|H(j)|$. Let $N_1, N_2 \in \mathbf{Z}_+$ be arbitrary but fixed. Invoking (3.6) we may write successively

$$\sum_{i=0}^{N_1} \left| \sum_{j=0}^{N_2} E[\Phi(t,t_0)H(j)\Phi^T(t,t_0)\chi_{\{\eta(t)=i\}}|\eta(t_0)=j] \right| \le$$

$$\sum_{j=0}^{N_2} \sum_{i=0}^{N_1} E\left[|\Phi(t,t_0)|^2\chi_{\{\eta(t)=i\}}|\eta(t_0)=j\right]|H(j)| \le$$

$$\sum_{j=0}^{N_2} E\left[|\Phi(t,t_0)|^2|\eta(t_0)=j\right]|H(j)| \le e^{\gamma(t-t_0)}\sum_{j=0}^{N_2}|H(j)|.$$

Taking the limits for $N_1 \to \infty, N_2 \to \infty$ we get

$$\sum_{i=0}^{\infty} \left| \sum_{j=0}^{\infty} E[\Phi(t,t_0)H(j)\Phi^T(t,t_0)\chi_{\{\eta(t)=i\}}|\eta(t_0)=j] \right| \le e^{\gamma(t-t_0)}\tilde{\|}\mathbf{H}\tilde{\|}_1 \quad (3.10)$$

with $\tilde{\|} \cdot \tilde{\|}_1$ being introduced by (2.11). Thus from (3.9) and (3.10) we obtain that

$$\tilde{\|}\Upsilon\mathbf{H}\tilde{\|}_1 \le e^{\gamma(t-t_0)}\tilde{\|}\mathbf{H}\tilde{\|}_1. \quad (3.11)$$

This allows us to conclude that Υ defined via (3.9) lies in $\mathbf{B}(\ell^1(\mathbf{Z}_+,\mathcal{S}_n))$. Now we are in position to prove a representation formula of the causal linear evolution operator $T(t,t_0)$ defined by the Lyapunov-type operator (2.140).

Proposition 3.1.5. *Assume that*

(a) The assumptions from Theorem 3.1.4 are fulfilled;
(b) The elements q_{ij} of the generator matrix Q satisfy the condition (2.149).

Under these conditions we have

$$(T(t,t_0)\mathbf{H})(i) = \sum_{j=0}^{\infty} E\left[\Phi(t,t_0)H(j)\Phi^T(t,t_0)\chi_{\{\eta(t)=i\}}|\eta(t_0)=j\right] \quad (3.12)$$

for all $i \in \mathbf{Z}_+, \mathbf{H} \in \ell^1(\mathbf{Z}_+,\mathcal{S}_n), t \ge t_0 \ge 0$.

Proof. Let $\mathbf{H} = \{H(i)\}_{i\in\mathbf{Z}_+} \in \ell^1(\mathbf{Z}_+,\mathcal{S}_n)$ and $\mathbf{G} = \{G(i)\}_{i\in\mathbf{Z}_+} \in \ell^2(\mathbf{Z}_+,\mathcal{S}_n)$ be arbitrary but fixed. Invoking Proposition 2.1.3 and Theorem 2.6.7 (ii) we may write

$$\langle\mathbf{G}, T(t,t_0)\mathbf{H}\rangle_2 = \langle T^*(t,t_0)\mathbf{G},\mathbf{H}\rangle_2.$$

Using Corollary 2.6.8 we may infer that

$$\langle\mathbf{G}, T(t,t_0)\mathbf{H}\rangle_2 = \langle T^a(t_0,t)\mathbf{G},\mathbf{H}\rangle_2. \quad (3.13)$$

Based on (2.13) and (3.1) we deduce

$$\langle T^a(t_0,t)\mathbf{G}, \mathbf{H}\rangle_2 = \sum_{j=0}^{\infty} Tr[(T^a(t_0,t)\mathbf{G})(j)H(j)] =$$

$$\sum_{j=0}^{\infty} Tr\left[E[\Phi^T(t,t_0)G(\eta(t))\Phi(t,t_0)|\eta(t_0)=j]H(j)\right]. \tag{3.14}$$

Similarly with the case $\mathcal{D} = \{1,2,\ldots,d\}$ we can show that

$$\begin{aligned}\Sigma_{j=0}^{\infty} Tr\left[E[\Phi^T(t,t_0)G(\eta(t))\Phi(t,t_0)|\eta(t_0)=j]H(j)\right] = \\ \Sigma_{i=0}^{\infty} Tr[G(i)(\Upsilon\mathbf{H})(i)] = \langle \mathbf{G}, \Upsilon\mathbf{H}\rangle_2.\end{aligned} \tag{3.15}$$

Combining (3.13)–(3.15) one obtains

$$\langle \mathbf{G}, T(t,t_0)\mathbf{H}\rangle_2 = \langle \mathbf{G}, \Upsilon\mathbf{H}\rangle_2.$$

Since $\mathbf{G}$ is arbitrary in $\ell^2(\mathbf{Z}_+, \mathcal{S}_n)$ it follows that

$$T(t,t_0)\mathbf{H} = \Upsilon\mathbf{H} \tag{3.16}$$

for all $\mathbf{H} \in \ell^1(\mathbf{Z}_+, \mathcal{S}_n)$. The conclusion follows from (3.9) and (3.16). Thus the proof is complete.   □

Let ζ be a random vector or a random matrix with the property that $E[|\zeta|] < \infty$. Let us define $\tilde{E}_t[\zeta] \triangleq \{E[\zeta\chi_{\{\eta(t)=i\}}]\}_{i\in\mathbf{Z}_+}$. From the inequality $\sum_{i\in\mathbf{Z}_+} |E[\zeta\chi_{\{\eta(t)=i\}}]| \leq E[|\zeta|]$ we deduce that $\tilde{E}_t[\zeta] \in \ell^1(\mathbf{Z}_+, \mathbf{R}^n)$ or $\tilde{E}_t[\zeta] \in \ell^1(\mathbf{Z}_+, \mathbf{R}^{n\times n})$, respectively.

Particularly, for $\zeta = x(t;t_0,x_0)x^T(t;t_0,x_0)$, we have $E[|\zeta|] = E[|x(t;t_0,x_0)|^2] < +\infty$. Hence, $\tilde{E}_t[x(t;t_0,x_0)x^T(t;t_0,x_0)] \in \ell^1(\mathbf{Z}_+, \mathcal{S}_n)$.

Now we prove a representation theorem for the operator $T(t,t_0)$ in the absence of the additional condition (2.149).

Theorem 3.1.6. *Under the assumptions of Theorem 3.1.4 the solution of the system (1.22) satisfies:*

$$\tilde{E}_t[x(t;t_0,x_0)x^T(t;t_0,x_0)] = T(t,t_0)\tilde{E}_{t_0}[x_0 x_0^T], \ \forall t \geq t_0 \geq 0, \ x_0 \in \mathfrak{X}_{t_0}$$

where $\mathfrak{X}_{t_0}$ is the set of n-dimensional random vectors x_0, $\mathcal{H}_{t_0}$-measurable and $E[|x_0|^2] < \infty$.

Proof. Applying the Itô-type formula (1.35) for $x(t;t_0,x_0) = \Phi(t,t_0)x_0$, $x_0 \in \mathfrak{X}_{t_0}, K(t) = H \in \mathcal{S}_n^\infty, k(t) = 0, k_0(t) = 0$, one gets

$$\begin{aligned}E\left[x^T(t;t_0,x_0)H(\eta(t))x(t;t_0,x_0)|\eta(t_0)=i\right] = E[x_0^T H(i)x_0|\eta(t_0)=i] \\ + \int_{t_0}^t E\left[x^T(s;t_0,x_0)(\mathcal{L}(s)H(s))(\eta(s))x(s;t_0,x_0)|\eta(t_0)=i\right]ds\end{aligned} \tag{3.17}$$

for all $i \in \mathbf{Z}_+$, $t \geq t_0 \geq 0$, $x_0 \in \mathfrak{X}_{t_0}$, where $\mathfrak{L}(s) : \mathcal{S}_n^\infty \to \mathcal{S}_n^\infty$ is defined as in (2.141).

Taking the expectation in (3.17) it follows that

$$
\begin{aligned}
E\left[x^T(t;t_0,x_0)H(\eta(t))x(t;t_0,x_0)\right] &= E[x_0^T H(\eta(t_0))x_0] \\
&+ \int_{t_0}^t E\left[x^T(s;t_0,x_0)(\mathfrak{L}(s)H)(\eta(s))x(s;t_0,x_0)\right] ds
\end{aligned}
\tag{3.18}
$$

for all $t \geq t_0 \geq 0$, $H \in \mathcal{S}_n^\infty$, $x_0 \in \mathfrak{X}_{t_0}$.

Let $x \in \mathbf{R}^n$, $i_0 \in \mathbf{Z}_+$ be arbitrary but fixed. Let $H_{i_0} \in \mathcal{S}_n^\infty$ be defined by $H_{i_0}(i) = 0$ if $i \neq i_0$ and $H_{i_0}(i) = xx^T$ if $i = i_0$. In this case, (3.18) becomes,

$$
\begin{aligned}
x^T E\left[x(t;t_0,x_0)x^T(t;t_0,x_0)\chi_{\{\eta(t)=i_0\}}\right]x &= x^T E[x_0 x_0^T \chi_{\{\eta(t_0)=i_0\}}]x \\
&+ \int_{t_0}^t E[x^T(s;t_0,x_0)(\mathfrak{L}(s)H_{i_0})(\eta(s))x(s;t_0,x_0)]ds.
\end{aligned}
\tag{3.19}
$$

By direct calculations, based on (2.140)–(2.141) as well as by the special form of H_{i_0} one obtains that

$$
E\left[x^T(s;t_0,x_0)(\mathfrak{L}(s)H_{i_0})(\eta(s))x(s;t_0,x_0)\right] = x^T\left(\mathcal{L}(s)\tilde{E}_s\left[x(s;t_0,x_0)x^T(s;t_0,x_0)\right]\right)(i_0)x.
\tag{3.20}
$$

Plugging (3.20) in (3.19) and taking into account that x, i_0 were arbitrarily chosen, we obtain:

$$
\tilde{E}_t[x(t;t_0,x_0)x^T(t;t_0,x_0)] = \tilde{E}_{t_0}[x_0 x_0^T] + \int_{t_0}^t \mathcal{L}(s)\tilde{E}_s[x(s;t_0,x_0)x^T(s;t_0,x_0)]ds.
$$

This shows that $t \to \tilde{E}_t[x(t;t_0,x_0)x^T(t;t_0,x_0)]$ solves the linear differential equation (2.147). Thus it is the representation formula from the statement and therefore the proof is complete. $\qquad \square$

3.1.3 The Third Representation Theorem

In this subsection $\mathcal{D} = \{1,2,\ldots,d\}$.

Throughout this monograph $(\mathbf{R}^n)^d$ stands for the direct product

$$
(\mathbf{R}^n)^d := \underbrace{\mathbf{R}^n \times \cdots \times \mathbf{R}^n}_{d},
$$

that is $y \in (\mathbf{R}^n)^d$ if and only if $y = (y(1),\ldots,y(d))$, $y(i) \in \mathbf{R}^n$, $i \in \mathcal{D}$.

We consider the inner product on $(\mathbf{R}^n)^d$

$$\langle y, z \rangle = \sum_{i=1}^{d} y^T(i) z(i)$$

for all $y = (y(1), \ldots, y(d))$ and $z = (z(1), \ldots, z(d))$ in $(\mathbf{R}^n)^d$.

By $\|y\|$ we denote the norm defined by:

$$\|y\|^2 = \langle y, y \rangle = \sum_{i=1}^{d} |y(i)|^2.$$

If $T : (\mathbf{R}^n)^d \to (\mathbf{R}^n)^d$ is a linear operator, then $\|T\|$ stands for the operator norm induced by the considered norm in $(\mathbf{R}^n)^d$.

Let $A : \mathbf{R}_+ \to \mathcal{M}_n^d$ be a bounded and continuous function, that is

$$A(t) = (A(t,1), \ldots, A(t,d)), t \in \mathbf{R}_+.$$

For each $t \geq 0$ we define the linear operator $M(t) : (\mathbf{R}^n)^d \to (\mathbf{R}^n)^d$ by

$$(M(t)y)(i) = A(t,i)y(i) + \sum_{j=1}^{d} q_{ji} y(j), i \in \mathcal{D} \tag{3.21}$$

$y = (y(1), \ldots, y(d)) \in (\mathbf{R}^n)^d$, $Q = (q_{ij}) \in \mathbf{R}^{d \times d}$ satisfies the conditions $q_{ij} \geq 0$ for $i \neq j$ and $\sum_{j=1}^{d} q_{ij} = 0$. It is easy to check that for each $t \geq 0$, $M(t)$ is a linear and bounded operator on the Hilbert space $(\mathbf{R}^n)^d$ and $t \to \|M(t)\|$ is a bounded function.

Let us consider the linear differential equation on $(\mathbf{R}^n)^d$:

$$\frac{d}{dt} y(t) = M(t)y(t). \tag{3.22}$$

Let $R(t,t_0)$ be the linear evolution operator associated with (3.22), that is

$$\frac{d}{dt} R(t,t_0) = M(t)R(t,t_0), R(t_0,t_0)y = y$$

for all $t, t_0 \geq 0, y \in (\mathbf{R}^n)^d$.

By $M^*(t)$ and $R^*(t,t_0)$ we denote the adjoint operators of $M(t)$ and $R(t,t_0)$, respectively, on $(\mathbf{R}^n)^d$. One can easily see that

$$(M^*(t)y)(i) = A^T(t,i)y(i) + \sum_{j=1}^{d} q_{ij} y(j), i \in \mathcal{D}, y \in (\mathbf{R}^n)^d$$

$$\frac{d}{dt} R^*(t,t_0) = R^*(t,t_0)M^*(t) \tag{3.23}$$

$$\frac{d}{dt} R^*(s,t) = -M^*(t)R^*(s,t)$$

for all $t, s \in \mathbf{R}_+$. The operator $R(t,t_0)$ will be termed the *evolution operator on* $(\mathbf{R}^n)^d$ *defined by the pair* (A, Q).

The next result provides the relationship between the evolution operator $R(t, t_0)$ and the fundamental matrix solution $\Phi(t, t_0)$ of the stochastic system (1.23).

Proposition 3.1.7. *Under the assumptions given at the beginning of the section, the following equality holds*

$$(R^*(t, t_0)y)(i) = E\left[\Phi^T(t, t_0)y(\eta(t)) \mid \eta(t_0) = i\right], t \geq t_0 \geq 0$$

$i \in \mathcal{D}, y = (y(1), \ldots, y(d)) \in (\mathbf{R}^n)^d.$

Proof. Let $t \geq t_0 \geq 0$ and the operator $V(t, t_0) : (\mathbf{R}^n)^d \to (\mathbf{R}^n)^d$ be defined by

$$(V(t, t_0)y)(i) = E\left[\Phi^T(t, t_0)y(\eta(t)) \mid \eta(t_0) = i\right],$$

$i \in \mathcal{D}, y = (y(1), \ldots, y(d)) \in (\mathbf{R}^n)^d.$ Let y be fixed and consider the function $v : \mathbf{R}^n \times \mathcal{D} \to \mathbf{R}$ by

$$v(x, i) = x^T y(i).$$

Applying the Itô-type formula (Theorem 1.10.2) to the function v and to the system (1.23), we obtain:

$$E[v(x(t), \eta(t)) \mid \eta(t_0) = i] - x_0^T y(i)$$

$$= E\left[\int_{t_0}^t x^T(s)\left(A_0^T(s, \eta(s))y(\eta(s)) + \sum_{j=1}^d q_{\eta(s)j}y(j)\right)ds \mid \eta(t_0) = i\right]$$

where $x(s) = \Phi(s, t_0)x_0$. Further, we write

$$x_0^T(V(t, t_0)y)(i) - x_0^T y(i) = x_0^T \int_{t_0}^t (V(s, t_0)M^*(s)y)(i)\,ds$$

for all $t \geq t_0 \geq 0$, $x_0 \in \mathbf{R}^n$, $i \in \mathcal{D}$. Therefore we may conclude that

$$V(t, t_0)y - y = \int_{t_0}^t V(s, t_0)M^*(s)y\,ds$$

for all $t \geq t_0$ and $y \in (\mathbf{R}^n)^d$.

By differentiation, we deduce that

$$\frac{d}{dt}V(t, t_0)y = V(t, t_0)M^*(t)y$$

for all $y \in (\mathbf{R}^n)^d$. Hence

$$\frac{d}{dt}V(t, t_0) = V(t, t_0)M^*(t), t \geq t_0.$$

Since $V(t_0, t_0) = R^*(t_0, t_0)$, from (3.23) it results that $V(t, t_0) = R^*(t, t_0)$, for all $t \geq t_0 \geq 0$ and the proof ends. $\square$

3.2   Mean Square Exponential Stability

In this section we introduce the concept of mean square exponential stability of
the zero-solution of the stochastic linear differential equations of type (1.22) and
we also give necessary and sufficient conditions ensuring this kind of stability. The
results proved in this section extend the ones corresponding to the particular cases
for the systems (1.23) and (1.24), respectively.

For the beginning we do not separate the case $\mathcal{D} = \{1, 2, \ldots, d\}$ and the case
$\mathcal{D} = \mathbf{Z}_{+}$.

Definition 3.2.1. We say that the zero state equilibrium of the system of stochastic
linear differential equations (1.22) is

(i) *Exponentially stable in mean square with conditioning (ESMS-C), if there exist*
    $\beta \geq 1$, $\alpha > 0$ such that

$$E[|\Phi(t, t_0)x_0|^2 \,|\, \eta(t_0) = i] \leq \beta e^{-\alpha(t-t_0)} |x_0|^2 \qquad (3.24)$$

   for all $t \geq t_0 \geq 0$, $x_0 \in \mathbf{R}^n$, $i \in \mathcal{D}$ and for every admissible initial distribution π_0
   of the Markov process;
(ii) *Exponentially stable in mean square (ESMS), if there exist* $\beta \geq 1$, $\alpha > 0$, *such
    that*

$$E[|\Phi(t, t_0)x_0|^2] \leq \beta e^{-\alpha(t-t_0)} |x_0|^2 \qquad (3.25)$$

   for all $t \geq t_0 \geq 0$, $x_0 \in \mathbf{R}^n$, and for any admissible initial distribution π_0 of the
   Markov process.

Remark 3.2.1. Since (1.22) and its special forms (1.23)–(1.25) are stochastic linear
differential equations, the ESMS of the zero state equilibrium is equivalent with
the exponential stability of every solution. Hence, both the properties of ESMS-C
and ESMS characterize the full system of differential equations, not only the zero
solution. Hence in the sequel we shall say that the system (1.22) is ESMS-C or
ESMS, respectively, if (3.24) or (3.25), respectively, are fulfilled.

Proposition 3.2.1. *The following are equivalent*

 (i) *The system (1.22) defines an ESMS-C evolution;*
(ii) *There exist* $\beta \geq 1$, $\alpha > 0$ *such that*

$$E\left[|\Phi(t, t_0)|^2 \,|\, \eta(t_0) = i\right] \leq \beta e^{-\alpha(t-t_0)}, \, t \geq t_0 \geq 0, \, i \in \mathcal{D} \,;$$

(iii) There exist $\beta_1 \geq 1$, $\alpha_1 > 0$ *such that*

$$E\left[|\Phi(t, t_0)|^2 \,|\, \eta(t_0)\right] \leq \beta_1 e^{-\alpha_1(t-t_0)}, \, a.s., \, t \geq t_0 \geq 0;$$

(iv) There exist $\tilde{\beta} \geq 1$, $\tilde{\alpha} > 0$ such that

$$E\left[|\Phi(t,t_0)\xi|^2 \mid \eta(t_0) = i\right] \leq \tilde{\beta}e^{-\tilde{\alpha}(t-t_0)}E\left[|\xi|^2 \mid \eta(t_0) = i\right],$$

$t \geq t_0 \geq 0$, $i \in \mathcal{D}$, and ξ is any random vector $\mathcal{H}_{t_0}$-measurable and $E\left[|\xi|^2\right] < \infty$.

Proof. (i)$\Longleftrightarrow$ (ii), (iii)$\Rightarrow$(ii) and (iv)$\Rightarrow$(i) are obvious.

We prove now the implication (i) $\Rightarrow$ (iii). Let $e_1,\ldots,e_n$ be the canonical basis in $\mathbf{R}^n$, that is $e_k = (0,\ldots,0,1,0,\ldots,0)^T$ with 1 being the k-th element. From the inequality

$$|\Phi(t,t_0)|^2 \leq \sum_{k=1}^{n} |\Phi(t,t_0)e_k|^2$$

we deduce that

$$E\left[|\Phi(t,t_0)|^2 \mid \eta(t_0)\right] \leq \sum_{k=1}^{n}\left[|\Phi(t,t_0)e_k|^2 \mid \eta(t_0)\right].$$

Since $\eta(t_0)$ has either a finite set of states or an infinite countable set of states we have

$$E\left[|\Phi(t,t_0)|^2 \mid \eta(t_0)\right] \leq \sum_{k=1}^{n}\sum_{j\in\mathcal{D}} \chi_{\eta(t_0)=j}E\left[|\Phi(t,t_0)e_k|^2 \mid \eta(t_0) = j\right] \text{ a.s.}$$

Using (3.24) we can write

$$E\left[|\Phi(t,t_0)|^2 \mid \eta(t_0)\right] \leq \beta\sum_{k=1}^{n}\sum_{j\in\mathcal{D}} \chi_{\eta(t_0)=j}e^{-\alpha(t-t_0)}|e_k|^2$$

$$= \beta n e^{-\alpha(t-t_0)} \text{ a.s.}$$

which shows that (iii) holds. Now we prove (iii) $\Rightarrow$(iv). Let ξ be an arbitrary random vector $\mathcal{H}_{t_0}$-measurable and $E\left[|\xi|^2\right] < \infty$. From the inequality

$$|\Phi(t,t_0)\xi|^2 \leq |\Phi(t,t_0)|^2|\xi|^2$$

we deduce that

$$E\left[|\Phi(t,t_0)\xi|^2 \mid \mathcal{H}_{t_0}\right] \leq E\left[|\Phi(t,t_0)|^2|\xi|^2 \mid \mathcal{H}_{t_0}\right]$$

$$= |\xi|^2 E\left[|\Phi(t,t_0)|^2 \mid \mathcal{H}_{t_0}\right].$$

Since the components of $\Phi(t,t_0)$ are measurable with respect to $\eta(s), w_j(s), t_0 \leq s \leq t, j = 1, \ldots, r$, it follows that we may apply Theorem 1.10.1 and we get

$$E\left[|\Phi(t,t_0)\xi|^2 \mid \mathcal{H}_{t_0}\right] \leq |\xi|^2 E\left[|\Phi(t,t_0)|^2 \mid \eta(t_0)\right] \text{ a.s.}$$

Using (iii) we deduce that

$$E\left[|\Phi(t,t_0)\xi|^2 \mid \mathcal{H}_{t_0}\right] \leq \beta_1 e^{-\alpha_1(t-t_0)} |\xi|^2, \text{ a.s., } t \geq t_0 \geq 0.$$

Further one easily deduces that

$$E\left[|\Phi(t,t_0)\xi|^2 \mid \eta(t_0) = i\right] \leq \beta_1 e^{-\alpha_1(t-t_0)} E\left[|\xi|^2 \mid \eta(t_0) = i\right]$$

for all $t \geq t_0 \geq 0$, $i \in \mathcal{D}$, and the proof ends. $\qquad \square$

Remark 3.2.2. (i) In the particular case when the considered system of stochastic differential equations is of type (1.24), the two types of mean square exponential stability introduced in Definition 3.2.1 reduce to

$$E\left[|\Phi(t,t_0)x_0|^2\right] \leq \beta e^{-\alpha(t-t_0)} |x_0|^2 \qquad (3.26)$$

for all $t \geq t_0$, $x_0 \in \mathbf{R}^n$.

(ii) From (3.24) and (3.25) it follows that if the system (1.22) is ESMS-C, then it is ESMS. However, we can notice that in the presence of Markovian perturbations in the system, the reverse implication is not always true. We shall see later (see Theorems 3.2.4 and 3.2.5) that ESMS-C is equivalent to ESMS in the case $\mathcal{D} = \{1,2,\ldots,d\}$ and the system (1.22) is either in the special case of (1.25) or it has periodic coefficients.

(iii) In the time-invariant case, based on Remark 3.1.1 (ii) we obtain that the system (1.25) defines an ESMS-C evolution if and only if there exist $\beta \geq 1$, $\alpha > 0$ such that

$$E\left[|\Phi(t,0)x_0|^2 \mid \eta(0) = i\right] \leq \beta e^{-\alpha t} |x_0|^2$$

for all $t \geq 0$, $i \in \mathcal{D}$, $x_0 \in \mathbf{R}^n$.

The next result emphasizes the relationship between the ESMS with conditioning ESMS-C of the system (1.22) and the exponential stability of Lyapunov-type differential equations on $\mathcal{S}_n^\mathcal{D}$.

Theorem 3.2.2. (a) If $\mathcal{D} = \{1,2,\ldots,d\}$, the following statements are equivalent:

(i) The system (1.22) defines an ESMS-C evolution;

(ii) The linear evolution operator $T(t,t_0)$ defined by the corresponding Lyapunov-type equation (2.125) satisfies (2.156) for some $\beta \geq 1$ and $\alpha > 0$, that is the system $(A_0, A_1, \ldots, A_r; Q)$ is stable.

(b) If $\mathcal{D} = \mathbf{Z}_+$, the following are equivalent:

(j) The system (1.22) defines an ESMS-C evolution;

(jj) The anticausal linear evolution operator $T^a(t,\tau)$ defined by the corresponding Lyapunov-type differential equation (2.147) associated with (1.22) satisfies an estimate of type (2.172) for some $\beta \geq 1, \alpha > 0$, that is the system $(A_0, A_1, \ldots, A_r; Q)$ is stable.

Proof. (a) follows from the representation Theorem 3.1.1 and Remark 2.7.1 (i). The equivalences (j) $\leftrightarrow$ (jj) from (b) follow from the representation Theorem 3.1.4 and Definition 2.8.1. □

Corollary 3.2.3. *The following are equivalent:*

(i) The system (1.22) defines an ESMS-C evolution;

(ii) There exists $\delta > 0$ not depending upon $t \in \mathbf{R}_+, x_0 \in \mathbf{R}^n$ and an initial distribution π_0, such that

$$E\left[\int_t^\infty |\Phi(s,t)x_0|^2 \, ds \mid \eta(t) = i\right] \leq \delta |x_0|^2$$

for all $t \geq 0$ and $x_0 \in \mathbf{R}^n$;

Proof. If $\mathcal{D} = \{1, 2, \ldots, d\}$ the equivalence follows combining Theorem 3.2.2 (a), Theorem 2.7.4 and the representation Theorem 3.1.1. In the case $\mathcal{D} = \mathbf{Z}_+$ one applies Theorem 3.2.2 (b), Theorem 2.8.2 together with the representation Theorem 3.1.4. □

The following result shows that in the time-invariant case the ESMS with conditioning is equivalent to the ESMS and with a type of attractivity of the zero solution.

Theorem 3.2.4. *Assume that $\mathcal{D} = \{1, 2, \ldots, d\}$, then the following assertions are equivalent:*

(i) The system (1.25) defines an ESMS-C evolution;

(ii) The system (1.25) defines an ESMS evolution;

(iii)

$$\lim_{t \to \infty} E\left[|x(t)|^2\right] = 0$$

for any solution $x(t)$ of the system (1.25) with $x(0) = x_0, x_0 \in \mathbf{R}^n$;

(iv)

$$\lim_{t \to \infty} E\left[x(t)x^T(t)\right] = 0$$

for all solution $x(t)$ of (1.25) as above;

(v)

$$\lim_{t \to \infty} E\left[\Phi^T(t,0)\Phi(t,0)\right] = 0.$$

Proof. The implications (i) $\Rightarrow$ (ii) $\Rightarrow$ (iii) directly follow from Remark 3.2.2 (ii) and (iii). (iii) $\Rightarrow$ (iv) follows from the inequality

$$0 \leq x(t)x^T(t) \leq |x(t)|^2 I_n.$$

(iv) $\Rightarrow$ (iii) follows from

$$|x(t)|^2 = Tr\left[x(t)x^T(t)\right].$$

(iii) $\Rightarrow$ (v) easily follows using the identity

$$E\left[x^T \Phi^T(t,0)\Phi(t,0)y\right] = \frac{1}{4}\left\{E\left[|\Phi(t,0)(x+y)|^2\right] - E\left[|\Phi(t,0)(x-y)|^2\right]\right\}$$
(3.27)

for all $x, y \in \mathbf{R}^n$.

It remains to prove that (v)$\Rightarrow$(i). Since $P(\eta(0) = i) > 0$, $i \in \mathcal{D}$, then from (v) we have

$$\lim_{t \to \infty} E\left[\Phi^T(t,0)\Phi(t,0) \mid \eta(0) = i\right] = 0, i \in \mathcal{D}.$$

Based on Theorem 3.1.1 and Remark 2.6.3 (i), the above equality gives:

$$\lim_{t \to \infty}\left(e^{\mathcal{L}^* t} J^d\right)(i) = 0, i \in \mathcal{D},$$

and therefore $\lim_{t \to \infty}\left|e^{\mathcal{L}^* t} J^d\right| = 0$. Applying Corollary 2.6.2 we conclude that $\lim_{t \to \infty}\left\|e^{\mathcal{L}^* t}\right\| = 0$. Further from (2.93) in the special case $\xi = J^d$ we obtain that

$$\lim_{t \to \infty}\left\|e^{\mathcal{L} t}\right\| = 0.$$
(3.28)

Since $\mathcal{L}$ is a linear operator on a finite dimensional Hilbert space, from (3.28) we deduce that the eigenvalues of the operator $\mathcal{L}$ are located in the half plane $\mathbf{C}^-$, and hence there exists $\beta \geq 1$, $\alpha > 0$ such that $\left\|e^{\mathcal{L} t}\right\| \leq \beta e^{-\alpha t}$. Invoking again (2.93) we get $\left\|e^{\mathcal{L}^* t}\right\| \leq \beta_1 e^{-\alpha t}$ for all $t \geq 0$. Combining Corollaries 2.1.7 (i) and 2.6.2 we deduce $\left|(e^{\mathcal{L}^* t} J^d)(i)\right| \leq \left|e^{\mathcal{L}^* t} J^d\right| \leq \beta_1 e^{-\alpha t}$ for all $t \geq 0$. Finally, applying Theorems 3.1.1 we obtain that (3.24) is fulfilled. $\square$

In the case of periodic coefficients we obtain the following analogous result:

Theorem 3.2.5. *Assume that $\mathcal{D} = \{1, 2, \ldots, d\}$ and $t \longmapsto A_k(t,i)$, $k = 0, \ldots, r$ are θ-periodic and continuous functions. Then the following are equivalent:*

(i) The system (1.22) defines an ESMS-C evolution;
(ii) The system (1.22) defines an ESMS evolution;

(iii)

$$\lim_{p\to\infty} E\left[|x(p\theta)|^2\right] = 0$$

for all solutions $x(t)$ of (1.22) with $x(0) = x_0$, $x_0 \in \mathbf{R}^n$;
(iv)

$$\lim_{p\to\infty} E\left[x(p\theta)x^T(p\theta)\right] = 0$$

for any solution $x(t)$ of (1.22) as above;
(v)

$$\lim_{p\to\infty} E\left[\Phi^T(p\theta,0)\Phi(p\theta,0)\right] = 0.$$

Proof. The implications (i) $\Rightarrow$ (ii) $\Rightarrow$ (iii) and the equivalence (iii) $\Longleftrightarrow$ (iv) are similar with the proof of Theorem 3.2.4;

(iii) $\Rightarrow$ (v) immediately follows from (3.27) and the Remark 3.1.1 (i). We prove (v) $\Rightarrow$ (i). If (v) is fulfilled then

$$\lim_{p\to\infty} E\left[\Phi^T(p\theta,0)\Phi(p\theta,0) \mid \eta(0) = i\right] = 0, i \in \mathcal{D}.$$

Using Theorem 3.1.1 we obtain

$$\lim_{p\to\infty} \left(T^*(p\theta,0)J^d\right)(i) = 0, i \in \mathcal{D}$$

and therefore

$$\lim_{p\to\infty} \left|\left(T^*(p\theta,0)J^d\right)(i)\right| = 0, i \in \mathcal{D}$$

which leads to

$$\lim_{p\to\infty} \left|T^*(p\theta,0)J^d\right| = 0.$$

Based on Corollary 2.6.2 we deduce that

$$\lim_{p\to\infty} \|T^*(p\theta,0)\| = 0.$$

Using (2.93) in the case of the Hilbert space $\mathcal{S}_n^d$ and $\xi = J^d$ we get

$$\lim_{p\to\infty} \|T(p\theta,0)\| = 0$$

which is equivalent with

$$\lim_{p \to \infty} \|(T(\theta,0))^p\| = 0, \qquad (3.29)$$

$T(\theta,0)$ being the monodromy operator associated with the differential equation (2.125). From (3.29) we deduce that the eigenvalues of $T(\theta,0)$ are inside of the unit disk $|\lambda| < 1$ or equivalently, $\rho(T\theta,0) < 1$. Using the implication (x) $\Rightarrow$ (i) from Theorem 2.7.6 we deduce that the zero solution of (2.125) is exponentially stable. This fact implies, via Theorem 3.2.2 (i), that (1.22) defines an ESMS-C evolution. The proof is complete.   □

In the case of systems of stochastic linear differential equations (1.22) with periodic coefficients but perturbed by a Markov chain with an infinite countable set of states, we have:

Theorem 3.2.6. *Assume that:*

(a) $\mathcal{D} = \mathbf{Z}_+$;
(b) *The assumptions from Theorem 3.1.4 are fulfilled;*
(c) *there exists $\theta > 0$ such that $A_k(t+\theta,i) = A_k(t,i)$, for all $0 \le k \le r$, $(t,i) \in \mathbf{R}_+ \times \mathbf{Z}_+$.*

Under these conditions, the following statements are equivalent:

(i) *The system (1.22) is ESMS-C;*
(ii) $\lim_{t \to \infty} E[|x(t;t_0,x_0)|^2|\eta_{t_0} = i] = 0$ *uniformly with respect to $i \in \mathbf{Z}_+$ for all $t_0 \ge 0$, $x_0 \in \mathbf{R}^n$ and every admissible distribution π_0;*
(iii) *There exists an admissible initial distribution $\tilde{\pi}_0$ such that for all $x_0 \in \mathbf{R}^n$*

$$\lim_{k \to \infty} E[|x(k\theta;0,x_0)|^2|\eta(0) = i] = 0$$

uniformly with respect to $i \in \mathbf{Z}_+$;
(iv) $\rho[T^a(0,\theta)] < 1$.

Proof. The implications (i) $\to$ (ii) $\to$ (iii) are obvious. Let us prove the implication (iii) $\to$ (iv). Using (3.1) we deduce that if (iii) is true then

$$\lim_{k \to \infty} x_0^T[(T^a(0,k\theta)\xi)(i)]x_0 = 0 \qquad (3.30)$$

uniformly with respect to $i \in \mathbf{Z}_+$ for all $x_0 \in \mathbf{R}^n$. Taking $x = e_l, y = e_j, l,j \in \{1,2,\ldots,n\}$ in the identity

$$4x^T[(T^a(0,k\theta)\xi)(i)]y = (x+y)^T[(T^a(0,k\theta)\xi)(i)](x+y) - (x-y)^T[(T^a(0,k\theta)\xi)(i)](x-y)$$

(e_l, e_j being vectors of the canonical base of $\mathbf{R}^n$) we deduce via (3.30) that

$$\lim_{k \to \infty} |(T^a(0,k\theta)\xi)(i)| = 0 \qquad (3.31)$$

uniformly with respect to $i \in \mathbf{Z}_+$. Combining (2.8) and Theorem 2.1.10 it follows that (3.31) yields

$$\lim_{k \to \infty} \|(T^a(0,k\theta))\| = 0. \tag{3.32}$$

Based on the uniqueness of the anticausal evolution operator as well as on the periodicity property of the coefficients one may prove inductively that

$$T^a(s+k\theta, t+k\theta) = T^a(s,t), \qquad t,s \in \mathbf{R}_+, \quad k \in \mathbf{Z}_+ \tag{3.33}$$

and

$$T^a(0,k\theta) = (T^a(0,\theta))^k, \quad k \in \mathbf{Z}_+. \tag{3.34}$$

Using (3.34) we obtain that (3.32) is equivalent with

$$\lim_{k \to \infty} \|(T^a(0,\theta))^k\| = 0. \tag{3.35}$$

Hence by [55] $\rho[T^a(0,\theta)] < 1$. This shows that (iv) is true. It remains to prove the implication (iv) $\to$ (i). If (iv) is fulfilled, then one shows, in a standard way (see the proof of Corollary 2.3.8), that there exist $\beta \geq 1$, $\alpha > 0$ such that

$$\|T^a(\tau,t)\| \leq \beta e^{-\alpha(t-\tau)}, \; \forall t \geq \tau \geq 0.$$

To this end, (3.33) and (3.34) are repeatedly used. Applying (jj) $\to$ (j) from Theorem 3.2.2 (b) we obtain that the system (1.22) is ESMS-C and thus the proof is complete. □

Remark 3.2.3. The statements (i)–(iii) of the previous Theorem are still valid (with $\theta = 1$) in the case $A_k(t,i) = A_k(i)$, $0 \leq k \leq r$, $(t,i) \in \mathbf{R}_+ \times \mathbf{Z}_+$. In this case, the statement (iv) is replaced with:

(iv') The spectrum of $\mathfrak{L}$ lies in the half plane $\mathbf{C}_- = \{z \in \mathbf{C}, \; Rez < 0\}$.

If $A_k(t,i) = 0, 1 \leq k \leq r$, $(t,i) \in \mathbf{R}_+ \times \mathbf{Z}_+$, we obtain the following special form of (1.22):

$$\frac{d}{dt}x(t) = A_0(t,\eta(t))x(t), \; t \geq 0. \tag{3.36}$$

Let $\mathfrak{L}_0(t) : S_n^\infty \to S_n^\infty$ be defined by

$$(\mathfrak{L}_0(t)H)(i) = A_0^T(t,i)H(i) + H(i)A_0(t,i) + \sum_{j=0}^\infty q_{ij}H(j) \tag{3.37}$$

for all $i \in \mathbf{Z}_+$, $H = \{H(i)\}_{i \in \mathbf{Z}_+} \in \mathcal{S}_n^\infty$. Since $\mathfrak{L}_0(t)$ is obtained from $\mathfrak{L}(t)$ taking $A_k(t,i) = 0$, $1 \leq k \leq r$, we deduce that $\mathfrak{L}_0(t)$ defines also an anticausal positive evolution on $\mathcal{S}_n^\infty$. Moreover $\mathfrak{L}_0(t) \leq \mathfrak{L}(t)$, $\forall t \geq 0$. Hence, we deduce, via Proposition 2.3.11 (ii), that the operator valued function $\mathfrak{L}_0(\cdot)$ generates an anticausal exponentially stable evolution if the operator valued function $\mathfrak{L}(\cdot)$ generates an anticausal exponentially stable evolution.

If $\Phi_0(t,t_0)$, $t \geq t_0 \geq 0$ is the fundamental random matrix solution of the system (3.36), then the following representation formula holds:

$$E[\Phi_0^T(t,t_0)H(\eta(t))\Phi_0(t,t_0)|\eta(t_0) = i] = (T_0^a(t_0,t)H)(i) \tag{3.38}$$

for all $t \geq t_0 \geq 0$, $i \in \mathbf{Z}_+$, $H \in \mathcal{S}_n^\infty$, $T_0^a(t,\tau)$ being the anticausal evolution operator on $\mathcal{S}_n^\infty$ defined by the operator valued function $\mathfrak{L}_0(\cdot)$.

The concepts of ESMS-C and ESMS introduced via Definition 3.2.1 for the system (1.22) can be specialized to the system (3.36), replacing $\Phi(t,t_0)$ by $\Phi_0(t,t_0)$ in (3.24), (3.25), respectively.

Based on Theorem 3.2.2 (b) and Proposition 2.3.11 (ii) one obtains:

Corollary 3.2.7. *Assume that: (a)* $\mathcal{D} = \mathbf{Z}_+$;
(b) The assumptions in Theorem 3.1.4 are fulfilled.
Then the system (3.36) is ESMS-C if the system (1.22) is ESMS-C.

The converse implication of the ones from Corollary 3.2.7 is not always true.

To obtain conditions under which these assertions become equivalent, one may use the developments from Sect. 2.5 applied to the operator valued functions $\mathfrak{L}_0(\cdot)$ and $\mathfrak{L}(\cdot)$.

Further we consider the case of system (1.22) affected by a Markov chain with a finite number of states.

Theorem 3.2.8. *Assume that the system (1.23) defines an ESMS-C evolution; then there exist* $\beta \geq 1$ *and* $\alpha > 0$ *such that* $\|R(t,t_0)\| \leq \beta e^{-\alpha(t-t_0)}$ *for all* $t \geq t_0 > 0$, $R(t,t_0)$ *being the linear evolution operator on* $(\mathbf{R}^n)^d$ *defined by the differential equation (3.22).*

Proof. Let $y = (y(1),\ldots,y(d)) \in (\mathbf{R}^n)^d$; then we have

$$\left| E\left[\Phi^T(t,t_0)y(\eta(t)) \mid \eta(t_0) = i\right]\right|^2 \tag{3.39}$$

$$\leq E\left[\left|\Phi^T(t,t_0)\right|^2 \mid \eta(t_0) = i\right] E\left[\left|y(\eta(t))\right|^2 \mid \eta(t_0) = i\right]$$

$t \geq t_0 \geq 0$. On the other hand

$$E\left[\left|y(\eta(t))\right|^2 \mid \eta(t_0) = i\right]$$

$$= \sum_{j=1}^d E\left[\chi_{\eta(t)=j} \mid \eta(t_0) = i\right]\left|y(j)\right|^2$$

$$= \sum_{j=1}^{d} p_{ij}(t-t_0)|y(j)|^2 \le \sum_{j=1}^{d}|y(j)|^2 = \|y\|^2.$$

Thus (3.39) leads to

$$\left| E\left[\Phi^T(t,t_0)y(\eta(t)) \mid \eta(t_0)=i\right]\right|^2$$

$$\le E\left[\left|\Phi^T(t,t_0)\right|^2 \mid \eta(t_0)=i\right]^2 \|y\|^2.$$

Since the system (1.23) defines an ESMS-C evolution and $\left|\Phi^T(t,t_0)\right| = \left|\Phi(t,t_0)\right|$, there exist $\beta \ge 1$, $\alpha > 0$ such that

$$E\left[\left|\Phi^T(t,t_0)\right|^2 \mid \eta(t_0)=i\right]^2 \le \beta e^{-\alpha(t-t_0)}.$$

Therefore

$$\left| E\left[\Phi^T(t,t_0)y(\eta(t)) \mid \eta(t_0)=i\right]\right|^2 \le \beta e^{-\alpha(t-t_0)}\|y\|^2$$

for all $t \ge t_0 \ge 0$. Based on Proposition 3.1.7 we deduce that

$$\left|(R^*(t,t_0)y)(i)\right|^2 \le \beta e^{-\alpha(t-t_0)}\|y\|^2.$$

Hence

$$\|R^*(t,t_0)y\|^2 = \sum_{i=1}^{d}\left|(R^*(t,t_0)y)(i)\right|^2$$

$$\le d\beta e^{-\alpha(t-t_0)}\|y\|^2$$

which gives

$$\|R^*(t,t_0)\| \le \sqrt{d\beta}\, e^{-\frac{\alpha}{2}(t-t_0)}$$

for all $t \ge t_0 \ge 0$. Since $\|R^*(t,t_0)\| = \|R(t,t_0)\|$ we conclude that

$$\|R(t,t_0)\| \le \sqrt{d\beta}\, e^{-\frac{\alpha}{2}(t-t_0)}.$$

Thus the proof ends.   □

Corollary 3.2.9. *If the system (1.23) defines an ESMS-C evolution, then for all $h : \mathbf{R}_+ \to (\mathbf{R}^n)^d$ continuous and bounded, the affine differential equation*

$$\frac{d}{dt}y(t) + M^T(t)y(t) + h(t) = 0$$

has a unique bounded on $\mathbf{R}_+$ solution, $M(t)$ being defined by (3.21).

The next result is the counterpart of Corollary 3.2.7 in the case $\mathcal{D} = \{1, 2, \ldots, d\}$.

Corollary 3.2.10. *If the system (1.22) defines an ESMS-C evolution, then the linear system*

$$\dot{x}(t) = A_0(t, \eta(t)) x(t)$$

obtained by ignoring the white noise perturbations in (1.22), defines an ESMS-C evolution, too.

3.3 Lyapunov-Type Criteria for Mean Square Exponential Stability in the Case $\mathcal{D} = \{1, 2, \ldots, d\}$

The results derived in Sect. 2.7 allow us to obtain useful criteria for ESMS of systems of stochastic linear differential equations (1.22)–(1.25) affected by a standard homogeneous Markov process with a finite number of states. Based on Theorems 2.7.4 and 3.2.2 we obtain the next result.

Theorem 3.3.1. *The following are equivalent:*

(i) The system (1.22) defines an ESMS-C evolution;

(ii) The system of linear differential equations:

$$\frac{d}{dt} K(t, i) + A_0^T(t, i) K(t, i) + K(t, i) A_0(t, i) + \sum_{k=1}^{r} A_k^T(t, i) K(t, i) A_k(t, i)$$

$$+ \Sigma_{j=1}^{d} q_{ij} K(t, j) + I_n = 0$$

$i \in \mathcal{D}, t \geq 0$, *has a bounded solution* $K \gg 0$

$$K(t) = (K(t, 1), \ldots, K(t, d));$$

(iii) There exists a bounded uniform positive and continuous function $H : \mathbf{R}_+ \to \mathcal{S}_n^d$, $H(t) = (H(t, 1), \ldots, H(t, d))$ such that the system of linear differential equations

$$\frac{d}{dt} K(t, i) + A_0^T(t, i) K(t, i) + K(t, i) A_0(t, i) + \sum_{k=1}^{r} A_k^T(t, i) K(t, i) A_k(t, i)$$

$$+ \sum_{j=1}^{d} q_{ij} K(t, j) + H(t, i) = 0$$

$$(3.40)$$

has a bounded and uniform positive solution $K(t) = (K(t, 1), \ldots, K(t, d))$;

(iv) For every bounded uniform positive and continuous function $H : \mathbf{R}_+ \to \mathcal{S}_n^d$, the system (3.40) has a bounded solution $K_0(t) = (K_0(t, 1), \ldots, K_0(t, d))$ with $K_0(t, i) \geq 0$ for all $(t, i) \in \mathbf{R}_+ \times \mathcal{D}$;

(v) For each $H(t)$ as above, there exists a C^1 function $K : \mathbf{R}_+ \to \mathcal{S}_n^d$, bounded with bounded derivative, $K \gg 0$ which solves the following system of linear differential inequalities

$$\frac{d}{dt}K(t,i) + A_0^T(t,i)K(t,i) + K(t,i)A_0(t,i) + \sum_{k=1}^{r} A_k^T(t,i)K(t,i)A_k(t,i)$$
$$+ \sum_{j=1}^{d} q_{ij}K(t,j) + H(t,i) < 0$$

$i \in \mathcal{D}$, *uniformly with respect to t, with* $t \geq 0$;

(vi) *There exists a* C^1*function* $K : \mathbf{R}_+ \to \mathcal{S}_n^d$, *bounded with bounded derivative,* $K \gg 0$ *which solves the following system of linear differential inequalities*

$$\frac{d}{dt}K(t,i) + A_0^T(t,i)K(t,i) + K(t,i)A_0(t,i) + \sum_{k=1}^{r} A_k^T(t,i)K(t,i)A_k(t,i)$$
$$+ \sum_{j=1}^{d} q_{ij}K(t,j) < 0$$

$i \in \mathcal{D}$, *uniformly with respect to t, with* $t \geq 0$.

Combining the results of Theorems 3.1.1 and 2.7.7 we obtain the following result for the time-invariant case.

Theorem 3.3.2. *The following are equivalent:*

(i) *The system (1.25) defines an ESMS evolution;*
(ii) *The system of linear matrix equalities (LME)*

$$A_0^T(i)X(i) + X(i)A_0(i) + \sum_{k=1}^{r} A_k^T(i)X(i)A_k(i) + \sum_{j=1}^{d} q_{ij}X(j) + I_n = 0,$$

$i \in \mathcal{D}$, *has a solution* $X = (X(1),\ldots,X(d))$ *with* $X(i) > 0, i \in \mathcal{D}$;
(iii) *There exists* $H = (H(1),\ldots,H(d)) \in \mathcal{S}_n^d$ *with* $H(i) > 0$ *such that the system of LME*

$$A_0^T(i)X(i) + X(i)A_0(i) + \sum_{k=1}^{r} A_k^T(i)X(i)A_k(i) + \sum_{j=1}^{d} q_{ij}X(j) + H(i) = 0,$$

(3.41)

$i \in \mathcal{D}$, *has a positive semidefinite solution* $X = (X(1),\ldots,X(d))$;
(iv) *For every* $H = (H(1),\ldots H(d)) \in \mathcal{S}_n^d$ *with* $H > 0$, *the system of LME (3.41) has a positive solution* $X = (X(1),\ldots,X(d))$;
(v) *For each* $H = (H(1),\ldots H(d)) \in \mathcal{S}_n^d$ *with* $H > 0$, *the system of LMI*

$$A_0^T(i)X(i) + X(i)A_0(i) + \sum_{k=1}^{r} A_k^T(i)X(i)A_k(i) + \sum_{j=1}^{d} q_{ij}X(j) + H(i) < 0,$$

$i \in \mathcal{D}$, *has a positive solution* $X = (X(1),\ldots,X(d))$;

(vi) The system of LMI:

$$A_0^T (i) X (i) + X (i) A_0 (i) + \sum_{k=1}^{r} A_k^T (i) X (i) A_k (i) + \sum_{j=1}^{d} q_{ij} X (j) < 0,$$

$i \in \mathcal{D}$ has a positive solution $X = (X (1), \ldots, X (d))$.

Similarly we have the next result.

Theorem 3.3.3. *The following are equivalent:*

(i) The system (1.25) defines an ESMS evolution;
(ii) The system of linear matrix equalities (LME)

$$A_0 (i) Y (i) + Y (i) A_0^T (i) + \sum_{k=1}^{r} A_k (i) Y (i) A_k^T (i) + \sum_{j=1}^{d} q_{ji} Y (j) + I_n = 0,$$

$i \in \mathcal{D}$, has a solution $Y = (Y (1), \ldots, Y (d))$ with $Y (i) > 0, i \in \mathcal{D}$;
(iii) There exists $H = (H (1), \ldots H (d)) \in \mathcal{S}_n^d$ with $H (i) > 0$ such that the system of LME:

$$A_0 (i) Y (i) + Y (i) A_0^T (i) + \sum_{k=1}^{r} A_k (i) Y (i) A_k^T (i) + \sum_{j=1}^{d} q_{ji} Y (j) + H (i) = 0,$$

(3.42)

$i \in \mathcal{D}$, has a positive semidefinite solution $Y = (Y (1), \ldots, Y (d))$;
(iv) For every $H = (H (1), \ldots H (d)) \in \mathcal{S}_n^d$ with $H > 0$, the system of LME (3.42) has a positive solution $Y = (Y (1), \ldots, Y (d))$;
(v) For each $H = (H (1), \ldots H (d)) \in \mathcal{S}_n^d$ with $H > 0$, the system of LMI

$$A_0 (i) Y (i) + Y (i) A_0^T (i) + \sum_{k=1}^{r} A_k (i) Y (i) A_k^T (i) + \sum_{j=1}^{d} q_{ji} Y (j) + H (i) < 0,$$

$i \in \mathcal{D}$ has a positive solution $Y = (Y (1), \ldots, Y (d))$;
(vi) The system of LMI:

$$A_0 (i) Y (i) + Y (i) A_0^T (i) + \sum_{k=1}^{r} A_k (i) Y (i) A_k^T (i) + \sum_{j=1}^{d} q_{ji} Y (j) < 0,$$

$i \in \mathcal{D}$ has a positive solution $Y = (Y (1), \ldots, Y (d))$.

In the following we consider the cases when the stochastic system (1.22) is subject only either to Markov jumping or to multiplicative white noise. Thus, in the case of system (1.23), Theorem 3.3.1 becomes as follows.

Theorem 3.3.4. *The following assertions are equivalent:*

(i) The system (1.23) defines an ESMS-C evolution;

(ii) The system of linear differential equations

$$\frac{d}{dt}K(t,i) + A^T(t,i)K(t,i) + K(t,i)A(t,i)$$
$$+ \sum_{j=1}^{d} q_{ij}K(t,j) + I_n = 0$$

$i \in \mathcal{D}, t \geq 0$, has a bounded and uniform positive solution

$$K(t) = (K(t,1), \ldots, K(t,d));$$

(iii) There exists a bounded uniform positive and continuous function $H : \mathbf{R}_+ \to \mathcal{S}_n^d, H(t) = (H(t,1), \ldots, H(t,d))$ such that the system of linear differential equations

$$\frac{d}{dt}K(t,i) + A^T(t,i)K(t,i) + K(t,i)A(t,i) \qquad (3.43)$$
$$+ \sum_{j=1}^{d} q_{ij}K(t,j) + H(t,i) = 0$$

has a bounded solution $K(t) = (K(t,1), \ldots, K(t,d))$ with $K(t,i) \geq 0$, for all $(t,i) \in \mathbf{R}_+ \times \mathcal{D}$;

(iv) For every bounded uniform positive and continuous function $H : \mathbf{R}_+ \to \mathcal{S}_n^d$, the system (3.43) has a bounded and uniform positive solution;

(v) For each $H(t)$ as above, there exists a C^1 function $K : \mathbf{R}_+ \to \mathcal{S}_n^d$, bounded with bounded derivative, $K \gg 0$ which solves the following system of linear differential inequalities

$$\frac{d}{dt}K(t,i) + A^T(t,i)K(t,i) + K(t,i)A(t,i)$$
$$+ \sum_{j=1}^{d} q_{ij}K(t,j) + H(t,i) < 0$$

$i \in \mathcal{D}$, uniformly with respect to t, with $t \geq 0$;

(vi) There exists a C^1 function $K : \mathbf{R}_+ \to \mathcal{S}_n^d$, bounded with bounded derivative, $K \gg 0$ which solves the following system of linear differential inequalities

$$\frac{d}{dt}K(t,i) + A^T(t,i)K(t,i) + K(t,i)A(t,i) + \sum_{j=1}^{d} q_{ij}K(t,j) < 0$$

$i \in \mathcal{D}$, uniformly with respect to t, with $t \geq 0$.

Remark 3.3.1. If the system (1.23) is in "time-invariant" case that is $A(t,i) = A(i)$, for all $t \geq 0$, $i \in \mathcal{D}$, similar results with the ones in Theorems 3.3.2 and 3.3.3 can also be formulated. In this case one obtains the well-known results concerning the ESMS of linear systems with jump Markov perturbations.

Let us consider now the case when the system (1.22) is subject only to white noise perturbations, that is the system under consideration is of the form (1.24). In this case from Theorem 3.3.1 one obtains some known results concerning the exponential stability of linear systems described by Itô differential equations [92].

Theorem 3.3.5. *The following assertions are equivalent:*

 (i) *The system (1.24) defines an ESMS evolution;*
 (ii) *The affine differential equation over the space of symmetric matrices*

$$\frac{d}{dt}X(t) + A_0^T(t)X(t) + X(t)A_0(t) + \sum_{k=1}^{r} A_k^T(t)X(t)A_k(t) + I_n = 0$$

has a bounded and uniform positive solution $X(t)$;
(iii) *There exists $H : \mathbf{R}_+ \to \mathcal{S}_n$ bounded and continuous function $H(t) \gg 0$ such that the affine differential equation*

$$\frac{d}{dt}X(t) + A_0^T(t)X(t) + X(t)A_0(t) + \sum_{k=1}^{r} A_k^T(t)X(t)A_k(t) + H(t) = 0 \quad (3.44)$$

has a bounded solution $X(t)$ with $X(t) \geq 0, t \in \mathbf{R}_+$;
(iv) *For each $H : \mathbf{R}_+ \to \mathcal{S}_n$ bounded, continuous and $H \gg 0$, the affine differential equation (3.44) has a bounded solution $X \gg 0$;*
 (v) *For each $H : \mathbf{R}_+ \to \mathcal{S}_n$ bounded, continuous function, $H \gg 0$, the linear differential inequality*

$$\frac{d}{dt}X(t) + A_0^T(t)X(t) + X(t)A_0(t) + \sum_{k=1}^{r} A_k^T(t)X(t)A_k(t) + H(t) < 0$$

uniformly with respect to $t \geq 0$, has a solution $X(t)$ bounded with bounded derivative $X \gg 0$;
(vi) *The linear differential inequality*

$$\frac{d}{dt}X(t) + A_0^T(t)X(t) + X(t)A_0(t) + \sum_{k=1}^{r} A_k^T(t)X(t)A_k(t) < 0$$

uniformly with respect to $t \geq 0$, has a C^1 solution $X : \mathbf{R}_+ \to \mathcal{S}_n$ which is bounded with bounded derivative and $X(t) \gg 0$.

Remark 3.3.2. If the system (1.24) is in "time-invariant" case, similar results with the ones in Theorems 3.3.2 and 3.3.3 can also be stated.

The next result is proved in a more general situation in [100].

Theorem 3.3.6. *The linear system of stochastic differential equations*

$$dx(t) = Ax(t)\,dt + bc^T x(t)\,dw_1(t); \, b, c \in \mathbf{R}^n \tag{3.45}$$

has an ESMS evolution if and only if A is stable and $\int_0^\infty |c^T e^{At} b|^2\,dt < 1$.

Proof. From Theorem 3.3.5 and Remark 3.3.2 it follows that (3.45) has an ESMS evolution if and only if there exists $X > 0$ such that

$$A^T X + XA + cb^T Xbc^T = -I_n$$

or equivalently,

$$A^T X + XA + cb^T Xbc^T + I_n = 0. \tag{3.46}$$

Assume that (3.46) is fulfilled for $X > 0$. Then it follows that A is stable and therefore we can define the linear operator $\mathcal{G} : \mathcal{S}_n \to \mathcal{S}_n$ by

$$\mathcal{G}(G) = \int_0^\infty e^{A^T t} G e^{At}$$

and $H = \mathcal{G}(G)$ is the unique solution of the Lyapunov equation

$$A^T H + HA = -G. \tag{3.47}$$

If $G > 0$, then $\mathcal{G}(G) > 0$; applying the operator $\mathcal{G}$ to the matrix from the left side of (3.46) and using (3.47) we obtain that

$$-X + b^T Xb\mathcal{G}(cc^T) + \mathcal{G}(I_n) = 0.$$

Hence

$$-b^T Xb + (b^T Xb)\, b^T \mathcal{G}(cc^T)\, b + b^T \mathcal{G}(I_n) b = 0$$

and therefore

$$b^T Xb\left(1 - b^T \mathcal{G}(cc^T)\, b\right) = b^T \mathcal{G}(I_n) b$$

which implies that $1 - b^T \mathcal{G}(cc^T)\, b > 0$, since if $b = 0$ the inequality is obvious and if $b \neq 0$ we have $b^T Xb > 0$, $b^T \mathcal{G}(I_n) b > 0$. Taking into account that

$$b^T \mathcal{G}(cc^T)\, b = \int_0^\infty |c^T e^{At} b|^2\,dt,$$

the inequality in the statement directly follows.

The condition in the statement is sufficient. Indeed, assume that A is stable and that $\int_0^\infty \left| c^T e^{At} b \right|^2 dt < 1$, namely $b^T \mathcal{G}\left(cc^T\right) b < 1$. Let

$$X = \mathcal{G}\left(I_n\right) + \frac{b^T \mathcal{G}\left(I_n\right) b}{1 - b^T \mathcal{G}\left(cc^T\right) b} \mathcal{G}\left(cc^T\right).$$

It is obvious that $X > 0$ and a direct calculation using (3.47) shows that X verifies (3.46) and the proof is complete.   □

Remark 3.3.3. From the Parseval's formula one easily obtains that:

$$\int_0^\infty \left| c^T e^{At} b \right|^2 dt = \frac{1}{2\pi} \int_{-\infty}^\infty \left| c^T \left(A - i\lambda I_n\right)^{-1} b \right|^2 d\lambda.$$

where $i = \sqrt{-1}$.

For each $i \in \mathcal{D}$ we can consider the following system subject only to white noise perturbations

$$dx_i(t) = \left(A_0(t,i) + \frac{1}{2} q_{ii} I_n\right) x_i(t)\, dt + \sum_{k=1}^r A_k(t,i) x_i(t)\, dw_k(t) \qquad (3.48)$$

$t \geq 0$, $i \in \mathcal{D}$. In this case one obtains

Corollary 3.3.7. *If the system (1.22) defines an ESMS evolution, then:*

(i) The system (3.48) defines an ESMS evolution for each $i \in \mathcal{D}$;
(ii) For each $i \in \mathcal{D}$ the deterministic system

$$\dot{x}_i(t) = \left(A_0(t,i) + \frac{1}{2} q_{ii} I_n\right) x_i(t)$$

defines an exponentially stable evolution.

At the end of this section we prove the following result under the assumptions in Sect. 2.7.

Theorem 3.3.8. *Assume that there exist a bounded and uniform positive function $K : \mathbf{R}_+ \to \mathcal{S}_n^d$, $K(t) = (K(t,1), \dots, K(t,d))$ and the constants $\tau > 0$, $\delta \in (0,1)$ such that*

$$\left(T^*(t+\tau,t) K(t+\tau)\right)(i) \leq \delta K(t,i)$$

for all $t \geq 0$, $i \in \mathcal{D}$. Then the system $(A_0, A_1, \dots, A_r; Q)$ is stable.

Proof. From the statement of the theorem it follows that

$$T^*(t+\tau,t) K(t+\tau) \leq \delta K(t), \ t \geq 0.$$

Let $t_0 \geq 0$ be fixed; since $T^*(t,t_0)$ is a positive operator, we obtain by induction that

$$T^*(t_0 + m\tau, t_0) K(t_0 + m\tau) \leq \delta^m K(t_0)$$

for all $m \geq 1$. Since K is bounded and uniformly positive there exist $\beta_i > 0, i = 1, 2$ such that $\beta_1 J^d \leq K(t) \leq \beta_2 J^d$ therefore

$$T^*(t_0 + m\tau, t_0) J^d \leq \beta \delta^m J^d$$

which leads to

$$\left| T^*(t_0 + m\tau, t_0) J^d \right| \leq \beta \delta^m, \quad m \geq 1.$$

Based on Corollary 2.6.2 we obtain:

$$\left\| T^*(t_0 + m\tau, t_0) J^d \right\| \leq \beta \delta^m.$$

Since $\sup_{t \geq 0} \|\mathcal{L}^*(t)\| < \infty$, we easily deduce (using (2.127) together with Remark 2.2.1 (i)) that $\|T^*(t,s)\| \leq \beta_1$ for all $0 \leq t - s \leq \tau$. Using (2.126) we deduce that $\|T^*(t,t_0)\| \leq \beta_2 e^{-\alpha(t-t_0)}$ for all $t \geq t_0 \geq 0$ for some $\beta_2 > 0$ and $\alpha = -\frac{1}{\tau} \ln \delta$. The proof ends. $\square$

3.4 Lyapunov-Type Criteria for Mean Square Exponential Stability in the Case $\mathcal{D} = \mathbf{Z}_+$

Combining the results of Theorems 2.8.2 and 3.2.2 (b), one obtains the following Lyapunov-type criteria for the property of ESMS-C of system (1.22) affected by a standard homogeneous Markov process with an infinite countable set of states.

Theorem 3.4.1. *Under the assumptions of Theorem 3.1.4 the following statements are equivalent:*

(i) *The system (1.22) is ESMS-C;*
(ii) *There exists a C^1 function $X : \mathbf{R}_+ \to \mathcal{S}_n^\infty$ bounded with bounded derivative, satisfying the affine differential equation on $\mathcal{S}_n^\infty$:*

$$\frac{d}{dt} X(t) + \mathcal{L}(t)X(t) + J^\infty = 0 \qquad (3.49)$$

and $X(t,i) \geq \mu I_n$, for all $(t,i) \in \mathbf{R}_+ \times \mathbf{Z}_+$, $\mu > 0$ is a constant;
(iii) *There exists a C^1 function $Y : \mathbf{R}_+ \to \mathcal{S}_n^\infty$ bounded with bounded derivative and the scalars $\mu_j > 0$, $j = 1, 2$ satisfying*

$$\frac{d}{dt} Y(t) + \mathcal{L}(t)Y(t) + \mu_1 J^\infty \leq 0 \qquad (3.50)$$

$t \in \mathbf{R}_+$,

$$Y(t,i) \geq \mu_2 I_n \qquad (3.51)$$

for all $(t,i) \in \mathbf{R}_+ \times \mathbf{Z}_+$.

From Corollary 2.8.3 and Theorem 3.2.2 (b) the next result directly follows.

Corollary 3.4.2. *Assume that*

(a) The assumptions of Theorem 3.1.4 are fulfilled;
(b) There exists $\theta > 0$ such that $A_k(t + \theta, i) = A_k(t,i), 0 \leq k \leq r, t \in \mathbf{R}_+, i \in \mathbf{Z}_+$;

Under these assumptions the following are equivalent

(i) The system (1.22) is ESMS-C;
(ii) The affine differential equation on $\mathcal{S}_n^\infty$ (3.49) has a periodic solution $\tilde{X}(t) = \{\tilde{X}(t,i)\}_{i \in \mathbf{Z}_+}, t \geq 0$ with the same period θ, having the property $\tilde{X}(t,i) \geq \mu I_n$, for all $(t,i) \in [0,\theta] \times \mathbf{Z}_+$, where $\mu > 0$ is a constant;
(iii) There exists a C^1 function which is periodic with period θ, $\tilde{Y}(t) = \{\tilde{Y}(t,i)\}_{i \in \mathbf{Z}_+}, t \geq 0$, and the scalars $\mu_j > 0, j = 1,2$, which satisfy (3.50), (3.51) for $t \in [0,\theta]$.

Based on the Remark 2.3.7 and Corollary 2.3.9 one obtains the following list of Lyapunov-type criteria for the property of ESMS in the invariant case of differential systems of type (1.25).

Corollary 3.4.3. *Assume that $A_k(t,i) = A_k(i), 0 \leq k \leq r, (t,i) \in \mathbf{R}_+ \times \mathbf{Z}_+$. Then the following statements are equivalent*

(i) The system (1.25) is ESMS-C;
(ii) There exist $X = \{X(i)\}_{i \in \mathbf{Z}_+} \in \mathcal{S}_n^\infty$ and the scalar $\mu > 0$ such that $\mathcal{L}X + J^\infty = 0$, $X(i) \geq \mu I_n, i \in \mathbf{Z}_+$;
(iii) There exist $Y = \{Y(i)\}_{i \in \mathbf{Z}_+} \in \mathcal{S}_n^\infty$ and the scalars $\mu_j > 0, j = 1,2$, satisfying

$$\mathcal{L}Y + \mu_1 J^\infty \leq 0, \qquad (3.52)$$

$Y(i) \geq \mu_2 I_n, i \in \mathbf{Z}_+$.

At the end of this section we provide a result which is a simple consequence of the representation Theorem 3.1.6.

Proposition 3.4.4. *Under the assumptions of Theorem 3.1.4 the system (1.22) is ESMS if the zero state equilibrium of the linear differential equation on $\ell^1(\mathbf{Z}_+, \mathcal{S}_n)$*

$$\frac{d}{dt}X(t) = \mathcal{L}(t)X(t) \qquad (3.53)$$

is exponentially stable.

Proof. If $x(t;t_0,x_0)$ is a solution of (1.22) we write successively

$$E[|x(t;t_0,x_0)|^2] = \sum_{i=0}^{\infty} E[|x(t;t_0,x_0)|^2 \chi_{\eta(t)=i}] = \sum_{i=0}^{\infty} Tr\{E[x(t;t_0,x_0)x^T(t;t_0,x_0)\chi_{\eta(t)=i}]\}$$

$$\leq n \sum_{i=0}^{\infty} |E[x(t;t_0,x_0)x^T(t;t_0,x_0)\chi_{\eta(t)=i}]| \|\tilde{E}_t[x(t;t_0,x_0)x^T(t;t_0,x_0)]\|_1.$$

Applying Theorem 3.1.6 we obtain:

$$E[|x(t;t_0,x_0)|^2] \leq n\|T(t,t_0)\| \, |x_0|^2. \tag{3.54}$$

The conclusion follows now from (3.54) and thus the proof is complete. □

In the case of systems of type (1.22) with $A_k(t,i) = 0, 1 \leq k \leq r$ and $A_0(t,i) = A(i)$ for all $(t,i) \in \mathbf{R}_+ \times \mathbf{Z}_+$ the inequality (3.54) was proved in Lemma 4.7 in [66].

3.5 Illustrative Examples

For the beginning we consider the case $\mathcal{D} = \{1,2,\ldots,d\}$.

Example 3.5.1. Let us consider the particular case $n = 1$ in which situation the system (1.25) reduces to the linear differential equation

$$dx(t) = a(\eta(t))x(t)\,dt + \sum_{k=1}^{r} g_k(\eta(t))x(t)\,dw_k(t), t \geq 0. \tag{3.55}$$

We shall prove that if

$$2a(i) + \sum_{k=1}^{r} g_k^2(i) < 0, \ i \in \mathcal{D}, \tag{3.56}$$

then (3.55) defines an ESMS evolution.

Indeed, taking $K = (1,\ldots,1)$ and using the fact that $\sum_{j=1}^{d} q_{ij} = 0$, we get

$$2a(i)K(i) + \sum_{k=1}^{r} g_k^2(i)K(i) + \sum_{j=1}^{d} q_{ij}K(j) = 2a(i) + \sum_{j=1}^{d} g_k^2(i),$$

$i \in \mathcal{D}$. Since the left side in the above equation coincides with L^*K and $K > 0$, from Theorem 3.3.2 it follows that if (3.56) is fulfilled then the system (3.55) defines an ESMS evolution.

Remark 3.5.1. (i) The above example shows that (3.56) are sufficient conditions
under which (3.55) defines an ESMS evolution. As we shall see in the next
example, these conditions are not necessary.

(ii) Using Theorem 3.3.5 and Remark 3.3.2, it is easy to check that (3.56) is a
necessary and sufficient condition for ESMS for the Itô equation

$$dx(t) = a(i)x(t)dt + \sum_{k=1}^{r} g_k(i)x(t)dw_k(t),$$

with $i \in \mathcal{D}$ fixed.

Example 3.5.2. Assume that in (3.55) we have $d = 2, r = 1$ and

$$Q = \begin{bmatrix} -\alpha & \alpha \\ \alpha & -\alpha \end{bmatrix}$$

with $\alpha > 0$. From Theorem 3.3.2 it results that (3.55) defines an ESMS evolution if
and only if there exists $K = (K_1, K_2), K_i > 0$ such that

$$2a_iK_i + g_i^2 K_i + \sum_{j=1}^{2} q_{ij}K_j = -\alpha, \ i = 1, 2,$$

where we denoted $a_i = a(i), g_i = g(i)$ and $K_i = K(i), i = 1, 2$. Then from the above
equations we obtain:

$$\left(2a_1 + g_1^2 - \alpha\right)K_1 + \alpha K_2 = -\alpha \tag{3.57}$$

$$\left(2a_2 + g_2^2 - \alpha\right)K_2 + \alpha K_1 = -\alpha,$$

from which yields the necessary conditions for stability

$$2a_i + g_i^2 - \alpha < 0, \ i = 1, 2.$$

Further, solving (3.57) we get

$$K_1 = \frac{\alpha\left(2a_2 + g_2^2 - 2\alpha\right)}{\alpha\left(2a_1 + g_1^2 + 2a_2 + g_2^2\right) - \left(2a_1 + g_1^2\right)\left(2a_2 + g_2^2\right)}$$

$$K_2 = \frac{\alpha\left(2a_1 + g_1^2 - 2\alpha\right)}{\alpha\left(2a_1 + g_1^2 + 2a_2 + g_2^2\right) - \left(2a_1 + g_1^2\right)\left(2a_2 + g_2^2\right)}.$$

Since $2a_i + g_i^2 - 2\alpha < 0$, it follows that

$$\alpha\left(2a_1 + g_1^2 + 2a_2 + g_2^2\right) - \left(2a_1 + g_1^2\right)\left(2a_2 + g_2^2\right) < 0. \tag{3.58}$$

Then the following cases can occur:

Case a. If $2a_1 + g_1^2 + 2a_2 + g_2^2 < 0$, the condition (3.58) is accomplished for

$$\alpha > \frac{\left(2a_1 + g_1^2\right)\left(2a_2 + g_2^2\right)}{2a_1 + g_1^2 + 2a_2 + g_2^2};$$

Case b. If $2a_1 + g_1^2 + 2a_2 + g_2^2 > 0$, then (3.58) holds for

$$\alpha < \frac{\left(2a_1 + g_1^2\right)\left(2a_2 + g_2^2\right)}{2a_1 + g_1^2 + 2a_2 + g_2^2}. \tag{3.59}$$

The case (b) implies $2a_i + g_i^2 > 0$, $i = 1, 2$ then (3.59) contradicts the necessary condition $\alpha > 2a_1 + g_1^2$. Therefore we conclude that the *Case (b)* from above must be excluded.

Summarizing, the stochastic system (3.55) with $d = 2$ and $r = 1$ considered in this example defines an ESMS evolution if and only if:

$$2a_1 + g_1^2 < 0 \text{ and } 2a_2 + g_2^2 < 0$$

(situation considered in Example 3.5.1), or if

$$2a_1 + g_1^2 + 2a_2 + g_2^2 < 0 \text{ and}$$

$$\alpha > \max\left\{2a_1 + g_1^2, 2a_2 + g_2^2, \frac{\left(2a_1 + g_1^2\right)\left(2a_2 + g_2^2\right)}{2a_1 + g_1^2 + 2a_2 + g_2^2}\right\}.$$

Example 3.5.3. Consider the stochastic system with jump Markov perturbations in which $n = d = 2$:

$$\frac{dx(t)}{dt} = A(\eta(t))x(t), t \geq 0 \tag{3.60}$$

where

$$A_1 := A(1) = \begin{bmatrix} -a\alpha & 0 \\ \alpha & -a\alpha \end{bmatrix}$$

$$A_2 := A(2) = \begin{bmatrix} -a\alpha & \alpha \\ 0 & -a\alpha \end{bmatrix}$$

with $a > 0$ and

$$Q = \begin{bmatrix} -\alpha & \alpha \\ \alpha & -\alpha \end{bmatrix}$$

with $\alpha > 0$. Then, according to Theorem 3.3.2, (3.60) defines an ESMS evolution if and only if there exist

$$X_1 := X(1) = \begin{bmatrix} x_1 & y_1 \\ y_1 & z_1 \end{bmatrix} \text{ and } X_2 := X(2) = \begin{bmatrix} x_2 & y_2 \\ y_2 & z_2 \end{bmatrix}$$

such that $X_1 > 0$, $X_2 > 0$ and

$$A_1^T X_1 + X_1 A_1 + \sum_{j=1}^{2} q_{1j} X_j = -\alpha I_2$$

$$A_2^T X_2 + X_2 A_2 + \sum_{j=1}^{2} q_{2j} X_j = -\alpha I_2$$

which are equivalent with

$$\beta x_1 - 2y_1 - x_2 = 1$$
$$\beta y_1 - z_1 - y_2 = 0$$
$$\beta z_1 - z_2 = 1$$
$$\beta x_2 - x_1 = 1$$
$$\beta y_2 - x_2 - y_1 = 0$$
$$\beta z_2 - 2y_2 - z_1 = 1,$$

where we denoted $\beta := 2a + 1$. By solving the above system of algebraic equations it follows that

$$z_1 = \frac{\beta + 1}{(\beta^3 - \beta^2 - \beta - 1)(\beta^3 + \beta^2 - \beta + 1)}.$$

Then for $a \to 0$ one obtains that $z_1 \to -\frac{1}{2}$. This shows that although $A(1)$ and $A(2)$ have their eigenvalues in $\mathbf{C}_-$, that is they are stable in the deterministic sense, the stochastic system (3.60) defines an unstable evolution.

Example 3.5.4. We consider now the case $n = d = 2$ and $r = 1$, namely the situation when the stochastic system is subject both to Markovian jumping and to multiplicative white noise:

$$dx(t) = A_0(\eta(t))x(t)dt + A_1(\eta(t))x(t)dw_1(t), t \geq 0 \qquad (3.61)$$

where

$$A_0(1) = \begin{bmatrix} -1 & 0 \\ 1 & -1 \end{bmatrix}, A_0(2) = \begin{bmatrix} -1 & 1 \\ 0 & -1 \end{bmatrix},$$

$$A_1(1) = \begin{bmatrix} a & 0 \\ 0 & 0 \end{bmatrix}, A_1(2) = \begin{bmatrix} 0 & 0 \\ 0 & a \end{bmatrix}$$

and

$$Q = \begin{bmatrix} -1 & 1 \\ 1 & -1 \end{bmatrix}.$$

According to Theorem 3.3.2, the necessary and sufficient condition such that $(A_0, A_1; Q)$ defines an ESMS evolution is that the equations

$$A_0^T(i)X(i) + X(i)A_0(i) + A_1^T(i)X(i)A_1(i) + \sum_{j=1}^{2} q_{ij}X(j) = -I_2,$$

$i = 1, 2$, have the solution $X(i) > 0$ with

$$X(i) = \begin{bmatrix} x_i & y_i \\ y_i & z_i \end{bmatrix}, i = 1, 2.$$

The above equation leads to

$$\begin{align} (3 - a^2)x_1 - 2y_1 - x_2 &= 1 \tag{3.62}\\ 3y_1 - z_1 - y_2 &= 0 \\ 3z_1 - z_2 &= 1 \\ 3x_2 - x_1 &= 1 \\ 3y_2 - x_2 - y_1 &= 0 \\ (3 - a^2)z_2 - 2y_2 - z_1 &= 1, \end{align}$$

from we deduce that

$$\begin{align} (24 - 9a^2)x_2 + (3a^2 - 10)z_1 &= 8 - 2a^2 \tag{3.63}\\ (3a^2 - 10)x_2 + (24 - 9a^2)z_1 &= 8 - 2a^2. \end{align}$$

For $a^2 = \frac{17}{6}$ we obtain that $x_2 + z_1 = -\frac{14}{9}$ which is not admissible since $X(i) > 0$, $i = 1, 2$ imply that $x_2 > 0$ and $z_1 > 0$.

On the other hand, if $a^2 = \frac{7}{3}$, the system (3.63) is incompatible and if $a^2 \neq \frac{17}{6}$ and $a^2 \neq \frac{7}{3}$, this system has the unique solution

$$x_2 = z_1 = \frac{a^2 - 4}{3a^2 - 7}$$

which gives in (3.62)

$$x_1 = z_2 = -\frac{5}{3a^2 - 7} \text{ and } y_1 = y_2 = \frac{a^2 - 4}{2(3a^2 - 7)}.$$

Therefore $X(1) > 0$ and $X(2) > 0$ if and only if $a^2 < \frac{7}{3}$, from which we conclude that $(A_0, A_1; Q)$ defines an ESMS evolution if and only if $a^2 < \frac{7}{3}$.

The next examples will illustrate the applicability of the Lyapunov-type criteria derived in Sect. 3.4 in the investigation of the ESMS of stochastic linear differential equations (1.22) affected by a standard homogeneous Markov chain with an infinite countable set of states.

Example 3.5.5. Consider the system (1.22) in the special case $n = 1$, $A_0(t, i) = a(i), A_k(t, i) = 0, 1 \leq k \leq r, (t, i) \in \mathbf{R}_+ \times \mathbf{Z}_+$. We have the differential equation:

$$\frac{d}{dt}x(t) = a(\eta(t))x(t), \, t \geq 0. \tag{3.64}$$

Here $\{\eta(t)\}_{t \geq 0}$ is an homogeneous Markov process having the state space $\mathbf{Z}_+$ and the generator matrix Q with the elements $\{q_{ij}\}_{(i,j) \in \mathbf{Z}_+ \times \mathbf{Z}_+}$ such that for each $i \in \mathbf{Z}_+, q_{ii} = -\lambda, q_{i,i+1} = \lambda, \lambda > 0$ and $q_{ij} = 0$ if $j \in \mathbf{Z}_+ \setminus \{i, i+1\}$. Hence $\{\eta(t)\}_{t \geq 0}$ is an homogeneous Poisson process with parameter λ. Assume that the sequence $\{a(i)\}_{i \in \mathbf{Z}_+}$ has the properties:

α) the sequence $\{a(i)\}_{i \in \mathbf{Z}_+}$ is bounded and $a(i) \leq \lambda/4$ for all $i \in \mathbf{Z}_+$.

β) $\overline{\lim}_{i \to \infty} a(i) < 0$.

We show that in this case the zero state equilibrium of (3.64) is ESMS-C. To this end, we shall use the equivalence (i)$\leftrightarrow$(iii) from the Corollary 3.4.3. Let $\delta > 0$ be defined by:

$$\overline{\lim_{i \to \infty}} a(i) = -2\delta. \tag{3.65}$$

Therefore there exists $i_\delta \geq 1$ such that $a(i) < -\delta$ for all $i \geq i_\delta$. The system of LMIs (3.52) becomes

$$(2a(i) - \lambda)y(i) + \lambda y(i+1) + \mu_1 \leq 0, i \in \mathbf{Z}_+. \tag{3.66}$$

Taking $y(i) = 1, i \geq i_\delta$ we obtain from (3.66) that $2a(i) + \mu_1 \leq 0$ for all $i \geq i_\delta$. If we take into account the choice of i_δ one obtains that (3.66) is verified by $y(i) = 1, i \geq i_\delta$ if $\mu_1 \in (0, 2\delta)$.

For $i \in \{0, 1, \ldots, i_\delta - 1\}$ we construct recursively $\hat{y}(i)$ from the equation

$$\hat{y}(i) = \frac{\lambda}{\lambda - 2a(i)}(\hat{y}(i+1) + 1),$$

$i \leq i_\delta - 1, \hat{y}(i_\delta) = 1$. So (3.66) is solved by $\{y(i)\}_{i \in \mathbf{Z}_+}$ where $y(i) = 1$ if $i \geq i_\delta$, $y(i) = \hat{y}(i), i < i_\delta$, and μ_1, μ_2 satisfying

$$0 < \mu_1 < min\{\lambda, 2\delta\}, \qquad \mu_2 = min\left\{1, min_{0 \leq i < i_\delta} \frac{\lambda}{\lambda - 2a(i)}\right\}.$$

Example 3.5.6. Consider the special case of the system (1.22) with $n = 1, A_0(t, i) = m(t)a(i)$, $A_k(t, i) = 0, 1 \leq k \leq r, (t, i) \in \mathbf{R}_+ \times \mathbf{Z}_+$. We have

$$\frac{d}{dt}x(t) = m(t)a(\eta_t)x(t). \tag{3.67}$$

As in the previous example $\{\eta(t)\}_{t \geq 0}$ is an homogeneous Poisson process with parameter $\lambda > 0$. Assume:

(α') $m : \mathbf{R}_+ \to \mathbf{R}_+$ is a continuous function with the property that $0 < m_0 \leq m(t) \leq m_1$ for all $t \geq 0$, where m_0, m_1 are positive constants.

(β') The sequence $a(i), i \in \mathbf{Z}_+$ is bounded and additionally $m(t)a(i) \leq \lambda/4$ for all $(t, i) \in \mathbf{R}_+ \times \mathbf{Z}_+$.

(γ') $\overline{lim}_{i \to \infty} a(i) < 0$.

To show that the zero state equilibrium of (3.67) is ESMS-C we apply the equivalence (i)$\leftrightarrow$(iii) from Theorem 3.4.1. The system of linear differential inequalities (3.50)–(3.51) becomes

$$\frac{d}{dt}y(t, i) + (2m(t)a(i) - \lambda)y(t, i) + \lambda y(t, i+1) + \mu_1 \leq 0$$

$$y(t, i) \geq \mu_2, (t, i) \in \mathbf{R}_+ \times \mathbf{Z}_+. \tag{3.68}$$

Let $i_\delta \geq 1$ be defined as in the previous example. One sees that for $i \geq i_\delta$ $y(t, i) = 1$ verifies (3.68) if $0 < \mu_1 < 2m_0\delta$. For $i \leq i_\delta - 1, y(t, i)$ may be constructed recursively as the unique bounded solution of the linear differential equation

$$\frac{d}{dt}y(t, i) + (2m(t)a(i) - \lambda)y(t, i) + \lambda(y(t, i+1) + 1) = 0, \quad i \leq i_\delta - 1, \tag{3.69}$$

$$y(t, i_\delta) = 1, t \geq 0.$$

Take

$$\hat{y}(t, i) = \lambda \int_t^\infty e^{\int_t^s (2m(\Sigma)a(i) - \lambda)d\Sigma}(1 + \hat{y}(s, i+1))ds \tag{3.70}$$

$0 \leq i \leq i_\delta - 1, \hat{y}(t, i_\delta) = 1, t \geq 0$. From ($\beta$) we have $\int_t^s (2m(\Sigma)a(i) - \lambda)d\Sigma \leq -\frac{\lambda}{2}(s - t)$ for all $s \geq t \geq 0$. This allows us to conclude that the integral from (3.70) is absolutely convergent and $t \to \hat{y}(t, i)$ is bounded. Applying Lemma 2.3.4 for the special case $\mathcal{X} = \mathbf{R}, \mathcal{X}^+ = \mathbf{R}_+, \zeta = 1$, we deduce that there exists $\tau_0 > 0$ such that $e^{\int_t^s (2m(\Sigma)a(i)-\lambda)d\Sigma} \geq \frac{1}{2}$ for all $t \leq s \leq t + \tau_0, t \geq 0$. This allows us to deduce that $\hat{y}(t,i) \geq \lambda \int_t^\infty e^{\int_t^s (2m(\Sigma)a(i)-\lambda)d\Sigma} ds \geq \frac{1}{2}\lambda\tau_0$ for all $t \geq 0, 0 \leq i \leq i_\delta - 1$. We have shown that (3.70) is solvable by $y(t,i) = 1$ if $i \geq i_\delta$, $y(t,i) = \hat{y}(t,i)$ if $0 \leq i \leq i_\delta - 1$ and $\mu_1 \in (0, min\{\lambda, 2m_0\delta\}), \mu_2 = min\{1, \frac{1}{2}\lambda\tau_0\}$.

Example 3.5.7. Consider the special case of the system (1.22) with $n = 1$, i.e.

$$dx(t) = a_0(t,\eta(t))dt + \sum_{k=1}^r a_k(t,\eta(t))x(t)dw_k(t) \tag{3.71}$$

where $t \to a_k(t,i) : \mathbf{R}_+ \to \mathbf{R}$ are continuous function uniformly with respect to $i \in \mathbf{Z}_+$; $\{\eta(t)\}_{t\geq 0}$ is an homogeneous Markov process and $\{w(t)\}_{t\geq 0}$ is a standard Wiener process. Assume

$$\sup_{t\geq 0}\sup_{i\in\mathbf{Z}_+}\left(\sum_{k=1}^r a_k^2(t,i) + 2a_0(t,i)\right) < 0. \tag{3.72}$$

Under these conditions the zero state equilibrium of (3.71) is ESMS-C. One sees that if (3.72) is fulfilled, then $y(t,i) = 1, t \geq 0, i \in \mathbf{Z}_+$ verifies (3.50)–(3.51) written for (3.71).

3.6   Affine Systems

Throughout this section we assume that $\mathcal{D} = \{1, 2, \ldots, d\}$. Consider the system

$$dx(t) = [A_0(t,\eta(t))x(t) + f_0(t)]dt + \sum_{k=1}^r [A_k(t,\eta(t))x(t) + f_k(t)]dw_k(t) \tag{3.73}$$

where $A_k(t,i), 0 \leq k \leq r$ are bounded on $\mathbf{R}_+$ and continuous matrix valued functions. Denote

$$u(t) = (f_0^T(t), f_1^T(t), \ldots, f_r^T(t))^T.$$

If $t_0 \geq 0, x_0 \in \mathbf{R}^n$ and $f_k \in L^2_{\eta,w}([t_0, T], \mathbf{R}^n), 0 \leq k \leq r$ for all $T > t_0$ by Theorem 1.11.1 it follows that there exists a unique solution $x_u(t, t_0, x_0)$ of the system (3.73) with $x_u(t_0, t_0, x_0) = x_0$ and $x_u(\cdot, t_0, x_0) \in L^2_{\eta,w}([t_0, T], \mathbf{R}^n), T > t_0$, that is all components of the vector x_u are in $L^2_{\eta,w}([t_0, T])$.

Unfortunately the representation formula (1.29) cannot be used to obtain some useful estimates for solutions of system (3.73) as in the deterministic case. Such estimations are obtained in an indirect way using some techniques based on Lyapunov functions.

Theorem 3.6.1. *Assume that the system* $(A_0, A_1, \ldots, A_r; Q)$ *is stable. Then*

(i) There exist $c \geq 1, \alpha > 0$ *such that*

$$E\left[|x_u(t, t_0, x_0)|^2 | \eta(t_0) = i\right] \leq c\left(e^{-\alpha(t-t_0)}|x_0|^2\right.$$
$$\left. + \sum_{k=0}^{r} E\left[\int_{t_0}^{t} e^{-\alpha(t-s)}|f_k(s)|^2 ds | \eta(t_0) = i\right]\right)$$

for all $t \geq t_0 \geq 0, x_0 \in \mathbf{R}^n, i \in \mathcal{D}$ *and all* $f_k \in L^2_{\eta,w}([t_0, \infty), \mathbf{R}^n), 0 \leq k \leq r;$
(ii) There exists $\beta > 0$ *such that*

$$E\left[\int_{t_0}^{\infty} |x_u(t, t_0, x_0)|^2 | \eta(t_0) = i\right] \leq \beta\left(|x_0|^2\right.$$
$$\left. + \sum_{k=0}^{r} E\left[\int_{t_0}^{\infty} |f_k(s)|^2 ds | \eta(t_0) = i\right]\right)$$

for all $t_0 \geq 0, x_0 \in \mathbf{R}^n, f_k \in L^2_{\eta,w}([t_0, \infty), \mathbf{R}^n), 0 \leq k \leq r, i \in \mathcal{D}.$
(iii)

$$\lim_{t \to \infty} E|x_u(t, t_0, x_0)|^2 = 0$$

for all $t_0 \geq 0, x_0 \in \mathbf{R}^n, f_k \in L^2_{\eta,w}([t_0, \infty), \mathbf{R}^n), 0 \leq k \leq r.$

Proof. Since $(A_0, A_1, \ldots, A_r; Q)$ is stable then by Theorem 2.7.4 the Lyapunov-type equation (2.157) has a unique bounded on $\mathbf{R}_+$ and uniformly positive solution $\tilde{K}(t) = (\tilde{K}(t, 1), \ldots, \tilde{K}(t, d))$. Therefore there exist $\alpha_1 > 0, \alpha_2 > 0$ such that

$$\alpha_1 J^d \leq \tilde{K}(t) \leq \alpha_2 J^d, \quad t \geq 0.$$

Let $x_u(t) = x_u(t, t_0, 0), t \geq t_0$. Applying the Itô-type formula (1.6) to the function $v(t, x, i) = x^T \tilde{K}(t, i)x$ and to the system (3.73), taking into account (2.157) for $\tilde{K}(t)$ we obtain:

$$E\left[v(t, x_u(t), \eta(t)) | \eta(t_0) = i\right] = E\left[\int_{t_0}^{t} \left\{-|x_u(s)|^2 + 2x_u^*(s)\left[\tilde{K}(s, \eta(s))f_0(s)\right.\right.\right.$$
$$\left.\left.\left. + \sum_{k=1}^{r} A_k^*(s, \eta(s))\tilde{K}(s, \eta(s))f_k(s)\right] + \sum_{k=1}^{r} f_k^*(s)\tilde{K}(s, \eta(s))f_k(s)\right\} ds | \eta(t_0) = i\right].$$

Denote

$$h_i(t) = E\left[v(t, x_u(t), \eta(t)) | \eta(t_0) = i\right], i \in \mathcal{D}$$

$$m_i(t) = \sqrt{E\left[|x_u(t)|^2 | \eta(t_0) = i\right]}, i \in \mathcal{D}$$

$$g_i(t) = \sqrt{\sum_{k=0}^{r} E\left[|f_k(t)|^2 | \eta(t_0) = i\right]}, i \in \mathcal{D}.$$

Then we may write

$$h_i'(t) = E\Bigg[\Big\{-|x_u(t)|^2$$

$$+2x_u^T(t)\left[\tilde{K}(t,\eta(t))f_0(t) + \sum_{k=1}^{r} A_k^T(t,\eta(t))\tilde{K}(t,\eta(t))f_k(t)\right]$$

$$+\sum_{k=1}^{r} f_k^T(t)\tilde{K}(t,\eta(t))f_k(t)\Big\} | \eta(t_0) = i\Bigg]$$

a.e. $t \geq t_0, i \in \mathcal{D}$.

Since $A_k, \tilde{K}$ are bounded, there exist $\gamma > 0, \delta > 0$ such that

$$h_i'(t) \leq -m_i^2(t) + \gamma\left[m_i(t)g_i(t) + g_i^2(t)\right] \leq -\frac{1}{2}m_i^2(t) + \delta g_i^2(t).$$

Taking into account that $\alpha_1 I_n \leq \tilde{K}(t,\eta(t)) \leq \alpha_2 I_n$ it follows that

$$\alpha_1 m_i^2(t) \leq h_i(t) \leq \alpha_2 m_i^2(t).$$

Hence $h_i'(t) \leq -\frac{1}{2\alpha_2}h_i(t) + \delta g_i^2(t)$. Since $h_i(t_0) = 0$ we obtain

$$\alpha_1 m_i^2(t) \leq h_i(t) \leq \delta \int_{t_0}^{t} e^{-\alpha(t-s)} g_i^2(s)ds, \ t \geq t_0, i \in \mathcal{D} \qquad (3.74)$$

with $\alpha = \frac{1}{2\alpha_2}$. On the other hand,

$$x_u(t,t_0,x_0) = x_u(t,t_0,0) + \Phi(t,t_0)x_0. \qquad (3.75)$$

Combining (3.74) and (3.75), (i) is proved. The assertion (ii) follows by (i) and Fubini theorem. We prove now (iii). Since

$$\sum_{i=1}^{d} E\left[\int_{t_0}^{\infty} \sum_{k=0}^{r} |f_k(t)|^2 dt | \eta(t_0) = i\right] < \infty,$$

it follows that for every $\varepsilon > 0$ there exists $t_\varepsilon > t_0$ such that

$$\sum_{i=1}^{d} \int_{t_\varepsilon}^{\infty} g_i^2(t)dt < \varepsilon.$$

For each $t \geq t_\varepsilon$ we have

$$\int_{t_0}^{t} e^{-\alpha(t-s)} g_i^2(s) ds = e^{-\alpha(t-t_\varepsilon)} \int_{t_0}^{t_\varepsilon} e^{-\alpha(t_\varepsilon-s)} g_i^2(s) ds + \int_{t_\varepsilon}^{t} e^{-\alpha(t-s)} g_i^2(s) ds$$

$$\leq e^{-\alpha(t-t_\varepsilon)} \int_{t_0}^{\infty} g_i^2(s) ds + \varepsilon.$$

From this inequality and (3.74) we conclude

$$\lim_{t \to \infty} E\left[|x_u(t,t_0,0)|^2 | \eta(t_0) = i\right] = 0.$$

Finally, using (3.75) we obtain

$$\lim_{t \to \infty} E\left[|x_u(t,t_0,x_0)|^2 | \eta(t_0) = i\right] = 0$$

and the proof is complete. □

Remark 3.6.1. If we do not know that the system $(A_0, A_1, \ldots, A_r; Q)$ is stable, then the estimation from Theorem 3.6.1 (i) is not uniform with respect to $t, t_0 \in \mathbf{R}_+$. In general we may prove that for any compact interval $[t_0, t_1]$ there exists a positive constant c depending upon $t_1 - t_0$ such that

$$E\left[|x_u(t,t_0,x_0)|^2 | \eta(t_0) = i\right] \leq c \left(|x_0|^2 + \sum_{k=0}^{r} E\left[\int_{t_0}^{t_1} |f_k(s)|^2 ds | \eta(t_0) = i \right] \right)$$

for all $t \in [t_0, t_1], x_0 \in \mathbf{R}^n, i \in \mathcal{D}$ and all $f_k \in L_{\eta,w}^2([t_0, t_1], \mathbf{R}^n), 0 \leq k \leq r$.

To this end we notice that since $A_k(t,i), 0 \leq k \leq r, i \in \mathcal{D}$ are bounded on $\mathbf{R}_+$, from (3.73) and Theorem 1.9.7 it follows easily that there exists an absolute constant $\gamma > 1$ such that for all $t \in [t_0, t_1], i \in \mathcal{D}$ we have

$$E\left[|x_u(t,t_0,x_0)|^2 | \eta(t_0) = i\right] \leq \gamma \left\{ |x_0|^2 + E\left[\int_{t_0}^{t} |x_u(s,t_0,x_0)|^2 ds | \eta(t_0) = i \right] \right.$$
$$\left. \times ((t_1 - t_0) + 1) + \sum_{k=0}^{r} E\left[\int_{t_0}^{t_1} |f_k(s)|^2 ds | \eta(t_0) = i \right] ((t_1 - t_0) + 1) \right\}.$$

By using the Gronwall Lemma we get

$$\sup_{t_0 \leq t \leq t_1} E\left[|x_u(t,t_0,x_0)|^2 \eta(t_0) = i\right] \leq c \left(|x_0|^2 + \sum_{k=0}^{r} E\left[\int_{t_0}^{t_1} |f_k(s)|^2 | \eta(t_0) = i \right] \right),$$

$i \in \mathcal{D}$, where $c > 0$ depends only on $t_1 - t_0$.

Notes and References

In the control literature one can find a large number of works devoted to the stability of Itô-type differential equations systems. For this reason it is impossible to give an exhaustive bibliography for this topics. We shall limit ourselves to indicate the monographs of [6,7,14,22,26,92,98,99,148] where one can found many references concerning this subject. Theorem 3.3.6 has been proved in [100] for a larger class of systems of linear stochastic differential equations.

ESMS for stochastic systems of differential equations with Markov perturbations has been introduced and studied for the first time in [91] in which characterizations using Lyapunov-type equations are given.

The results in this chapter concerning time-varying linear differential systems with jump Markov perturbations have been proved in [112]. The mean square exponential stability for time-invariant differential systems with jump Markov perturbations has been investigated in [61,64,86,103,106,108,109].

The ESMS problem for differential equations subject both with Markov perturbations and with multiplicative white noise has also been considered in [104]. In that paper sufficient conditions for stability are given in terms of some M-matrices and it is proved that ESMS implies almost sure stability. Results concerning the stability and the boundedness of solutions of nonlinear Itô differential systems subject to Markov perturbations can be also found in [101].

The most results included in Sects. 3.2–3.3 have been proved in [41], while the ones from Sect. 3.4 were proved in [52].

Chapter 4
Structural Properties of Linear Stochastic Systems

In this chapter we present the stochastic version of some basic concepts in control theory, namely the stabilizability, detectability, observability and controllability. All these concepts are defined both in Lyapunov operators terms and in stochastic systems terms. The definitions given in this chapter extend the corresponding definitions from the deterministic time-varying systems. Some examples will show that the stochastic observability does not always imply stochastic detectability and stochastic controllability does not necessarily imply stochastic stabilizability. As in the deterministic case the concepts of stochastic detectability and observability are used in some criteria of exponential stability in mean square. Throughout this chapter we will assume that $\mathcal{D} = \{1, 2, \ldots, d\}$ even if some of its developments remain true when $\mathcal{D} = \mathbf{Z}_+$ (see, e.g., [63, 144]).

4.1 Stabilizability and Detectability of Stochastic Linear Systems

Let us consider the following stochastic input–output system:

$$dx(t) = [A_0(t, \eta(t))x(t) + B_0(t, \eta(t))u(t)] dt$$

$$+ \sum_{k=1}^{r} [A_k(t, \eta(t))x(t) + B_k(t, \eta(t))u(t)] dw_k(t) \qquad (4.1)$$

$$y(t) = C_0(t, \eta(t))x(t)$$

$t \in \mathbf{R}_+$ with the inputs $u \in \mathbf{R}^m$ and the outputs $y \in \mathbf{R}^p$. We denote $\mathbf{A} = (A_0, A_1, \ldots, A_r)$ and $\mathbf{B} = (B_0, B_1, \ldots, B_r)$.

Definition 4.1.1. (a) We say that the system (4.1) is *stochastically stabilizable* or equivalently, *the triple* $(\mathbf{A}, \mathbf{B}; Q)$ *is stabilizable* if there exists $F : \mathbf{R}_+ \to \mathcal{M}_{m,n}^d$ bounded and continuous function such that the zero solution of the system obtained by taking $u(t) = F(t, \eta(t))x(t)$, namely

$$dx(t) = [A_0(t, \eta(t)) + B_0(t, \eta(t))F(t, \eta(t))]x(t)dt$$

$$+ \sum_{k=1}^{r} [A_k(t, \eta(t)) + B_k(t, \eta(t))F(t, \eta(t))]x(t)dw_k(t),$$

$t \geq 0$ is ESMS-C.

(b) We say that the system (4.1) is *stochastically detectable*, or equivalently, the triple $(C_0, \mathbf{A}; Q)$ is detectable if there exists $K : \mathbf{R}_+ \to \mathcal{M}_{n,p}^d$ continuous and bounded function such that the zero solution of the system

$$dx(t) = [A_0(t, \eta(t)) + K(t, \eta(t))C_0(t, \eta(t))]x(t)dt + \sum_{k=1}^{r} A_k(t, \eta(t))x(t)dw_k(t)$$

is ESMS-C.

Remark 4.1.1. (a) The above definition of the stochastic detectability could also be stated if the output of the system (4.1) is of the form

$$dy(t) = C_0(t, \eta(t))x(t)dt + \sum_{k=1}^{r} C_k(t, \eta(t))x(t)dw_k(t).$$

(b) The function $F(t) = (F(t,1), F(t,2), \dots F(t,d))$ and the function $K(t) = (K(t,1), K(t,2), \dots, K(t,d))$, respectively, from the above definition will be termed stabilizing feedback gain and stabilizing injection, respectively.

The concepts of stochastic stabilizability and stochastic detectability in the particular cases when the system (4.1) is subject only either to Markovian jumping (i.e., $A_k = 0, B_k = 0, 1 \leq k \leq r$) or to multiplicative white noise (i.e., $\mathcal{D} = \{1\}$) are defined obviously in the same way. In the case of the systems with Markovian jumping only we shall say that $(A_0, B_0; Q)$ is stabilizable and $(C_0, A_0; Q)$ is detectable and in the case of Itô systems we shall say that $(\mathbf{A}, \mathbf{B})$ is stabilizable and $(C_0, \mathbf{A})$ is detectable.

Remark 4.1.2. If the system (4.1) is in "stationary case", then the stabilizing feedback gain and the stabilizing injection are supposed to be of the form $F = (F(1), \dots, F(d)), K = (K(1), \dots, K(d))$.

In the next chapter we shall show that in the case when the coefficients of the system (4.1) are θ-periodic functions with respect to their first argument, then this system is stochastically stabilizable (stochastically detectable), if and only if there exists a θ-periodic stabilizing feedback gain (a θ-periodic stabilizing injection, respectively). Moreover, if the system (4.1) is in the time invariant case, then it is stochastically stabilizable (stochastically detectable) if and only if there exists a stabilizing feedback gain $F = (F(1), F(2), \dots, F(d))$ (a stabilizing injection $K = (K(1), K(2), \dots, K(d))$, respectively).

Let us consider the following illustrative example with $n = 2$, $d = 2$ and $r = 1$ where

$$Q = \begin{bmatrix} -1 & 1 \\ 1 & -1 \end{bmatrix}, A_0(1) = \begin{bmatrix} -1 & 0 \\ \alpha & \beta \end{bmatrix}, A_0(2) = \begin{bmatrix} \gamma & \delta \\ 0 & -1 \end{bmatrix},$$

$$A_1(1) = \begin{bmatrix} a & 0 \\ 0 & 0 \end{bmatrix}, A_1(2) = \begin{bmatrix} 0 & 0 \\ 0 & a \end{bmatrix}, B(1) = \begin{bmatrix} 0 \\ 1 \end{bmatrix}, B(2) = \begin{bmatrix} 1 \\ 0 \end{bmatrix}$$

with $a^2 < 7/3$ and $\alpha, \beta, \gamma, \delta \in \mathbf{R}$. The system $(A_0, A_1, B; Q)$ is stabilizable. Indeed, let $F(1) = \begin{bmatrix} 1 - \alpha & -1 - \beta \end{bmatrix}$, $F(2) = \begin{bmatrix} -1 - \gamma & 1 - \delta \end{bmatrix}$. Then

$$A_0(1) + B(1)F(1) = \begin{bmatrix} -1 & 0 \\ 1 & -1 \end{bmatrix} \text{ and } A_0(2) + B(2)F(2) = \begin{bmatrix} -1 & 1 \\ 0 & -1 \end{bmatrix}$$

from which we deduce, according to Example 3.5.4 that $(A_0 + BF, A_1; Q)$ is stable. Let us remark that the pairs $(A_0(1), B(1))$ and $(A_0(2), B(2))$ are not controllable. One can also note that if $\beta \geq 1/2$ or $\gamma \geq 1/2$ then the system $(A_0, A_1; Q)$ is not stable since it does not satisfy the necessary conditions of stability, namely the matrices $A_0(i) + \frac{1}{2}q_{ii}I_2$, $i = 1, 2$ be stable.

The next result immediately follows.

Proposition 4.1.1. (i) *The system (4.1) is stochastically stabilizable if and only if there exists a continuous and bounded function* $F : \mathbf{R}_+ \to \mathcal{M}_{m,n}^d$ *such that the system* $(A_0 + B_0F, A_1 + B_1F, \ldots, A_r + B_rF; Q)$ *is stable.*

(ii) *The system (4.1) is stochastically detectable if and only if there exists a continuous and bounded function* $K : \mathbf{R}_+ \to \mathcal{M}_{n,p}^d$ *such that the system* $(A_0 + KC_0, A_1, \ldots A_r; Q)$ *is stable.*

From Theorems 3.3.1, 3.3.4, and 3.3.5 the following result can be obtained.

Proposition 4.1.2. (i) *If the system (4.1) is stochastically stabilizable (stochastically detectable, respectively), then the system with Markovian jumping:*

$$\dot{x}(t) = A_0(t, \eta(t))x(t) + B_0(t, \eta(t))u(t)$$

$$y(t) = C_0(t, \eta(t))x(t)$$

is stochastically stabilizable (stochastically detectable, respectively).

(ii) *If the system (4.1) is stochastically stabilizable (stochastically detectable, respectively) then, for each* $i \in \mathcal{D}$, *the system described by Itô differential equations:*

$$dx_i(t) = [\tilde{A}_0(t, i)x_i(t) + B_0(t, i)u(t)]dt$$

$$+ \sum_{k=1}^r [A_k(t, i)x_i(t) + B_k(t, i)u(t)]dw_k(t)$$

$$y_i(t) = C_0(t, i)x_i(t)$$

is stochastically stabilizable (stochastically detectable, respectively) where $\tilde{A}_0(t, i) = A_0(t, i) + \frac{1}{2}q_{ii}I_n$.

Remark 4.1.3. It is not difficult to see that the definition of the stochastic stabilizability and stochastic detectability can be stated for triplets $(\mathbf{A}, \mathbf{B}; Q)$ and $(\mathbf{C}, \mathbf{A}; Q)$ in the case when the elements of the matrix Q verify only condition (2.123); $\mathbf{C} = (C_0, C_1, \ldots, C_r)$ and A_k, B_k, C_k are continuous matrix valued functions on an right unbounded interval $\mathcal{I} \subseteq \mathbf{R}$.

More precisely:

Definition 4.1.2. (a) The triple $(\mathbf{A}, \mathbf{B}; Q)$ is *stabilizable* if there exists a bounded and continuous function $F : \mathcal{I} \to \mathcal{M}^d_{m,n}$ such that

$$\|T_F(t,s)\| \le \beta e^{-\alpha(t-s)}, \ \forall t \ge s \in \mathcal{I},$$

($\alpha > 0, \beta > 0$ being constants), $T_F(\cdot, \cdot)$ is the linear evolution operator defined by the linear differential equation over $\mathcal{S}^d_n$:

$$\frac{d}{dt} S(t) = \mathcal{L}_F(t) S(t),$$

where $\mathcal{L}_F(t) : \mathcal{S}^d_n \to \mathcal{S}^d_n$ by

$$(\mathcal{L}_F(t)S)(i) = [A_0(t,i) + B_0(t,i)F(t,i)] S(i) + S(i) [A_0(t,i) + B_0(t,i)F(t,i)]^T$$

$$+ \sum_{k=1}^{r} [A_k(t,i) + B_k(t,i)F(t,i)] S(i) \tag{4.2}$$

$$\times [A_k(t,i) + B_k(t,i)F(t,i)]^T + \sum_{j=1}^{d} q_{ji} S(j)$$

$i \in \mathcal{D}, S \in \mathcal{S}^d_n$.

(b) The triple $(\mathbf{C}, \mathbf{A}; Q)$ is *detectable* if there exists a bounded and continuous function $K : \mathcal{I} \to \mathcal{M}^d_{n,p}$, such that $\|T^K(t,s)\| \le \beta e^{-\alpha(t-s)}, \ \forall t \ge s \in \mathcal{I}, \beta > 0, \alpha > 0$ being constants. $T^K(t,s)$ is the linear evolution operator defined by the linear differential equation:

$$\frac{d}{dt} S(t) = \mathcal{L}^K(t) S(t)$$

where $\mathcal{L}^K(t) : \mathcal{S}^d_n \to \mathcal{S}^d_n$ by

$$[\mathcal{L}^K(t)S](i) = [A_0(t,i) + K(t,i)C_0(t,i)] S(i) + S(i) [A_0(t,i) + K(t,i)C_0(t,i)]^T$$

$$+ \sum_{k=1}^{r} [A_k(t,i) + K(t,i)C_k(t,i)] S(i) [A_k(t,i) + K(t,i)C_k(t,i)]^T$$

$$+ \sum_{j=1}^{d} q_{ji} S(j) \tag{4.3}$$

$i \in \mathcal{D}, S \in \mathcal{S}^d_n$.

The next result easily follows from Theorem 3.3.3.

Proposition 4.1.3. *Assume that the system (4.1) is in the time-invariant case. Then the following are equivalent:*

(i) *The system (4.1) is stochastically stabilizable;*
(ii) *There exists* $F = (F(1), F(2), \ldots, F(d)) \in \mathcal{M}_{m,n}^d$ *such that the affine Lyapunov equation over* $\mathcal{S}_n^d$ *:*

$$\mathcal{L}_F X + J^d = 0$$

has a solution $X > 0$;
(iii) *The linear matrix inequalities*

$$\begin{bmatrix} \mathcal{L}(X,\Gamma)(i) & \mathcal{P}(X,\Gamma)(i) \\ \mathcal{P}^T(X,\Gamma)(i) & \mathcal{R}(X)(i) \end{bmatrix} < 0 \tag{4.4}$$

have a solution $(X,\Gamma) \in \mathcal{S}_n^d \times \mathcal{M}_{m,n}^d, X > 0$, *where*

$$\mathcal{L}(X,\Gamma)(i) = A_0(i)X(i) + X(i)A_0^T(i) + B_0(i)\Gamma(i) + \Gamma^T(i)B_0^T(i) + \sum_{j=1}^d q_{ji}X(j)$$

$$\mathcal{P}(X,\Gamma)(i) = (A_1(i)X(i) + B_1(i)\Gamma(i) \quad A_2(i)X(i) + B_2(i)\Gamma(i) \quad \ldots \quad A_r(i)X(i) + B_r(i)\Gamma(i))$$

$$\mathcal{R}(X)(i) = \begin{bmatrix} -X(i) & 0 & 0 & \ldots & 0 \\ 0 & -X(i) & 0 & \ldots & 0 \\ 0 & 0 & -X(i) & \ldots & 0 \\ \ldots & \ldots & \ldots & \ldots & \ldots \\ 0 & 0 & 0 & \ldots & -X(i) \end{bmatrix} \in \mathcal{S}_{rn}.$$

Moreover, if $(X,\Gamma) \in \mathcal{S}_n^d \times \mathcal{M}_{m,n}^d$ *is a solution of the linear matrix inequalities (4.4) with* $X > 0$, *then* $F = (F(1), F(2), \ldots F(d))$, *with*

$$F(i) = \Gamma(i)X(i)^{-1} \tag{4.5}$$

$i \in \mathcal{D}$ *is a stabilizing feedback gain.*

In the particular case with $B_k = 0, k = 1, 2 \ldots, r$ we have the following result.

Proposition 4.1.4. *Assume that the system (4.1) is in the time-invariant case and* $B_k(i) = 0, i \in \mathcal{D}, k = 1, .., r$, *then the following are equivalent:*

(i) *The system (4.1) is stochastically stabilizable;*
(ii) *The system of linear matrix equations*

$$A_0(i)X(i) + X(i)A_0^T(i) + B_0(i)\Gamma(i) + \Gamma^T(i)B_0^T(i)$$
$$+ \sum_{k=1}^r A_k(i)X(i)A_k^T(i) + \sum_{j=1}^d q_{ji}X(j) + I_n = 0 \tag{4.6}$$

$i \in \mathcal{D}$, has a solution $(X, \Gamma) \in \mathcal{S}_n^d \times \mathcal{M}_{m,n}^d, X > 0$. Moreover, if $(X, \Gamma) \in \mathcal{S}_n^d \times \mathcal{M}_{m,n}^d$ is a solution of the system (4.6) with $X > 0$ then a stabilizing feedback gain may be obtained as in (4.5).

The next result follows easily from Theorem 3.3.2.

Proposition 4.1.5. *Assume that the system (4.1) is in the time-invariant case, then the following are equivalent:*

 (i) *The system (4.1) is stochastically detectable;*
(ii) *The system of linear matrix equations*

$$
\begin{aligned}
& A_0^T(i)Y(i) + Y(i)A_0(i) + \Lambda(i)C_0(i) + C_0(i)^T \Lambda^T(i) \\
& + \sum_{k=1}^r A_k^T(i)Y(i)A_k(i) + \sum_{j=1}^d q_{ij}Y(j) + I_n = 0
\end{aligned}
\tag{4.7}
$$

$i \in \mathcal{D}$ has a solution $(Y, \Lambda) \in \mathcal{S}_n^d \times \mathcal{M}_{n,p}^d, Y > 0$. Moreover, if (Y, Λ) is a solution of the system (4.7), then $K = (K(1), \ldots, K(d))$, with

$$
K(i) = Y^{-1}(i)\Lambda(i)
\tag{4.8}
$$

$i \in \mathcal{D}$ is a stabilizing injection;
(iii) *The system of linear matrix inequalities*

$$
\begin{aligned}
& A_0^T(i)Y(i) + Y(i)A_0(i) + \Lambda(i)C_0(i) + C_0^T(i)\Lambda^T(i) \\
& + \sum_{k=1}^r A_k^T(i)Y(i)A_k(i) + \sum_{j=1}^d q_{ij}Y(j) < 0
\end{aligned}
\tag{4.9}
$$

$i \in \mathcal{D}$ has a solution $(Y, \Lambda) \in \mathcal{S}_n^d \times \mathcal{M}_{n,p}^d, Y > 0$. Moreover if (Y, Λ) is a solution of the system (4.9) with $Y > 0$, then a stabilizing injection is obtained as in (4.8).

Based on the Remark 4.1.3 we can establish a duality relationship between the stabilizability and detectability in this stochastic framework.

Proposition 4.1.6. *Assume that*

 (i) *$A_k : \mathbf{R} \to \mathcal{M}_n^d, B_k : \mathbf{R} \to \mathcal{M}_{n,m}^d$ are continuous and bounded functions, $k = 0, 1, \ldots, r$;*
(ii) *The elements of the matrix Q, verify (2.123).*

Then the triple $(\mathbf{A}, \mathbf{B}; Q)$ is stabilizable if and only if the triple $(\mathbf{B}^\sharp, \mathbf{A}^\sharp; Q^\sharp)$ is detectable, where

$$
\mathbf{A}^\sharp = \left(A_0^\sharp, A_1^\sharp, \ldots, A_r^\sharp \right), \; B^\sharp = \left(B_0^\sharp, B_1^\sharp, \ldots, B_r^\sharp \right),
$$

$$
A_k^\sharp(t) = \left(A_k^\sharp(t,1), A_k^\sharp(t,2), \ldots, A_k^\sharp(t,d) \right),
$$

$$
B_k^\sharp(t) = \left(B_k^\sharp(t,1), B_k^\sharp(t,2), \ldots, B_k^\sharp(t,d) \right)
$$

$$A_k^\sharp(t,i) \; : \; = A_k^T(-t,i),$$

$$B_k^\sharp(t,i) \; : \; = B_k^T(-t,i)$$

$$Q^\sharp = Q^T.$$

$t \in \mathbf{R}, i \in \mathcal{D}, k = 0, 1, \dots, r.$

Proof. If $(\mathbf{A}, \mathbf{B}; Q)$ is stabilizable, then there exists a bounded and continuous function $F : \mathbf{R} \to \mathcal{M}_{m,n}^d$ such that:

$$\|T_F(t,s)\| \leq \beta e^{-\alpha(t-s)} \tag{4.10}$$

for all $t \geq s, t, s \in \mathbf{R}, \beta > 0, \alpha > 0$ being positive constants $T_F(\cdot, \cdot)$ being the linear evolution operator defined by linear differential equation over $\mathcal{S}_n^d$,

$$\frac{d}{dt} S(t) = \mathcal{L}_F(t) S(t) \tag{4.11}$$

$\mathcal{L}_F(t)$ being defined as in (4.2).

It is easy to see that $S(t)$ is a solution of (4.11) if and only if $t \to S(-t)$ is a solution of the equation

$$\frac{d}{dt} X(t) + (\mathcal{L}^\sharp(t))^* X(t) = 0 \tag{4.12}$$

where $\mathcal{L}^\sharp(t) : \mathcal{S}_n^d \to \mathcal{S}_n^d$ is defined by

$$(\mathcal{L}^\sharp(t)S)(i) = \left[A_0^\sharp(t,i) + K^\sharp(t,i) B_0^\sharp(t,i) \right] S(i)$$

$$+ S(i) \left[A_0^\sharp(t,i) + K^\sharp(t,i) B_0^\sharp(t,i) \right]^T$$

$$+ \sum_{k=1}^r \left[A_k^\sharp(t,i) + K^\sharp(t,i) B_k^\sharp(t,i) \right] S(i)$$

$$\times \left[A_k^\sharp(t,i) + K^\sharp(t,i) B_k^\sharp(t,i) \right]^T$$

$$+ \sum_{j=1}^d q_{ji}^\sharp S(j), i \in \mathcal{D}, S \in \mathcal{S}_n^d,$$

where $A_k^\sharp, B_k^\sharp$ were defined in the statement and $K^\sharp(t,i) = F^T(-t,i), q_{ji}^\sharp = q_{ij}, i, j \in \mathcal{D}$. If $T^\sharp(t,s)$ stands for the linear evolution operator over $\mathcal{S}_n^d$ defined by the differential equation

$$\frac{d}{dt} S(t) = \mathcal{L}^\sharp(t) S(t)$$

then we obtain from (4.12) that $S(-t) = (T^\sharp(s,t))^* S(-s)$ for all $t \leq s$, hence $S(t) = (T^\sharp(-s,-t))^* S(s)$ for all $t \geq s$.

On the other hand $S(t) = T_F(t,s)S(s), t \geq s$. Hence we have $T^{\sharp}(t,s) = T_F^*(-s,-t)$, $(\forall)t \geq s$. Finally invoking (4.10) we deduce that

$$||T^{\sharp}(t,s)|| \leq \beta e^{-\alpha(t-s)}, \ (\forall) \ t \geq s$$

which shows that $(\mathbf{B}^{\sharp}, \mathbf{A}^{\sharp}; Q^{\sharp})$ is detectable and the proof ends. $\square$

Remark 4.1.4. (i) In the same way may prove that $(\mathbf{C}, \mathbf{A}; Q)$ is detectable, if and only if $(\mathbf{A}^{\sharp}, \mathbf{C}^{\sharp}; Q^{\sharp})$ is stabilizable.
(ii) From Proposition 4.1.6 it follows immediately that in the time invariant case $(\mathbf{A}, \mathbf{B}; Q)$ is stabilizable if and only if the triple $(\mathbf{B}^T, \mathbf{A}^T; Q^T)$ is detectable.

Now we prove the following theorem, which extends a well-known result from the deterministic framework.

Theorem 4.1.7. *Suppose that*

 (i) $(C_0, \mathbf{A}; Q)$ *is stochastically detectable.*
(ii) *The differential equation*

$$\frac{d}{dt}K(t) + \mathcal{L}^*(t)K(t) + \tilde{C}(t) = 0 \tag{4.13}$$

has a bounded solution $\tilde{K} : \mathbf{R}_+ \to \mathcal{S}_n^d, \tilde{K}(t) = \big(\tilde{K}(t,1), \ldots, \tilde{K}(t,d)\big), \tilde{K}(t,i) \geq 0, t \geq 0, i \in \mathcal{D}$ *where* $\tilde{C}(t) = \big(\tilde{C}(t,1), \ldots, \tilde{C}(t,d)\big), \tilde{C}(t,i) = C_0^T(t,i)C_0(t,i).$ *Then the solution of the system* (1.22) *is ESMS-C (or equivalently the system* $((A_0, A_1, \ldots, A_r); Q)$ *is stable.*

Proof. Consider $v : \mathbf{R}_+ \times \mathbf{R}^n \times \mathcal{D} \to \mathbf{R}, v(t,x,i) = x^T \tilde{K}(t,i)x.$ Let $x(t) = x(t,t_0,x_0)$ be a solution of the system (1.22). Applying the identity (1.6) to the function v and to the system (1.22) and taking into account (4.13) we get for all $t \geq t_0$ and $i \in \mathcal{D}$

$$E\left[v(t,x(t),\eta(t))|\eta(t_0)=i\right] - x_0^T \tilde{K}(t_0,i)x_0 = -E\left[\int_{t_0}^t |C_0(s,\eta(s))x(s)|^2 ds|\eta(t_0) = i\right].$$

Hence

$$E\left[\int_{t_0}^{\infty} |C_0(t,\eta(t))x(t)|^2 dt|\eta(t_0) = i\right] \leq x_0^T K(t_0,i)x_0 \leq \gamma|x_0|^2 \tag{4.14}$$

$t_0 \geq 0, x_0 \in \mathbf{R}^n, i \in \mathcal{D}.$
We may write

$$dx(t) = \{[A_0(t,\eta(t)) + H(t,\eta(t))C_0(t,\eta(t))]x(t) + f_0(t)\}dt$$

$$+ \sum_{k=1}^r A_k(t,\eta(t))x(t)dw_k(t)$$

where $f_0(t) = -H(t,\eta(t))C_0(t,\eta(t))x(t).$

Since the system $(A_0 + HC_0, A_1, \ldots, A_r; Q)$ is stable and $f_0 \in L^2_{\eta,w}([t_0, \infty) \times \mathbf{R}^n)$ (see (4.14)) we may use the Theorem 3.6.1 (ii) to obtain;

$$E\left[\int_{t_0}^{\infty} |\Phi(t,t_0)x_0|^2 dt \,|\, \eta(t_0) = i\right] \leq \delta[|x_0|^2 + E\left[\int_{t_0}^{\infty} |f_0(t)|^2 dt \,|\, \eta(t_0) = i\right]$$

$$\leq \beta|x_0|^2$$

for all $t_0 \geq 0, x_0 \in \mathbf{R}^n, i \in \mathcal{D}$.

Using Corollary 3.2.3 we conclude that the system $(A_0, A_1, \ldots, A_r; Q)$ is stable and the proof is complete. $\qquad\qquad\square$

Remark 4.1.5. If $(\mathbf{C}, \mathbf{A}; Q)$ is detectable, then it follows based on a similar proof that the result proved in Theorem 4.1.7 remains valid if one replaces $\widetilde{C}(t)$ with

$$\widetilde{C}(t,i) = \sum_{k=0}^{r} C_k^T(t,i) C_k(t,i).$$

4.2 Stochastic Observability

Definition 4.2.1. We say that the system (4.1) is *stochastically uniformly observable* (or equivalently $(C_0, \mathbf{A}; Q)$ *is uniformly observable*) if there exist $\tau > 0, \beta > 0$ such that

$$\int_t^{t+\tau} T^*(s,t)\widetilde{C}(s)ds \geq \beta J^d \tag{4.15}$$

for all $t \geq 0$, where $\widetilde{C}(s) = (\widetilde{C}(s,1), \widetilde{C}(s,2), \ldots, \widetilde{C}(s,d)), \widetilde{C}(s,i) = C_0^T(s,i)C_0(s,i)$, $i \in \mathcal{D}, s \geq 0$. In the time invariant case we shall say that the system (4.1) is *stochastically observable*, or the triple $(C_0, \mathbf{A}; Q)$ is *observable*.

Remark 4.2.1. (a) If in the system (4.1) we have $A_k(t,i) = 0, k = 1, \ldots, r, \mathcal{D} = \{1\}$, then the Lyapunov operator (2.124) is the Lyapunov operator of deterministic framework. In this case (4.15) becomes

$$\int_t^{t+\tau} \Phi_0^T(s,t)C_0^T(s)C_0(s)\Phi_0(s,t)ds \geq \beta I_n, \forall t \geq 0$$

where $\Phi_0(\cdot, \cdot)$ is the fundamental matrix solution of the differential equation $\dot{x}(t) = A_0(t)x(t)$.

This shows that the above definition of stochastic uniform observability is a natural extension of the uniform observability used for linear time-varying deterministic systems (see [89]).

(b) If the system (4.1) is subject only to Markovian jumping, then the condition (4.15) becomes $\int_t^{t+\tau} \widetilde{T}^*(s,t)\widetilde{C}(s)ds \geq \beta J^d$. If this is fulfilled we shall say that the triple $(C_0, A_0; Q)$ is uniformly observable.

(c) If the system (4.1) is subject only to multiplicative white noise and the corresponding inequality (4.15) is fulfilled, then we shall say that $(C_0, A_0, A_1, \ldots, A_r)$ or shortly $(C_0, \mathbf{A})$ is *uniformly observable*.

The following result follows immediately from Theorem 3.1.1.

Proposition 4.2.1. *The system (4.1) is stochastically uniformly observable if and only if there exist* $\beta > 0, \tau > 0$ *such that*

$$E\left[\int_t^{t+\tau} \Phi^T(s,t)C_0^T(s,\eta(s))C_0(s,\eta(s))\Phi(s,t)ds|\eta(t) = i\right] \geq \beta I_n$$

for all $t \geq 0, i \in \mathcal{D}, \Phi(\cdot,\cdot)$ *being the fundamental matrix solution of the system (1.22).*

The proof of the next result is based on some preliminary results that develop the ones presented in Sect. 2.7. First, remark that since

$$\Phi_i(t,t_0) = e^{\frac{1}{2}q_{ii}(t-t_0)}\widehat{\Phi}_i(t,t_0)$$

where $\Phi_i(t,t_0)$ is the fundamental matrix solution for fixed $i \in \mathcal{D}$ of the deterministic linear differential equation on $\mathbf{R}^n$

$$\frac{dx}{dt} = \left[A_0(t,i) + \frac{1}{2}q_{ii}I_n\right]x(t),$$

and $\widehat{\Phi}_i(t,t_0)$ is the fundamental matrix solution for fixed $i \in \mathcal{D}$ of the deterministic linear differential equation

$$\frac{dx}{dt} = A_0(t,i)x(t),$$

it follows that for each $i \in \mathcal{D}$ the pair $\left(C_0(\cdot,i), \widetilde{A}_0(\cdot,i)\right)$ is uniformly observable if and only if the pair $(C_0(\cdot,i), A_0(\cdot,i))$ is uniformly observable, where

$$\widetilde{A}_0(t,i) = A_0(t,i) + \frac{1}{2}q_{ii}I_n.$$

Further, for each $i \in \mathcal{D}$, let $\mathcal{L}^i(t) : \mathcal{S}_n \to \mathcal{S}_n$ be the Lyapunov-type linear operator defined by

$$\mathcal{L}^i(t)M = \widetilde{A}_0(t,i)M + M\widetilde{A}_0^T(t,i) + \sum_{j=1}^r A_j(t,i)MA_j^T(t,i), \; M \in \mathcal{S}_n$$

and let $T^i(t,t_0)$ be the linear evolution operator on $\mathcal{S}_n$ associated with the operator $\mathcal{L}^i(t)$.

Let $\overline{\mathcal{L}}(t) : \mathcal{S}_n^d \to \mathcal{S}_n^d$ defined by

$$\left(\overline{\mathcal{L}}(t)H\right)(i) = \mathcal{L}^i(t)H(i), \ H \in \mathcal{S}_n^d, i \in \mathcal{D}$$

and let $\overline{T}(t,t_0)$ be the linear evolution operator on $\mathcal{S}_n^d$ associated with the linear operator $\overline{\mathcal{L}}(t)$. It is easy to prove that

$$\left(\overline{T}(t,t_0)H\right)(i) = T^i(t,t_0)H(i), \ H \in \mathcal{S}_n^d, i \in \mathcal{D}.$$

From the definitions of $\widehat{T}(t,t_0)$, $T_1(t,t_0)$ (see Sect. 2.6.2) easily follows

$$T(t,t_0) \geq \widehat{T}(t,t_0) \geq T_1(t,t_0), \tag{4.16}$$

$$T(t,t_0) \geq \overline{T}(t,t_0).$$

From (4.16) one obtains the following result.

Proposition 4.2.2. *(i) If for each $i \in \mathcal{D}$, the pair $(C_0(\cdot,i),A_0(\cdot,i))$ is uniformly observable then the triple $(C_0,A_0;Q)$ is uniformly observable.*
(ii) If $(C_0,A_0;Q)$ is uniformly observable, then $(C_0,\mathbf{A};Q)$ is uniformly observable.
(iii) If for every $i \in \mathcal{D}$, the system $\left(C_0(\cdot,i),\widetilde{A}_0(\cdot,i),A_1(\cdot,i),\ldots,A_r(\cdot,i)\right)$ is uniformly observable then the system $(C_0,\mathbf{A};Q)$ is uniformly observable, too.

Proposition 4.2.3. *Assume that the system (4.1) is in the time-invariant case. Then the following are equivalent.*

(i) The system (4.1) is stochastically observable.
(ii) There exists $\tau > 0$ such that

$$\int_0^\tau e^{\mathcal{L}^* s}\widetilde{C}ds > 0.$$

(iii) There exists $\tau > 0$ such that $X_0(\tau) > 0$, where $X_0(t)$ is the solution of the problem with initial value:

$$\frac{d}{dt}X_0(t) = \mathcal{L}^*X_0(t) + \widetilde{C}, \quad X_0(0) = 0.$$

Proof. (i) $\Longleftrightarrow$ (ii) follows from (2.132).

Since $X_0(t) = \displaystyle\int_0^t e^{\mathcal{L}^*(t-s)}\widetilde{C}ds = \int_0^t e^{\mathcal{L}^* s}\widetilde{C}ds$, $t \geq 0$ it follows that (iii) $\Longleftrightarrow$ (ii). The proof is complete. $\qquad\square$

Proposition 4.2.4. *Assume that the system (4.1) is in the time-invariant case. Let $X_0(t)$ be the solution of the Cauchy problem on $\mathcal{S}_n^d$*

$$\frac{d}{dt}X_0(t) = \mathcal{L}^*X_0(t) + \widetilde{C}, \ t \geq 0, \quad X_0(0) = 0.$$

If there exists $\tau > 0$, such that $X_0(\tau) > 0$ then $X_0(t) > 0$ for all $t > 0$.

Proof. For each $t > 0$, we write the representation

$$X_0(t) = (X_0(t,1), X_0(t,2),\ldots,X(t,d)) = \int_0^t e^{\mathcal{L}^*(t-s)}\tilde{C}ds.$$

Since $e^{\mathcal{L}^*(t-s)} : \mathcal{S}_n^d \to \mathcal{S}_n^d$ is a positive operator, we deduce that $X_0(t) \geq 0$, for all $t \geq 0$. Moreover if $t \geq \tau$ we have $X_0(t) \geq X_0(\tau)$, therefore if $X_0(\tau) > 0$, we have $X_0(t) > 0$ for all $t \geq \tau$. It remains to show that $X_0(t) > 0, 0 < t < \tau$. To this end we show that $detX_0(t,i) > 0$, $0 < t < \tau$, $i \in \mathcal{D}$. Indeed, since $detX_0(t,i) = det\left\{(\int_0^t e^{\mathcal{L}^*(t-s)}\tilde{C}ds)(i)\right\}$, we deduce that $t \to detX_0(t,i)$ is an analytic function.

The set of its zeros on $[0,\tau]$ has no accumulation point. In this way it will follow that there exists $\tau_1 > 0$ such that $detX_0(t,i) > 0$ for all $t \in (0,\tau_1]$. Invoking again the monotonicity of the function $t \to X_0(t)$ we conclude that $X_0(t) > 0$ for all $t \geq \tau_1$, and the proof ends. $\square$

Remark 4.2.2. From Propositions 4.2.3 and 4.2.4 it follows that the stochastic observability for a system (4.1) in the time invariant case may be checked by using a numerical procedure to compute the solution $X_0(t)$ through an enough long interval of time.

The following two results can be considered as Barbashin–Krasovskii type theorems [74].

Theorem 4.2.5. *Assume that $(C_0,\mathbf{A};Q)$ is uniformly observable and the affine differential equation*

$$\frac{d}{dt}X(t) + \mathcal{L}^*(t)X(t) + \tilde{C}(t) = 0 \tag{4.17}$$

has a bounded and positive semidefinite solution $\tilde{X}(t), t \geq 0$. Then

(i)   *The system $(A_0, A_1,\ldots,A_r; Q)$ is stable;*
(ii) *$\tilde{X}(t) \gg 0$;*
(iii) *Equation (4.17) has only one bounded solution which is uniform positive.*

Proof. From (2.127) it follows that

$$\tilde{X}(t) = T^*(s,t)\tilde{X}(s) + \int_t^s T^*(u,t)\tilde{C}(u)du, \quad s \geq t. \tag{4.18}$$

Since $0 \leq \tilde{X}(s) \leq \beta_0 J^d$ with some $\beta_0 > 0$ and $T(s,t) \geq 0$ one gets $0 \leq \int_t^s T^*(u,t)\tilde{C}(u)du \leq \tilde{X}(t) \leq \beta_0 J^d$ for all $s \geq t \geq 0$. Hence the integral $\hat{X}(t) = \int_t^\infty T^*(s,t)\tilde{C}(s)ds$ is convergent and $0 \leq \hat{X}(t) \leq \beta_0 J^d$, $t \geq 0$. By (2.127) it follows directly that $\hat{X}$ is a solution of (4.17). Since $(C_0; A_0,\ldots,A_r, Q)$ is uniformly observable it follows that $\hat{X}$ is uniform positive. Since $T^*(t+\tau,t)T^*(s,t+\tau) = T^*(s,t)$ we have

$$T^*(t+\tau,t)\hat{X}(t+\tau) = \int_{t+\tau}^\infty T^*(s,t)\tilde{C}(s)ds = \hat{X}(t) - \int_t^{t+\tau} T^*(s,t)\tilde{C}(s)ds$$

Hence $T^*(t+\tau)\widehat{X}(t+\tau) \le \widehat{X}(t) - \beta J^d \le \left(1 - \frac{\beta}{\beta_0}\right)\widehat{X}(t), t \ge 0$. Thus by Theorem 3.3.8 it follows that the system $(A_0,\dots,A_r,Q)$ is stable. Hence by Theorem 2.7.4 (ii), $\|T^*(s,t)\| \le \gamma e^{-\alpha(s-t)}, s \ge t$. Taking $s \to \infty$ in (4.18) one gets $\widecheck{X}(t) = \widehat{X}(t), t \ge 0$ and thus the proof is complete. $\qquad\square$

Corollary 4.2.6. *Suppose that* $A_k(t,i) = A_k(i)$, $C_0(t,i) = C(i), t \ge 0, i \in \mathcal{D}, 0 \le k \le r$. *Assume that* $(C_0; A_0, \dots, A_r, Q)$ *is observable and the algebraic equation on* $\mathcal{S}_n^d$

$$\mathcal{L}^*X + \widetilde{C} = 0 \tag{4.19}$$

has a solution $\widetilde{X} \ge 0$. *Then*

(i) *The system* $(A_0, A_1, \dots, A_r; Q)$ *is stable.*
(ii) $\widetilde{X} > 0$.
(iii) *Equation (4.19) has a unique positive semidefinite solution.*

The next result gives sufficient conditions concerning the observability of the system $(C_0, A_0, \dots A_r; Q)$.

Theorem 4.2.7. *Under the assumption of Proposition 4.2.3 if the system* $(C_0; A_0, \dots, A_r, Q)$ *is not observable then there exist* $x_0 \in \mathbf{R}^n, x_0 \ne 0$ *and* $i_0 \in \mathcal{D}$ *such that*

(i) $C_0(i_0)x_0 = 0$.
(ii) $q_{i_0 i} C_0(i)x_0 = 0$ *for all* $i \in \mathcal{D}$.
(iii) $C_0(i_0)(A_0(i_0))^m x_0 = 0$ *for all* $m \ge 1$.
(iv) $q_{i_0 i} q_{ij} C_0(j)x_0 = 0$ *for all* $i \ne i_0, j \in \mathcal{D}$.
(v) $C_0(i_0)A_k(i_0)x_0 = 0$, $1 \le k \le r$.

Proof. Suppose that $(C_0, A_0, \dots, A_r; Q)$ is not observable. From Proposition 4.2.3 it follows that there exist $x_0 \in \mathbf{R}^n, x_0 \ne 0$ and $i_0 \in \mathcal{D}$ such that $x_0^T \int_0^1 (e^{\mathcal{L}^*t}\widetilde{C})(i_0)dt x_0 = 0$.

Hence $x_0^T(e^{\mathcal{L}^*t}\widetilde{C})(i_0)x_0 = 0$ for all $t \in [0,1]$. Since $e^{\mathcal{L}^*t} \ge e^{\widehat{\mathcal{L}}^*t} \ge e^{\mathcal{L}_1^*t}$ (see 4.16 and Remark 2.6.3) one gets $x_0^T(e^{\widehat{\mathcal{L}}^*t}\widetilde{C})(i_0)x_0 = 0, x_0^T(e^{\mathcal{L}_1^*t}\widetilde{C})(i_0)x_0 = 0, t \in [0,1]$. From the last equality we get $C_0(i_0)e^{A_0(i_0)t}x_0 = 0, t \in [0,1)$.

Hence differentiating successively we have

$$x_0^T((\mathcal{L}^*)^m\widetilde{C})(i_0)x_0 = 0, \quad m \ge 0 \tag{4.20}$$

$$C_0(i_0)(A_0(i_0))^m x_0 = 0, \quad m \ge 0 \tag{4.21}$$

$$x_0^T((\widehat{\mathcal{L}}^*)^m\widetilde{C})(i_0)x_0 = 0, x_0^T((\mathcal{L}_1^*)^m\widetilde{C})(i_0)x_0 = 0 \tag{4.22}$$

for all $m \ge 0$. Thus (i) and (iii) follow from (4.21).

Now, from (4.20) and (4.22) we have

$$0 = x_0^T(\mathcal{L}^*\widetilde{C})(i_0)x_0 = x_0^T(\mathcal{L}_2^*\widetilde{C})(i_0)x_0 + x_0^T(\widehat{\mathcal{L}}^*\widetilde{C})(i_0)x_0$$

$$= x_0^T (\mathcal{L}_2^* \tilde{C})(i_0)x_0 = x_0^T \sum_{k=1}^{r} A_k^T(i_0)C_0^T(i_0)C_0(i_0)A_k(i_0)x_0$$

and thus (v) follows.

Further, by (4.22) we can write

$$0 = x_0^T (\hat{\mathcal{L}}^* \tilde{C})(i_0)x_0 = x_0^T (\mathcal{L}_1^* \tilde{C})(i_0)x_0 + x_0^T (\mathcal{L}_3^* \tilde{C})(i_0)x_0$$

$$= x_0^T (\mathcal{L}_3^* \tilde{C})(i_0)x_0 = x_0^T \sum_{j \neq i_0} q_{i_0 j} C_0^T(j)C_0(j)x_0$$

where $\mathcal{L}$, $\hat{\mathcal{L}}$, $\mathcal{L}_1$ are defined in Sect. 2.6.2 and $\mathcal{L}_2 = \mathcal{L} - \hat{\mathcal{L}}$ and $\mathcal{L}_3 = \hat{\mathcal{L}} - \mathcal{L}_1$. Then since $q_{ij} \geq 0$ if $i \neq j$ one gets (ii). From (4.22) it also follows that

$$0 = x_0^T ((\hat{\mathcal{L}}^*)^2 \tilde{C})(i_0)x_0 =$$

$$= x_0^T \left\{ \left[\left((\mathcal{L}_1^*)^2 + \mathcal{L}_1^* \mathcal{L}_3^* + \mathcal{L}_3^* \mathcal{L}_1^* + (\mathcal{L}_3^*)^2 \right) \tilde{C} \right] (i_0) \right\} x_0 =$$

$$= x_0^T \left[(\mathcal{L}_1^* \mathcal{L}_3^* \tilde{C})(i_0) + (\mathcal{L}_3^* \mathcal{L}_1^* \tilde{C})(i_0) + ((\mathcal{L}_3^*)^2 \tilde{C})(i_0) \right] x_0.$$

Further, using (ii) we can write

$$x_0^T (\mathcal{L}_1^* \mathcal{L}_3^* \tilde{C})(i_0)x_0 = 2x_0^T \left[A_0^T(i_0) + \frac{1}{2} q_{i_0 i_0} I_n \right] \sum_{i \neq i_0} q_{i_0 i} C_0^T(i)C_0(i)x_0 = 0$$

$$x_0^T (\mathcal{L}_3^* \mathcal{L}_1^* \tilde{C})(i_0)x_0 = 2x_0^T \sum_{i \neq i_0} q_{i_0 i} \left(A_0^T(i) + \frac{1}{2} q_{ii} I_n \right) C_0^T(i)C_0(i)x_0 = 0.$$

Hence one gets

$$0 = x_0^T ((\mathcal{L}_3^*)^2 \tilde{C})(i_0)x_0 = x_0^T \sum_{i \neq i_0} \sum_{j \neq i} q_{i_0 i} q_{ij} C_0^T(j)C_0(j)x_0$$

and since $q_{i_0 i} q_{ij} \geq 0$ for $i \neq i_0, j \neq i$ one obtains $q_{i_0 i} q_{ij} C(j)x_0 = 0$ for all $i \neq i_0$ and $j \neq i$ and thus by (ii) it follows that (iv) holds and hence the proof is complete. $\square$

Corollary 4.2.8. *Under the assumption of Proposition 4.2.3 if for every $i \in \mathcal{D}$, rank $M(i) = n$, where*

$$M(i) = \left[C_0^T(i), A_0^T(i)C_0^T(i), \ldots, (A_0^T(i))^{n-1}C_0^T(i), \right.$$

$$\left. q_{i1}C_0^T(1), \ldots, q_{id}C_0^T(d), A_1^T(i)C_0^T(i), \ldots, A_r^T(i)C_0^T(i) \right]$$

then the system $(C_0, A_0, A_1, \ldots, A_r; Q)$ is observable.

In the following examples, the stochastic observability used in this paper is compared with other types of stochastic observability, for example the one introduced in [86, 109, 110]. We also show that the stochastic observability used in this paper doesn't imply the stochastic detectability as we would expect.

Example 4.2.1. The case of a system with Markovian jumping with $d = 2, n = 2, p = 1$.

Take

$$A_0(1) = A_0(2) = \begin{bmatrix} \alpha & 0 \\ 0 & \alpha \end{bmatrix}$$

$$C_0(1) = \begin{bmatrix} 1 & 0 \end{bmatrix}, C_0(2) = \begin{bmatrix} 0 & 1 \end{bmatrix}, Q = \begin{bmatrix} -q & q \\ q & -q \end{bmatrix}, \alpha \in \mathbf{R}, q > 0.$$

It is obvious that the pairs $(C_0(1), A_0(1)), (C_0(2), A_0(2))$ are not observable. Therefore this system is not stochastically observable, in the sense of [109]. We shall show that this system is stochastically observable in the sense of Definition 4.2.1.

To this end we use the implication (iii) $\Longrightarrow$ (i) in Proposition 4.2.3. We show that there exists $\tau > 0$ such that $X_1(\tau) > 0, X_2(\tau) > 0$, where $X_i(t), i = 1, 2$ is the solution of the Cauchy problem:

$$\frac{d}{dt} X_i(t) = A_0^T(i) X_i(t) + X_i(t) A_0(i) + \sum_{j=1}^{2} q_{ij} X_j(t) \qquad (4.23)$$

$$+ C_0^T(i) C_0(i),$$

$$X_i(0) = 0, \quad i = 1, 2.$$

From the representation formula

$$(X_1(t), X_2(t)) = \int_0^t e^{\mathcal{L}_0^*(t-s)} \tilde{C} ds$$

it follows that $X_i(t) \geq 0$ for all $t \geq 0$. Therefore it is sufficient to show that there exists $\tau > 0$ such that $det X_i(\tau) > 0$.

Set $X_i(t) = \begin{pmatrix} x_i(t) & y_i(t) \\ y_i(t) & z_i(t) \end{pmatrix}, i = 1, 2$ and obtain from (4.23) the following system of affine differential equations:

$$x_1'(t) = (2\alpha - q)x_1(t) + qx_2(t) + 1$$

$$x_2'(t) = qx_1(t) + (2\alpha - q)x_2(t)$$

$$y_1'(t) = (2\alpha - q)y_1(t) + qy_2(t)$$

$$y_2'(t) = \alpha y_1(t) + (2\alpha - q)y_2(t)$$

$$z_1'(t) = (2\alpha - q)z_1(t) + qz_2(t)$$

$$z_2'(t) = qz_1(t) + (2\alpha - q)z_2(t) + 1$$

$$x_i(0) = y_i(0) = z_i(0) = 0, i = 1, 2.$$

Hence $y_1(t) = y_2(t) = 0, t \geq 0$.

From the uniqueness of the solution of a Cauchy problem it follows that $x_1(t) = z_2(t) = \tilde{x}(t)$ and $x_2(t) = z_1(t) = \tilde{z}(t)$ where $t \to (\tilde{x}(t), \tilde{z}(t))$ is the solution of the problem

$$\frac{d}{dt}\tilde{x}(t) = (2\alpha - q)\tilde{x}(t) + q\tilde{z}(t) + 1$$

$$\frac{d}{dt}\tilde{z}(t) = q\tilde{x}(t) + (2\alpha - q)\tilde{z}(t)$$

$$\tilde{x}(0) = \tilde{z}(0) = 0.$$

We have $det X_i(t) = x_i(t)z_i(t) - y_i^2(t) = x_i(t)z_i(t) = \tilde{x}(t)\tilde{z}(t), t \geq 0$.
But

$$\tilde{x}(t) = \frac{1}{2}\int_0^t [e^{\alpha s} + e^{(2\alpha - q)s}]ds$$

$$\tilde{z}(t) = \frac{1}{2}\int_0^t [e^{2\alpha s} - e^{2(\alpha - q)s}]ds.$$

It is easy to see that for every $\alpha \in \mathbf{R}, q > 0$ we have $\lim_{t \to \infty} \tilde{x}(t)\tilde{z}(t) > 0$.

Remark 4.2.3. Let us consider the system of type (4.1) with $n = 2, d = 2, p = 1, r = 1$ and $A_0(1) = A_0(2) = \alpha I_2, C_0(1) = [1 \quad 0], C_0(2) = [0 \quad 1], A_1(i) \, a \, 2 \times 2$ arbitrary matrix, $Q = \begin{bmatrix} -q & q \\ q & -q \end{bmatrix}, \alpha \in \mathbf{R}, q > 0$. Combining the conclusion of Example 4.2.1 with Proposition 4.2.2 it follows that the system $(C_0, (A_0, A_1); Q)$ is observable.

Example 4.2.2. The stochastic observability does not imply always stochastic detectability. Let us consider the system with Markovian jumping with $d = 2$, $n = 2, p = 1$,

$$A_0(1) = A_0(2) = \frac{q}{2}I_2, C_0(1) = [1 \quad 0], C_0(2) = [0 \quad 1], Q = \begin{bmatrix} -q & q \\ q & -q \end{bmatrix}. \quad (4.24)$$

From the previous example we conclude that the system $(C_0, A_0; Q)$ is observable. Invoking (i) $\Leftrightarrow$ (ii) from Proposition 4.1.5 we deduce that if the system (4.24) would be stochastically detectable, then would exist the matrices $X(i) > 0$, and $\Lambda(i) = \begin{bmatrix} \lambda_1(i) \\ \lambda_2(i) \end{bmatrix}, i = 1, 2$ which verify the following system of linear equations:

$$A_0^T(i)X(i) + X(i)A_0(i) + \Lambda(i)C_0(i) + C_0^T(i)\Lambda^T(i) + \sum_{j=1}^{2} q_{ij}X(j) + I_2 = 0, \ i = 1, 2$$

which implies $I_2 + \begin{bmatrix} 2\lambda_1(1) & \lambda_2(1) \\ \lambda_2(1) & 0 \end{bmatrix} < 0$ which is a contradiction.

Example 4.2.3. Let us consider the stochastic system

$$dx(t) = A_0(\eta(t))x(t)dt + A_1(\eta(t))x(t)dw_1(t) \qquad (4.25)$$

$$y(t) = C_0(\eta(t))x(t)$$

with $n = 2$, $d = 2$, $r = 1$, $p = 1$, $A_0(1) = A_0(2) = \alpha I_2$, $C_0(1) = [1 \quad 0]$, $C_0(2) = [0 \quad 1]$, $A_1(1) = \beta I_2$, $A_1(2)$ is a 2×2 arbitrary matrix, $Q = \begin{bmatrix} -q & q \\ q & -q \end{bmatrix}$, $\alpha \in \mathbf{R}$, $\beta \in \mathbf{R}$, $q > 0$ which satisfy $2\alpha - q + \beta^2 = 0$.

From Remark 4.2.3 it follows that the system described by (4.25) is stochastically observable. We show that it is not stochastically detectable. If by contrary the system (4.25) is stochastically detectable, then, using again Proposition 4.1.5, we deduce that there exist matrices $X(i) > 0, \Lambda(i) = \begin{bmatrix} \lambda_1(i) \\ \lambda_2(i) \end{bmatrix}, \lambda_k(i) \in \mathbf{R}$ which verify the following system of linear equations

$$A_0^T(i)X(i) + X(i)A_0(i) + \Lambda(i)C_0(i) + C_0^T(i)\Lambda^T(i) + A_1^T(i)X(i)\Lambda_1(i)$$

$$+ \Sigma_{j=1}^2 q_{ij}X(j) + I_2 = 0$$

which leads to the same contradiction as in the previous example.

Remark 4.2.4. One can see that the system

$$dx(t) = A_0(\eta(t))x(t)dt + \sum_{k=1}^r A_k(\eta(t))x(t)dw_k(t) \qquad (4.26)$$

$$y(t) = C_0(\eta(t))x(t)$$

with $A_0(i), C_0(i)$ as in (4.24) and $A_k(i), k = 1, 2, \ldots, r$, 2×2 arbitrary matrices, is stochastically observable but it is not stochastically detectable. If, by contrary, (4.26) would be stochastically detectable, then by Proposition 4.1.2 (i) could follow that the system described by (4.24) would be stochastically detectable, which contradicts the conclusion of Example 4.2.2.

From the representation formula in Theorem 3.1.1 it follows the next result.

Proposition 4.2.9. *Assume that the system (4.1) is in the time-invariant case. Then the triple $(C_0, \mathbf{A}; Q)$ is observable if and only if they do not exist $\tau > 0, i \in \mathcal{D}$ and $x_0 \neq 0$ such that*

$$E\left[|y(t, 0, x_0)|^2 | \eta(0) = i\right] = 0$$

$\forall t \in [0, \tau]$ *where $y(t, 0, x_0) = C_0(\eta(t))x(t, 0, x_0)$, $x(t, 0, x_0)$ being the solution of (4.1) for $u(t) = 0$ and having the initial condition $x(0, 0, x_0) = x_0$.*

In the deterministic framework the analogous of the above statement is one of the usual definitions of observability.

Remark 4.2.5. In Definition 4.2.1 of observability no condition on Q is imposed. All the results proved above except Propositions 4.2.2 and 4.2.9 require only the condition $q_{ij} \geq 0$ for $i \neq j$. The additional condition $\sum_{j=1}^{d} q_{ij} = 0$ is used only in the proof of the two aforementioned propositions.

4.3 Stochastic Controllability

In this section the controllability of stochastic systems will be introduced. For simplicity we shall consider only the time-invariant case.

Let $A_k(i) \in \mathbf{R}^{n \times n}, 0 \leq k \leq r, i \in \mathcal{D}, B(i) \in \mathbf{R}^{n \times m}, Q = [q_{ij}], i, j \in \mathcal{D}$ with $q_{ij} \geq 0$ for $i \neq j$.

Definition 4.3.1. We say that the system $(A_0, A_1, \ldots, A_r, B; Q)$ is *controllable* if it exists $\tau > 0$ such that

$$\int_0^{\tau} e^{\mathcal{L}t} \tilde{B} dt > 0$$

where $\mathcal{L}$ is defined by (2.130) and $\tilde{B} \in \mathcal{S}_n^d, \tilde{B}(i) = B(i) B^T(i), i \in \mathcal{D}$.

Remark 4.3.1. One can easily see that in the deterministic case, namely if $\mathcal{D} = \{1\}, q_{11} = 0$ and $A_k(1) = 0, 1 \leq k \leq r$, the above definition reduces to the definition of controllability of the pair $(A_0(1), B(1))$.

The following result can be directly proved.

Proposition 4.3.1. *The system* $(A_0, A_1, \ldots, A_r, B; Q)$ *is controllable if and only if the system* $(B^T, A_0^T, A_1^T, \ldots, A_r^T; Q^T)$ *is observable.*

From the above proposition and from Propositions 4.2.3, 4.2.4 and Remark 4.3.1, immediately follows.

Proposition 4.3.2. *The following assertions are equivalent*

(i) *The system* $(A_0, A_1, \ldots, A_r, B; Q)$ *is controllable.*
(ii) *It exists* $\tau > 0$ *such that* $K_0(\tau) > 0$ *where* $K_0(t)$ *denotes the solution of the affine equation in the space* $\mathcal{S}_n^d$:

$$\frac{d}{dt} K_0(t) = \mathcal{L} K_0(t) + \tilde{B}$$

with $K_0(0) = 0$.
(iii) *For any* $t > 0, K_0(t) > 0$.

In the following we shall consider the situation when the system is subject only to white noise perturbations, namely $\mathcal{D} = \{1\}$, $q_{11} = 0$, $A_k(1) = A_k$, $B(1) = B$. The inequality in Definition 4.3.1 becomes

$$\int_0^\tau e^{\overline{\mathcal{L}}t}\hat{B}dt > 0$$

where $\overline{\mathcal{L}}$ denotes the linear operator defined on $\mathcal{S}_n$ by (2.137) and $\hat{B} = BB^T$. If this inequality is fulfilled for some $\tau > 0$ we shall say that the system $(A_0, A_1, \ldots, A_r, B)$ *is controllable*. Therefore in the case of systems with multiplicative white noise the above proposition gives the following result.

Proposition 4.3.3. *The following assertions are equivalent*

(i) *The system* $(A_0, A_1, \ldots, A_r, B)$ *is controllable;*
(ii) *It exists* $\tau > 0$ *such that* $\tilde{K}(\tau) > 0$ *where*

$$\frac{d}{dt}\tilde{K}(t) = A_0\tilde{K}(t) + \tilde{K}(t)A_0^T + \sum_{k=1}^r A_k\tilde{K}(t)A_k^T + BB^T \text{ with } \tilde{K}(0) = 0; \quad (4.27)$$

(iii) $\tilde{K}(t) > 0$ *for all* $t > 0$.

From Proposition 3.1.3 immediately follows that

$$e^{\overline{\mathcal{L}}t}H = E\left[\Phi(t,0)H\Phi^T(t,0)\right], t \geq 0, H \in \mathcal{S}_n,$$

where $\Phi(t,t_0)$, $t \geq t_0$ denotes the fundamental matrix solution associated with the linear Itô system

$$dx(t) = A_0 x(t)dt + \sum_{k=1}^r A_k x(t)dw_k(t).$$

Therefore the next result directly follows.

Proposition 4.3.4. *The system* $(A_0, A_1, \ldots, A_r, B)$ *is controllable if and only if it exists* $\tau > 0$ *such that* $E\int_0^\tau \left[\Phi(t,0)BB^T\Phi^T(t,0)\right]dt > 0$.

We shall give now another characterization, in stochastic terms, of the controllability of the system $(A_0, A_1, \ldots, A_r, B)$. Consider the affine Itô system:

$$dx(t) = A_0 x(t)dt + \sum_{k=1}^r A_k x(t)dw_k(t) + Bdv(t), t \geq 0 \quad (4.28)$$

where $(w(t), v(t))^T$ is a standard $r + m$-dimensional Wiener process. Let $\tilde{x}(t), t \geq 0$ be the solution of (4.28) with $\tilde{x}(0) = 0$. Using the Itô's formula (Theorem 1.9.9) one can easily verify that $\tilde{K}(t) = E\left[\tilde{x}(t)\tilde{x}^T(t)\right]$, $\tilde{K}$ being defined in Proposition 4.3.3.

Then the following result is immediately obtained.

Proposition 4.3.5. *The system* $(A_0, A_1, \ldots, A_r, B)$ *is controllable if and only if*

$$E\left[\tilde{x}(t)\tilde{x}^T(t)\right] > 0$$

for all $t > 0$.

The above characterization has been considered as a definition of controllability of the system $(A_0, A_1, \ldots, A_r, B)$ in [13].

The next result proved in [13] characterizes the controllability of the system $(A_0, A_1, \ldots, A_r, B)$ in terms of invariant subspaces as in the deterministic case ($A_k = 0$, $1 \leq k \leq r$).

Theorem 4.3.6. *The system* $(A_0, A_1, \ldots, A_r, B)$ *is controllable if and only if no invariant subspace exists with the dimension less than n of the collection A_k, $0 \leq k \leq r$ containing all columns of B.*

For the proof of the above theorem we need the following lemma.

Lemma 4.3.7. *The following two assertions are equivalent*

  (i) *It exists an invariant subspace with the dimension less than n of the matrices A_k, $0 \leq k \leq r$ containing all columns of B;*
 (ii) *It exists $\xi \in \mathbf{R}^n$, $\xi \neq 0$ such that $\xi^T MB = 0$ for all $M = A_{i_1}^{s_1} A_{i_2}^{s_2} \ldots A_{i_p}^{s_p}$ where $0 \leq i_j \leq r$ and $s_j \geq 0$, $1 \leq j \leq p$, $p \geq 1$ are natural numbers.*

Proof. (i) $\Rightarrow$ (ii) Let S be an invariant subspace of the matrices A_k, $0 \leq k \leq r$ with the dimension less than n containing all columns of B. Denote by $S^\perp$ the orthogonal subspace of S. Since $S^\perp \neq \{0\}$, consider $\xi \in S^\perp$ such that $\xi \neq 0$. Since all the columns of the matrices MB with M as in the statement are included in S it follows that $\xi^T MB = 0$.

(ii) $\Rightarrow$ (i) Assume that it exists $\xi \neq 0$ satisfying (ii). Let S be the subspace generated by the columns of all matrices MB, M being defined as in the statement. Since $\xi \neq 0$ it follows that $S \neq \mathbf{R}^n$. On the other hand it is easy to check that if $x \in S$ then $A_k x \in S$ for all $0 \leq k \leq r$. Thus the proof is complete. $\square$

Proof of Theorem 4.3.6. Necessity. Assume that the system $(A_0, A_1, \ldots, A_r, B)$ is controllable. It follows that $B \neq 0$ and therefore if $n = 1$ the condition in the statement is automatically accomplished. We consider now the case $n \geq 2$ and that it exists a subspace S, $S \neq \{0\}$, $S \neq \mathbf{R}^n$ invariant of A_k, $0 \leq k \leq r$ containing all columns of B. Then it follows that it exists a basis in $\mathbf{R}^n$ with respect to which the matrices A_k have the structure

$$\tilde{A}_k = \begin{bmatrix} A_{1k} & A_{2k} \\ 0 & A_{3k} \end{bmatrix}, 0 \leq k \leq r$$

and B has the form

$$\tilde{B} = \begin{bmatrix} B_0 \\ 0 \end{bmatrix},$$

where A_{1k} are $s \times s$ matrices with $1 \leq s < n$. Let $\overline{K}(t), t \geq 0$ be the solution of (4.27) corresponding to the matrices $\tilde{A}_k$ and $\tilde{B}$ and $\overline{K}(0) = 0$. It is easy to check that if

$$\overline{K}(t) = \begin{bmatrix} \overline{K}_{11}(t) & \overline{K}_{12}(t) \\ \overline{K}_{21}(t) & \overline{K}_{22}(t) \end{bmatrix}$$

then $\overline{K}_{22}(t)$ verifies a linear equation. Since $\overline{K}(0) = 0$ it follows that $\overline{K}_{22}(t) = 0$ for all $t \geq 0$ and therefore $\overline{K}_{22}(t)$ is not positive definite for all $t \geq 0$. Taking into account that $\tilde{K}(t) = T\overline{K}(t)T^T$ with T nonsingular it follows that $\tilde{K}(t)$ is not positive definite which fact contradicts the assumption (see Proposition 4.3.3).

Sufficiency. We prove that $\tilde{K}(t) > 0$ for all $t > 0$. Indeed, assume that it exists $\tau > 0$ and $\xi \in \mathbf{R}^n$, $\xi \neq 0$ such that $\xi^T \tilde{K}(t)\xi = 0$. Then one can easily check that

$$\tilde{K}(t) = \sum_{k=1}^{r} \int_0^t e^{A_0(t-s)} A_k \tilde{K}(s) A_k^T e^{A_0^T(t-s)} ds + \int_0^t e^{A_0 s} BB^T e^{A_0^T s} ds. \tag{4.29}$$

Since $\tilde{K}(t) \geq 0$ from (4.29) we successively obtain:

$$\tilde{K}(t) \geq \int_0^t e^{A_0 s} BB^T e^{A_0^T s} ds.$$

$$\tilde{K}(t) \geq \sum_{i_1=1}^{r} \int_0^t \left(\int_0^{s_1} e^{A_0(t-s_1)} A_{i_1} e^{A_0 s_0} BB^T A_{i_1}^T e^{A_0^T(t-s_1)} ds \right) ds_1,$$

$$\vdots$$

$$\tilde{K}(t) \geq \sum_{i_p, i_{p-1}, \ldots, i_1} \int_0^t \int_0^{s_p} \int_0^{s_{p-1}} \cdots \int_0^{s_1} e^{A_0(t-s_p)} A_{i_p} e^{A_0(s_p-s_{p-1})} A_{i_{p-1}} \cdots$$

$$\cdots e^{A_0(s_2-s_1)} A_{i_1} e^{A_0 s_0} BB^T A_{i_1}^T e^{A_0^T(s_2-s_1)} \ldots A_{i_p}^T e^{A_0^T(t-s_p)} ds_0 \cdots ds_p.$$

Therefore $\xi^T e^{A_0 s} B = 0$ for all $0 \leq s \leq \tau$ and

$$\xi^T e^{A_0(\tau-s_p)} A_{i_p} e^{A_0(s_p-s_{p-1})} \ldots e^{A_0(s_2-s_1)} A_{i_1} e^{A_0 s_0} B = 0$$

for all $\tau > s_p > s_{p-1} > \cdots > s_2 > s_1 > s_0 \geq 0$ and for all $1 \leq i_j \leq r, 1 \leq s \leq p$. It follows that $\xi^T A_0^k B = 0, k \geq 0$ and

$$\xi^T A_0^{k_1} A_{i_p} A_0^{k_2} A_{i_{p-1}} \cdots A_{i_1} A_0^{k_p} B = 0$$

for all $1 \leq i_j \leq r, 1 \leq j \leq p$ and $k_s \geq 0, 0 \leq s \leq p$. Therefore $\xi^T MB = 0$ for all M as in the statement of the Lemma 4.3.7 and according to this lemma, we obtained a contradiction. Thus the proof of the theorem is complete. $\qquad \square$

From the above theorem immediately follows.

Corollary 4.3.8. *If a pair (A_k, B) is controllable for a certain $k \in \{0, 1, \ldots, r\}$, then the system $(A_0, A_1, \ldots, A_r, B)$ is controllable.*

We shall show below that the converse of the corollary is not usually true. However in the case $n = 2$, $m = 1$, $r = 1$ such implication is valid, namely one can prove:

Proposition 4.3.9. *If $n = 2$, $m = 1$, $r = 1$ and the pairs (A_0, B) and (A_1, B) are not controllable, then the system (A_0, A_1, B) is not controllable.*

Proof. Let

$$A_0 = \begin{bmatrix} a & b \\ c & d \end{bmatrix}, A_1 = \begin{bmatrix} \alpha & \beta \\ \gamma & \delta \end{bmatrix} \text{ and } B = \begin{bmatrix} b_1 \\ b_2 \end{bmatrix}$$

such that (A_0, B) and (A_1, B) are not controllable, that is

$$b_1 b_2 (d - a) = bb_2^2 - b_1^2 c \text{ and } b_1 b_2 (\delta - \alpha) = \beta b_2^2 - b_1^2 \gamma \qquad (4.30)$$

According to Proposition 4.3.3 the considered system (A_0, A_1, B) is controllable if and only if $\tilde{K}(t) > 0$ for all $t > 0$ where $\tilde{K}$ verifies (4.27) written for this particular case. Taking

$$\tilde{K} = \begin{bmatrix} x & y \\ y & z \end{bmatrix}$$

(4.27) gives

$$\frac{dx}{dt} = (2a + \alpha^2) x + 2(b + \alpha\beta) y + \beta^2 z + b_1^2$$

$$\frac{dy}{dt} = (c + \alpha\gamma) x + (a + d + \gamma\beta + \alpha\delta) y + (b + \beta\delta) z + b_1 b_2 \qquad (4.31)$$

$$\frac{dz}{dt} = \gamma^2 x + 2(c + \gamma\delta) y + (2d + \delta^2) z + b_2^2.$$

If $b_1 = 0$ and $b_2 = 0$ immediately follows that $x(t) = y(t) = z(t) = 0$ for all $t \in \mathbf{R}$.

If $b_1 \neq 0$ and $b_2 = 0$ from (4.30), one obtains that $c = 0$ and $\gamma = 0$ and therefore $z(t) = 0$ for all $t \in \mathbf{R}$.

If $b_1 = 0$ and $b_2 \neq 0$, then (4.30) gives $b = \beta = 0$ and hence $x(t) = 0$ for all $t \in \mathbf{R}$.

Assume that $b_1 \neq 0$ and $b_2 \neq 0$. Using (4.30) one can easily check that $(\bar{x}, \bar{y}, \bar{z})$ verifies (4.31) where

$$\bar{x}(t) = \frac{b_1}{b_2}\bar{y}(t), \ \bar{z}(t) = \frac{b_2}{b_1}\bar{y}(t)$$

and $\bar{y}(t)$ is the solution of the equation

$$\frac{d\bar{y}}{dt} = \left[(c+\alpha\gamma)\frac{b_1}{b_2} + (a+d+\gamma\beta+\alpha\delta) + (b+\beta\delta)\frac{b_2}{b_1}\right]\bar{y} + b_1 b_2$$

and $\bar{y}(0) = 0$. From the uniqueness of the solution it follows that $x(t) = \bar{x}(t)$, $y(t) = \bar{y}(t)$, $z(t) = \bar{z}(t)$ and therefore $x(t)z(t) - (y(t))^2 = 0$ for all $t \in \mathbf{R}$, and therefore by Proposition 4.3.3 (A_0, A_1, B) is not controllable.   □

The next example shows that the converse of the Corollary 4.3.8 is not generally true, namely it is possible to have a controllable system (A_0, A_1, B) but with the pairs (A_0, B) and (A_1, B) not controllable.

Example 4.3.1. Consider the case $n = 3$, $m = 1$ and $r = 1$ in which

$$A_0 = \begin{bmatrix} 1 & 0 & 0 \\ 0 & -1 & 3 \\ 0 & 0 & 2 \end{bmatrix}, A_1 = \begin{bmatrix} 3 & 0 & 0 \\ 2 & 1 & 0 \\ 0 & 0 & -1 \end{bmatrix}, B = \begin{bmatrix} 1 \\ 1 \\ 1 \end{bmatrix}.$$

It is easy to check that (A_0, B) and (A_1, B) are not controllable. In this case (4.27) gives for

$$\tilde{K} = \begin{bmatrix} x & y & z \\ y & u & v \\ z & v & q \end{bmatrix},$$

$$\frac{dx}{dt} = 11x + 1$$

$$\frac{dy}{dt} = 3y + 6x + 3z + 1$$

$$\frac{dz}{dt} = 1$$

$$\frac{du}{dt} = -u + 6v + 4x + 4y + 1$$

$$\frac{dv}{dt} = 3q - 2z + 1$$

$$\frac{dq}{dt} = 5q + 1$$

with $x(0) = y(0) = z(0) = u(0) = v(0) = q(0) = 0$. One can directly check that the solution of the above system is given by

$$x(t) = \frac{1}{11}\left(e^{11t} - 1\right)$$

$$y(t) = \frac{3}{44}e^{11t} + \frac{5}{12}e^{3t} - t - \frac{16}{33}$$

$$z(t) = t$$

$$v(t) = \frac{3}{25}\left(e^{5t} - 1\right) - t^2 + \frac{2}{5}t$$

$$q(t) = \frac{1}{5}\left(e^{5t} - 1\right)$$

and $u(t)$ has the form:

$$u(t) = \frac{17}{132}e^{11t} + \alpha_1 e^{5t} + \alpha_2 e^{3t} + \alpha_3 e^{-t} + \alpha_4 t^2 + \alpha_5 t + \alpha_6.$$

Then it follows that $\lim_{t \to \infty} \det \tilde{K}(t) = \infty$, which fact implies that $\tilde{K}(t) > 0$ for some $t \geq 0$ and therefore, according to Proposition 4.3.3 the system (A_0, A_1, B) is controllable.

Remark 4.3.2. We have previously shown that by contrast with the deterministic case, the stochastic controllability of Markovian systems does not imply their stochastic stabilizability. A similar affirmation is valid for the stochastic systems subject to Itô multiplicative noise.

Indeed the system (A_0, A_1, B) in the above example is controllable but it is not stabilizable since in such situation, according to Proposition 4.1.4 applied in this case $(D = \{1\}, q_{11} = 0)$ it exists $(X, \Delta), X > 0$,

$$X = \begin{bmatrix} x & y & z \\ y & u & v \\ z & v & q \end{bmatrix} \text{ and } \Delta = \begin{bmatrix} 2f_1 & f_1 + f_2 & f_1 + f_3 \\ f_1 + f_2 & 2f_2 & f_2 + f_3 \\ f_1 + f_3 & f_2 + f_3 & 2f_3 \end{bmatrix}$$

such that

$$A_0 X + X A_0^* + A_1 X A_1^* + I_3 + \Delta = 0.$$

Therefore

$$11x + 1 + 2f_1 = 0$$

$$3y + 6x + 3z + f_1 + f_2 = 0$$

$$f_1 + f_3 = 0$$

$$-u + 6v + 4x + 4y + 1 + 2f_2 = 0$$

$$3q - 2z + f_2 + f_3 = 0$$

$$5q + 1 + 2f_3 = 0$$

Since $x > 0$ and $q > 0$ it follows $f_1 < 0$, $f_3 < 0$ which contradicts $f_1 + f_3 = 0$. Hence (A_0, A_1, B) is not stabilizable.

Notes and References

Stochastic controllability for Itô differential equations has been introduced in [13]. Theorem 4.3.6 can be found in [13]. The numerical example and Remark 4.3.2 appear for the first time in this book.

Other concepts of stochastic controllability have been studied in terms of control which generalize recurrence notions of stochastic processes (see, e.g., [15, 16, 56, 93, 94, 131, 132, 155]) for Itô systems and [86] for jump linear Markovian systems. In the present book the concept of stochastic controllability is not used and therefore a reduced space is devoted to this concept.

The stochastic uniform observability has been defined in [111] for Itô systems and in [112] for systems with jump Markovian perturbations. Stochastic uniform observability of linear differential equations with multiplicative noise were studied also in [142, 143] and for systems driven by Markov processes with infinite number of states in [63, 144].

These concepts have been used to solve the linear quadratic problem with infinite horizon for these corresponding systems. The results in this chapter devoted to stochastic stabilizability, detectability and observability can be found in [41, 45] and in [46].

Chapter 5
A Class of Nonlinear Differential Equations on an Ordered Linear Space of Symmetric Matrices with Applications to Riccati Differential Equations of Stochastic Control

In many control problems, both in deterministic and in stochastic framework, a crucial role is played by a class of nonlinear matrix differential equations or nonlinear matrix algebraic equations known as *matrix Riccati equations*.

In this chapter we deal with a class of systems of matrix differential equations as well as systems of nonlinear algebraic equations arising in connection with the solution of several control problems as: linear quadratic optimization, H^2 control and H^∞ control problem for stochastic systems. These will be called *stochastic generalized Riccati differential equations* (SGRDE) or *stochastic generalized Riccati algebraic equations* (SGRAE). It is easy to see that the systems of matrix Riccati differential equations considered in this chapter contain as particular cases many types of matrix Riccati equations which are known both in the deterministic and in the stochastic framework. The results derived in the general case considered in this chapter are also applicable to the aforementioned particular cases.

In the first part of this chapter we shall study global solutions of a class of nonlinear differential equations on the Banach space $\mathcal{S}_n^\mathcal{D}$ which will be called *generalized Riccati differential equations* (GRDEs). These kinds of GRDEs are regarded as mathematical objects with interest in themselves and the proofs are done avoiding any connection with some optimization problems. The proofs are mainly based on positivity properties of linear evolution operators defined by the Lyapunov differential equations. We provide conditions which guarantee the existence and the uniqueness of some global solutions of GRDE as maximal solution, minimal solution, and stabilizing solution.

In the second part of this chapter, the general results obtained for GRDEs are specialized to obtain necessary and sufficient conditions for the existence of the maximal solution, stabilizing solution, and minimal solution of SGRDEs. We prove that if the coefficients of SGRDE are periodic functions, then the maximal solution, the minimal solution, and the stabilizing solution are also periodic functions. Moreover, if the coefficients of the SGRDE are not depending on the time parameter t, then the above-mentioned special solutions are constant and they solve the corresponding SGRAE. The necessary and sufficient conditions which guarantee the existence of the maximal solution, of the minimal, and of the stabilizing solution,

V. Dragan et al., *Mathematical Methods in Robust Control of Linear Stochastic Systems*, DOI 10.1007/978-1-4614-8663-3_5, © Springer Science+Business Media New York 2013

respectively, are expressed in terms of solvability of a class of suitable systems of linear matrix inequalities. Finally we shall provide an iterative procedure which allows to compute these special solutions to the SGRDE and to the SGRAE.

5.1   Generalized Riccati Differential Equations: Preliminaries

On the Banach space $\mathcal{S}_n^{\mathcal{D}}$ we consider the nonlinear differential equation often named GRDE

$$\frac{d}{dt}X(t)+A^T(t)X(t)+X(t)A(t)+\Pi_1(t)[X(t)]-\{X(t)B(t)+\Pi_{12}(t)[X(t)]+L(t)\}\times$$

$$\{R(t)+\Pi_2(t)[X(t)]\}^{-1}\{X(t)B(t)+\Pi_{12}(t)[X(t)]+L(t)\}^T+M(t)=0 \quad (5.1)$$

where $A:\mathcal{I}\to\mathcal{M}_n^{\mathcal{D}}$, $B:\mathcal{I}\to\mathcal{M}_{nm}^{\mathcal{D}}$, $M:\mathcal{I}\to\mathcal{S}_n^{\mathcal{D}}$, $L:\mathcal{I}\to\mathcal{M}_{nm}^{\mathcal{D}}$, $R:\mathcal{I}\to\mathcal{S}_m^{\mathcal{D}}$ are bounded and continuous functions and $\Pi_1:\mathcal{I}\to\mathbf{B}(\mathcal{S}_n^{\mathcal{D}})$, $\Pi_{12}:\mathcal{I}\to\mathbf{B}(\mathcal{S}_n^{\mathcal{D}},\mathcal{M}_{nm}^{\mathcal{D}})$, $\Pi_2:\mathcal{I}\to\mathbf{B}(\mathcal{S}_n^{\mathcal{D}},\mathcal{S}_m^{\mathcal{D}})$ are operator valued functions, $\mathcal{I}\subset\mathbb{R}$ is a right unbounded interval.

In (5.1) we have used the conventions of notation established in Sect. 2.6.1. So, the i-th component of the GRDE (5.1) is:

$$\frac{d}{dt}X(t,i)+A^T(t,i)X(t,i)+X(t,i)A(t,i)+\Pi_1(t)[X(t)](i)-\{X(t,i)B(t,i)+$$

$$\Pi_{12}(t)[X(t)](i)+L(t,i)\}\{\Pi_2(t)[X(t)](i)+R(t,i)\}^{-1}\{X(t,i)B(t,i)+$$

$$\Pi_{12}(t)[X(t)](i)+L(t,i)\}^T+M(t,i)=0,$$

$i\in\mathcal{D}$, $t\in\mathcal{I}$, where $X\to\Pi_1(t)[X](i)\in\mathbf{B}(\mathcal{S}_n^{\mathcal{D}},\mathcal{S}_n)$, $X\to\Pi_{12}(t)[X](i)\in\mathbf{B}(\mathcal{S}_n^{\mathcal{D}},\mathbf{R}^{n\times m})$, $X\to\Pi_2(t)[X](i)\in\mathbf{B}(\mathcal{S}_n^{\mathcal{D}},\mathcal{S}_m)$.

The GRDE (5.1) contains as special cases various types of Riccati differential equations arising in connection with the linear quadratic optimization control problems, both in the deterministic and in the stochastic framework. Thus, if $\mathcal{D}=\{1\}$ and $\Pi_1(t)[X]=0$, $\Pi_{12}(t)[X]=0$, $\Pi_2(t)[X]=0$ for all $(t,X)\in\mathcal{I}\times\mathcal{S}_n$, (5.1) reduces to the well-known matrix Riccati equation intensively investigated starting with the pioneering work of Kalman [89].

In this chapter we will especially deal with systems of coupled Riccati differential equations (SGRDE) of the form

$$\frac{d}{dt}X(t,i)+\left[A_0(t,i)+\frac{1}{2}q_{ii}I_n\right]^TX(t,i)+X(t,i)\left[A_0(t,i)+\frac{1}{2}q_{ii}I_n\right]+$$

$$\sum_{k=1}^{r} A_k^T(t,i)X(t,i)A_k(t,i) + \sum_{j \in \mathcal{D}\backslash\{i\}} q_{ij}X(t,j) - \{X(t,i)B_0(t,i) +$$

$$\sum_{k=1}^{r} A_k^T(t,i)X(t,i)B_k(t,i) + L(t,i)\} \left\{ \sum_{k=1}^{r} B_k^T(t,i)X(t,i)B_k(t,i) + \quad (5.2)\right.$$

$$R(t,i)\}^{-1} \left\{ X(t,i)B_0(t,i) + \sum_{k=1}^{r} A_k^T(t,i)X(t,i)B_k(t,i) + L(t,i) \right\}^T + M(t,i) = 0$$

$i \in \mathcal{D}, t \in \mathbb{R}_+$.

The SGRDEs (5.2) occur in connection with the linear quadratic optimization problems described by controlled systems of type (4.1) and quadratic cost functionals of the form:

$$J = E\left[\int_{t_0}^{\infty} \begin{pmatrix} x(t) \\ u(t) \end{pmatrix}^T \begin{pmatrix} M(t,\eta(t)) & L(t,\eta(t)) \\ L^T(t,\eta(t)) & R(t,\eta(t)) \end{pmatrix} \begin{pmatrix} x(t) \\ u(t) \end{pmatrix} dt \right].$$

This problem of optimal control will be detailed studied in the next chapter, where the role of some global solutions of (5.2) will be clarified.

Further we introduce the notations $\mathcal{Q}(t) = \{Q(t,i)\}_{i \in \mathcal{D}}$ and $\Pi(t)[X] = \{\Pi(t)[X](i)\}_{i \in \mathcal{D}}$ where

$$\mathcal{Q}(t,i) = \begin{pmatrix} M(t,i) & L(t,i) \\ L^T(t,i) & R(t,i) \end{pmatrix} \in \mathcal{S}_{n+m} \qquad (5.3)$$

$$\Pi(t)[X](i) = \begin{pmatrix} \Pi_1(t)[X](i) & \Pi_{12}(t)[X](i) \\ (\Pi_{12}^T(t)[X](i))^T & \Pi_2(t)[X](i) \end{pmatrix}, \qquad (5.4)$$

$i \in \mathcal{D}$, $X \in \mathcal{S}_n^{\mathcal{D}}$, $t \in \mathcal{I}$. It is obvious that for $t \in \mathcal{I}$ we have $X \to \Pi(t)[X] \in \mathbf{B}(\mathcal{S}_n^{\mathcal{D}}, \mathcal{S}_{n+m}^{\mathcal{D}})$. Regarding the operator valued function, $\Pi(\cdot)$ we make the assumptions:

$(\mathbf{\Pi}_1)$: (i) $\Pi : \mathcal{I} \to \mathbf{B}(\mathcal{S}_n^{\mathcal{D}}, \mathcal{S}_{n+m}^{\mathcal{D}})$ is a bounded and continuous operator valued function;

(ii) For each $t \in \mathcal{I}$, $\Pi(t)$ is a positive operator, that is $\Pi(t)[X] \geq 0$ if $X \geq 0$.

Remark 5.1.1. In the special case of SGRDE (5.2) we have

$$\Pi(t)[X](i) = \sum_{k=1}^{r} \begin{pmatrix} A_k^T(t,i) \\ B_k^T(t,i) \end{pmatrix} X(i) \begin{pmatrix} A_k(t,i) & B_k(t,i) \end{pmatrix} + \sum_{j \in \mathcal{D}\backslash\{i\}} q_{ij} \begin{pmatrix} X(j) & 0 \\ 0 & 0 \end{pmatrix}. (5.5)$$

It is easy to see that if $\mathcal{D} = \{1, 2, \ldots, d\}$ then the operator valued function introduced by (5.5) satisfies the assumptions $(\mathbf{\Pi}_1)$, if the scalars $q_{ij} \geq 0$ for $i \neq j$. If $\mathcal{D} = \mathbb{Z}_+$

and the scalars q_{ij} satisfy the conditions (2.138) and (2.139), then reasoning as in the proof of Proposition 2.6.5 one may show that $\Pi(t)$ defined in (5.5) satisfies also the assumptions (Π_1).

If $W : \mathcal{I} \to \mathcal{M}_{mn}^{\mathcal{D}}$ is a bounded and continuous function, then $\Pi_W : \mathcal{I} \to \mathbf{B}(\mathcal{S}_n^{\mathcal{D}})$ is defined by

$$\Pi_W(t)[X] = \left(J^{\mathcal{D}} \ W^T(t) \right) \Pi(t)[X] \begin{pmatrix} J^{\mathcal{D}} \\ W(t) \end{pmatrix} \tag{5.6}$$

for all $X \in \mathcal{S}_n^{\mathcal{D}}$ and

$$M_W(t) = \left(J^{\mathcal{D}} \ W^T(t) \right) \mathcal{Q}(t) \begin{pmatrix} J^{\mathcal{D}} \\ W(t) \end{pmatrix}. \tag{5.7}$$

In (5.6) and (5.7) $J^{\mathcal{D}}$ is defined by $J^{\mathcal{D}} = \begin{cases} J^d \ if \ \mathcal{D} = \{1,2,\dots,d\} \\ J^\infty \ if \ \mathcal{D} = \mathbb{Z}_+. \end{cases}$

So, the i-th component of $\Pi_W(t)[X]$, $M_W(t)$, respectively, are given by $\Pi_W(t)[X](i) = \left(I_n \ W^T(t,i) \right) \Pi(t)[X](i) \begin{pmatrix} I_n \\ W(t,i) \end{pmatrix}$ and $M_W(t,i) = \left(I_n \ W^T(t,i) \right)$ $\begin{pmatrix} M(t,i) \ L(t,i) \\ L^T(t,i) \ R(t,i) \end{pmatrix} \begin{pmatrix} I_n \\ W(t,i) \end{pmatrix}$ for all $(t,i) \in I \times \mathcal{D}$.

Further we denote

$$Dom\mathfrak{R} = \left\{ (t,X) \in \mathcal{I} \times \mathcal{S}_n^{\mathcal{D}} ; \inf_{i \in \mathcal{D}} |det[\Pi_2(t)[X](i) + R(t,i)]| \geq \delta > 0 \right\} \tag{5.8}$$

and $\mathfrak{R} : Dom\mathfrak{R} \to \mathcal{S}_n^{\mathcal{D}}$ by

$$\mathfrak{R}(t,X) = A^T(t)X + XA(t) + \Pi_1(t)[X] - \{XB(t) + \Pi_{12}(t)[X] + L(t)\}\{\Pi_2(t)[X] +$$
$$R(t)\}^{-1}\{XB(t) + \Pi_{12}(t)[X] + L(t)\}^T + M(t). \tag{5.9}$$

In (5.8) δ may depend upon (t,X).
The differential equation (5.1) may be written in the following compact form

$$\frac{d}{dt}X(t) + \mathfrak{R}(t,X(t)) = 0. \tag{5.10}$$

Definition 5.1.1. A C^1-function $X : \mathcal{I}_1 \subset \mathcal{I} \to \mathcal{S}_n^{\mathcal{D}}$ is a *solution of (5.1)* if $(t,X(t)) \in Dom\mathfrak{R}$ for all $t \in \mathcal{I}_1$ and if X verifies the relation (5.1) on $\mathcal{I}_1$. If $\mathcal{I}_1 = \mathcal{I}$, then $X(\cdot)$ is named global solution of GRDE (5.1), or equivalently of (5.10).

The next results will be repeatedly used in the rest of this book.

Lemma 5.1.1. *If* $W : \mathcal{I} \to \mathcal{M}_{mn}^{\mathcal{D}}$ *then, for all* $(t, X) \in Dom\mathfrak{R}$ *we have*

$$\mathfrak{R}(t, X) = (A(t) + B(t)W(t))^T X + X(A(t) + B(t)W(t)) + \Pi_W(t)[X] + M_W(t) -$$
$$(W(t) - F^X(t))^T (\Pi_2(t)[X] + R(t))(W(t) - F^X(t)) \quad (5.11)$$

where Π_W *and* M_W *were introduced by (5.6) and (5.7) and* $F^X(t) \in \mathcal{M}_{mn}^{\mathcal{D}}$ *is defined by:*

$$F^X(t) = -\{\Pi_2(t)[X] + R(t)\}^{-1}\{XB(t) + \Pi_{12}(t)[X] + L(t)\}^T. \quad (5.12)$$

Proof is done by standard direct calculation. The details are omitted.

The next version of parameterized Schur complement technique will be frequently used in the next developments.

Lemma 5.1.2 (Schur Complement). *Let* $S : \mathcal{I} \to \mathcal{S}_{n+m}^{\mathcal{D}}$ *be a bounded function. For each* t *we consider the partition* $S(t) = \begin{pmatrix} S_{11}(t) & S_{12}(t) \\ S_{12}^T(t) & S_{22}(t) \end{pmatrix}$ *where* $S_{11}(t) \in \mathcal{S}_n^{\mathcal{D}}$ *and* $S_{22}(t) \in \mathcal{S}_m^{\mathcal{D}}$. *Assume that* $S_{22}(t) \gg 0$, $t \in \mathcal{I}$. *Under these conditions,* $S(t) \gg 0$, $t \subset \mathcal{I}$ $(S(t) \geq 0, t \in \mathcal{I})$ *if and only if* $S_{11}(t) - S_{12}(t)S_{22}^{-1}(t)S_{12}^T(t) \gg 0$, $t \in \mathcal{I}$ $(S_{11}(t) - S_{12}(t)S_{22}^{-1}(t)S_{12}^T(t) \geq 0, t \in \mathcal{I})$.

Proof. Under the considered assumptions there exist positive constants μ, ν such that $\|S(t)\|_\infty \leq \mu$ and $S_{22}(t, i) \geq \nu I_m$ for all $t \in \mathcal{I}, i \in \mathcal{D}$. The last inequality guarantees the invertibility of the matrices $S_{22}(t, i)$ and we have $|S_{22}^{-1}(t, i)| \leq \nu^{-1}$ for all $(t, i) \in \mathcal{I} \times \mathcal{D}$. For each $t \in \mathcal{I}, i \in \mathcal{D}$ we consider $U(t, i) \in \mathbf{R}^{n+m \times n+m}$ defined by

$$U(t, i) = \begin{pmatrix} I_n & -S_{12}(t, i)S_{22}^{-1}(t, i) \\ 0 & I_m \end{pmatrix}.$$

One obtains that $U(t, i)$ is invertible and we have

$$U^{-1}(t, i) = \begin{pmatrix} I_n & S_{12}(t, i)S_{22}^{-1}(t, i) \\ 0 & I_m \end{pmatrix}.$$

By direct calculation we deduce

$$U(t, i)S(t, i)U^T(t, i) = \begin{pmatrix} \tilde{S}(t, i) & 0 \\ 0 & S_{22}(t, i) \end{pmatrix} \quad (5.13)$$

for all $(t, i) \in \mathcal{I} \times \mathcal{D}$, where $\tilde{S}(t, i) = S_{11}(t, i) - S_{12}(t, i)S_{22}^{-1}(t, i)S_{12}^T(t, i)$. From $U(t, i) = I_{n+m} + \begin{pmatrix} 0 & -S_{12}(t, i)S_{22}^{-1}(t, i) \\ 0 & 0 \end{pmatrix}$, we infer $|U(t, i)| \leq 1 + |S_{12}(t, i)S_{22}^{-1}(t, i)|$. Since $|S_{12}(t, i)| \leq |S(t, i)| \leq \|S(t)\|_\infty$ we conclude that

$$|U(t,i)| \le \tilde{\mu} \tag{5.14}$$

for all $(t,i) \in \mathcal{I} \times \mathcal{D}$ where $\tilde{\mu} = 1 + \mu \nu^{-1}$. In a similar way one obtains

$$|U^{-1}(t,i)| \le \tilde{\mu} \tag{5.15}$$

for all $(t,i) \in \mathcal{I} \times \mathcal{D}$. Employing (5.14) and (5.15) we get

$$U(t,i)U^T(t,i) \ge \tilde{\mu}^{-2} I_{n+m}$$
$$(U^T(t,i)U(t,i))^{-1} \ge \tilde{\mu}^{-2} I_{n+m} \tag{5.16}$$

for all $(t,i) \in \mathcal{I} \times \mathcal{D}$. Now, we are in position to prove that $\tilde{S}(t) \gg 0, t \in \mathcal{I}$, if and only if $S(t) \gg 0, t \in \mathcal{I}$. Indeed, if $S(t) \gg 0, t \in \mathcal{I}$ there exists $\delta > 0$ such that $S(t,i) \ge \delta I_{n+m}$ for all $(t,i) \in \mathcal{I} \times \mathcal{D}$. From (5.13) we get $\begin{pmatrix} \tilde{S}(t,i) & 0 \\ 0 & S_{22}(t,i) \end{pmatrix} \ge$ $\delta U(t,i)U^T(t,i)$. Invoking the first inequality from (5.16) we deduce that $\begin{pmatrix} \tilde{S}(t,i) & 0 \\ 0 & S_{22}(t,i) \end{pmatrix} \ge \delta \tilde{\mu}^{-2} I_{n+m}$. This allows us to conclude that $\tilde{S}(t,i) \ge \delta \tilde{\mu}^{-2} I_n$ for all $(t,i) \in \mathcal{I} \times \mathcal{D}$. So, we have obtained that $\tilde{S}(t) \gg 0, t \in \mathcal{I}$.

Let us prove the converse implication. To this end, we assume that $\tilde{S}(t) \gg 0, t \in \mathcal{I}$. This means that there exists $\tilde{\delta} > 0$ such that $\tilde{S}(t,i) \ge \tilde{\delta} I_n$ for all $(t,i) \in \mathcal{I} \times \mathcal{D}$. Therefore, $\begin{pmatrix} \tilde{S}(t,i) & 0 \\ 0 & S_{22}(t,i) \end{pmatrix} \ge \hat{\delta} I_{n+m}$ for all $(t,i) \in \mathcal{I} \times \mathcal{D}$ where $\hat{\delta} = \min\{\tilde{\delta}, \nu\}$. Further (5.13) yields $S(t,i) = U^{-1}(t,i) \begin{pmatrix} \tilde{S}(t,i) & 0 \\ 0 & S_{22}(t,i) \end{pmatrix} (U^T(t,i))^{-1}$. Hence $S(t,i) \ge \hat{\delta}(U^T(t,i)U(t,i))^{-1}$. Invoking the second inequality (5.16) we deduce

$$S(t,i) \ge \hat{\delta} \tilde{\nu}^{-2} I_{n+m}$$

for all $(t,i) \in \mathcal{I} \times \mathcal{D}$. Thus we have proved that $S(t) \gg 0, t \in \mathcal{I}$. The equivalence $S(t) \ge 0, t \in \mathcal{I} \leftrightarrow \tilde{S}(t) \ge 0, t \in \mathcal{I}$ may be proved in a similar way. Thus the proof is complete. $\qquad\square$

The case when the Banach space of the sequences of symmetric matrices is replaced by a Banach space of the sequences of self-adjoint linear operators on a Hilbert space may be viewed in [146].

As usual, $\tilde{S}(t) = S_{11}(t) - S_{12}(t)S_{22}^{-1}(t)S_{12}^T(t)$ is named *Schur complement of* $S_{22}(t)$ *in* $S(t)$.

One sees that the operator $\mathcal{R}$ and consequently equation (5.10) are associated with the quadruple $\Sigma = (A, B, \Pi, \mathcal{Q})$ where (A, B, Π) are as before and $\mathcal{Q}$ is defined by (5.3). We introduce the so-called *generalized dissipation operator* $\Lambda^\Sigma : C^1(\mathcal{I}, \mathcal{S}_n^\mathcal{D}) \to \mathcal{S}_{n+m}^\mathcal{D}$ associated with the quadruple Σ by

$$\Lambda^{\Sigma}(t)[X(t)]=\begin{pmatrix} \Lambda_1(t) & X(t)B(t)+\Pi_{12}(t)[X(t)]+L(t) \\ \{X(t)B(t)+\Pi_{12}(t)[X(t)]+L(t)\}^T & R(t)+\Pi_2(t)[X(t)] \end{pmatrix},$$

where

$$\Lambda_1(t) = \frac{d}{dt}X(t)+A^T(t)X(t)+X(t)A(t)+\Pi_1(t)[X(t)]+M(t)$$

and $C^1(\mathcal{I},\mathcal{S}_n^{\mathcal{D}})$ is the space of C^1-functions defined on the interval $\mathcal{I}$ taking values in $\mathcal{S}_n^{\mathcal{D}}$. Notice that $\frac{d}{dt}X(t)+\mathfrak{R}(t,X(t))$ is the Schur complement of $R(t)+\Pi_2(t)[X(t)]$ in $\Lambda^{\Sigma}(t)[X(t)]$.

The following two subsets of

$$C_b^1(\mathcal{I},\mathcal{S}_n^{\mathcal{D}}) = \left\{ X \in C^1(\mathcal{I},\mathcal{S}_n^{\mathcal{D}}) \;\middle|\; X, \frac{d}{dt}X \text{ are bounded} \right\}$$

will play an important role in the next developments:

$$\Gamma^{\Sigma} = \left\{ X \in C_b^1(\mathcal{I},\mathcal{S}_n^{\mathcal{D}}) \mid \Lambda^{\Sigma}(t)[X(t)] \geq 0,\, R(t)+\Pi_2(t)[X(t)] \gg 0,\, t \in \mathcal{I} \right\},$$

$$\tilde{\Gamma}^{\Sigma} = \left\{ X \in C_b^1(\mathcal{I},\mathcal{S}_n^{\mathcal{D}}) \mid \Lambda^{\Sigma}(t)[X(t)] \gg 0,\, t \in \mathcal{I} \right\}.$$

Remark 5.1.2. (a) In the case when $\Pi_2(t)$ is the zero operator then in the definition of the set Γ^{Σ} we ask $R(t) \gg 0$ which is the usual condition used in the case of Riccati differential equations of deterministic and stochastic control. If $\Pi_2(t)$ is not the zero operator it is not necessary to make any assumptions concerning the sign of $R(t)$.

(b) We shall see later that if A, B, Π, $\mathcal{Q}$ are θ-periodic functions and if Γ^{Σ} is not empty ($\tilde{\Gamma}^{\Sigma}$ is not empty, respectively), then Γ^{Σ} contains also a θ-periodic function ($\tilde{\Gamma}^{\Sigma}$ contains also a θ-periodic function). Moreover, we shall show that if $A(t) = A$, $B(t) = B$, $\Pi(t) = \Pi$, $\mathcal{Q}(t) = \mathcal{Q}$ for all $t \in \mathbf{R}$ and if Γ^{Σ} is not empty, ($\tilde{\Gamma}^{\Sigma}$ is not empty, respectively), then there exists $X \in \mathcal{S}_n^{\mathcal{D}}$ which lies in Γ^{Σ} ($X \in \tilde{\Gamma}^{\Sigma}$, respectively).

(c) Based on the Schur complement technique one obtains that Γ^{Σ} contains in particular all bounded solutions of (5.1) verifying the additional condition $R(t)+\Pi_2(t)[X(t)] \gg 0$.

Now we introduce the following linear operators $\mathfrak{L}_{A+BW}(t)$ and $\mathfrak{L}_{A+BW,\Pi_W}(t)$ by

$$\mathfrak{L}_{A+BW}(t)[X] = (A(t)+B(t)W(t))^T X + X(A(t)+B(t)W(t)) \qquad (5.17)$$

$$\mathfrak{L}_{A+BW,\Pi_w}(t)[X]=(A(t)+B(t)W(t))^T X+X(A(t)+B(t)W(t))+\Pi_W(t)[X] \quad (5.18)$$

for all $X \in \mathcal{S}_n^{\mathcal{D}}$.

Remark 5.1.3. (a) If $W : \mathcal{I} \to \mathcal{M}_{mn}^{\mathcal{D}}$ is a bounded and continuous function, then under the assumption $(\mathbf{\Pi}_1)$, $\Pi_W(t)$ introduced via (5.6) is a positive operator.

(b) Combining (2.122) written for $A(t)$ replaced by $A(t) + B(t)W(t)$ and Corollary 2.2.6 (ii) we obtain that the operator valued function $\mathcal{L}_{A+BW,\Pi_W}(\cdot)$ generates an anticausal positive evolution on the space $\mathcal{S}_n^{\mathcal{D}}$.

(c) The equality proved in Lemma 5.1.1 may be rewritten in the form

$$\mathfrak{R}(t,X) = \mathcal{L}_{A+BW,\Pi_W}(t) + M_W(t) - (W(t) - F^X(t))^T (\Pi_2(t)[X] + R(t))(W(t) - F^X(t)). \quad (5.19)$$

At the end of this section let us remark that if $(t,X) \in Dom\mathfrak{R}$ and $Y \in \mathcal{S}_n^{\mathcal{D}}$ then there exists $\tilde{h} > 0$ such that $(t, X + hY) \in Dom\mathfrak{R}$ for all h, with $|h| < \tilde{h}$. This allows us to compute the Fréchet derivative $\mathfrak{R}'(t,X)$ of the operator valued function $X \to \mathfrak{R}(t,X)$ when $(t,X) \in Dom\mathfrak{R}$.

Lemma 5.1.3. *If $(t,X) \in Dom\mathfrak{R}$, then the Fréchet derivative $\mathfrak{R}'(t,X)$ is given by*

$$\mathfrak{R}'(t,X)[Y] = (A(t) + B(t)F^X(t))^T Y + \quad (5.20)$$

$$Y(A(t) + B(t)F^X(t)) + \Pi_{F^X}(t)[Y] = \mathcal{L}_{A+BF^X,\Pi_{F^X}}(t)[Y], \quad \forall \ Y \in \mathcal{S}_n^{\mathcal{D}}$$

where $F^X(t)$ is introduced by (5.12).

Proof. From (5.8) and (5.9) one sees that $\mathfrak{R}(t,\cdot)$ is Fréchet differentiable in X if $(t,X) \in Dom\mathfrak{R}$. The formula (5.20) is obtained by direct calculation taking into account that $\mathfrak{R}'(t,X)[Y] = \lim_{h \to 0} \frac{1}{h}[\mathfrak{R}(t,X + hY) - \mathfrak{R}(t,X)]$.

5.2   A Comparison Theorem and Several Consequences

First we prove the following important result concerning the monotonic dependence of the solutions of (5.1) with respect to the data.

Theorem 5.2.1 (Comparison   Theorem). *Let $\hat{\mathfrak{R}}$ be the operator (5.9) associated with the quadruple $\hat{\Sigma} = (A,B,\Pi,\hat{Q})$ and $\tilde{\mathfrak{R}}$ be the operator of type (5.9) associated with the quadruple $\tilde{\Sigma} = (A,B,\Pi,\tilde{Q})$ where A, B, Π are as before and $\hat{Q}(t) = \{\hat{Q}(t)\}_{i \in \mathcal{D}}, \tilde{Q}(t) = \{\tilde{Q}(t,i)\}_{i \in \mathcal{D}}, \ \hat{Q}(t,i) = \begin{pmatrix} \hat{M}(t,i) & \hat{L}(t,i) \\ \hat{L}(t,i)^T & \hat{R}(t,i) \end{pmatrix},$ $\tilde{Q}(t,i) = \begin{pmatrix} \tilde{M}(t,i) & \tilde{L}(t,i) \\ \tilde{L}(t,i)^T & \tilde{R}(t,i) \end{pmatrix}$ with $\hat{L}(t,i), \tilde{L}(t,i) \in \mathbf{R}^{n \times m}, \ \hat{M}(t,i), \tilde{M}(t,i) \in \mathcal{S}_n$ and $\hat{R}(t,i), \tilde{R}(t,i) \in \mathcal{S}_m$. Let $X_i : \mathcal{I}_1 \subset \mathcal{I} \to \mathcal{S}_n^{\mathcal{D}}, \ i = 1,2$, be the solutions of*

$$\frac{d}{dt}X_1(t) + \hat{\mathfrak{R}}(t,X_1(t)) = 0, \qquad \frac{d}{dt}X_2(t) + \tilde{\mathfrak{R}}(t,X_2(t)) = 0.$$

Assume that: (a) $\hat{Q}(t,i) \geq \tilde{Q}(t,i)$ for all $(t,i) \in \mathcal{I} \times \mathcal{D}$; (b) $\tilde{R}(t,i) + \Pi_2(t)[X_2(t)](i) > 0$ for $(t,i) \in \mathcal{I}_1 \times \mathcal{D}$; (c) There exists $\tau \in \mathcal{I}_1$ such that $X_1(\tau,i) \geq X_2(\tau,i)$ for all $i \in \mathcal{D}$. Under these conditions we have $X_1(t,i) \geq X_2(t,i)$ for all $(t,i) \in ((-\infty,\tau] \cap \mathcal{I}_1) \times \mathcal{D}$.

Proof. Let

$$F_1(t) := -\left\{\hat{R}(t) + \Pi_2(t)[X_1(t)]\right\}^{-1}\left\{X_1(t)B(t) + \Pi_{12}(t)[X_1(t)] + \hat{L}(t)\right\}^T$$

and

$$F_2(t) := -\left\{\tilde{R}(t) + \Pi_2(t)[X_2(t)]\right\}^{-1}\left\{X_2(t)B(t) + \Pi_{12}(t)[X_2(t)] + \tilde{L}(t)\right\}^T.$$

Applying Lemma 5.1.1 with $W(t) = F_1(t)$ both for $\hat{\mathfrak{R}}(t, X_1(t))$ and $\tilde{\mathfrak{R}}(t, X_2(t))$, one obtains

$$\frac{d}{dt}X_1(t) + \mathcal{L}_{A+BF_1}(t)[X_1(t)] + \Pi_{F_1}(t)[X_1(t)] + \hat{M}_{F_1}(t) = 0$$

and

$$\frac{d}{dt}X_2(t) + \mathcal{L}_{A+BF_1}(t)[X_2(t)] + \Pi_{F_1}(t)[X_2(t)] + \tilde{M}_{F_1}(t)$$
$$- \left[F_1(t) - F_2(t)\right]^T\left\{\tilde{R}(t) + \Pi_2(t)[X_2(t)]\right\}\left[F_1(t) - F_2(t)\right] = 0.$$

This leads to

$$\frac{d}{dt}\left[X_1(t) - X_2(t)\right] + \mathcal{L}_{A+BF_1}(t)\left[X_1(t) - X_2(t)\right] + \Pi_{F_1}(t)\left[X_1(t) - X_2(t)\right] + H(t) = 0,$$

and

$$H(t) := \left[F_1(t) - F_2(t)\right]^T\left\{\tilde{R}(t) + \Pi_2(t)[X_2(t)]\right\}\left[F_1(t) - F_2(t)\right]$$
$$+ \hat{M}_{F_1}(t) - \tilde{M}_{F_1}(t).$$

Since

$$\hat{M}_{F_1}(t,i) - \tilde{M}_{F_1}(t,i) = \begin{pmatrix} I_n \\ F_1(t,i) \end{pmatrix}^T \left[\hat{Q}(t,i) - \tilde{Q}(t,i)\right] \begin{pmatrix} I_n \\ F_1(t,i) \end{pmatrix} \geq 0$$

it follows that $H(t,i) \geq 0$ for all $(t,i) \in \mathcal{I}_1 \times \mathcal{D}$.

Since $\Pi_{F_1}(t)$ is a positive operator, applying Corollary 2.2.6 (ii), to the operator $X \to \mathcal{L}_{A+BF_1}(t)[X] + \Pi_{F_1}(t)[X]$ one gets that $X_1(t) - X_2(t) \geq 0$ for all $t \in (-\infty, \tau] \cap \mathcal{I}_1$ and thus the proof is complete. $\square$

Using the above theorem we prove the following result concerning the maximal interval of definition of a solution of (5.1) with given terminal conditions.

Theorem 5.2.2. *Assume that* $\Sigma = (A, B, \Pi, \mathcal{Q})$ *satisfies* $\Gamma^{\Sigma} \neq \emptyset$. *Let*

$$\tilde{\mathcal{D}}(\mathfrak{R}) := \left\{(\tau, X) \in Dom(\mathfrak{R}); \exists \hat{X} \in \Gamma^{\Sigma} \text{ such that } X \geq \hat{X}(\tau)\right\}$$

and let $X(\cdot, \tau, X_0)$ *be the solution of (5.1) with* $X(\tau, \tau, X_0) = X_0$.

If $(\tau, X_0) \in \tilde{\mathcal{D}}(\mathfrak{R})$ then the solution $X(\cdot, \tau, X_0)$ is well defined on $(-\infty, \tau] \cap \mathcal{I}$.

Proof. Let $\mathcal{I}_{\tau, X_0} \subset (-\infty, \tau]$ be the maximal interval on which $X(\cdot, \tau, X_0)$ is defined and let $\hat{X} \in \Gamma^\Sigma$ be such that $X_0 \geq \hat{X}(\tau)$. Obviously there exists a bounded and continuous function $M_\ell : \mathcal{I} \to \mathcal{S}_n^{\mathcal{D}}$ such that $M_\ell(t) \leq 0$ and

$$\frac{d}{dt}\hat{X}(t) + \mathfrak{R}(t, \hat{X}(t)) + M_\ell(t) = 0.$$

Applying Theorem 5.2.1 for the quadruples $\hat{\Sigma} = \Sigma$ and $\tilde{\Sigma} = (A, B, \Pi, \tilde{\mathcal{Q}})$ with

$$\tilde{\mathcal{Q}}(t,i) := \begin{pmatrix} M(t,i) + M_\ell(t,i) & L(t,i) \\ L^T(t,i) & R(t,i) \end{pmatrix} = \mathcal{Q}(t,i) + \begin{pmatrix} M_\ell(t,i) & 0 \\ 0 & 0 \end{pmatrix}$$

we conclude that

$$X(t, \tau, X_0) \geq \hat{X}(t) \quad \text{for all} \quad t \in \mathcal{I}_{\tau, X_0}. \tag{5.21}$$

Let Y be the solution of the terminal value problem

$$\frac{d}{dt}Y(t) + A^T(t)Y(t) + Y(t)A(t) + \Pi_1(t)[Y(t)] + M(t) = 0, \quad Y(\tau) = X_0. \tag{5.22}$$

Since $Y \to \Pi_1(t)[Y]$ is a linear operator it follows that $Y(t)$ is well defined for all $t \in \mathcal{I}$.

By direct calculation we obtain that

$$\frac{d}{dt}\big[Y(t) - X(t)\big] + \mathfrak{L}_A(t)\big[Y(t) - X(t)\big] + \Pi_1(t)\big[Y(t) - X(t)\big] + \hat{H}(t) = 0 \tag{5.23}$$

for $t \in \mathcal{I}_{\tau, X_0}$ where $X(t) = X(t, \tau, X_0)$. Here

$$\hat{H}(t) := F^T(t)\big\{R(t) + \Pi_2(t)[X(t)]\big\}F(t)$$

with $F(t) := F^X(t)$. From (5.21) and assumptions (Π_1) we deduce that

$$R(t) + \Pi_2(t)[X(t)] \geq R(t) + \Pi_2(t)[\hat{X}(t)] \gg 0.$$

So $\hat{H}(t) \geq 0$ for $t \in \mathcal{I}_{\tau, X_0}$. Corollary 2.2.6 (ii) and (2.44), applied in the case of (5.23), gives

$$X(t) \leq Y(t) \tag{5.24}$$

for all $t \in \mathcal{I}_{\tau, X_0}$.

From (5.21), (5.24), $\hat{X} \in \Gamma^\Sigma$ and [129] vol. 2, it follows easily that $X(t, \tau, X_0)$ is defined for all $t \in (-\infty, \tau] \cap \mathcal{I}$ and thus the proof ends. $\square$

The proof of Theorem 5.2.2 shows that—as a consequence of the Comparison Theorem—each element $\hat{X} \in \Gamma^{\Sigma}$ is providing a lower bound for the solution $X(\cdot, \tau, X_0)$ of (5.1) (see (5.21), whereas the solution Y of (5.22) gives an upper bound).

Corollary 5.2.3. *Assume that* $0 \in \Gamma^{\Sigma}$. *Then for all* $(\tau, X_0) \in \mathcal{I} \times \mathcal{S}_{n+}^{\mathcal{D}}$ *the solution* $X(\cdot, \tau, X_0)$ *of the GRDE (5.1) is defined on the whole interval* $(-\infty, \tau] \cap \mathcal{I}$ *and fulfills the inequality*

$$0 \leq X(t, \tau, X_0) \leq Y(t)$$

where Y *is the solution of (5.22). Moreover if* $\mathcal{D} = \{1, 2, \ldots, d\}$ *and* $X_0 = \{X_0(i)\}_{i \in \mathcal{D}}$ *with* $X_0(i) > 0$, *then* $X(t, \tau, X_0) > 0$ *for all* $t \in \mathcal{I}$ *with* $t \leq \tau$.

Proof. Since $0 \in \Gamma^{\Sigma}$ it follows that $\mathcal{I} \times \mathcal{S}_{n+}^{\mathcal{D}} \subset \tilde{\mathcal{D}}(\mathfrak{R})$. So from the above theorem one obtains that $X(t, \tau, X_0)$ is well defined for all $t \in (-\infty, \tau] \cap \mathcal{I}$ for arbitrary $(\tau, X_0) \in \mathcal{I} \times \mathcal{S}_{n+}^{\mathcal{D}}$.

The inequality $X(t, \tau, X_0) \geq 0$ is just (5.21) for $\hat{X}(t) \equiv 0$. On account of (5.24) it remains to prove that $X(t, \tau, X_0) > 0$ if $X_0 > 0$. To this end we set $X(t) = X(t, \tau, X_0)$ and $F(t) = F^X(t)$ for $t \in \mathcal{I}$ with $t \leq \tau$. Applying Lemma 5.1.1 for $W(t) = F(t)$ one obtains

$$\frac{d}{dt}X(t) + \mathcal{L}_{A+BF}(t)[X(t)] + \Pi_F(t)[X(t)] + M_F(t) = 0.$$

Further we write the representation formula

$$X(t) = \Phi_{A+BF}^T(\tau, t)X_0\Phi_{A+BF}(\tau, t) + \int_t^{\tau} \Phi_{A+BF}^T(s, t)H(s)\Phi_{A+BF}(s, t)\, ds,$$

where $\Phi_{A+BF}(s, t)$ is the fundamental matrix solution defined by

$$\frac{d}{ds}\Phi_{A+BF}(s, t) = [A(s) + B(s)F(s)]\Phi_{A+BF}(s, t), \quad \Phi_{A+BF}(t, t) = I$$

and where

$$H(s) = \Pi_F(s)[X(s)] + M_F(s) \quad \text{for } s \in \mathcal{I} \text{ with } s \leq \tau.$$

The assumption $0 \in \Gamma^{\Sigma}$ is equivalent to

$$R(t) \gg 0 \quad \text{and} \quad \begin{pmatrix} M(t) & L(t) \\ L^T(t) & R(t) \end{pmatrix} \geq 0. \tag{5.25}$$

From (5.25) and the monotonicity of $\Pi(s)[\cdot]$ we conclude that $H(s) \geq 0$ for all $s \in \mathcal{I}$ with $s \leq \tau$. From the representation formula we obtain that

$$X(t) \geq \Phi_{A+BF}^T(\tau,t)X_0\Phi_{A+BF}(\tau,t) > 0.$$

For the last inequality we have taken into account that $X_0 > 0$ and $\Phi_{A+BF}(\tau,t)$ is invertible, thus the proof is complete.   □

5.3   The Maximal Solution of GRDE

First we introduce the concept of maximal solution of the GRDE (5.1).

Definition 5.3.1. A solution $\check{X} \colon \mathcal{I} \to \mathcal{S}_n^{\mathcal{D}}$ of the GRDE (5.1) is said to be the *maximal solution with respect to* Γ^Σ (or *maximal solution* for shortness) if $\check{X}(t) \geq \hat{X}(t)$ for arbitrary $\hat{X} \in \Gamma^\Sigma$.

Remark 5.3.1. If Γ^Σ is not empty and $\check{X}$ is a bounded and maximal solution of (5.1), then from $\check{X}(t) \geq \hat{X}(t)$ for arbitrary $\hat{X} \in \Gamma^\Sigma$, it follows that $\check{X} \in \Gamma^\Sigma$. Therefore the bounded and maximal solution of (5.1) (if it exists) is unique (under the assumption $\Gamma^\Sigma \neq \emptyset$).

In this section we prove a result concerning the existence of the bounded and maximal solution with respect to Γ^Σ of (5.1). First, we give a definition which will play a crucial role in the next developments. That is the concept of stabilizability for the triple (A,B,Π).

Definition 5.3.2. We say that the triple (A,B,Π) is *stabilizable* if there exists a bounded and continuous function $F \colon \mathcal{I} \to \mathcal{M}_{mn}^{\mathcal{D}}$ such that the operator valued function $\mathcal{L}_{A+BF,\Pi_F}(\cdot)$ generates an anticausal exponentially stable evolution. The function F will be termed a *stabilizing feedback gain*.

We shall show later (Corollary 5.4.9) that if A, B, Π are periodic functions with period θ and if the triple (A,B,Π) is stabilizable then there exists a stabilizing feedback gain which is a periodic function with period θ. Moreover if $A(t) \equiv A$, $B(t) \equiv B$, $\Pi(t) \equiv \Pi$ for $t \in \mathbf{R}$, and if the triple (A,B,Π) is stabilizable, then there exists a stabilizing feedback gain which is constant.

In the particular case when $\Pi(t)$ is of the form (5.5) then the above definition of stabilizability reduces to the standard definition of stabilizability for stochastic systems (mean-square stabilizability—see also Chap. 4).

Applying Theorem 2.3.12 we have the following result.

Corollary 5.3.1. *The triple (A,B,Π) is stabilizable if and only if there exists some $X \in C_b^1(\mathcal{I},\mathcal{S}_n^{\mathcal{D}})$ with $X(t) \gg 0, t \in \mathcal{I}$ and a bounded and continuous function $F \colon \mathcal{I} \to \mathcal{M}_{mn}^{\mathcal{D}}$ such that*

$$\frac{d}{dt}X(t) + \mathcal{L}_{A+BF,\Pi_F}(t)[X(t)] \ll 0 \quad \text{for all} \quad t \in \mathcal{I}.$$

Remark 5.3.2. In the case $\mathcal{D} = \{1, 2, \ldots, d\}$ we may define the linear operators $\mathcal{L}_{A+BF,\Pi_F^*} : \mathcal{S}_n^d \to \mathcal{S}_n^d$, by $\mathcal{L}_{A+BF,\Pi_F^*}[X] = (A(t) + B(t)F(t))X + X(A(t) + B(t)F(t))^T + \Pi_F^*(t)[X]$ where $\Pi_F^*(t) : \mathcal{S}_n^d \to \mathcal{S}_n^d$ is the adjoint of the operator $\Pi_F(t)$ with respect to the inner product (2.16) and $F : \mathcal{I} \to \mathcal{M}_{mn}^d$ is a given bounded and continuous function. By direct calculation, based on the definition of the adjoint operator we obtain the equality

$$\mathcal{L}_{A+BF,\Pi_F}(t) = \mathcal{L}_{A+BF,\Pi_F^*}^*(t) \tag{5.26}$$

for all $t \in \mathcal{I}$ and all bounded and continuous function $F : \mathcal{I} \to \mathcal{M}_{mn}^d$.

Applying Proposition 2.4.1 in the case of operator valued function $\mathcal{L}_{A+BF,\Pi_F^*}(\cdot)$ and using (5.26) we obtain that in the case $\mathcal{D} = \{1, 2, \ldots, d\}$ the triple (A, B, Π) is stabilizable, if and only if there exists a bounded and continuous function $F : \mathcal{I} \to \mathcal{M}_{mn}^d$ such that the operator valued function $\mathcal{L}_{A+BF,\Pi_F^*}(\cdot)$ defines an exponentially stable evolution on $\mathcal{S}_n^d$.

In the case $\mathcal{I} = \mathbf{R}$, using Theorem 2.3.6, we get

Corollary 5.3.2. *For $\mathcal{I} = \mathbf{R}$ and $\mathcal{D} = \{1, 2, \ldots, d\}$ the following are equivalent: (a) The triple (A, B, Π) is stabilizable. (b) There exists a bounded C^1-function $X : \mathbf{R} \to \mathcal{S}_{n+}^{\mathcal{D}}$ with bounded derivative, $X(t) \gg 0$ and a bounded and continuous function $F : \mathbf{R} \to \mathcal{M}_{mn}^{\mathcal{D}}$ which satisfy*

$$\frac{d}{dt}X(t) - \mathcal{L}_{A+BF,\Pi_F^*}(t)[X(t)] \gg 0. \tag{5.27}$$

Remark 5.3.3. In the particular case when the coefficients do not depend on t and Π takes the special form of (5.5), (5.27) can be converted in a system of LMIs which can be solved using an LMI solver (see [124]).

Now we state an auxiliary result which together with Lemma 5.1.1 plays a crucial role in the proof of the main result of this section and follows directly from (5.18).

Lemma 5.3.3. *If $W : \mathcal{I} \to \mathcal{M}_{mn}^{\mathcal{D}}$ is a continuous function and if X is a solution of the differential equation*

$$\frac{d}{dt}X(t) + \mathcal{L}_{A+BW,\Pi_W}(t)[X(t)] + M_W(t) = 0$$

and if $\det\{R(t, i) + \Pi_2(t)[X(t)](i)\} \neq 0$, then X verifies also the following differential equation:

$$\frac{d}{dt}X(t) + \mathcal{L}_{A+BF,\Pi_F}(t)[X(t)] + M_F(t)$$

$$+ [F(t) - W(t)]^T \Theta(t, X(t)) [F(t) - W(t)] = 0$$

where $F(t) := F^X(t)$ and $\Theta(t, X(t)) := R(t) + \Pi_2(t)[X(t)]$.

Regarding the operator valued function Π we make the assumption

$(\mathbf{\Pi}_2)$ for each $t \in \mathcal{I}$ and for any bounded and monotone sequence $\{X_k\}_{k\geq 0} \subset \mathcal{S}_n^{\mathcal{D}}$ we have

$$\lim_{k\to\infty} \Pi(t)[X_k](i) = \Pi(t)[X](i) \tag{5.28}$$

for all $i \in \mathcal{D}$ where $X = \{X(i)\}_{i\in\mathcal{D}} \in \mathcal{S}_n^{\mathcal{D}}$ is such that

$$X(i) = \lim_{k\to\infty} X_k(i) \tag{5.29}$$

Remark 5.3.4. (a) Since $\mathcal{S}_{n+m}$ and $\mathcal{S}_n$, respectively, are finite dimensional linear spaces, the limits from (5.28) and (5.29) hold under any norms of these linear spaces, as well as component wise.

(b) If in (5.29) the convergence is uniform with respect to $i \in \mathcal{D}$, then (5.28) is automatically satisfied because $\Pi(t)$ is a linear and bounded operator. Therefore, the operator $\Pi(t)$ satisfies the assumption $(\mathbf{\Pi}_2)$ if it satisfies (5.28) even if the limit (5.29) is not uniform with respect to $i \in \mathcal{D}$.

(c) From (a) we deduce that in the case $\mathcal{D} = \{1, 2, \ldots, d\}$ any linear operator $\Pi(t)$ satisfies the assumption $(\mathbf{\Pi}_2)$.

(d) Following step by step the proof of Lemma 2.8.5 (i) one obtains that the operator $\Pi(t)$ described in (5.5) satisfies the assumption $(\mathbf{\Pi}_2)$ both in the case $\mathcal{D} = \{1, 2, \ldots, d\}$ and in the case $\mathcal{D} = \mathbf{Z}_+$.

Lemma 5.3.4. *Assume that:*

(a) $A(\cdot), B(\cdot), Q(\cdot)$ are bounded and continuous functions;
(b) The operator valued function $\Pi(\cdot)$ satisfies the assumptions $(\mathbf{\Pi}_1)$, $(\mathbf{\Pi}_2)$.

Let $\hat{X} \in \mathcal{S}_n^{\mathcal{D}}$ be such that $R(t,i) + \Pi_2(t)[\hat{X}](i) \geq \delta I_m$ for all $i \in \mathcal{D}$, where $\delta > 0$ may depend on $t \in \mathcal{I}$. Under these conditions for any sequence $\{X_k\}_{k\geq 0} \subset \mathcal{S}_n^{\mathcal{D}}$ which satisfy $X_k(i) \geq X_{k+1}(i) \geq \hat{X}(i)$, for all $k \geq 0$, $i \in \mathcal{D}$ we have

$$\lim_{k\to\infty} \mathfrak{R}(t, X_k)(i) = \mathfrak{R}(t, X)(i) \tag{5.30}$$

for all $i \in \mathcal{D}$, where $X = \{X(i)\}_{i\in\mathcal{D}}$ is such that $X(i) = \lim_{k\to\infty} X_k(i), i \in \mathcal{D}$.

Proof. First, let us remark that based on the Remark 5.3.4 (a) we infer that

$$\lim_{k\to\infty} \Pi_l(t)[X_k](i) = \Pi_l(t)[X](i), \quad l = 1, 2. \tag{5.31}$$

and

$$\lim_{k\to\infty} \Pi_{12}(t)[X_k](i) = \Pi_{12}(t)[X](i) \tag{5.32}$$

$i \in \mathcal{D}$, if X is defined by (5.29).

Employing (5.31) for $l = 1$ we deduce that (5.30) is true if and only if

$$\lim_{k\to\infty} \mathbf{G}(t, X_k)(i) = \mathbf{G}(t, X)(i) \tag{5.33}$$

for all $i \in \mathcal{D}$, where

$$\mathbf{G}(t,Y) = (YB(t) + \Pi_{12}(t)[Y] + L(t))(\Pi_2(t)[Y] + \\ + R(t))^{-1}(YB(t) + \Pi_{12}(t)[Y] + L(t))^T, \quad Y = \{Y(i)\}_{i \in \mathcal{D}} \subset \mathcal{S}_n^{\mathcal{D}}. \quad (5.34)$$

Before showing that (5.33) holds we prove that

$$\lim_{k \to \infty} x^T \mathbf{G}(t, X_k)(i)x = x^T \mathbf{G}(t,X)(i)x \quad (5.35)$$

for all $i \in \mathcal{D}$, $x \in \mathbf{R}^n$, is true.

We have

$$x^T \mathbf{G}(t, X_k)(i)x - x^T \mathbf{G}(t,X)(i)x = y_k^T(t,i)[(\Pi_2(t)[X_k](i) + R(t,i))^{-1} - \quad (5.36)$$

$$-(\Pi_2(t)[X](i) + R(t,i))^{-1}]y_k(t,i) + x^T[(X_k(i) - X(i))B(t,i) + \Pi_{12}(t)[X_k](i) - \quad (5.37)$$

$$-\Pi_{12}(t)[X](i)][\Pi_2(t)[X](i) + R(t,i)]^{-1}[(X_k(i) + X(i))B(t,i) + \Pi_{12}(t)[X_k](i) + \\ +\Pi_{12}(t)[X](i) + 2L(t,i)]^T x$$

where $y_k(t,i) = [X_k(i)B(t,i) + \Pi_{12}(t)[X_k](i) + L^T(t,i)]x \in \mathbf{R}^m$. Since $\{\|X_k\|_\infty\}_{k \geq 0}$ is a bounded sequence we deduce via (5.32) that

$$|y_k(t,i)| \leq c_1|x| \quad (5.38)$$

for all $k \geq 0, t \in \mathcal{I}, i \in \mathcal{D}$.

On the other hand, since

$$\Pi_2(t)[X_k](i) + R(t,i) \geq \Pi_2(t)[X](i) + R(t,i) \geq \Pi_2(t)[\hat{X}](i) + R(t,i) \geq \delta I_m$$

we get

$$|(\Pi_2(t)[X_k](i) + R(t,i))^{-1}| \leq |(\Pi_2(t)[X](i) + R(t,i))^{-1}| \leq \delta^{-1} \quad (5.39)$$

for all $k \geq 0, i \in \mathcal{D}$.

Further, we write

$$y_k^T(t,i)[(\Pi_2(t)[X_k](i) + R(t,i))^{-1} - (\Pi_2(t)[X](i) + R(t,i))^{-1}]y_k(t,i) = \\ y_k^T(t,i)(\Pi_2(t)[X_k](i) + R(t,i))^{-1}(\Pi_2(t)[X](i) \\ -\Pi_2(t)[X_k](i))(\Pi_2(t)[X](i) + R(t,i))^{-1}y_k(t,i).$$

Employing (5.31) for $l = 2$ together with (5.38) and (5.39) we obtain

$$\lim_{k \to \infty} y_k^T(t,i)[(\Pi_2(t)[X_k](i) + R(t,i))^{-1} - (\Pi_2(t)[X](i) + R(t,i))^{-1}]y_k(t,i) = 0 \quad (5.40)$$

for all $i \in \mathcal{D}, x \in \mathbf{R}^n$.

On the other hand, (5.39) together with the boundedness of the sequence $\{\|X_k\|_\infty\}_{k\geq 0}$ yields

$$|(\Pi_2(t)[X](i)+R(t,i))^{-1}[(X_k(i)+X(i))B(t,i)+\Pi_{12}(t)[X_k](i)+\Pi_{12}(t)[X](i)+2L(t,i)]^T x|\leq c_2|x|$$

for all $k \geq 0$, $i \in \mathcal{D}$, $x \in \mathbf{R}^n$, where $c_2 > 0$ is a constant. So, combining (5.29) and (5.32) we may infer

$$\begin{aligned}
\lim_{k\to\infty} x^T &[(X_k(i) - X(i))B(t,i) + \Pi_{12}(t)[X_k](i) - \Pi_{12}(t)[X](i)] \\
&\times [\Pi_2(t)[X](i) + R(t,i)]^{-1} \\
\times &[(X_k(i)+X(i))B(t,i) + \Pi_2(t)[X_k](i) + \Pi_{12}(t)[X](i) + 2L(t,i)]^T x = 0
\end{aligned} \tag{5.41}$$

for all $i \in \mathcal{D}$, $x \in \mathbf{R}^n$. Now, (5.36), (5.40), (5.41) allow us to conclude that (5.35) is true. Further, invoking the identity $x^T My = \frac{1}{4}[(x+y)^T M(x+y) - (x-y)^T M(x-y)]$ for all $x,y \in \mathbf{R}^n$ ($M \in \mathcal{S}_n$) we deduce that (5.35) yields

$$\lim_{k\to\infty} x^T \mathbf{G}(t,X_k)(i)y = x^T \mathbf{G}(t,X)(i)y \tag{5.42}$$

for all $i \in \mathcal{D}$, $x,y \in \mathbf{R}^n$.

Replacing x and y by the vectors of the canonical basis of $\mathbf{R}^n$ in (5.42) we obtain that (5.33) holds component wise. Thus the proof is complete.   □

The main result of this section is the following theorem.

Theorem 5.3.5. *Assume that:*

(a) The assumptions of Lemma 5.3.4 are fulfilled,
(b) The triple (A,B,Π) is stabilizable.

Then the following are equivalent: (i) $\mathbf{\Gamma}^\Sigma \neq \emptyset$. (ii) The GRDE (5.1) has a maximal and bounded solution $\tilde{X}: \mathcal{I} \to \mathcal{S}_n^{\mathcal{D}}$ which verifies $R(t) + \Pi_2(t)[\tilde{X}(t)] \gg 0$, $t \in \mathcal{I}$. If A, B, Π, $\mathcal{Q}$ are θ-periodic functions, then $\tilde{X}$ is also a θ- periodic function.

Moreover, if $A(t) \equiv A \in \mathcal{M}_n^{\mathcal{D}}$, $B(t) \equiv B \in \mathcal{M}_{nm}^{\mathcal{D}}$, $\Pi(t) \equiv \Pi \in \mathbf{B}(\mathcal{S}_n^{\mathcal{D}}, \mathcal{S}_{n+m}^{\mathcal{D}})$, $\mathcal{Q}(t) \equiv \mathcal{Q} \in \mathcal{S}_{n+m}^{\mathcal{D}}$, then the maximal solution of (5.1) is constant and it solves the nonlinear algebraic equation

$$A^T X + XA + \Pi_1[X] - [XB + \Pi_{12}[X] + L][\Pi_2[X] + R]^{-1}[XB + \Pi_{12}[X] + L]^T + M = 0. \tag{5.43}$$

Proof. (ii) $\Rightarrow$ (i) is obvious, since $\tilde{X} \in \mathbf{\Gamma}^\Sigma$.

It remains to prove the implication (i) $\Rightarrow$ (ii). Since (A,B,Π) is stabilizable it follows that there exists a bounded and continuous function $F_0: \mathcal{I} \to \mathcal{M}_{mn}^{\mathcal{D}}$ such that the operator $\mathcal{L}_{A+BF_0,\Pi_{F_0}}$ generates an exponentially stable anticausal evolution.

Let $\varepsilon > 0$ be fixed. Using Theorem 2.3.13 one obtains that the backward affine differential equation

$$\frac{d}{dt}X(t) + \mathcal{L}_{A+BF_0,\Pi_{F_0}}(t)[X(t)] + M_{F_0}(t) + \varepsilon J^{\mathcal{D}} = 0 \tag{5.44}$$

has a unique bounded solution $X_1 : \mathcal{I} \to \mathcal{S}_n^{\mathcal{D}}$. We shall show that $X_1(t) \gg \hat{X}(t)$ for arbitrary $\hat{X} \in \Gamma^{\Sigma}$. If $\hat{X} \in \Gamma^{\Sigma}$, then we obtain immediately via Lemma 5.1.2 that $t \mapsto \hat{X}(t)$ verifies

$$\frac{d}{dt}\hat{X}(t) + \mathcal{R}(t,\hat{X}(t)) \geq 0 \quad \text{for} \quad t \in \mathcal{I};$$

consequently $t \mapsto \hat{X}(t)$ solves the equation

$$\frac{d}{dt}\hat{X}(t) + \mathcal{R}(t,\hat{X}(t)) - \hat{M}(t) = 0, \qquad (5.45)$$

where $\hat{M}(t) = \frac{d}{dt}\hat{X}(t) + \mathcal{R}(t,\hat{X}(t)) \geq 0$. Applying Lemma 5.1.1, (5.45) may be written as

$$\frac{d}{dt}\hat{X}(t) + \mathcal{L}_{A+BF_0,\Pi_{F_0}}(t)[\hat{X}(t)] + M_{F_0}(t)$$
$$- [F_0(t) - \hat{F}(t)]^T \Theta(t,\hat{X}(t)) [F_0(t) - \hat{F}(t)] - \hat{M}(t) = 0, \qquad (5.46)$$

where $\hat{F}(t) := F^{\hat{X}}(t)$. From (5.44) and (5.45) we deduce that $t \mapsto X_1(t) - \hat{X}(t)$ is a bounded solution of the differential equation

$$\frac{d}{dt}Y(t) + \mathcal{L}_{A+BF_0,\Pi_{F_0}}(t)[Y(t)] + H_1(t) = 0$$

with

$$H_1(t) = \varepsilon J^{\mathcal{D}} + [F_0(t) - \hat{F}(t)]^T \Theta(t,\hat{X}(t))[F_0(t) - \hat{F}(t)] + \hat{M}(t).$$

Clearly $H_1(t) \geq \varepsilon J^{\mathcal{D}} > 0$. Hence Theorem 2.3.13 (iv) implies $X_1(t) - \hat{X}(t) \gg 0, t \in \mathcal{I}$. Therefore $R(t) + \Pi_2(t)[X_1(t)] \geq R(t) + \Pi_2(t)[\hat{X}(t)] \gg 0$ for $t \in \mathcal{I}$. Thus we obtain that $F_1(t) := F^{X_1}(t)$ is well defined. We show that F_1 is a stabilizing feedback gain for the triple (A, B, Π). To this end, based on Lemma 5.3.3, we rewrite (5.44) as

$$\frac{d}{dt}X_1(t) + \mathcal{L}_{A+BF_1,\Pi_{F_1}}(t)[X_1(t)] + M_{F_1}(t) + \varepsilon J^{\mathcal{D}}$$
$$+ [F_1(t) - F_0(t)]^T \Theta(t,X_1(t)) [F_1(t) - F_0(t)] = 0. \qquad (5.47)$$

On the other hand, based on Lemma 5.1.1, (5.45) can be rewritten as

$$\frac{d}{dt}\hat{X}(t) + \mathcal{L}_{A+BF_1,\Pi_{F_1}}(t)[\hat{X}(t)] + M_{F_1}(t) - \hat{M}(t)$$
$$- [F_1(t) - \hat{F}(t)]^T \Theta(t,\hat{X}(t)) [F_1(t) - \hat{F}(t)] = 0.$$

Subtracting the last equation from (5.47) we obtain

$$\frac{d}{dt}\big[X_1(t) - \hat{X}(t)\big] + \mathcal{L}_{A+BF_1,\Pi_{F_1}}(t)\big[X_1(t) - \hat{X}(t)\big] + \tilde{H}(t) = 0$$

where

$$\tilde{H}(t) = \varepsilon J^{\mathcal{D}} + \big[F_1(t) - F_0(t)\big]^T \Theta(t, X_1(t))\big[F_1(t) - F_0(t)\big]$$
$$+ \big[F_1(t) - \hat{F}(t)\big]^T \Theta(t, \hat{X}(t))\big[F_1(t) - \hat{F}(t)\big] + \hat{M}(t) \gg 0, \ t \in \mathcal{I}.$$

Applying the implication (vi) $\Rightarrow$ (i) in Theorem 2.3.12 we infer that $\mathcal{L}_{A+BF_1,\Pi_{F_1}}$ generates an anticausal exponentially stable evolution. This means that $F_1 = F^{X_1}$ *is a stabilizing feedback gain*; notice that, as a consequence of Theorem 2.3.13, F_1 is constant (or periodic) if the coefficients of (5.1) are constant (or periodic, respectively).

Taking X_1, F_1 as a first step we construct two sequences $\{X_k\}_{k\geq 1}$ and $\{F_k\}_{k\geq 1}$, where $X_k(t) = \{X_k(t,i)\}_{i\in\mathcal{D}}, t \in \mathcal{I}$ is the unique bounded solution of the differential equation

$$\frac{d}{dt}X_k(t) + \mathcal{L}_{A+BF_{k-1},\Pi_{F_{k-1}}}(t)[X_k(t)] + M_{F_{k-1}}(t) + \frac{\varepsilon}{k}J^{\mathcal{D}} = 0 \qquad (5.48)$$

and $F_k(t) := F^{X_k}(t)$. We show inductively that the following items hold

(a_k) $X_k(t,i) - \hat{X}(t,i) > \mu_k I_n$ for all $(t,i) \in \mathcal{I} \times \mathcal{D}$, and for arbitrary $\hat{X}(t) = \{\hat{X}(t,i)\}_{i\in\mathcal{D}} \in \mathbf{\Gamma}^{\Sigma}, \mu_k > 0$ independent of $\hat{X}$.
(b_k) F_k is a stabilizing feedback gain for the triple (A,B,Π).
(c_k) $X_k(t,i) \geq X_{k+1}(t,i)$ for $(t,i) \in \mathcal{I} \times \mathcal{D}$.

For $k = 1$, items (a_1), (b_1) were proved before.

To prove (c_1) we subtract (5.48), written for $k = 2$, from (5.47) and get

$$\frac{d}{dt}\big[X_1(t) - X_2(t)\big] + \mathcal{L}_{A+BF_1,\Pi_{F_1}}(t)\big[X_1(t) - X_2(t)\big] + \Delta_1(t) = 0,$$

where

$$\Delta_1(t) := \frac{\varepsilon}{2}J^{\mathcal{D}} + \big[F_1(t) - F_0(t)\big]^T \Theta(t, X_1(t))\big[F_1(t) - F_0(t)\big] \gg 0, t \in \mathcal{I}.$$

Invoking Theorem 2.3.13 (iv), one obtains that $X_1(t) - X_2(t) \gg 0$ and thus (c_1) is fulfilled. Let us assume that (a_i), (b_i), (c_i) are fulfilled for $i \leq k - 1$ and let us prove them for $i = k$.

Based on (b_{k-1}) and Theorem 2.3.13 (i) we deduce that (5.48) has a unique bounded solution $X_k: \mathcal{I} \to \mathcal{S}_n^{\mathcal{D}}$. Applying Lemma 5.1.1 with $W(t) := F_{k-1}(t)$ one obtains that (5.45) may be rewritten as

$$\frac{d}{dt}\hat{X}(t) + \mathfrak{L}_{A+BF_{k-1},\Pi_{F_{k-1}}}(t)[\hat{X}(t)] + M_{F_{k-1}}(t)$$

$$- [F_{k-1}(t) - \hat{F}(t)]^T \Theta(t, \hat{X}(t)) [F_{k-1}(t) - \hat{F}(t)] - \hat{M}(t) = 0.$$

Subtracting this equation from (5.48) one obtains that $t \mapsto X_k(t) - \hat{X}(t)$ is a bounded solution of the equation

$$\frac{d}{dt}X(t) + \mathfrak{L}_{A+BF_{k-1},\Pi_{F_{k-1}}}(t)[X(t)] + H_k(t) = 0,$$

where

$$H_k(t) = \frac{\varepsilon}{k}J^{\mathcal{D}} + [F_{k-1}(t) - \hat{F}(t)]^T \Theta(t, \hat{X}(t)) [F_{k-1}(t) - \hat{F}(t)] + \hat{M}(t) \geq \frac{\varepsilon}{k}J^{\mathcal{D}} \gg 0.$$

Since $\mathfrak{L}_{A+BF_{k-1},\Pi_{F_{k-1}}}$ generates an anticausal exponentially stable evolution, we obtain from Theorem 2.3.13 (iv), that there exist

$$\mu_k > 0 \quad \text{such that} \quad X_k(t) - \hat{X}(t) \geq \mu_k J^{\mathcal{D}} \quad \text{for} \quad t \in \mathcal{I}, \qquad (5.49)$$

thus (a_k) is fulfilled.

Let us show that (b_k) is fulfilled. Firstly from (5.49) we have

$$R(t) + \Pi_2(t)[X_k(t)] \gg 0,$$

therefore F_k is well defined. Applying Lemma 5.3.3 to (5.48), one obtains that X_k solves the backward affine differential equation

$$\frac{d}{dt}X_k(t) + \mathfrak{L}_{A+BF_k,\Pi_{F_k}}(t)[X_k(t)] + M_{F_k}(t) + \frac{\varepsilon}{k}J^{\mathcal{D}}$$

$$+ [F_k(t) - F_{k-1}(t)]^T \Theta(t, X_k(t)) [F_k(t) - F_{k-1}(t)] = 0. \qquad (5.50)$$

On the other hand, Lemma 5.1.1 applied to (5.45) gives

$$\frac{d}{dt}\hat{X}(t) + \mathfrak{L}_{A+BF_k,\Pi_{F_k}}(t)[\hat{X}(t)] + M_{F_k}(t)$$

$$- [F_k(t) - \hat{F}(t)]^T \Theta(t, \hat{X}(t)) [F_k(t) - \hat{F}(t)] - \hat{M}(t) = 0.$$

From the last two equations one obtains

$$\frac{d}{dt}[X_k(t) - \hat{X}(t)] + \mathfrak{L}_{A+BF_k,\Pi_{F_k}}(t)[X_k(t) - \hat{X}(t)] + \tilde{H}_k(t) = 0$$

with

$$\tilde{H}_k(t) = \frac{\varepsilon}{k} J^\mathcal{D} + \left[F_k(t) - F_{k-1}(t)\right]^T \Theta(t, X_k(t)) \left[F_k(t) - F_{k-1}(t)\right]$$

$$+ \left[F_k(t) - \hat{F}(t)\right]^T \Theta(t, \hat{X}(t)) \left[F_k(t) - \hat{F}(t)\right] + \hat{M}(t) \geq \frac{\varepsilon}{k} J^\mathcal{D} \gg 0.$$

Implication (vi) $\Rightarrow$ (i) of Theorem 2.3.12 allows us to conclude that $\mathfrak{L}_{A+BF_k, \Pi_{F_k}}$ generates an anticausal exponentially stable evolution which shows that (b_k) is fulfilled.

It remains to prove that (c_k) holds. To this end we subtract (5.48), written for $k+1$ instead of k, from (5.50) and get

$$\frac{d}{dt}\left[X_k(t) - X_{k+1}(t)\right] + \mathfrak{L}_{A+BF_k, \Pi_{F_k}}(t)\left[X_k(t) - X_{k+1}(t)\right] + \frac{\varepsilon}{k(k+1)} J^\mathcal{D} \quad (5.51)$$

$$+ \left[F_k(t) - F_{k-1}(t)\right]^T \Theta(t, X_k(t))\left[F_k(t) - F_{k-1}(t)\right] = 0.$$

Since $\mathfrak{L}_{A+BF_k, \Pi_{F_k}}$ generates an anticausal exponentially stable evolution one obtains, via Theorem 2.3.13 (iv), that (5.51) has a unique bounded solution which additionally is uniformly positive. Therefore $X_k(t) - X_{k+1}(t) \gg 0$ and thus (c_k) holds.

Now, from (a_k) and (c_k) we obtain that for each $(t, i) \in \mathcal{I} \times \mathcal{D}$ we have

$$X_k(t, i) \geq X_{k+1}(t, i) \geq \hat{X}(t, i) \quad (5.52)$$

for all $\hat{X}(t) = \{\hat{X}(t, i)\}_{i \in \mathcal{D}} \in \mathbf{\Gamma}^\Sigma$.

Let $\tilde{X}(t, i)$ be defined by $\tilde{X}(t, i) = \lim_{k \to \infty} X_k(t, i)$ for all $t, i \in \mathcal{I} \times \mathcal{D}$. We show that $t \to \tilde{X}(t) = \{\tilde{X}(t, i)\}_{i \in \mathcal{D}} : \mathcal{I} \to \mathcal{S}_n^\mathcal{D}$ is just the maximal and bounded solution of GRDE (5.1). First we show that for all $t \in \mathcal{I}$, $(t, \tilde{X}(t)) \in Dom\mathfrak{R}$.

Let $\hat{X}(\cdot) \in \mathbf{\Gamma}^\Sigma$ be arbitrary but fixed, this means that there exists $\nu > 0$ such that $R(t, i) + \Pi_2(t)[\hat{X}(t)](i) \geq \nu I_m$, $\forall (t, i) \in \mathcal{I} \times \mathcal{D}$. Invoking again (a_k) we obtain that

$$R(t, i) + \Pi_2(t)[X_k(t)](i) \geq R(t, i) + \Pi_2(t)[\hat{X}(t)](i) \geq \nu I_m \quad (5.53)$$

$\forall (t, i) \in \mathcal{I} \times \mathcal{D}$. Taking the limit for $k \to \infty$ we obtain via (5.31) for $l = 2$ and $X_k(i)$ replaced by $X_k(t, i)$: $R(t, i) + \Pi_2(t)[\tilde{X}(t)](i) \geq \nu I_m$, $\forall (t, i) \in \mathcal{I} \times \mathcal{D}$. Thus we have shown that both $(t, X_k(t))$ and $(t, \tilde{X}(t))$ lie in $Dom\mathfrak{R}$.

Let us show that $t \to \tilde{X}(t)$ is differentiable and satisfies the (GRDE) (5.1). To this end we rewrite (5.48) in the integral form

$$X_k(\tau) - X_k(t) = \int_t^\tau \left(\mathfrak{L}_{A+BF_{k-1}, \Pi_{F_{k-1}}}(s)[X_k(s)] + M_{F_{k-1}}(s) + \frac{\varepsilon}{k} J^\mathcal{D}\right) ds$$

for all $t \leq \tau$, $t, \tau \in \mathcal{I}$, $k \geq 1$. Further, using (5.19) for X replaced by $X_k(t)$ and $W(t)$ replaced by $F_{k-1}(t)$ we obtain $X_k(\tau) - X_k(t) = \int_t^\tau [\Re(s, X_k(s)) + (F_k(s) - F_{k-1}(s))^T (R(s) + \Pi_2(s)[X_k(s)])(F_k(s) - F_{k-1}(s)) + \frac{\varepsilon}{k} J^{\mathcal{D}}] ds$.

This may be written as

$$X_k(\tau, i) - X_k(t, i) = \int_t^\tau [\Re(s, X_k(s))(i) + (F_k(s, i) - F_{k-1}(s, i))^T \tag{5.54}$$

$$(R(s, i) + \Pi_2(s)[X_k(s)](i))(F_k(s, i) - F_{k-1}(s, i)) + \frac{\varepsilon}{k} I_n] ds$$

for all $t \leq \tau$, $t, \tau \in \mathcal{I}$, $i \in \mathcal{D}$.

Proceeding as in the proof of Lemma 5.3.4 one may prove that

$$\lim_{k \to \infty} (F_k(s, i) - F_{k-1}(s, i)) = 0. \tag{5.55}$$

Let $X_k^{j,l}(t, i)$ be the scalar component, jl-th of the matrix $X_k(t, i)$. We have $X_k^{jl}(\tau, i) - X_k^{jl}(t, i) = \int_t^\tau \varphi_k^{jl}(s, i) ds$ where

$$\varphi_k^{jl}(s, i) = \{\Re(s, X_k(s))(i) + (F_k(s, i) - F_{k-1}(s, i))^T \tag{5.56}$$

$$(R(s, i) + \Pi_2(s)[X_k(s)](i))(F_k(s, i) - F_{k-1}(s, i)) + \frac{\varepsilon}{k} I_n\}_{jl}$$

is the jl-th component of the integrant from (5.54). Employing (5.35) and (5.52) we deduce that there exists a positive constant $\tilde{\Gamma}$ such that $|\varphi_k^{jl}(s, i)| \leq \tilde{\Gamma}$ for all $t \leq s \leq \tau \in \mathcal{I}$. On the other hand, (5.30) written for $X_k(i), X(i)$ replaced by $X_k(s, i)$, $\tilde{X}(s, i)$, respectively, together with (5.55) yield $\lim_{k \to \infty} \varphi_k^{jl}(s, j) = \Re_{jl}(s, \tilde{X}(s))(i)$. Thus, applying Lebesques's Theorem for each component jl, we finally obtain

$$\lim_{k \to \infty} \int_t^\tau \{\Re(s, X_k(s))(i) + (F_k(s, i) - F_{k-1}(s, i))^T (R(s, i) +$$

$$\Pi_2(s, i)[X_k(s)](i))(F_k(s, i) - F_{k-1}(s, i)) + \frac{\varepsilon}{k} I_n\} ds = \int_t^\tau \Re(s, \tilde{X}(s))(i) ds$$

for all $i \in \mathcal{D}$. Taking the limit for $k \to \infty$ in the both sides of (5.54) one gets $\tilde{X}(\tau, i) - \tilde{X}(t, i) = \int_t^\tau \Re(s, \tilde{X}(s))(i) ds$, or, in a compact form

$$\tilde{X}(\tau) - \tilde{X}(t) = \int_t^\tau \Re(s, \tilde{X}(s)) ds \tag{5.57}$$

for all $t < \tau$, $t, \tau \in \mathcal{I}$. On the other hand, from (5.35) and (5.53) we deduce that there exists a constant $\Gamma > 0$ such that $\|\mathfrak{R}(s, \tilde{X}(s))\|_\infty < \Gamma$, $\forall$ $s \in \mathcal{I}$. Therefore $\|\tilde{X}(\tau) - \tilde{X}(t)\|_\infty \leq \gamma(\tau - t)$, $\forall$ $t \leq \tau$, $t, \tau \in \mathcal{I}$. This allows us to obtain that $s \to \mathfrak{R}(s, \tilde{X}(s)) : \mathcal{I} \to \mathcal{S}_n^{\mathcal{D}}$ is a continuous function. So, we may conclude that the right-hand side of (5.57) is differentiable with respect to $t \in \mathcal{I}$. Hence, $t \to \tilde{X}(t)$ is also differentiable. Differentiating (5.57) we obtain that

$$\frac{d}{dt}\tilde{X}(t) + \mathfrak{R}(t, \tilde{X}(t)) = 0, \; t \in \mathcal{I},$$

which means $\tilde{X}(t)$ is a global solution of GRDE (5.1). Taking the limit for $k \to \infty$ in (5.53) we obtain that $\tilde{X}(t) \geq \hat{X}(t)$, $t \in \mathcal{I}$, $\forall \hat{X}(\cdot) \in \Gamma^\Sigma$. Thus we have obtained $\tilde{X}(\cdot)$ is just the maximal solution of (5.1) and thus (i) $\Rightarrow$ (ii) is proved.

To complete the proof, let us remark that if A, B, Π, $\mathcal{Q}$ are periodic functions with the same period θ then via Corollary 5.4.9, (i), it follows that there exists a stabilizing feedback gain which is a θ-periodic function. Applying Theorem 2.3.13 (ii), one obtains that X_k, F_k are θ-periodic functions for all k and thus $\tilde{X}$ will be a θ-periodic function, too. Also if $A(t) \equiv A$, $B(t) \equiv B$, $\Pi(t) \equiv \Pi$, $\mathcal{Q}(t) \equiv \mathcal{Q}$ and (A, B, Π) is stabilizable, then from Corollary 5.4.9 (ii), we obtain that there exists a stabilizing feedback gain which is constant. Applying again Theorem 2.3.13 (iii), one obtains that X_k and F_k are constant functions for all $k \geq 1$ and therefore $\tilde{X}$ is constant. Thus the proof is complete.   □

In order to facility the statement of the next result we introduce several notations. Let $\Sigma^j = (A, B, \Pi, \mathcal{Q}^j)$, $j = 0, 1, \ldots$ where $A(\cdot), B(\cdot), \Pi(\cdot)$ are as in the case of the GRDE (5.1) and $\mathcal{Q}^j : \mathcal{I} \to \mathcal{S}_{n+m}^{\mathcal{D}}$ are bounded and continuous functions. For each $t \in \mathcal{I}$, $\mathcal{Q}^j(t) = \{\mathcal{Q}^j(t, i)\}_{i \in \mathcal{D}}$, $\mathcal{Q}^j(t, i) = \begin{pmatrix} M^j(t, i) & L^j(t, i) \\ (L^j(t, i))^T & R^j(t, i) \end{pmatrix}$ where $M^j(t, i) \in \mathcal{S}_n$ and $R^j(t, i) \in \mathcal{S}_n$, $\forall i \in \mathcal{D}$. Set $M^j(t) = \{M^j(t, i)\}_{i \in \mathcal{D}} \in \mathcal{S}_n^{\mathcal{D}}$, $R^j(t) = \{R^j(t, i)\}_{i \in \mathcal{D}} \in \mathcal{S}_n^{\mathcal{D}}$ and $L^j(t) = \{L^j(t, i)\}_{i \in \mathcal{D}} \in \mathcal{M}_{nm}^{\mathcal{D}}$. For each quadruple Σ^j we define the operator $\mathfrak{R}^j : Dom\mathfrak{R}^j \to \mathcal{S}_n^{\mathcal{D}}$ where $Dom\mathfrak{R}^j$ is defined as in (5.8) and $\mathfrak{R}^j(t, X)$ is defined as in (5.35) replacing $M(t), L(t), R(t)$ by $M^j(t), L^j(t), R^j(t)$. To the quadruple Σ^j we associate the GRDE

$$\frac{d}{dt}X(t) + \mathfrak{R}^j(t, X(t)) = 0. \tag{5.58}$$

Theorem 5.3.6. *Let* $\Sigma^j = (A, B, \Pi, \mathcal{Q}^j)$, $j = 0, 1, \ldots$ *be as before and* $\Sigma = (A, B, \Pi, \mathcal{Q})$ *be the quadruple defining the GRDE (5.1). We assume that*

(a) (A, B, Π) *is stabilizable and* Γ^Σ *is not empty.*
(b) For each $t \in \mathcal{I}$ *the operator* $\Pi(t)$ *satisfies the assumptions* (Π_1) *and* (Π_2).
(c) $\mathcal{Q}^j(t, i) \geq \mathcal{Q}^{j+1}(t, i) \geq \mathcal{Q}(t, i)$, $\forall$ $j \geq 0$, $t, i \in \mathcal{I} \times \mathcal{D}$.

Under these conditions, the following hold

(i) $\Gamma^\Sigma \subset \Gamma^{\Sigma^j} \subset \Gamma^{\Sigma^{j-1}}$, $\forall$ $j \geq 1$;

(ii) *If $\tilde{X}^j(t) = \{\tilde{X}^j(t,i)\}_{i \in \mathcal{D}}$, $t \in \mathcal{I}$, is the bounded maximal solution of the GRDE (5.58) satisfying the condition*

$$R^j(t) + \Pi_2(t)[\tilde{X}^j(t)] \gg 0, \quad t \in \mathcal{I} \tag{5.59}$$

then:

$$\tilde{X}^j(t,i) \geq \tilde{X}^{j+1}(t,i) \geq \tilde{X}(t,i), \quad \forall \, j \geq 0, (t,i) \in \mathcal{I} \times \mathcal{D}, \tag{5.60}$$

where $\tilde{X}(t) = \{\tilde{X}(t,i)\}_{i \in \mathcal{D}}$, $t \in \mathcal{I}$, is the maximal and bounded solution of the GRDE (5.1) which satisfies the condition $R(t) + \Pi_2(t)[\tilde{X}(t)] \gg 0$, $t \in \mathcal{I}$. Furthermore if

$$\lim_{j \to \infty} \mathcal{Q}^j(t,i) = \mathcal{Q}(t,i) \tag{5.61}$$

for all $(t,i) \in \mathcal{I} \times \mathcal{D}$ then $\lim_{j \to \infty} \tilde{X}^j(t,i) = \tilde{X}(t,i)$, for all $(t,i) \in \mathcal{I} \times \mathcal{D}$.

Proof. (i) If $\Lambda^{\Sigma^j}(t)$ and $\Lambda^{\Sigma}(t)$, respectively, are the generalized dissipation operators associated with Σ^j and Σ, respectively, we obtain:

$$\Lambda^{\Sigma^j}(t)[\hat{X}(t)] = \Lambda^{\Sigma}(t)[\hat{X}(t)] + \mathcal{Q}^j(t) - \mathcal{Q}(t). \tag{5.62}$$

Let $\hat{X}(\cdot) \in \Gamma^{\Sigma}$, this means that $\hat{X}(\cdot) \in C_b^1(\mathcal{I}, \mathcal{S}_n^{\mathcal{D}})$ which satisfies $\Lambda^{\Sigma}(t)[\hat{X}(t)] \geq 0$, $t \in \mathcal{I}$, and $R(t,i) + \Pi_2(t)[\hat{X}(t)](i) \geq v I_m$, $\forall (t,i) \in \mathcal{I} \times \mathcal{D}$, $v > 0$ being a constant not depending upon t, i.

From (5.62), we obtain that $\Lambda^{\Sigma^j}(t)[\hat{X}(t)] \geq 0$, $\forall t \in \mathcal{I}$. On the other hand, $R^j(t,i) + \Pi_2(t)[\hat{X}(t)](i) \geq R(t,i) + \Pi_2(t)[\hat{X}(t)](i) \geq v I_m$, $\forall (t,i) \in \mathcal{I} \times \mathcal{D}$, thus we may conclude that $\Gamma^{\Sigma} \subset \Gamma^{\Sigma^j}$. The fact that $\Gamma^{\Sigma^j} \subset \Gamma^{\Sigma^{j-1}}$ may be proved in a similar way.

Since (A, B, Π) is stabilizable and Γ^{Σ^j} is not empty we may apply Theorem 5.3.5 to (5.58) to obtain the existence of the bounded and maximal solution $\tilde{X}^j(t)$ satisfying the condition (5.59). From $\tilde{X}^{j+1}(\cdot) \in \Gamma^{\Sigma^{j+1}} \subset \Gamma^{\Sigma^j}$ we obtain that

$$\tilde{X}^{j+1}(t,i) \leq \tilde{X}^j(t,i) \text{ for all } j \geq 0, t, i \in \mathcal{I} \times \mathcal{D}. \tag{5.63}$$

Thus the first part of (ii) is proved. Let $Y(t,i) = \lim_{j \to \infty} \tilde{X}^j(t,i)$, $(t,i) \in \mathcal{I} \times \mathcal{D}$. We show that if (5.61) is satisfied, then $Y(t) = \{Y(t,i)\}_{i \in \mathcal{D}}$ coincides with the maximal and bounded solution $\tilde{X}(t)$ of the GRDE (5.1).

To this end, we write (5.58) in the form

$$\tilde{X}^j(\tau,i) - \tilde{X}^j(t,i) = \int_t^\tau \mathfrak{R}^j(s, \tilde{X}^j(s))(i) ds, \quad \forall t \leq \tau, t, \tau \in \mathcal{I}, i \in \mathcal{D}. \tag{5.64}$$

Let us remark that $\mathfrak{R}^j(s,\tilde{X}(s))(i) = \mathfrak{R}(s,\tilde{X}^j(s))(i) + (\tilde{\mathbf{G}}^j(s,\tilde{X}^j(s)))(i)$ where $(\tilde{\mathbf{G}}^j(s,\tilde{X}^j(s)))(i) = M^j(s,i) - M(s,i) + (\mathbf{G}^j(s,\tilde{X}^j(s)))(i) - (\mathbf{G}(s,\tilde{X}^j(s)))(i)$ with $\mathbf{G}(s,\tilde{X}^j(s))$ defined as in (5.34), with Y replaced by $\tilde{X}^j(s)$ and $\mathbf{G}^j(s,\tilde{X}^j(s))$ is defined as in (5.34) with Y replaced by $\tilde{X}(s)$ and $L(s), R(s)$ replaced by $L^j(s), R^j(s)$.

Proceeding as in the proof of Lemma 5.3.4 one may show that $\lim_{j\to\infty}(\tilde{\mathbf{G}}^j(s,\tilde{X}^j(s)))(i) = 0$, if (5.61) holds. This allows us to deduce that $\lim_{j\to\infty}(\mathfrak{R}^j(s,\tilde{X}^j(s)))(i) = (\mathfrak{R}(s,Y(s)))(i), i \in \mathcal{D}, s \in \mathcal{I}$. Also, one obtains that there exists $\gamma > 0$, not depending upon s and j, such that $|(R^j(s,\tilde{X}^j(s)))(i)| \le \gamma, \quad \forall s, j$. By a standard reasoning based on Lebesque's Theorem of the convergence of integrals we obtain that $\lim_{j\to\infty}\int_t^\tau(\mathfrak{R}^j(s,\tilde{X}^j(s)))(i)ds = \int_t^\tau\mathfrak{R}(s,Y(s))ds$. Taking the limit for $j \to \infty$ in (5.64) we finally obtain $Y(\tau) - Y(t) = \int_t^\tau \mathfrak{R}(s,Y(s))ds, \forall t \le \tau, t, \tau \in \mathcal{I}$. This allows us to conclude that $t \to Y(t)$ is differentiable and it is a bounded solution of the GRDE (5.1).

From $\tilde{X}^j(t,i) \ge \tilde{X}(t,i), \quad (\forall) \quad j \ge 0$ we deduce that $Y(t,i) \ge \tilde{X}(t,i), \forall (t,i) \in \mathcal{I} \times \mathcal{D}$. Based on the uniqueness of the maximal solution we deduce that $Y(t)$ coincides with $\tilde{X}(t)$. This completes the proof. $\qquad\qquad\square$

5.4   The Stabilizing Solution of the GRDE

In this section we deal with the stabilizing solutions of the GRDE (5.1) in the case when $X \to \Pi(t)[X]$ satisfies the assumptions $(\mathbf{\Pi}_1)$ and $(\mathbf{\Pi}_2)$. We shall prove the uniqueness of a bounded and stabilizing solution and we shall provide a necessary and sufficient condition for the existence of a bounded and stabilizing solution of the GRDE 5.1.

Definition 5.4.1. Let $X_s \colon \mathcal{I} \to \mathcal{S}_n^{\mathcal{D}}$ be a solution of the GRDE (5.1) and denote by $F_s(t) := F^{X_s}(t)$ the corresponding feedback matrix. Then X_s is called a *stabilizing solution* if the operator $\mathcal{L}_{A+BF_s,\Pi_{F_s}}$ generates an anticausal exponentially stable evolution where Π_{F_s} is defined as in (5.6) for $W(t) = F_s(t)$. This means that there exist $\beta \ge 1, \alpha > 0$ such that

$$\|T_{F_s}^a(t,t_0)\| \le \beta e^{\alpha(t-t_0)}, \quad \forall \ t \le t_0, t, t_0 \in \mathcal{I} \qquad (5.65)$$

where $T_{F_s}^a(t,t_0)$ is the anticausal linear evolution operator on $\mathcal{S}_n^{\mathcal{D}}$ defined by the backward linear differential equation

$$\frac{d}{dt}Y(t) + \mathcal{L}_{A+BF_s,\Pi_{F_s}}(t)[Y(t)] = 0.$$

We recall that for each $t \in \mathcal{I}$ $F_s(t) = \{F_s(t,i)\}_{i\in\mathcal{D}} \in \mathcal{M}_{mn}^{\mathcal{D}}$ where

$$F_s(t,i) = -[\Pi_2(t)[X_s(t)](i) + R(t,i)]^{-1}[X_s(t,i)B(t,i) + \Pi_{12}(t)[X_s(t](i) + L(t,i)]^T \quad (5.66)$$

Remark 5.4.1. (a) According to (5.20) we obtain that the global solution $X_s(\cdot)$ is a stabilizing solution of GRDE (5.1) if and only if the backward differential equation

$$\frac{d}{dt}Y(t) + \mathfrak{R}'(t,X_s(t))[Y(t)] = 0$$

defines an anticausal exponentially stable evolution.

(b) If $\mathcal{D} = \{1,2,\ldots,d\}$, then the equality (5.26) written for $F(t) = F_s(t)$ yields via Proposition 2.4.1 that $X_s(\cdot)$ is a stabilizing solution of GRDE (5.1) if and only if the operator valued function $\mathcal{L}_{A+BF_s,\Pi_{F_s}^*}(\cdot)$ defines an exponentially stable evolution on $\mathcal{S}_n^d$. This means that there exist $\beta_1 \geq 1, \alpha > 0$ such that

$$\|T_{F_s}(t,t_0)\| \leq \beta_1 e^{-\alpha(t-t_0)}, \quad \forall\ t \geq t_0, t, t_0 \in \mathcal{I} \tag{5.67}$$

$T_{F_s}(t,t_0)$ being the linear evolution operator on $\mathcal{S}_n^d$ defined by the linear differential equation:

$$\frac{d}{dt}X(t) = \mathcal{L}_{A+BF_s,\Pi_{F_s}^*}(t)[X(t)].$$

In fact the equivalence between (5.65) and (5.67) is obtained using the identity $T_{F_s}^a(\tau,t) = T_{F_s}^*(t,\tau)$.

Theorem 5.4.1. *Let* $\Sigma = (A,B,\Pi,\mathcal{Q})$ *be such that* $\Gamma^\Sigma \neq \emptyset$. *If* $X_s: \mathcal{I} \to \mathcal{S}_n^{\mathcal{D}}$ *is a bounded and stabilizing solution of the GRDE (5.1) then* X_s *coincides with the maximal solution with respect to* Γ^Σ *of (5.1).*

Proof. Applying Lemma 5.1.1 we deduce that X_s verifies the differential equation

$$\frac{d}{dt}X_s(t) + \mathcal{L}_{A+BF_s,\Pi_{F_s}}(t)[X_s(t)] + M_{F_s}(t) = 0 \tag{5.68}$$

where F_s is as in (5.66). Let $\hat{X}$ be arbitrary in Γ^Σ. As in the proof of Theorem 5.3.5 one obtains that there exists $\hat{M}(t) \geq 0$ such that $\hat{X}$ verifies a differential equation of the form (5.45). Applying Lemma 5.1.1 to (5.45) we get

$$\frac{d}{dt}\hat{X}(t) + \mathcal{L}_{A+BF_s,\Pi_{F_s}}(t)[\hat{X}(t)] + M_{F_s}(t) - \hat{M}(t)$$

$$- [F_s(t) - \hat{F}(t)]^T \Theta(t,\hat{X}(t))[F_s(t) - \hat{F}(t)] = 0.$$

Subtracting the last two equations we obtain that $t \mapsto X_s(t) - \hat{X}(t)$ is a bounded solution of the backward differential equation

$$\frac{d}{dt}X(t) + \mathfrak{L}_{A+BF_s,\Pi_{F_s}}(t)[X(t)] + H_s(t) = 0$$

where

$$H_s(t) = [F_s(t) - \hat{F}(t)]^T \Theta(t, \hat{X}(t))[F_s(t) - \hat{F}(t)] + \hat{M}(t) \geq 0.$$

Since $\mathfrak{L}_{A+BF_s,\Pi_{F_s}}$ generates an anticausal exponentially stable evolution one obtains, using Theorem 2.3.13 (iv), that $X_s(t) - \hat{X}(t) \geq 0, t \in \mathcal{I}$ and thus the proof is complete. $\square$

Remark 5.4.2. From Theorem 5.4.1 it follows that if Γ^Σ is not empty then a bounded and stabilizing solution of the GRDE (5.1) (if it exists) will verify the condition

$$R(t) + \Pi_2(t)[X_s(t)] \gg 0, t \in \mathcal{I}.$$

The next result follows directly from Theorem 5.4.1 and Remark 5.3.1.

Corollary 5.4.2. *If Γ^Σ is not empty, then the differential equation (5.1) has at most one bounded and stabilizing solution.*

In Sect. 5.6 we will show that in the particular case when $\Pi(t)$ is of the form (5.5), the uniqueness of the bounded and stabilizing solution of (5.1) follows in the absence of any assumption concerning Γ^Σ. In that case $R(t) + \Pi_2(t)[X_s(t)]$ has not a definite sign.

Theorem 5.4.3. *Assume that A, B, Π, Q are periodic functions with period θ and that Γ^Σ is not empty. Then the bounded and stabilizing solution of (5.1) (if it exists) is θ-periodic.*

Proof. Let $X_s : \mathcal{I} \to \mathcal{S}_n^{\mathcal{D}}$ be a bounded and stabilizing solution of (5.1). We define $\tilde{X}(t) := X_s(t + \theta), t \in \mathcal{I}$. By direct computation we obtain that $\tilde{X}$ is a solution of (5.1) too. We shall prove that $\tilde{X}$ is also a stabilizing solution of (5.1).

Set $\tilde{F}(t) := F^{\tilde{X}}(t)$. We show that the operator valued function $\mathfrak{L}_{A+B\tilde{F},\Pi_{\tilde{F}}}(\cdot)$ generates an anticausal exponentially stable evolution. Let $\tilde{T}^a(t, t_0)$ be the anticausal linear evolution operator defined by the backward linear differential equation

$$\frac{d}{dt}S(t) + \mathfrak{L}_{A+B\tilde{F},\Pi_{\tilde{F}}}(t)[S(t)] = 0. \tag{5.69}$$

Because of the periodicity we obtain that

$$\mathfrak{L}_{A+B\tilde{F},\Pi_{\tilde{F}}}(t) = \mathfrak{L}_{A+BF_s,\Pi_{F_s}}(t + \theta) \quad \text{for} \quad t \in \mathcal{I}.$$

If $\tilde{S}(t, t_0, H)$ is the solution of (5.69) with $\tilde{S}(t_0, t_0, H) = H$, then we have

$$\frac{d}{dt}\tilde{S}(t,t_0,H) + \mathfrak{L}_{A+BF_s,\Pi^*_{F_s}}(t+\theta)\tilde{S}(t,t_0,H) = 0.$$

From the uniqueness of the solution of this initial value problem we infer that

$$\tilde{S}(t,t_0,H) = S(t+\theta,t_0+\theta,H),$$

where $t \mapsto S(t,\tau,H)$ is the solution of the problem

$$\frac{d}{dt}S(t) + \mathfrak{L}_{A+BF_s,\Pi_{F_s}}(t)[S(t)] = 0, \quad S(\tau,\tau,H) = H. \tag{5.70}$$

Thus we get $\tilde{T}^a(t,t_0) = T^a_{F_s}(t+\theta,t_0+\theta)$ where $T^a_{F_s}(t,t_0)$ is the anticausal linear evolution operator defined by (5.70). The last equality leads to

$$\|\tilde{T}^a(t,t_0)\| = \|T^a_{F_s}(t+\theta,t_0+\theta)\| \leq \beta e^{\alpha(t-t_0)} \quad \text{for} \quad t \leq t_0$$

with $\alpha,\beta > 0$, which shows that $\tilde{X}$ is also a bounded and stabilizing solution of GRDE (5.1).

Applying Corollary 5.4.2 one obtains that $\tilde{X}(t) = X_s(t)$ for $t \in \mathcal{I}$, that means $X_s(t+\theta) = X_s(t)$ for all t, which shows that X_s is a θ-periodic function and thus the proof ends. □

Corollary 5.4.4. *If* $\Gamma^\Sigma \neq \emptyset$ *and* $A(t) \equiv A$, $B(t) \equiv B$, $\Pi(t) \equiv \Pi$, $\mathcal{Q}(t) \equiv \mathcal{Q}$, $t \in \mathbf{R}$, *then the stabilizing solution of (5.1) (if it exists) is constant and solves the algebraic equation (5.43).*

Proof. Since the matrix coefficients of (5.1) are constant functions they may be viewed as periodic functions with arbitrary period. Applying Theorem 5.4.3 it follows that the bounded and stabilizing solution of (5.1) is a periodic function with arbitrary period. Therefore it is a constant function and thus the proof ends. □

The following lemma which is a generalization of the invariance under feedback transformations of standard Riccati differential equations will be useful in the next developments. Since

$$[W(t,i) - F^X(t,i)]^T \{R(t,i) + \Pi_2(t)[X(t)](i)\}$$

$$= X(t,i)B(t,i) + \Pi_{12}(t)[X(t)](i) + W^T(t,i)\Pi_2(t)[X(t)](i) + L(t,i) + W^T(t,i)R(t,i)$$

the conclusion of this lemma follows immediately from Lemma 5.1.1.

Lemma 5.4.5. *Let* $W: \mathcal{I} \to \mathcal{M}^{\mathcal{D}}_{mn}$ *be a bounded and continuous function. Then* $X: \mathcal{I}_1 \subset \mathcal{I} \to \mathcal{S}^{\mathcal{D}}_n$ *is a solution of the differential equation (5.1) associated with the quadruple* $\Sigma = (A,B,\Pi,\mathcal{Q})$ *if and only if* X *is a solution of the differential equation of type (5.1) associated with the quadruple* $\Sigma^W = (A + BW, B, \Pi^W, \mathcal{Q}^W)$, *where* $\Pi^W(t): \mathcal{S}^{\mathcal{D}}_n \to \mathcal{S}^{\mathcal{D}}_{n+m}$ *is given by*

$$\Pi^W(t)[X](i) = \begin{pmatrix} I_n & 0 \\ W(t,i) & I_m \end{pmatrix}^T \begin{pmatrix} \Pi_1(t)[X](i) & \Pi_{12}(t)[X](i) \\ \{\Pi_{12}(t)[X](i)\}^T & \Pi_2(t)[X](i) \end{pmatrix} \begin{pmatrix} I_n & 0 \\ W(t,i) & I_m \end{pmatrix}$$

$$= \begin{pmatrix} \Pi_W(t)[X](i) & \Pi_{12}(t)[X](i) + W^T(t,i)\Pi_2(t)[X](i) \\ \{\Pi_{12}(t)[X](i) + W^T(t,i)\Pi_2(t)[X](i)\}^T & \Pi_2(t)[X](i) \end{pmatrix}$$

and

$$Q^W(t,i) = \begin{pmatrix} I_n & 0 \\ W(t,i) & I_m \end{pmatrix}^T \begin{pmatrix} M(t,i) & L(t,i) \\ L^T(t,i) & R(t,i) \end{pmatrix} \begin{pmatrix} I_n & 0 \\ W(t,i) & I_m \end{pmatrix}$$

$$= \begin{pmatrix} M_W(t,i) & L(t,i) + W^T(t,i)R(t,i) \\ L^T(t,i) + R(t,i)W(t,i) & R(t,i) \end{pmatrix}.$$

Theorem 5.4.6. *Under the considered assumptions the following assertions are equivalent*

(i) (A,B,Π) is stabilizable and the set $\tilde{\Gamma}^\Sigma$ is not empty;
(ii) The GRDE (5.1) has a stabilizing and bounded solution $X_s : \mathcal{I} \to \mathcal{S}_n^\mathcal{D}$ satisfying

$$R(t,i) + \Pi_2(t)[X_s(t)](i) \geq \nu I_m, \quad \forall \ (t,i) \in \mathcal{I} \times \mathcal{D}, \tag{5.71}$$

$\nu > 0$ *is a constant.*

Proof. First we notice that according to the adopted convention of notations the sign conditions (5.71) may be written in the following compact form

$$R(t) + \Pi_2(t)[X_s(t)] \gg 0, t \in \mathcal{I}.$$

(i) $\Rightarrow$ (ii). If (i) holds, then Theorem 5.3.5 yields that (5.1) has a bounded maximal solution $\tilde{X} : \mathcal{I} \to \mathcal{S}_n^\mathcal{D}$ satisfying the sign condition $R(t) + \Pi_2(t)[\tilde{X}(t)] \gg 0$. We show that $\tilde{X}$ is just the stabilizing solution. If $\tilde{F}$ is the feedback matrix associated with $\tilde{X}$, then (5.45) may be written as

$$\frac{d}{dt}\hat{X}(t) + \mathfrak{L}_{A+B\tilde{F},\Pi_{\tilde{F}}}(t)[\hat{X}(t)] + M_{\tilde{F}}(t) - \hat{M}(t)$$

$$- [\tilde{F}(t) - \hat{F}(t)]^T \Theta(t,\hat{X}(t))[\tilde{F}(t) - \hat{F}(t)] = 0.$$

Since $\hat{X} \in \tilde{\Gamma}^\Sigma$ then it is a solution of an equation of type (5.45) with $\hat{M}(t) \gg 0$. Using again Lemma 5.1.1 one obtains that $t \mapsto \tilde{X}(t) - \hat{X}(t)$ is a bounded and positive semi-definite solution of the backward differential equation

$$\frac{d}{dt}X(t) + \mathfrak{L}_{A+B\tilde{F},\Pi_{\tilde{F}}}(t)[X(t)] + H(t) = 0,$$

where

$$H(t) := \hat{M}(t) + \left[\tilde{F}(t) - \hat{F}(t)\right]^T \Theta(t, \hat{X}(t)) \left[\tilde{F}(t) - \hat{F}(t)\right].$$

Since $\hat{M}(t) \gg 0, t \in \mathcal{I}$ it follows that $H(t) \gg 0, t \in \mathcal{I}$. Applying implication (vi) $\Rightarrow$ (i) of Theorem 2.3.12 one gets that $\mathfrak{L}_{A+B\tilde{F}, \Pi_{\tilde{F}}}$ generates an anticausal exponentially stable evolution which shows that $\hat{X}$ is a stabilizing solution of (5.1).

We prove now (ii) $\Rightarrow$ (i). If (5.1) has a bounded and stabilizing solution $X_s : \mathcal{I} \to \mathcal{S}_n^{\mathcal{D}}$, then $F_s := F^{X_s}$ is a stabilizing feedback gain and therefore (A, B, Π) is stabilizable.

Applying Lemma 5.4.5 with $W(t) = F_s(t)$ we rewrite (5.1) as

$$\frac{d}{dt} X(t) + \mathfrak{L}_{A+BF_s, \Pi_{F_s}}(t)[X(t)] + M_{F_s}(t)$$

$$- \mathcal{P}_{F_s}^T(t, X(t)) \Theta(t, X(t))^{-1} \mathcal{P}_{F_s}(t, X(t)) = 0,$$

where $X \mapsto \mathcal{P}_{F_s}(t, X) : \mathcal{S}_n^{\mathcal{D}} \to \mathcal{M}_{mn}^{\mathcal{D}}$ is given by

$$\mathcal{P}_{F_s}(t, X)(i) = x\{X(i)B(t, i) + \Pi_{12}(t)[X](i) + F_s^T(t, i)\Pi_2(t)[X](i) + L(t, i)$$

$$+ F_s^T(t, i)R(t, i)\}^T$$

$i \in \mathcal{D}$ and $\Theta(t, X)$ is as in Lemma 5.3.3. Let $T_{F_s}^a(t, t_0)$ be the anticausal linear evolution operator defined by

$$\frac{d}{dt} S(t) + \mathfrak{L}_{A+BF_s, \Pi_{F_s}}(t)[S(t)] = 0.$$

Since F_s is a stabilizing feedback gain it follows that there exist $\alpha, \beta > 0$ such that (5.65) holds.

Let $C_b(\mathcal{I}, \mathcal{S}_n^{\mathcal{D}})$ be the Banach space of bounded and continuous functions $X : \mathcal{I} \to \mathcal{S}_n^{\mathcal{D}}$. Since $\Theta(t, X_s(t)) \gg 0$ for $t \in \mathcal{I}$, it follows that there exist an open set $\mathcal{U} \subset C_b(\mathcal{I}, \mathcal{S}_n^{\mathcal{D}})$ such that $X_s \in \mathcal{U}$ and $\Theta(t, X(t)) \gg 0$ for all $X \in \mathcal{U}$. Let $\Psi : \mathcal{U} \times \mathbf{R} \to C_b$ be defined by

$$\Psi(X, \delta)(t) = \int_t^\infty T_{F_s}^a(t, \sigma) \left[M_{F_s}(\sigma) + \delta J^{\mathcal{D}}\right.$$

$$\left. - \mathcal{P}_{F_s}^T(\sigma, X(\sigma)) \Theta^{-1}(\sigma, X(\sigma)) \mathcal{P}_{F_s}(\sigma, X(\sigma))\right] d\sigma - X(t), t \in \mathcal{I}.$$

We apply the implicit functions theorem to the equation

$$\Psi(X, \delta) = 0 \tag{5.72}$$

in order to obtain that there exists a function $X_\delta \in \mathcal{U}$ such that

$$X_\delta(t) = \int_t^\infty T_{F_s}^a(t,\sigma)\left[M_{F_s}(\sigma) + \delta J^{\mathcal{D}}\right.$$
$$\left. - \mathcal{P}_{F_s}^T(\sigma, X_\delta(\sigma))\Theta^{-1}(\sigma, X_\delta(\sigma))\mathcal{P}_{F_s}(\sigma, X_\delta(\sigma))\right] d\sigma$$

for $|\delta|$ small enough.

It is clear that $(X_s, 0)$ is a solution of (5.72). We show now that

$$d_1\Psi(X_s(\cdot), 0) \colon C_b(\mathcal{I}, \mathcal{S}_n^{\mathcal{D}}) \to C_b(\mathcal{I}, \mathcal{S}_n^{\mathcal{D}})$$

is an isomorphism, $d_1\Psi$ being the derivative of Ψ with respect to its first argument. Since

$$d_1\Psi(X_s, 0)Y = \lim_{\varepsilon \to 0} \frac{1}{\varepsilon}\left[\Psi(X_s + \varepsilon Y, 0) - \Psi(X_s, 0)\right]$$

and $\mathcal{P}_{F_s}(\sigma, X_s(\sigma)) \equiv 0$ we obtain that $d_1\Psi(X_s, 0)Y = -Y$ for all $Y \in C_b(\mathcal{I}, \mathcal{S}_n^{\mathcal{D}})$. Therefore $d_1\Psi(X_s, 0) = -I_{C_b}$, where I_{C_b} is the identity operator of $C_b(\mathcal{I}, \mathcal{S}_n^{\mathcal{D}})$ which is an isomorphism. Also we see that $d_1\Psi(X, \delta)$ is continuous in $(X, \delta) = (X_s, 0)$. Applying the implicit function theorem (see [129] vol. 1) we deduce that there exists $\tilde{\delta} > 0$ and a smooth function $X_\delta(\cdot) \colon (-\tilde{\delta}, \tilde{\delta}) \to \mathcal{U}$ which satisfies $\Psi(X_\delta(\cdot), \delta) = 0$ for all $\delta \in (-\tilde{\delta}, \tilde{\delta})$. It is easy to see that if $\delta \in (-\tilde{\delta}, 0)$ then $X_\delta(\cdot) \in \tilde{\mathbf{\Gamma}}^\Sigma$ and the proof is complete.   □

Corollary 5.4.7. *Assume that A, B, Π, $\mathcal{Q}$ are periodic functions with period $\theta > 0$. Under these conditions the following are equivalent*

(i) (A, B, Π) is stabilizable and $\tilde{\mathbf{\Gamma}}^\Sigma$ is not empty.

(ii) The GRDE (5.1) has a bounded, θ-periodic and stabilizing solution $X_s \colon \mathcal{I} \to \mathcal{S}_n^{\mathcal{D}}$ which verifies $R(t) + \Pi_2(t)[X_s(t)] \gg 0, t \in \mathcal{I}$.

(iii) (A, B, Π) is stabilizable and $\tilde{\mathbf{\Gamma}}^\Sigma$ contains at least a θ-periodic function $\check{X}$.

Proof. (i) ⇔ (ii) follows from Theorems 5.4.6 and 5.4.3.

(iii) ⇒ (i) is obvious.

It remains to prove (ii) ⇒ (iii). In the proof of the implication (ii) ⇒ (i) in Theorem 5.4.6 we have shown that there exists $\tilde{\delta} > 0$ and a smooth function $\delta \mapsto X_\delta(\cdot) \colon (-\tilde{\delta}, \tilde{\delta}) \to C_b(\mathcal{I}, \mathcal{S}_n^{\mathcal{D}})$ which satisfies

$$\frac{d}{dt}X_\delta(t) + \mathfrak{R}(t, X_\delta(t)) + \delta J^{\mathcal{D}} = 0.$$

Choose $\delta_1 \in (-\tilde{\delta}, 0)$ and set $\Sigma_1 := (A, B, \Pi, \mathcal{Q}_1)$ with

$$\mathcal{Q}_1(t, i) := \begin{pmatrix} M(t,i) + \delta_1 I_n & L(t,i) \\ L^T(t,i) & R(t,i) \end{pmatrix}, i \in \mathcal{D}.$$

It is easy to see that if $\delta \in (-\tilde{\delta}, \delta_1)$ then $X_\delta(\cdot) \in \tilde{\mathbf{\Gamma}}^{\Sigma_1}$. Applying implication (i) $\Rightarrow$ (ii) of Theorem 5.4.6 one obtains that the equation

$$\frac{d}{dt}X(t) + \mathfrak{R}(t, X(t)) + \delta_1 J^{\mathcal{D}} = 0$$

has a bounded and stabilizing solution $\hat{X}_{\delta_1}$. Based on Theorem 5.4.3 one obtains that $\hat{X}_{\delta_1}$ is a periodic function. The conclusion follows since $\hat{X}_{\delta_1}(\cdot) \in \tilde{\mathbf{\Gamma}}^{\Sigma}$ and the proof is complete. $\square$

With the similar proof based on Corollary 5.4.4 and Theorem 5.4.6 we obtain:

Corollary 5.4.8. *Assume that* $A(t) \equiv A \in \mathcal{M}_n^{\mathcal{D}}$, $B(t) \equiv B \in \mathcal{M}_{nm}^{\mathcal{D}}$, $\Pi(t) \equiv \Pi \in \mathbf{B}(\mathcal{S}_n^{\mathcal{D}}, \mathcal{S}_{n+m}^{\mathcal{D}})$ *and* $\mathcal{Q}(t) \equiv \mathcal{Q} \in \mathcal{S}_{n+m}^{\mathcal{D}}$. *Then the following are equivalent:*

(i) (A, B, Π) is stabilizable and $\tilde{\mathbf{\Gamma}}^{\Sigma}$ is not empty;

(ii) The GRDE (5.1) has a bounded and stabilizing solution X_s which is constant and solves the algebraic equation (5.43).

(iii) (A, B, Π) is stabilizable and there exists at least a sequence of symmetric matrices $\hat{X} = \{\hat{X}(i)\}_{i \in \mathcal{D}}$ such that $\hat{X} \in \tilde{\mathbf{\Gamma}}^{\Sigma}$.

As a simple consequence of Theorem 5.4.6 we have:

Corollary 5.4.9. *Assume that* (A, B, Π) *is stabilizable. Then:*

(i) If A, B, Π are periodic functions with period θ, then there exists a stabilizing feedback gain $F: \mathbf{R} \to \mathcal{M}_{mn}^{\mathcal{D}}$ which is a periodic function with period θ.

(ii) If $A(t) \equiv A$, $B(t) \equiv B$, $\Pi(t) \equiv \Pi$ for all $t \in \mathbf{R}$, then there exists a stabilizing feedback gain $F \in \mathcal{M}_{mn}^{\mathcal{D}}$.

Proof. Consider the differential equation

$$\frac{d}{dt}X(t) + A^T(t)X(t) + X(t)A(t) + J^{\mathcal{D}} + \Pi_1(t)[X(t)]$$

$$- \left\{X(t)B(t) + \Pi_{12}(t)[X(t)]\right\}\left\{R_J + \Pi_2(t)[X(t)]\right\}^{-1}$$

$$\times \left\{X(t)B(t) + \Pi_{12}(t)[X(t)]\right\}^T = 0, \tag{5.73}$$

where $R_J \in \mathcal{S}_m^{\mathcal{D}}, R_J = \{R_J(i)\}_{i \in \mathcal{D}} \in \mathcal{S}_m^{\mathcal{D}}, R_J(i) = I_m, i \in \mathcal{D}$. Equation (5.73) is a GRDE of type (5.1) corresponding to the quadruple $\Sigma_0 := (A, B, \Pi, \mathcal{Q}_0)$ where A, B, Π are as in (5.1) and $\mathcal{Q}_0(t, i) = \begin{pmatrix} I_n & 0 \\ 0 & I_m \end{pmatrix}$.

It is seen that $\Lambda^{\Sigma_0}(t)[\mathbf{0}](i) = \begin{pmatrix} I_n & 0 \\ 0 & I_m \end{pmatrix} \gg 0, \forall \ i \in \mathcal{D}$ and hence $\mathbf{0} \in \tilde{\mathbf{\Gamma}}^{\Sigma_0}$.

Therefore (5.73) has a bounded and stabilizing solution X_s with the corresponding stabilizing feedback gain

$$F_s(t, i) = -\left\{I_m + \Pi_2(t)[X_s(t)](i)\right\}^{-1}\left\{X_s(t, i)B(t, i) + \Pi_{12}(t)[X_s(t)](i)\right\}^T.$$

If the matrix coefficients of (5.73) are periodic functions with period θ, then by Theorem 5.4.3 we obtain that F_s is a periodic function with the same period θ. If the matrix coefficients of (5.73) are constants, then by Corollary 5.4.4 one obtains that F_s is constant and thus the proof is complete.   □

The result of Corollary 5.4.9 shows that if A, B, Π are periodic functions, then, without loss of generality, we may restrict the definition of stabilizability working only with periodic stabilizing feedback gains. Also, if A, B, Π are constant functions, then, without loss of generality the definition of stabilizability may be restricted only to the class of stabilizing feedback gains which are constant functions.

5.5   The Minimal Solution of the GRDE

In this section we focus our attention on those equations (5.1) associated with the quadruple $\Sigma = (A, B, \Pi, \mathcal{Q})$ where the operator valued function $\Pi(\cdot)$ satisfies the assumptions $(\Pi_1), (\Pi_2)$. Additionally, we assume that $0 \in \Gamma^\Sigma$. This means that $R(t,i) \geq \nu I_m$ and $\begin{pmatrix} M(t,i) & L(t,i) \\ L^T(t,i) & R(t,i) \end{pmatrix} \geq 0$, for all $(t,i) \in \mathcal{I} \times \mathcal{D}$. Applying Lemma 5.1.2 we deduce that these are equivalent to

$$
\begin{aligned}
&R(t,i) \geq \nu I_m \\
&M(t,i) - L(t,i)R^{-1}(t,i)L^T(t,i) \geq 0, \quad \forall\ (t,i) \in \mathcal{I} \times \mathcal{D}
\end{aligned}
\tag{5.74}
$$

$\nu > 0$ being a constant.

Remark 5.5.1. Let $W : \mathcal{I} \to \mathcal{M}_{mn}^{\mathcal{D}}$ be a continuous and bounded function. Let $X : \mathcal{I} \to \mathcal{S}_n^{\mathcal{D}}$ be a solution of the backward differential equation

$$
\frac{d}{dt}X(t) + \mathfrak{L}_{A+BW,\Pi_W}(t)[X(t)] + H(t) = 0
$$

where $H : \mathcal{I} \to \mathcal{S}_n^{\mathcal{D}}$ is a continuous function such that $H(t) \geq 0$, $t \in \mathcal{I}$. If there exists $\tau \in \mathcal{I}$ such that $X(\tau) \geq 0$, then $X(t) \geq 0$ for all $t \in (-\infty, \tau] \cap \mathcal{I}$. Indeed we have the following representation formula:

$$
X(t) = T_W^a(t,\tau)X(\tau) + \int_t^\tau T_W^a(t,s)H(s)ds
\tag{5.75}
$$

$T_W^a(t,s)$ being the anticausal linear evolution operator on $\mathcal{S}_n^{\mathcal{D}}$ defined by the linear differential equation

$$
\frac{d}{dt}X(t) + \mathfrak{L}_{A+BW,\Pi_W}(t)[X(t)] = 0.
$$

Since $T_W^a(t,s)$ is a positive operator for $t \le s$ we obtain via (5.75) that $X(t) \ge 0$ for all $t \le \tau, t \in \mathcal{I}$.

First we prove the following result.

Theorem 5.5.1. *Assume that the quadruple* $\Sigma = (A, B, \Pi, \mathcal{Q})$ *verifies the following assumptions: (a)* (A, B, Π) *is stabilizable;*

(b) $0 \in \Gamma^\Sigma$. *Then the GRDE (5.1) has two bounded solutions* $\tilde{X} : \mathcal{I} \to \mathcal{S}_n^\mathcal{D}$, $\tilde{\tilde{X}} : \mathcal{I} \to \mathcal{S}_n^\mathcal{D}$ *with the property* $\tilde{X}(t) \ge \bar{X}(t) \ge \tilde{\tilde{X}}(t) \ge 0$ *for all* $t \in \mathcal{I}$ *and for arbitrary bounded and positive semi-definite solution* $\bar{X} : \mathcal{I} \to \mathcal{S}_n^\mathcal{D}$ *of (5.1).*

Moreover if A, B, Π, $\mathcal{Q}$ are periodic functions with period $\theta > 0$, then both $\tilde{X}$ and $\tilde{\tilde{X}}$ are periodic functions with period θ. If $A(t) \equiv A$, $B(t) \equiv B$, $\Pi(t) \equiv \Pi$ and $\mathcal{Q}(t) \equiv \mathcal{Q}$, then both $\tilde{X}$ and $\tilde{\tilde{X}}$ are constant and solve the algebraic equation (5.43).

Proof. The existence of the solution $\tilde{X}$ is guaranteed by Theorem 5.3.5. Now we prove the existence of $\tilde{\tilde{X}}$. For each τ from the interior of $\mathcal{I}$ we consider $X_\tau(t) := X(t, \tau, 0)$ the solution of the following problem with given terminal conditions

$$\frac{d}{dt}X(t) + \mathfrak{R}(t, X(t)) = 0$$
$$X(\tau, \tau, 0) = 0. \tag{5.76}$$

Applying Corollary 5.2.3 we infer that $X_\tau(t)$ is well defined for all $t \in (-\infty, \tau] \cap \mathcal{I}$, and additionally we have, $X_\tau(t, i) \ge 0$ for all $t \in \mathcal{I}, t \le \tau, i \in \mathcal{D}$. We set $F_\tau(t) = F^{X_\tau}(t)$. Based on (5.74) it follows that $F_\tau(t)$ is well defined for all $t \in (-\infty, \tau] \cap \mathcal{I}$. Let $\tau_1 < \tau_2$ be arbitrary in the interior of $\mathcal{I}$. We show that

$$X_{\tau_1}(t, i) \le X_{\tau_2}(t, i) \tag{5.77}$$

for all $t \in \mathcal{I}, t \le \tau_1, i \in \mathcal{D}$. Applying Lemma 5.1.1 with $W(t) = F_{\tau_2}(t)$ we obtain that $X_{\tau_2}(t)$ solves the following backward differential equation:

$$\frac{d}{dt}X_{\tau_2}(t) + \mathcal{L}_{A+BF_{\tau_2}, \Pi_{F_{\tau_2}}}(t)[X_{\tau_2}(t)] + M_{F_{\tau_2}}(t) = 0, t \le \tau_2, t \in \mathcal{I}.$$

Applying again Lemma 5.1.1 taking $W(t) = F_{\tau_2}(t)$ we obtain that $X_{\tau_1}(t)$ satisfies the following backward differential equation

$$\frac{d}{dt}X_{\tau_1}(t) + \mathcal{L}_{A+BF_{\tau_2}, \Pi_{F_{\tau_2}}}(t)[X_{\tau_1}(t)] + M_{F_{\tau_2}}(t) - (F_{\tau_1}(t) - F_{\tau_2}(t))^T \Theta(t, X_{\tau_1}(t))(F_{\tau_1}(t)$$
$$- F_{\tau_2}(t)) = 0, t \le \tau_1, t \in \mathcal{I}$$

with

$$\Theta(t, X_{\tau_1}(t))(i) = \Pi_2(t)[X_{\tau_1}(t)](i) + R(t, i) \ge \nu I_m.$$

Subtracting the last two equations we deduce that $t \to X_{\tau_2}(t) - X_{\tau_1}(t)$ satisfies the backward differential equation:

$$\frac{d}{dt}Y(t) + \mathcal{L}_{A+BF_{\tau_2}, \Pi_{F_{\tau_2}}}(t)[Y(t)] + \Delta(t) = 0$$

where $\Delta(t,i) = [F_{\tau_1}(t,i) - F_{\tau_2}(t,i)]^T \Theta(t, X_{\tau_1}(t))(i)[F_{\tau_1}(t,i) - F_{\tau_2}(t,i)] \geq 0$. Since $X_{\tau_2}(\tau_1) - X_{\tau_1}(\tau_1) = X_{\tau_2}(\tau_1) \geq 0$ we may conclude based on Remark 5.5.1 that $X_{\tau_2}(t) - X_{\tau_1}(t) \geq 0$ for all $t \leq \tau_1, t \in \mathcal{I}$. Therefore (5.77) is true. Further, if (A, B, Π) is stabilizable, then there exists a bounded and continuous function $F: \mathcal{I} \to \mathcal{M}_{mn}^{\mathcal{D}}$ such that the corresponding operator $\mathcal{L}_{A+BF, \Pi_F}$ generates an anticausal exponentially stable evolution. Based on Theorem 2.3.13 we deduce that the equation

$$\frac{d}{dt}Y(t) + \mathcal{L}_{A+BF, \Pi_F}(t)[Y(t)] + M_F(t) = 0 \qquad (5.78)$$

has a unique bounded solution $\tilde{Y}(t) \geq 0$ on $\mathcal{I}$.

Let X_τ be the solution of the problem with given terminal value (5.76). Applying Lemma 5.1.1 equation (5.76) verified by X_τ can be rewritten as:

$$\frac{d}{dt}X_\tau(t) + \mathcal{L}_{A+BF_\tau, \Pi_{F_\tau}}(t)[X_\tau(t)] + M_{F_\tau}(t) = 0. \qquad (5.79)$$

On the other hand, applying Lemma 5.3.3 with $W(t) = F_\tau(t)$ to (5.78) one obtains

$$\frac{d}{dt}\tilde{Y}(t) + \mathcal{L}_{A+BF_\tau, \Pi_{F_\tau}}(t)[\tilde{Y}(t)] + M_{F_\tau}(t)$$

$$+ [F_\tau(t) - F(t)]^T \{R(t) + \Pi_2(t)[\tilde{Y}(t)]\}[F_\tau(t) - F(t)] = 0 \qquad (5.80)$$

for $t \in \mathcal{I}, t \leq \tau$. From (5.79) and (5.80) one obtains

$$\frac{d}{dt}[\tilde{Y}(t) - X_\tau(t)] + \mathcal{L}_{A+BF_\tau, \Pi_{F_\tau}}(t)[\tilde{Y}(t) - X_\tau(t)]$$

$$+ [F_\tau(t) - F(t)]^T \{R(t) + \Pi_2(t)[\tilde{Y}(t)]\}[F_\tau(t) - F(t)] = 0.$$

Since $\tilde{Y}(\tau) - X_\tau(\tau) = \tilde{Y}(\tau) \geq 0$, then invoking again Remark 5.5.1 we conclude that

$$\tilde{Y}(t) - X_\tau(t) \geq 0 \quad \text{for all} \quad t \in (-\infty, \tau] \cap \mathcal{I}. \qquad (5.81)$$

Inequality (5.81) together with (5.77) shows that the sequence $\{X_\tau(t)\}_{\tau \in \mathcal{I}}$ is monotonically increasing and bounded, hence it is convergent. Define

$$\tilde{X}(t,i) := \lim_{\tau \to \infty} X_\tau(t,i) \quad \text{for} \quad (t,i) \in \mathcal{I} \times \mathcal{D}. \qquad (5.82)$$

First we show that $t \to \tilde{X}(t) = \{\tilde{X}(t,i)\}_{i \in \mathcal{D}} : \mathcal{I} \to \mathcal{S}_n^{\mathcal{D}}$ is a solution of the GRDE (5.1). From (5.82) we obtain that $\tilde{X}(t,i) \geq X_\tau(t,i) \geq 0$. Therefore, $R(t,i) + \Pi_2(t)[\tilde{X}(t)](i) \geq \nu I_m$. Hence $(t, \tilde{X}(t)) \in Dom\mathfrak{R}$ for all $t \in \mathcal{I}$. Let τ_0 be arbitrary but fixed in the interior of $\mathcal{I}$. Let $\{\tau_k\}_{k \geq 0} \subset \mathcal{I}$ be a strictly increasing sequence with the properties: $\tau_k \geq \tau_0$ for all $k \geq 0$ and $\lim_{k\to\infty} \tau_k = +\infty$. Let $X_{\tau_k}(t)$ be the solution of the problem with given terminal conditions (5.76) for $\tau = \tau_k$. We have

$$X_{\tau_k}(\tau_0, i) - X_{\tau_k}(t, i) = \int_t^{\tau_0} \mathfrak{R}(s, X_{\tau_k}(s))(i)ds \qquad (5.83)$$

for all $t \leq \tau_0, t \in \mathcal{I}, i \in \mathcal{D}$. Following step by step the proof of Lemma 5.3.4 one shows that $\lim_{k\to\infty} \mathfrak{R}(s, X_{\tau_k}(s))(i) = \mathfrak{R}(s, \tilde{\tilde{X}}(s))(i)$ for all $i \in \mathcal{D}, t \leq \tau_0, t \in \mathcal{I}$. Further, reasoning as in the proof of Theorem 5.3.5 one obtains based on the convergence theorem of Lebesque that $\lim_{k\to\infty} \int_t^{\tau_0} \mathfrak{R}(s, X_{\tau_k}(s))(i)ds = \int_t^{\tau_0} \mathfrak{R}(s, \tilde{\tilde{X}}_{\tau_k}(s))(i)ds$ for all $i \in \mathcal{D}, t \leq \tau_0, t \in \mathcal{I}$. Taking the limit for $k \to \infty$ in (5.83) we get

$$\tilde{\tilde{X}}(\tau_0) - \tilde{\tilde{X}}(t) = \int_t^{\tau_0} \mathfrak{R}(s, \tilde{\tilde{X}}(s))ds$$

for all $t \leq \tau_0, t, \tau_0 \in \mathcal{I}$. This allows us to conclude that $t \to \tilde{\tilde{X}}(t)$ is differentiable and satisfies the differential equation (5.1). Let now $\bar{X} : \mathcal{I} \to \mathcal{S}_n^{\mathcal{D}}$ be an arbitrary global solution of (5.1) such that $\bar{X}(t) \geq 0$ for all $t \in \mathcal{I}$. Applying Theorem 5.2.1 we obtain that $\bar{X}(t) \geq X_\tau(t)$ for all $t \leq \tau, t, \tau \in \mathcal{I}$. Taking the limit for $\tau \to \infty$ we may conclude that $\tilde{\tilde{X}}(t) \leq \bar{X}(t)$ for all $t \in \mathcal{I}$. Thus we have shown that $\tilde{\tilde{X}}$ is the minimal positive semidefinite solution of the GRDE (5.1).

Let us assume that $A, B, \Pi, \mathcal{Q}$ are periodic functions with period $\theta > 0$. For each $\tau \in \mathcal{I}$ we define $Y_\tau(t)$ by

$$Y_\tau(t) = X_{\tau+\theta}(t + \theta), t \in \mathcal{I}.$$

One obtains that $Y_\tau(t)$ is a solution of GRDE (5.1). Additionally we have $Y_\tau(\tau) = 0 = X_\tau(\tau)$. From the uniqueness of the solution of the problem (5.76) we deduce that $Y_\tau(t, i) = X_\tau(t, i)$ for all $i \in \mathcal{D}, t \leq \tau, t \in \mathcal{I}$. We have $\tilde{\tilde{X}}(t, i) = \lim_{\tau\to\infty} X_\tau(t, i) = \lim_{\tau\to\infty} Y_\tau(t, i) = \lim_{\tau\to\infty} X_{\tau+\theta}(t + \theta, i) = \tilde{\tilde{X}}(t + \theta, i)$ for all $(t, i) \in \mathcal{I} \times \mathcal{D}$. So, we have shown that $\tilde{\tilde{X}}(\cdot)$ is a θ periodic function if the coefficients of GRDE (5.1) are periodic functions of the same period θ. Thus the proof is complete. $\square$

In the rest of this section we restrict our attention to the case $\mathcal{D} = \{1, 2, \ldots, d\}$. First we introduce a concept of detectability which extends to more general framework the concept of detectability introduced by Definition 4.1.2 (b).

Definition 5.5.1. Let $A : \mathcal{I} \to \mathcal{M}_n^{\mathcal{D}}, C : \mathcal{I} \to \mathcal{M}_{pn}^{\mathcal{D}}, \hat{\Pi} : \mathcal{I} \to \mathbf{B}(\mathcal{S}_n^d)$ be bounded and continuous functions. We say that the triple $(C, A, \hat{\Pi})$ is detectable if there exists a bounded and continuous function $K : \mathcal{I} \to \mathcal{M}_{np}^d$ such that the operator valued function $\mathcal{L}_{A+KC, \hat{\Pi}^*}(\cdot)$ defines an exponentially stable evolution on $\mathcal{S}_n^d$.

Here $X \to \mathcal{L}_{A+KC, \hat{\Pi}^*}(t)[X](i) = (A(t, i) + K(t, i)C(t, i))X(i) + X(i)(A(t, i) + K(t, i)C(t, i))^T + \hat{\Pi}^*(t)[X](i), i \in \mathcal{D}$.

Lemma 5.5.2. *Assume that the quadruple* $\Sigma = (A, B, \Pi, \mathcal{Q})$ *satisfies (a)* $0 \in \Gamma^\Sigma$. *(b) The triple* $(C, A + BW, \Pi^*_W)$ *is detectable, where* $W(t,i) = -R^{-1}(t,i)L^T(t,i)$ *and* C *is such that* $C^T(t,i)C(t,i) = M(t,i) - L(t,i)R^{-1}(t,i)L^T(t,i), i \in \mathcal{D}$. *Under these assumptions any bounded and positive semi-definite solution of GRDE (5.1) is a stabilizing solution.*

Proof. The proof has two stages. Firstly, the proof of the lemma is made in the particular case $L(t,i) \equiv 0$. Secondly, we shall show that the general case may be reduced to the particular case of the first step.

(i) Assume that $L(t,i) \equiv 0$. In this case $W(t,i) \equiv 0$ and $\Pi_W(t) = \Pi_1(t)$ for $t \in \mathcal{I}$ and the assumption (b) in the statement is equivalent to the detectability of the triple (C, A, Π^*_1) where C is such that $C^T(t,i)C(t,i) = M(t,i)$ for $t \in \mathcal{I}, i \in \mathcal{D}$. Let $X(t) = \{X(t,i)\}_{i \in \mathcal{D}}$ be a bounded and positive semi-definite solution of (5.1), that is $X(t,i) \geq 0$ for all $(t,i) \in \mathcal{I} \times \mathcal{D}$. Set $F(t) := F^X(t)$. We have to show that $\mathcal{L}_{A+BF,\Pi^*_F}$ generates an exponentially stable evolution.

Let $(t_0, H) \in \mathcal{I} \times \mathcal{S}^d_{n+}$ be fixed and let S be the solution of the initial value problem

$$\frac{d}{dt}S(t) = \mathcal{L}_{A+BF,\Pi^*_F}(t)[S(t)], \quad S(t_0) = H. \tag{5.84}$$

We show that

$$\int_{t_0}^{\infty} \|S(t)\|_\infty \, dt \leq \delta \|H\|_\infty,$$

where $\delta > 0$ is constant independent of t_0 and H. By the detectability assumption it follows that there exists a bounded and continuous function K such that the operator $\mathcal{L}_{A+KC,\Pi_1}$ generates an exponentially stable evolution, where

$$\mathcal{L}_{A+KC,\Pi_1}(t)[X](i) = \left[A(t,i) + K(t,i)C(t,i)\right]X(i) + X(i)\left[A(t,i) + K(t,i)C(t,i)\right]^T$$
$$+ \Pi^*_1(t)[X](i). \tag{5.85}$$

Using (5.85) equation 5.84 may be written as

$$\frac{d}{dt}S(t) = \mathcal{L}_{A+KC,\Pi_1}(t)[S(t)] - K(t)C(t)S(t) - S(t)C^T(t)K^T(t)$$
$$+ \left[\Pi^*_F(t)[S(t)] - \Pi^*_1(t)[S(t)]\right] + B(t)F(t)S(t) + S(t)F^T(t)B^T(t). \tag{5.86}$$

We introduce the following perturbed operator $X \to \mathcal{L}_\varepsilon(t)[X]$ by

$$\mathcal{L}_\varepsilon(t)[X](i) := \mathcal{L}_{A+KC,\Pi_1}(t)[X](i) + 2\varepsilon^2 X(i) + \varepsilon^2 \Pi^*_1(t)[X](i).$$

Let $T(t,s)$ be the evolution operator on $\mathcal{S}^d_n$ defined by

$$\frac{d}{dt}S(t) = \mathcal{L}_{A+KC,\Pi_1}(t)[S(t)].$$

Since $\mathcal{L}_{A+KC,\Pi_1}$ generates an exponentially stable evolution, we have

$$\|T(t,s)\| \le \beta e^{-2\alpha(t-s)} \quad \text{for} \quad t \ge s, \, t,s \in \mathcal{I}$$

for some constants $\alpha, \beta > 0$. By a standard argument based on Gronwall's Lemma one obtains that for $\varepsilon > 0$ small enough

$$\|T_\varepsilon(t,s)\| \le \beta e^{-\alpha(t-s)} \quad \text{for} \quad t \ge s, \, t,s \in \mathcal{I}, \tag{5.87}$$

where $T_\varepsilon(t,s)$ is the linear evolution operator on $\mathcal{S}_n^d$ defined by the linear differential equation

$$\frac{d}{dt}Y(t) = \mathcal{L}_\varepsilon(t)[Y(t)].$$

Let $\varepsilon > 0$ be such that (5.87) is fulfilled and let $Y(t)$ be the solution of the following forward differential equation

$$\begin{aligned} \frac{d}{dt}Y(t) &= \mathcal{L}_\varepsilon(t)[Y(t)] + \frac{1}{\varepsilon^2}K(t)C(t)S(t)C^T(t)K^T(t) \\ &+ \left(1+\frac{1}{\varepsilon^2}\right)\hat{\Pi}_F^*(t)[S(t)] + \frac{1}{\varepsilon^2}B(t)F(t)S(t)F^T(t)B^T(t), \end{aligned} \tag{5.88}$$

with initial condition $Y(t_0) = H$, where

$$\hat{\Pi}_F(t)[X] = F^T(t)\Pi_2(t)[X]F(t).$$

Set $Z(t) := Y(t) - S(t)$. We obtain from (5.86) and (5.88) that

$$\frac{d}{dt}Z(t) = \mathcal{L}_\varepsilon(t)[Z(t)] + U(t), \quad Z(t_0) = 0,$$

where

$$\begin{aligned} U(t,i) &= \left[\varepsilon I_n + \frac{1}{\varepsilon}K(t,i)C(t,i)\right]S(t,i)\left[\varepsilon I_n + \frac{1}{\varepsilon}K(t,i)C(t,i)\right]^T \\ &+ \Pi_{\varepsilon,F}^*(t)[S(t)](i) + (\varepsilon I_n - \frac{1}{\varepsilon}B(t,i)F(t,i))S(t,i)(\varepsilon I_n - \frac{1}{\varepsilon}B(t,i)F(t,i))^T \end{aligned}$$

with

$$\Pi_{\varepsilon,F}(t)[X](i) = \begin{pmatrix} \varepsilon I_n \\ -\frac{1}{\varepsilon}F(t,i) \end{pmatrix}^T \Pi(t)[X](i) \begin{pmatrix} \varepsilon I_n \\ -\frac{1}{\varepsilon}F(t,i) \end{pmatrix}.$$

Taking into account that $S(t_0) = H \geq 0$ it follows that $S(t) \geq 0$ for $t \geq t_0$ and hence $U(t) \geq 0$ for $t \geq t_0$. On the other hand $X \mapsto 2\varepsilon^2 X + \varepsilon^2 \Pi_1^*(t)[X]$ is a positive linear operator.

Since $\mathcal{L}_{A+BK,\Pi_1}$ generates a positive evolution it follows from Corollary 2.2.6 (i) that $\mathcal{L}_\varepsilon$ generates a positive evolution. So, from the representation formula $Z(t) = \int_{t_0}^t T_\varepsilon(t,s)U(s)ds$ we may conclude that $Z(t) \geq 0$ for all $t \geq 0$, hence $0 \leq S(t) \leq Y(t)$ which leads to

$$0 \leq \|S(t)\|_\infty \leq \|Y(t)\|_\infty. \tag{5.89}$$

Applying the representation formula (2.43) to (5.88) we may write

$$Y(t) = T_\varepsilon(t,t_0)H + \int_{t_0}^t T_\varepsilon(t,s)U_1(s)ds \quad \text{for} \quad t \geq t_0, \tag{5.90}$$

where

$$U_1(s,i) = \frac{1}{\varepsilon^2}K(s,i)C(s,i)S(s,i)C^T(s,i)K^T(s,i)$$

$$+ \left(1+\frac{1}{\varepsilon^2}\right)\hat{\Pi}_F^*(s)[S(s)](i) + \frac{1}{\varepsilon^2}B(s,i)F(s,i)S(s,i)F^T(s,i)B^T(s,i).$$

Taking into account the definition of the adjoint operator we obtain

$$\hat{\Pi}_F^*(s)[S(s)](i) = \Pi_2^*(s)\left[F(s)S(s)F^T(s)\right](i).$$

This allows us to write $U_1(s)$ as

$$U_1(s,i) = \tfrac{1}{\varepsilon^2}[K(s,i)C(s,i)S(s,i)C^T(s,i)K^T(s,i)$$
$$+K_1(s,i)R^{\frac{1}{2}}(s,i)F(s,i)S(s,i)F^T(s,i)R^{\frac{1}{2}}(s,i)K_1^T(s,i)]$$
$$+ \left(1+\tfrac{1}{\varepsilon^2}\right)\check{\Pi}_2(s)\left[R^{1/2}(s)F(s)S(s)F^T(s)R^{1/2}(s)\right](i),$$

where $Y \mapsto \check{\Pi}_2(s)[Y]$ is defined by

$$\check{\Pi}_2(s)[Y] := \Pi_2^*(s)\left[R^{-1/2}(s)YR^{-1/2}(s)\right]$$

and $K_1(s,i) = B(s,i)R^{-\frac{1}{2}}(s,i)$.

Further we have

$$\|U_1(s)\|_\infty = max_{i\in\mathcal{D}|U_1(s,i)|} \leq \left(1+\frac{1}{\varepsilon^2}\right)\gamma[\|C(s)S(s)C^T(s)\|_\infty$$
$$+\|R^{1/2}(s)F(s)S(s)F^T(s)R^{1/2}(s)\|_\infty], \tag{5.91}$$

where $\gamma = \max\{\sup_{s\in\mathcal{I}}\|K(s)\|_\infty^2, \sup_{s\in\mathcal{I}}\|K_1(s)\|_\infty^2, \sup_{s\in\mathcal{I}}\|\check{\Pi}_2(s)\|\}$.

From Remark 2.1.4 we deduce

$$\|C(s)S(s)C^T(s)\|_\infty + \|R^{1/2}(s)F(s)S(s)F^T(s)R^{1/2}(s)\|_\infty \le$$
$$\|C(s)S(s)C^T(s)\|_1 + \|R^{1/2}(s)F(s)S(s)F^T(s)R^{1/2}(s)\|_1 =$$
$$\sum_{i=1}^d \{Tr[C(s,i)S(s,i)C^T(s,i)] + Tr[R^{1/2}(s,i)F(s,i)S(s,i)F^T(s,i)R^{1/2}(s,i)]\}.$$

Using the properties of the trace together with (2.16) we have

$$\|C(s)S(s)C^T(s)\|_\infty + \|R^{1/2}(s)F(s)S(s)F^T(s)R^{1/2}(s)\|_\infty$$
$$\le \langle C^T(s)C(s) + F^T(s)R(s)F(s), S(s)\rangle. \tag{5.92}$$

Applying Lemma 5.1.1 we may write (5.1), verified by the bounded and positive semi-definite solution X, in the form

$$\frac{d}{ds}X(s) + \mathcal{L}^*_{A+BF,\Pi^*_F}(s)[X(s)] + C^T(s)C(s) + F^T(s)R(s)F(s) = 0.$$

Thus we obtain

$$\begin{aligned}
&\langle C^T(s)C(s) + F^T(s)R(s)F(s), S(s)\rangle \\
&= -\langle \tfrac{d}{ds}X(s), S(s)\rangle - \langle \mathcal{L}^*_{A+BF,\Pi^*_F}(s)[X(s)], S(s)\rangle \\
&= -\langle \tfrac{d}{ds}X(s), S(s)\rangle - \langle X(s), \mathcal{L}_{A+BF,\Pi^*_F}(s)[S(s)]\rangle \\
&= -\tfrac{d}{ds}\langle X(s), S(s)\rangle.
\end{aligned} \tag{5.93}$$

From (5.92), (5.93) we get

$$\int_{t_0}^t \left[\|C(s)S(s)C^T(s)\|_\infty + \|R^{1/2}(s)F(s)S(s)F^T(s)R^{1/2}(s)\|_\infty \right] ds$$

$$\le \langle X(t_0), S(t_0)\rangle - \langle X(t), S(t)\rangle.$$

Taking into account that $\langle X(t), S(t)\rangle \ge 0$ for $t \ge t_0$ and $\|X(t)\|_\infty \le \rho$ for all $t \in \mathcal{I}$ where $\rho > 0$ is a constant not depending on t, we obtain

$$\int_{t_0}^t \left[\|C(s)S(s)C^T(s)\|_\infty + \|R^{1/2}(s)F(s)S(s)F^T(s)R^{1/2}(s)\|_\infty \right] ds \le \rho \|H\|_\infty \tag{5.94}$$

for $t \ge t_0$. From (5.87), (5.90), and (5.91) we have

$$\|Y(t)\|_\infty \le \beta e^{-\alpha(t-t_0)}\|H\|_\infty + \beta\gamma \left(1 + \frac{1}{\varepsilon^2}\right) \int_{t_0}^t e^{-\alpha(t-s)} [\|C(s)S(s)C^T(s)\|_\infty$$

$$+ \|R^{1/2}(s)F(s)S(s)F^T(s)R^{1/2}(s)\|_\infty] ds,$$

which leads to

$$\int_{t_0}^{\tau} \|Y(t)\|_\infty \, dt \leq \frac{\beta}{\alpha} \|H\|_\infty + \beta\gamma \left(1 + \frac{1}{\varepsilon^2}\right) \int_{t_0}^{\tau} \int_{t_0}^{t} e^{-\alpha(t-s)} \left[\|C(s)S(s)C^T(s)\|_\infty \right.$$

$$\left. + \|R^{1/2}(s)F(s)S(s)F^T(s)R^{1/2}(s)\|_\infty \right] ds \, dt.$$

Changing the order of integration and invoking (5.94) we obtain

$$\int_{t_0}^{\tau} \|Y(t)\|_\infty \, dt \leq \frac{\beta}{\alpha} \left[1 + \left(1 + \frac{1}{\varepsilon^2}\right) \gamma\rho \right] \|H\|_\infty =: \delta \|H\|_\infty.$$

Taking the limit for $\tau \to \infty$ we deduce

$$\int_{t_0}^{\infty} \|Y(t)\|_\infty \, dt \leq \delta \|H\|_\infty \quad \text{for all} \quad t_0 \in \mathcal{I}, \, H \in \mathcal{S}_{n+}^d,$$

where δ is independent of t_0 and H. From (5.89) it follows now that

$$\int_{t_0}^{\infty} \|S(t)\|_\infty \, dt \leq \delta \|H\|_\infty \quad \text{for all} \quad t_0 \in \mathcal{I}, \, H \in \mathcal{S}_{n+}^d. \tag{5.95}$$

Taking $H = J^d$ the last inequality becomes:

$$\int_{t_0}^{\infty} \|T(s,t_0)J^d\|_\infty ds \leq \delta$$

for all $t_0 \in \mathcal{I}, T(s,t_0)$ being the linear evolution operator on $\mathcal{S}_n^d$ defined by the linear differential equation (5.84). Further from Corollary 2.6.2 we obtain

$$\int_{t_0}^{\infty} \|T(s,t_0)\| ds \leq \delta \tag{5.96}$$

for all $t_0 \in \mathcal{I}$. Applying (2.93) we deduce from (5.96) that

$$\int_{t_0}^{\infty} \|T^*(s,t_0)\| ds \leq \delta_1$$

for all $t_0 \in \mathcal{I}$ where δ_1 is a constant not depending upon t_0. Using implication (iii) $\to$ (i) of Theorem 2.7.4 we deduce that the zero solution of (5.84) is exponentially stable. This completes the proof of part (a).

(ii) Let us consider the general case when $L(t,i) \not\equiv 0$. Let X be a bounded and positive semi-definite solution of the GRDE (5.1). Applying Lemma 5.4.5 for $W(t) = -R^{-1}(t)L^T(t)$ we obtain that X is a bounded and positive semi-definite solution of the equation

$$\frac{d}{dt}X(t) + \mathcal{L}^*_{A-BR^{-1}L^T}(t)[X(t)] + C^T(t)C(t) + \Pi_W(t)[X(t)]$$
$$- \{X(t)B(t) + \Pi_{12W}(t)[X(t)]\}\{R(t) + \Pi_2(t)[X(t)]\}^{-1} \quad (5.97)$$
$$\times \{X(t)B(t) + \Pi_{12W}(t)[X(t)]\}^T = 0,$$

where $\Pi_W(t)$ is defined as in (5.6),

$$\Pi_{12W}(t) := \Pi_{12}(t) - L(t)R^{-1}(t)\Pi_2(t)$$

and

$$C^T(t)C(t) = M(t) - L(t)R^{-1}(t)L^T(t).$$

Equation (5.97) is an equation of type (5.1) with $L(t) \equiv 0$. Applying the first part of the proof we deduce that X is a stabilizing solution of (5.97). Let

$$\hat{F}(t) := -\{R(t) + \Pi_2(t)[X(t)]\}^{-1}\{X(t)B(t) + \Pi_{12W}(t)[X(t)]\}^T$$

be the stabilizing feedback gain associated with the solution X regarded as a solution of (5.97). Then it is easy to see that $\hat{F}(t) - R^{-1}(t)L^T(t) = F(t)$, where $F(t) := F^X(t)$ is defined as in (5.12). Hence,

$$A(t) - B(t)R^{-1}(t)L^T(t) + B(t)\hat{F}(t) = A(t) + B(t)F(t),$$

and

$$\begin{pmatrix} I_n \\ \hat{F}(t,i) \end{pmatrix}^T \begin{pmatrix} \Pi_W(t)[X](i) & \Pi_{12W}(t)[X](i) \\ \{\Pi_{12W}(t)[X](i)\}^T & \Pi_2(t)[X](i) \end{pmatrix} \begin{pmatrix} I_n \\ \hat{F}(t,i) \end{pmatrix}$$

$$= \begin{pmatrix} I_n \\ \hat{F}(t,i) \end{pmatrix}^T \begin{pmatrix} I_n & -L(t,i)R^{-1}(t,i) \\ 0 & I_m \end{pmatrix} \Pi(t)[X](i) \begin{pmatrix} I_n & 0 \\ -R^{-1}(t,i)L^T(t,i) & I_m \end{pmatrix} \begin{pmatrix} I_n \\ \hat{F}(t,i) \end{pmatrix}$$

$$= \begin{pmatrix} I_n \\ F(t,i) \end{pmatrix}^T \Pi(t)[X](i) \begin{pmatrix} I_n \\ F(t,i) \end{pmatrix} = \Pi_F(t)[X](i).$$

These facts allow us to conclude that X is a stabilizing solution of (5.1) and the proof ends. $\square$

Remark 5.5.2. Assume that the quadruple $\Sigma = (A, B, \Pi, \mathcal{Q})$ satisfies $0 \in \tilde{\Gamma}^\Sigma$. Then any bounded and positive semi-definite solution $X : \mathcal{I} \to \mathcal{S}^d_{n+}$ of (5.1) is a stabilizing solution, and we have $X(t) \gg 0$ for $t \in \mathcal{I}$. Indeed by using (5.11) with $W = F^X$, together with Theorems 2.3.12 and 2.3.13, we conclude that the above assertions hold.

Theorem 5.5.3. *Assume that the quadruple $\Sigma = (A, B, \Pi, \mathcal{Q})$ satisfies the following assumptions*

(a) $0 \in \Gamma^{\Sigma}$;
(b) (A, B, Π) is stabilizable.
(c) $(C, A - BR^{-1}L^{T}, \Pi_{W}^{*})$ is detectable where Π_{W} is defined as in (5.6) for $W(t) = -R^{-1}(t)L^{T}(t)$ and C is such that $C^{T}(t)C(t) = M(t) - L(t)R^{-1}(t)L^{T}(t)$.

Then (5.1) has a unique bounded solution $X: \mathcal{I} \to \mathcal{S}_{n+}^{d}$ which is stabilizing.

Proof. Based on Theorem 5.5.1 we deduce that (5.1) has both a bounded maximal solution $\check{X}$ and a bounded minimal positive semi-definite solution $\tilde{X}$ such that $\check{X}(t) \geq \bar{X}(t) \geq \tilde{X}(t) \geq 0$ for all $t \in \mathcal{I}$, where $\bar{X}$ is an arbitrary bounded and positive semi-definite solution of (5.1). Applying Lemma 5.5.2 it follows that both $\check{X}$ and $\tilde{X}$ are stabilizing solutions.

From the uniqueness of the stabilizing and bounded solution of (5.1) we conclude that $\tilde{X}(t) = \check{X}(t)$ for all $t \in \mathcal{I}$ and thus the proof is complete.   □

Remark 5.5.3. As we have seen in Theorem 5.5.1 equation (5.1) has two remarkable bounded solutions, namely $\check{X}: \mathcal{I} \to \mathcal{S}_{n}^{d}$, which is the maximal solution, and $\tilde{X}: \mathcal{I} \to \mathcal{S}_{n}^{d}$, which is the minimal solution in the class of all bounded and positive semi-definite solutions of (5.1). Theorem 5.5.3 shows that under the assumption of detectability these two solutions coincide. However in the absence of detectability these two solutions may be different (see the illustrative example from Sect. 5.6.4). If in addition to the assumptions of Theorem 5.5.3 we assume that $0 \in \tilde{\Gamma}^{\Sigma}$—which is equivalent to $\mathcal{Q}(t) \gg 0$—then $\check{X} \gg 0$. This follows immediately from Theorem 2.3.13 (iv) and formula (5.11) with $X := \check{X}$ and $W := F^{\check{X}}$.

5.6   Systems of Generalized Riccati Equations on the Space $\mathcal{S}_{n}^{d}$

5.6.1   Preliminaries

In this section we study systems of nonlinear matrix differential equations of the form:

$$
\begin{aligned}
&\tfrac{d}{dt}X(t,i) + A_{0}^{T}(t,i)X(t,i) + X(t,i)A_{0}(t,i) + \sum_{k=1}^{r}A_{k}^{T}(t,i)X(t,i)A_{k}(t,i) \\
&+ \sum_{j=1}^{d}q_{ij}X(t,j) - (X(t,i)B_{0}(t,i) + \sum_{k=1}^{r}A_{k}^{T}(t,i)X(t,i)B_{k}(t,i) \\
&+ L(t,i))(R(t,i) + \sum_{k=1}^{r}B_{k}^{T}(t,i)X(t,i)B_{k}(t,i))^{-1}(B_{0}^{T}(t,i)X(t,i) + \\
&\sum_{k=1}^{r}B_{k}^{T}(t,i)X(t,i)A_{k}(t,i) + L^{T}(t,i)) + M(t,i) = 0.
\end{aligned}
\tag{5.98}
$$

where $t \to A_{k}(t,i): \mathcal{I} \to \mathbf{R}^{n \times n}$, $t \to B_{k}(t,i): \mathcal{I} \to \mathbf{R}^{n \times m}$, $0 \leq k \leq r$, $t \to M(t,i): \mathcal{I} \to \mathcal{S}_{n}$, $t \to L(t,i): \mathcal{I} \to \mathbf{R}^{n \times m}$, $t \to R(t,i): \mathcal{I} \to \mathcal{S}_{m}$, $i \in \mathcal{D}$ are bounded and continuous matrix valued functions. $\mathcal{I} \subset \mathbf{R}$ is a right unbounded interval. The elements q_{ij} of the matrix Q verify only the weaker assumption $q_{ij} \geq 0$ for $i \neq j$. The assumption $\sum_{j=1}^{d} q_{ij} = 0$ will be used only for the results referring to stochastic observability and detectability. If $A_k(t,i) = 0$, $B_k(t,i) = 0$, $1 \leq k \leq r$, $(t,i) \in \mathcal{I} \times \mathcal{D}$ the system (5.98) becomes the system of Riccati-type equations intensively investigated in connection with the linear quadratic problem for linear stochastic systems with Markovian jumping. In the particular case $\mathcal{D} = \{1\}$ the system (5.98) reduces to:

$$\begin{aligned}
&\tfrac{d}{dt}X(t) + A_0^T(t)X(t) + X(t)A_0(t) + \sum_{k=1}^{r} A_k^T(t)X(t)A_k(t) \\
&-(X(t)B_0(t) + \sum_{k=1}^{r} A_k^T(t)X(t)B_k(t)) \\
&+L(t))(R(t) + \sum_{k=1}^{r} B_k^T(t)X(t)B_k(t))^{-1}(B_0^T(t)X(t)+ \\
&\sum_{k=1}^{r} B_k^T(t)X(t)A_k(t) + L^T(t)) + M(t) = 0,\ t \in \mathcal{I}
\end{aligned} \tag{5.99}$$

where we denoted $A_0(t) = A_0(t,1) + \tfrac{1}{2}q_{11}I_n$, $A_k(t) = A_k(t,1)$, $1 \leq k \leq r$, $B_k(t) = B_k(t,1)$, $0 \leq k \leq r$, $M(t) = M(t,1)$, $L(t) = L(t,1)$, $R(t) = R(t,1)$. If $A_k(t) = 0$, $B_k(t) = 0$, $1 \leq k \leq r$, $t \in \mathcal{I}$, (5.99) becomes the well-known matrix Riccati differential equation intensively investigated in connection with various types of control problems in the deterministic framework.

In this book the system (5.98) and its particular form (5.99) will be called *SGRDE*.

Remark 5.6.1. (i) The SGRDE (5.98) is a special form of SGRDE (5.1) for the case $\mathcal{D} = \{1, 2, \ldots, d\}$. Therefore, in this section we shall apply the results derived in the previous sections to obtain necessary and sufficient conditions for the existence of the maximal solution, stabilizing solution and minimal solution of SGRDE (5.98) and its special form (5.99).

(ii) In the developments of the previous sections, the GRDEs (5.1) are defined by the quadruple $\Sigma = (A, B, \Pi, \mathcal{Q})$. In the special case of SGRDE (5.98) regarded as a special form of (5.1) the operator $\Pi(t)$ is the one described in (5.5) while

$$A(t,i) = A_0(t,i) + \frac{1}{2}q_{ii}I_n \tag{5.100}$$

$$B(t,i) = B_0(t,i), 1 \leq i \leq d, t \in \mathcal{I}.$$

Hence, in the special case of SGRDE (5.98) the triple (A, B, Π) may be identified with the triple $(\mathbf{A}, \mathbf{B}, \mathcal{Q})$ where $\mathbf{A} = (A_0, A_1, \ldots, A_r)$, $\mathbf{B} = (B_0, B_1, \ldots, B_r)$, $\mathcal{Q} = (q_{ij})_{i,j \in \mathcal{D}}$. Here $t \to A_k(t) = (A_k(t,1), A_k(t,2), \ldots, A_k(t,d))$: $\mathcal{I} \to \mathcal{M}_n^d$, $t \to B_k(t) = (B_k(t,1), B_k(t,2), \ldots, B_k(t,d))$: $\mathcal{I} \to \mathcal{M}_{nm}^d$. This allows us as every time when we refer to SGRDE (5.98) to say without loss of generality that this kind of Riccati differential equations are associated with a quadruple $\Sigma = (\mathbf{A}, \mathbf{B}, \mathcal{Q}, \mathcal{Q})$.

Lemma 5.6.1. *If (A, B, Π) is the triple described according to (5.5) and (5.100), then the following are equivalent*

 (i) *The triple (A, B, Π) is stabilizable in the sense of Definition 5.3.2.*
(ii) *The triple $(\mathbf{A}, \mathbf{B}, Q)$ is stabilizable in the sense of Definition 4.1.2 (a).*

Proof. According to Remark 5.3.2, the triple (A, B, Π) is stabilizable if and only if there exists a bounded and continuous function $F : \mathcal{I} \to \mathcal{M}_{mn}^d$ such that the operator valued function $\mathcal{L}_{A+BF,\Pi_F^*}(\cdot)$ generates an exponentially stable evolution on $\mathcal{S}_n^d$. We have

$$\mathcal{L}_{A+BF,\Pi_F^*}(t)[X](i) = (A_0(t,i) + B_0(t,i)F(t,i))X(i) + X(i)(A_0(t,i) + B_0(t,i)F(t,i))^T +$$

$$\sum_{j=1}^{d} q_{ji}X(j) + \sum_{k=1}^{r}(A_k(t,i) + B_k(t,i)F(t,i))X(i)(A_k(t,i) + B_k(t,i)F(t,i))^T = (\mathcal{L}_F(t)X)(i)$$

for all $i \in \mathcal{D}, X = (X(1), X(2), \ldots, X(d)) \in \mathcal{S}_n^d$. So, we have obtained that

$$\mathcal{L}_{A+BF,\Pi_F^*}(t) = \mathcal{L}_F(t) \tag{5.101}$$

$\mathcal{L}_F(t)$ being the linear operator introduced via (4.2).

The equivalence of the assertions in the statement follows employing (5.101), Remark 5.3.2 and Definition 4.1.2 (a).   □

Using the conventions of notations established in Sect. 2.6.1, the system of differential equations (5.98) may be written in the following compact form on the space $\mathcal{S}_n^d$

$$\frac{d}{dt}X(t) + \mathcal{L}^*(t)X(t) - \mathcal{P}^*(t,X(t))\mathcal{R}^{-1}(t,X(t))\mathcal{P}(t,X(t)) + M(t) = 0, \tag{5.102}$$

$\mathcal{L}^*(t)$ being the adjoint operator of $\mathcal{L}(t)$ defined as in (2.124)

$$X \to \mathcal{P}(t,X) : \mathcal{S}_n^d \to \mathcal{M}_{m,n}^d$$

$$\mathcal{P}(t,X) = (\mathcal{P}_1(t,X), \mathcal{P}_2(t,X), \ldots, \mathcal{P}_d(t,X)),$$

$$\mathcal{P}_i(t,X) = B_0^T(t,i)X(i) + \sum_{k=1}^{r} B_k^T(t,i)X(i)A_k(t,i) + L^T(t,i)$$

$$X \to \mathcal{R}(t,X) : \mathcal{S}_n^d \to \mathcal{S}_m^d \text{ by}$$

$$\mathcal{R}(t,X) = (\mathcal{R}_1(t,X), \mathcal{R}_2(t,X), \ldots, \mathcal{R}_d(t,X)),$$

$$\mathcal{R}_i(t,X) = R(t,i) + \sum_{k=1}^{r} B_k^T(t,i)X(i)B_k(t,i),$$

$$M(t) = (M(t,1), M(t,2), \ldots M(t,d)) \in \mathcal{S}_n^d.$$

If the coefficients of (5.98) do not depend upon t, then the operators $\mathcal{L}, \mathcal{P}, \mathcal{R}$ are also independent upon t. In this case we shall use the following algebraic nonlinear equation over $\mathcal{S}_n^d$

$$L^*X - \mathcal{P}^*(X)\mathcal{R}^{-1}(X)\mathcal{P}(X) + M = 0. \tag{5.103}$$

Let us remark that (5.102) is defined on the set

$$\Gamma = \left\{ (t,X) \in \mathcal{I} \times \mathcal{S}_n^d \mid \det \mathcal{R}_i(t,X) \neq 0, \ \forall \, i \in \mathcal{D} \right\}.$$

Now we recall the definitions of the dissipation operator Λ^Σ and of the sets $\Gamma^\Sigma, \tilde{\Gamma}^\Sigma$ introduced in Sect. 5.1 updated with the notations introduced in connection with (5.102).

If $X : \mathcal{I} \to \mathcal{S}_n^d$ is a C^1 function, we denote

$$\Lambda_i^\Sigma(t)[X(t)] = \begin{bmatrix} \frac{d}{dt}X(t,i) + \mathcal{L}_i^*(t)(X(t)) + M(t,i) & \mathcal{P}_i^*(t,X(t)) \\ \mathcal{P}_i(t,X(t)) & \mathcal{R}_i(t,X(t)) \end{bmatrix}$$

which will be called the *dissipation matrix*, where

$$\mathcal{L}_i^*(t)(X(t)) = (\mathcal{L}^*(t)X(t))(i),$$

$\mathcal{L}^*(t)$ being the adjoint operator of the Lyapunov operator $\mathcal{L}(t)$ with respect to the inner product (2.16) and $\mathcal{P}_i$, $\mathcal{R}_i$ are defined above related to (5.102). We shall also denote

$$\Lambda^\Sigma(t)[X(t)] = \left(\Lambda_1^\Sigma(t)[X(t)], \ldots, \Lambda_d^\Sigma(t)[X(t)] \right) \in \mathcal{S}_{n+m}^d.$$

To a quadruple $\Sigma = (\mathbf{A}, \mathbf{B}, Q, \mathcal{Q})$ we associate the following two sets of C^1 functions which will play an important role in the next developments:

$$\Gamma^\Sigma = \left\{ \hat{X} \in C_b^1\left(\mathcal{I}, \mathcal{S}_n^d\right) \mid \Lambda_i^\Sigma(t)[\hat{X}(t)] \geq 0, \ \mathcal{R}_i\left(t, \hat{X}(t)\right) \gg 0, \ t \in \mathcal{I}, i \in \mathcal{D} \right\} \tag{5.104}$$

and

$$\tilde{\Gamma}^\Sigma = \left\{ \hat{X} \in C_b^1\left(\mathcal{I}, \mathcal{S}_n^d\right) \mid \Lambda_i^\Sigma(t)[\hat{X}(t)] \gg 0, \ t \in \mathcal{I}, i \in \mathcal{D} \right\} \tag{5.105}$$

where $C_b^1\left(\mathcal{I}, \mathcal{S}_n^d\right) = \left\{ X \in C^1\left(\mathcal{I}, \mathcal{S}_n\right) \mid X, \frac{d}{dt}X \text{ are bounded functions} \right\}$.

Remark 5.6.2. Excepting some particular cases which will be displayed later, we do not make any assumption concerning the signature of the matrices $Q(t,i)$ in (5.3) and $R(t,i)$. As we shall see in the next developments an important role in the characterization of SGRDE (5.98) is played by the sign of the expression

$$\mathcal{R}_i(t, X(t)) = R(t,i) + \sum_{k=1}^{r} B_k^T(t,i)X(t,i)B_k(t,i).$$

In this chapter we consider only the case $\mathcal{R}_i(t, X(t)) > 0$ because this is the case required by the quadratic optimization problem. In Chap. 8 the case $\mathcal{R}_i(t, X(t)) < 0$ will be considered in connection with some Bounded Real Lemma type results.

5.6.2   The Maximal Solution of the SGRDE

The notion of the maximal solution of SGRDE (5.98) is similar to the one introduced for the GRDE (5.1) (see Definition 5.3.1).

Theorem 5.6.2. *Assume that* $(\mathbf{A}, \mathbf{B}; Q)$ *is stabilizable. Then the following are equivalent:*

(i) The set $\mathbf{\Gamma}^{\Sigma}$ *is not empty;*
(ii) The SGRDE (5.98) has a bounded maximal solution $\tilde{X} : \mathcal{I} \to \mathcal{S}_n^d$ *which verifies*

$$\mathcal{R}_i(t, \hat{X}(t)) \gg 0, t \in \mathcal{I}, i \in \mathcal{D}. \tag{5.106}$$

Moreover if the coefficients of the SGRDE (5.98) are θ-*periodic functions then the maximal solution* $\tilde{X}(t)$ *is a* θ-*periodic function too. If the coefficients of (5.98) do not depend upon t, then the maximal solution* $\tilde{X}(t)$ *is constant and it solves (5.103).*

Proof follows directly from Theorem 5.3.5.

Corollary 5.6.3. *Assume that*

(a) $(\mathbf{A}, \mathbf{B}; Q)$ is stabilizable;
(b) $R(t, i) \geq \rho^2 I_m$, $(t, i) \in \mathcal{I} \times \mathcal{D}$.
(c) $M(t, i) - L(t, i) R^{-1}(t, i) L^T(t, i) \geq 0$, $(t, i) \in \mathcal{I} \times \mathcal{D}$.

Under these conditions (5.102) has a bounded solution $\tilde{X}(t) \geq 0$. *Moreover* $\tilde{X}(t) \geq \hat{X}(t)$ *for any bounded and semipositive solution* $\hat{X}(t)$ *of (5.102).*

Proof. Under the considered assumptions $\hat{X}(t) = 0$ solves the differential inequality $\Lambda_i^{\Sigma}(t)[X(t)] \geq 0$, $(t, i) \in \mathcal{I} \times \mathcal{D}$ and thus we obtain $\mathbf{0} \in \mathbf{\Gamma}^{\Sigma}$. Therefore the assumptions of Theorem 5.6.2 are fulfilled.   □

With the same technique as in Theorem 5.6.2 we may prove the following dual result.

Theorem 5.6.4. *Assume that*

(a) $(\mathbf{A}, \mathbf{B}; Q)$ is stabilizable;
(b) The differential inequality

$$\Lambda^{\Sigma}(t)[X(t)] \leq 0, \tag{5.107}$$

$$\Lambda^{\Sigma}(t)[X(t)] = \begin{bmatrix} \frac{d}{dt}X(t) + \mathcal{L}^*(t)X(t) + M(t) & \mathcal{P}^*(t,X(t)) \\ \mathcal{P}(t,X(t)) & \mathcal{R}(t,X(t)) \end{bmatrix}$$

has a bounded solution $\hat{X}(t)$ which verifies

$$\mathcal{R}(t,\hat{X}(t)) \ll 0. \tag{5.108}$$

Under these conditions the differential equation (5.102) has a bounded solution $\tilde{X}(t)$ which verifies $\tilde{X}(t) \leq \check{X}(t)$ for any bounded solutions $\check{X}(t)$ of the inequality (5.107) which satisfies (5.108).

5.6.3 Stabilizing Solution of the SGRDE

In this subsection we investigate some aspects concerning the stabilizing solution of the SGRDE (5.98). First we show that the SGRDE (5.98) has at most one bounded and stabilizing solution. The uniqueness of the stabilizing solution is proved without any assumption concerning the sign of $\mathcal{R}_i(t,X(t))$. Further we provide a necessary and sufficient condition which guarantees the existence of the bounded and stabilizing solution of (5.98) satisfying the additional condition (5.106).

Definition 5.6.1. A solution $\tilde{X} : \mathcal{I} \rightarrow \mathcal{S}_n^d$ of (5.98) is called *stabilizing solution* if it has the following properties:

(a)

$$\inf_{t \in \mathcal{I}} |det[R(t,i) + \sum_{k=1}^r B_k^T(t,i)\tilde{X}(t,i)B_k(t,i)]| > 0, \ i \in \mathcal{D}.$$

(b) The system

$$(A_0 + B_0\tilde{F}, A_1 + B_1\tilde{F}, \ldots, A_r + B_r\tilde{F}; Q)$$

is stable in the sense of Definition 2.7.1, where

$$\tilde{F}(t) = (\tilde{F}(t,1), \ \tilde{F}(t,2), \ \ldots, \ \tilde{F}(t,d)), \tag{5.109}$$

$$\tilde{F}(t,i) = -[R(t,i) + \sum_{k=1}^r B_k^T(t,i)\tilde{X}(t,i)B_k(t,i)]^{-1}[B_0(t,i)\tilde{X}(t,i)$$

$$+ \sum_{k=1}^r B_k^T(t,i)\tilde{X}(t,i)A_k(t,i) + L^T(t,i)].$$

Remark 5.6.3. (a) The condition (a) in Definition 5.6.1 is assumed in order to be sure that the stabilizing feedback gain in (5.109) is bounded if $\tilde{X}(t)$ is bounded.

(b) If the scalars q_{ij} satisfy the additional condition $\sum_{j=1}^{d} q_{ij} = 0$ for all $i \in \mathcal{D}$, then the solution $\tilde{X}(t)$ of the system (5.98) is a stabilizing solution if the control $u(t) = \tilde{F}(t, \eta(t))x(t)$ stabilizes the system

$$dx(t) = [A_0(t, \eta(t))x(t) + B_0(t, \eta(t))u(t)]\,dt$$

$$+ \sum_{k=1}^{r} [A_k(t, \eta(t))x(t) + B_k(t, \eta(t))u(t)]\,dw_k(t).$$

(c) Employing Remark 5.4.1 (b) together with formula (5.101) written for $\tilde{F}(t)$ instead of $F(t)$ we obtain that the solution $\tilde{X}(t), t \in \mathcal{I}$ is stabilizing in the sense of Definition 5.6.1 if and only if it is a stabilizing solution of SGRDE (5.98) regarded as a special form of (5.1) in the sense of Definition 5.4.1. This fact allows us to use the general results from Sect. 5.4 to derive similar results for the stabilizing solution of SGRDE (5.98).

Theorem 5.6.5.   *(i) The system of generalized matrix Riccati differential equations (5.98) has at most one stabilizing and bounded on $\mathcal{I}$ solution.*

(ii) If the coefficients of the system (5.98) are θ-periodic functions, then the stabilizing and bounded solution $\tilde{X}(t)$ (if it exists) is θ-periodic function too.

(iii) If the coefficients of the system (5.98) do not depend upon t, then its stabilizing and bounded solution $\tilde{X}(t)$ is constant and solves the following system of nonlinear algebraic equations

$$A_0^T(i)X(i) + X(i)A_0(i) + \sum_{k=1}^{r} A_k^T(i)X(i)A_k(i) + \sum_{j=1}^{d} q_{ij}X(j)$$
$$-(X(i)B_0(i) + \sum_{k=1}^{r} A_k^T(i)X(i)B_k(i) + L(i))(R(i)$$
$$+ \sum_{k=1}^{r} B_k^T(i)X(i)B_k(i))^{-1}(B_0^T(i)X(i)$$
$$+ \sum_{k=1}^{r} B_k^T(i)X(i)A_k(i) + L^T(i)) + M(i) = 0, \; i \in \mathcal{D}$$

$$(5.110)$$

Proof. (i) Let us suppose that the differential equation (5.98) has two bounded and stabilizing solutions, $X_l : \mathcal{I} \to \mathcal{S}_n^d, l = 1,2$; hence, the systems $(A_0 + B_0F_l, A_1 + B_1F_l, \ldots, A_r + B_rF_l; Q), l = 1,2$ are stable, the stabilizing feedback gain being defined as in (5.109). By direct computation we obtain that:

$$\frac{d}{dt}X_l(t,i) + [A_0(t,i) + B_0(t,i)F_1(t,i)]^T X_l(t,i) + X_l(t,i)[A_0(t,i)$$
$$+ B_0(t,i)F_2(t,i)] + \sum_{k=1}^{r} [A_k(t,i) + B_k(t,i)F_1(t,i)]^T X_l(t,i)$$
$$\times [A_k(t,i) + B_k(t,i)F_2(t,i)] + \sum_{j=1}^{d} q_{ij}X_l(t,j) + F_1^T(t,i)R(t,i)F_2(t,i)$$
$$+ M(t,i) + L(t,i)F_2(t,i) + F_1^T(t,i)L^T(t,i) = 0$$

$l = 1,2; \; i \in \mathcal{D}, \; t \in \mathcal{I}$.

Set $\hat{X}(t,i) = X_1(t,i) - X_2(t,i), \quad i \in \mathcal{D}, \quad t \in \mathcal{I}$ and obtain that $\hat{X}(t) = (\hat{X}(t,1), \ldots, \hat{X}(t,d))$ is a bounded solution of the system:

$$\frac{d}{dt}\hat{X}(t,i) + [A_0(t,i) + B_0(t,i)F_1(t,i)]^T \hat{X}(t,i) + \hat{X}(t,i)[A_0(t,i)$$

$$+ B_0(t,i)F_2(t,i)] + \sum_{k=1}^r [A_k(t,i) + B_k(t,i)F_1(t,i)]^T \tag{5.111}$$

$$\times \hat{X}(t,i)[A_k(t,i) + B_k(t,i)F_2(t,i)] + \sum_{j=1}^d q_{ij}\hat{X}(t,j) = 0,$$

$i \in \mathcal{D}$, $t \in \mathcal{I}$. It is easy to see that (5.111) is equivalent to the following linear equation on $\mathcal{S}_{2n}^d$:

$$\frac{d}{dt}\hat{X}_e(t) + \mathcal{L}_e^*(t)\hat{X}_e(t) = 0 \tag{5.112}$$

where $\mathcal{L}_e(t) : \mathcal{S}_{2n}^d \to \mathcal{S}_{2n}^d$, $i \in \mathcal{D}, t \in \mathcal{I}$

$$A_{k,e}(t,i) = \begin{pmatrix} A_k(t,i) + B_k(t,i)F_1(t,i) & 0 \\ 0 & A_k(t,i) + B_k(t,i)F_2(t,i) \end{pmatrix},$$

$$k = 0, 1, \ldots r.$$

$$\hat{X}_e(t,i) = \begin{bmatrix} 0 & \hat{X}(t,i) \\ \hat{X}(t,i) & 0 \end{bmatrix}.$$

From Theorem 2.7.4 we deduce that there exist the $\mathbf{C}^1$ functions $K_j : \mathcal{I} \to \mathcal{S}_n^d, K_j(t) \gg 0$ which are bounded on $\mathcal{I}$ and verify the linear differential equations

$$\frac{d}{dt}K_j(t) + \mathcal{L}_j^*(t)K_j(t) + J^d = 0, \quad j = 1, 2.$$

where $\mathcal{L}_j$ are the Lyapunov operators associated with $(A_0 + B_0F_j, \ldots, A_r + B_rF_j; Q)$, $j = 1, 2$. Set $K_e(t) = \begin{pmatrix} K_1(t) & 0 \\ 0 & K_2(t) \end{pmatrix}$. It is easy to see that $K_e(t)$ is a solution of the linear differential equation on $\mathcal{S}_{2n}^d$

$$\frac{d}{dt}K_e(t) + \mathcal{L}_e^*(t)K_e(t) + J^{2d} = 0 \tag{5.113}$$

where, by definition $J^{2d} = (J^{2d}(1), \ldots, J^{2d}(d))$ with $J^{2d}(i) = \begin{pmatrix} I_n & 0 \\ 0 & I_n \end{pmatrix}$. From Theorem 2.7.4 (v)$\to$(i) we conclude that the augmented system $(A_{0,e}, \ldots, A_{r,e}; Q)$ is stable. Applying Theorem 2.7.5 we deduce that (5.112) has a unique bounded solution. Therefore $\hat{X}_e(t) = 0$ and hence $X_1(t,i) = X_2(t,i)$ for all $(t,i) \in \mathcal{I} \times \mathcal{D}$, and the proof of part (i) is complete.

(ii) Let $\tilde{X}(t) = (\tilde{X}(t,1), \ldots, \tilde{X}(t,d))$ be the bounded and stabilizing solution of (5.98). According to Remark 5.6.3 (c) it follows that $\tilde{X}(t)$ is the stabilizing solution of SGRDE (5.98) in the sense of Definition 5.4.1. Then, applying Theorem 5.4.3 we obtain that $\tilde{X}(t)$ is periodic of period θ.

The assertion (iii) follows directly from Corollary 5.4.4.   □

A result concerning the existence of a stabilizing solution of SGRDE (5.98) is given by the next theorem.

Theorem 5.6.6. *The following are equivalent:*

(i) The triple $(\mathbf{A}, \mathbf{B}; Q)$ *is stabilizable and it exists a* C^1 *function* $\widehat{X} : \mathcal{I} \to \mathcal{S}_n^d$ *bounded, with bounded derivative such that the differential inequality*

$$\Lambda^{\Sigma}(t)\,[\widehat{X}(t)] \gg 0; \tag{5.114}$$

is satisfied.

(ii) The differential equation on $\mathcal{S}_n^d$ *(5.98) has a bounded on* $\mathcal{I}$ *and stabilizing solution* $\tilde{X}(t)$ *which satisfies* $\mathcal{R}(t, \tilde{X}(t)) \gg 0, t \in \mathcal{I}.$

Proof follows directly from Theorem 5.4.6.

Corollary 5.6.7. *If the SGRDE (5.98) has a stabilizing and bounded on* $\mathcal{I}$ *solution* $\tilde{X}$ *which verifies (5.106), then* $\tilde{X}(t)$ *is the maximal solution with respect to* Γ^{Σ} *of (5.98).*

Proof. Suppose that (5.98) has a stabilizing and bounded on $\mathcal{I}$ solution $\tilde{X}$ which satisfies (5.106). Then by Theorem 5.6.6 it follows that the assumptions of Theorem 5.6.2 are fulfilled. Therefore there exists a bounded solution $\hat{X}$ of (5.98) with the maximality property as in Theorem 5.6.2. The conclusion follows now applying Theorem 5.4.1 and thus the proof is complete.   □

The counterpart of the above theorem for the periodic case is the following.

Theorem 5.6.8. *Assume that the coefficients of (5.98) are* θ-*periodic functions. Then the following are equivalent*

(i) $(\mathbf{A}, \mathbf{B}; Q)$ *is stabilizable and the differential inequality (5.114) has a* θ-*periodic solution.*

(ii) Equation (5.98) has a stabilizing θ-*periodic solution* $\tilde{X}(t)$ *which verifies (5.106).*

Proof. (i) → (ii) Applying Theorem 5.6.6 (i) ⇒ (ii) we deduce that (5.98) has a stabilizing and bounded on $\mathcal{I}$ solution $\tilde{X}(t)$ which verifies (5.106). Using Theorem 5.6.5 (ii) we conclude that $\tilde{X}(t)$ is a θ-periodic function too.

(ii) → (i) follows from the implication (ii) → (iii) of Corollary 5.4.7 together with Lemma 5.6.1.   □

With the same proof as in the previous theorem we get the time-invariant counterpart of Theorem 5.6.6.

Theorem 5.6.9. *Assume that the coefficients of (5.98) do not depend upon* t. *Then the following are equivalent*

(i) The triple $(\mathbf{A}, \mathbf{B}; Q)$ *is stabilizable and it exists* $\widehat{X} \in \mathcal{S}_n^d$ *such that* $\Lambda^{\Sigma}\left(\widehat{X}\right) > 0;$

(ii) The system of generalized Riccati algebraic equations (5.110) has a stabilizing solution $\widetilde{X}$ which verifies $\mathcal{R}_i\left(\widetilde{X}\right) > 0$ for all $i \in \mathcal{D}$.

Remark 5.6.4. (a) From Corollary 5.4.9 and Lemma 5.6.1 we may conclude that if $A_k(\cdot,i), B_k(\cdot,i)$, $k = 0,1,\ldots,r$, are continuous θ-periodic functions and the triple $(\mathbf{A},\mathbf{B};Q)$ is stabilizable, then there exists a stabilizing feedback gain $\tilde{F}(t) = (\tilde{F}(t,1), \ldots, \tilde{F}(t,d))$ which is θ-periodic function too. Also, if $A_k(t,i) = A_k(i), B_k(t,i) = B_k(i), k = 0,1,\ldots,r$, $(t,i) \in \mathcal{I} \times \mathcal{D}$, and $(\mathbf{A},\mathbf{B};Q)$ is stabilizable, then there exists a stabilizing feedback gain, $F = (F(1), \ldots F(d))$. Therefore we may infer, without loss of generality, that in the case of periodic coefficients the triple $(\mathbf{A},\mathbf{B};Q)$ is stabilizable, if and only if there exists a stabilizing feedback gain $F(t)$ which is θ-periodic function; in the time invariant case the triple $(\mathbf{A},\mathbf{B};Q)$ is stabilizable if and only if there exists a stabilizing feedback gain $F = (F(1), \ldots F(d))$ not depending upon t.

(b) Combining the result in Corollary 5.4.9, Lemma 5.6.1 and Remark 4.1.4 we may conclude that if $A_k(\cdot,i), C_k(\cdot,i), k = 0,1,\ldots r$, are θ-periodic functions defined on $\mathbf{R} \times \mathcal{D}$, then the triple $(\mathbf{C},\mathbf{A};Q)$ is detectable if and only if there exists a stabilizing injection $K(t)$ which is continuous θ-periodic function. Also, in the time invariant case the triple $(\mathbf{C},\mathbf{A};Q)$ is detectable, if and only if there exists a stabilizing injection $K = (K(1), K(2), \ldots, K(d)) \in \mathcal{M}_{n,p}^d$.

5.6.4 The Minimal Solution of the SGRDE

In the following we focus our attention to the case when the coefficients of the SGRDE (5.98) (and equivalently of (5.102)) satisfy the additional conditions of type (5.74).

From (5.104) we see that conditions (5.74) are equivalent with the fact that $\hat{X}(t) \equiv 0$ belongs to $\mathbf{\Gamma}^{\Sigma}$. We start presenting several results with interest in themselves which follows immediately from the general results proved in Sect. 5.2.

Lemma 5.6.10. *Assume that (5.74) holds. Then*

(i) Let $X : \mathcal{I}_1 \subseteq \mathcal{I} \to \mathcal{S}_n^d$ be a solution of (5.98). If there exists $\tau \in \mathcal{I}_1$ such that $X(\tau,i) \geq 0, i \in \mathcal{D}$, then $X(t,i) \geq 0$ for all $t \in \mathcal{I}_1 \cap (-\infty, \tau]$.

(ii) Let $\hat{X} : \mathcal{I}_1 \subset \mathcal{I} \to \mathcal{S}_n^d, \check{X} : \mathcal{I}_1 \subset \mathcal{I} \to \mathcal{S}_n^d$ be two solutions of (5.98).
   If there exists $\tau \in \mathcal{I}_1$ such that $\check{X}(\tau) \geq \hat{X}(\tau) \geq 0$, then $\check{X}(t) \geq \hat{X}(t)$ for all $t \in \mathcal{I}_1 \cap (-\infty, \tau]$.

Proof. (i) follows directly applying Theorem 5.2.2.

(ii) is obtained applying Theorem 5.2.1 to SGRDE (5.98) regarded as special case of (5.1) and taking $\tilde{Q}(t) \equiv Q(t)$. $\qquad\square$

For each $\tau \in \mathcal{I}$ we denote $X_\tau(\cdot)$ the solution of (5.98) which verifies the condition $X_\tau(\tau, i) = 0$, $i \in \mathcal{D}$.

Proposition 5.6.11. *Assume that* $(\mathbf{A}, \mathbf{B}; Q)$ *is stabilizable and the conditions (5.74) are fulfilled. Then*

(i) *For each* $\tau \in \mathcal{I}$, *the solution* $X_\tau(\cdot)$ *is defined on* $\mathcal{I} \cap (-\infty, \tau]$. *Moreover there exists* $c > 0$, *such that* $0 \le X_\tau(t) \le c J^d$, $\forall\, t \le \tau$, $t \in \mathcal{I}$.

(ii) $X_{\tau_1}(t) \le X_{\tau_2}(t)$ $\forall\, t \le \tau_1 < \tau_2$, $t \in \mathcal{I}$.

Proof. (i) The fact that $t \to X_\tau(t)$ is well defined and $X_\tau(t) \ge 0$ for all $t \le \tau, t \in \mathcal{I}$ follows from Corollary 5.2.3. Since $(\mathbf{A}, \mathbf{B}; Q)$ is stabilizable, there exists $F^0 : \mathcal{I} \to \mathcal{M}^d_{m,n}$ continuous and bounded function, such that the system $(A_0 + B_0 F^0, A_1 + B_1 F^0, \ldots, A_r + B_r F^0; Q)$ is stable. Let $X^0(t)$ be the unique bounded on $\mathcal{I}$ solution of the affine Lyapunov-type differential equation

$$\frac{d}{dt} X^0(t) + \mathcal{L}^*_{F^0}(t) X^0(t) + M^0(t) = O$$

where $M^0(t) = (M^0(t,1), M^0(t,2), \ldots, M^0(t,d))$,

$$M^0(t,i) = M(t,i) + L(t,i) F^0(t,i) + (F^0(t,i))^T L^T(t,i) + (F^0(t,i))^T R(t,i) F^0(t,i).$$

Since (5.74) is fulfilled we obtain that $M^0(t) \ge 0$, $t \in \mathcal{I}$. Hence by Theorem 2.7.5 there exists $c > 0$ such that $0 \le X^0(t) \le c J^d$ for all $t \in \mathcal{I}$. By direct computation we obtain that $X^0(t) - X_\tau(t)$ verifies the affine differential equation of Lyapunov-type

$$\frac{d}{dt}(X^0(t) - X_\tau(t)) + \mathcal{L}^*_{F^0}(t)(X^0(t) - X_\tau(t)) + \tilde{M}^0(t) = 0, \qquad (5.115)$$

$t \in \mathcal{I}, t \le \tau$ where $\tilde{M}^0(t) = (\tilde{M}^0(t,1), \tilde{M}^0(t,2), \ldots, \tilde{M}^0(t,d))$

$$\tilde{M}^0(t,i) = (F^0(t,i) - F_\tau(t,i))^T \mathcal{R}_i(t, X_\tau(t))(F^0(t,i) - F_\tau(t,i)),$$

$(t,i) \in \mathcal{I} \times \mathcal{D}, t \le \tau$ where $F_\tau(t) = F^{X_\tau}(t)$. Since $X_\tau(t) \ge 0$ we get $\tilde{M}^0(t) \ge 0$, $t \in \mathcal{I}, t \le \tau$.

From (5.115) we deduce that

$$X^0(t) - X_\tau(t) \ge 0 \qquad (5.116)$$

$\forall t \in \mathcal{I}, t \le \tau$ which leads to $0 \le X_\tau(t) \le X^0(t) \le c J^d$, $\forall\, t \in \mathcal{I}, t \le \tau$.

(ii) follows immediately from Lemma 5.6.10 and the proof is complete. $\square$

Now we have the following result.

Theorem 5.6.12. *Assume that* $(\mathbf{A}, \mathbf{B}; Q)$ *is stabilizable and the conditions* (5.74) *are fulfilled. Under these assumptions the SGRDE* (5.98) *has two bounded solutions* $\tilde{X} : \mathcal{I} \to \mathcal{S}_n^d$, $\tilde{\tilde{X}} : \mathcal{I} \to \mathcal{S}_n^d$ *with the property* $\tilde{X}(t) \geq \hat{X}(t) \geq \tilde{\tilde{X}}(t) \geq 0$ *for all* $t \in \mathcal{I}$, $\hat{X}(t)$ *being any bounded and positive semidefinite solution of* (5.98).

Proof. follows applying Theorem 5.5.1 to the SGRDE (5.98). □

To solve the linear quadratic problems, a crucial role is played by the minimal solution, stabilizing solution, respectively, of the following system of matrix nonlinear differential equations

$$
\begin{aligned}
&\tfrac{d}{dt}X(t,i) + A_0^T(t,i)X(t,i) + X(t,i)A_0(t,i) + \sum_{k=1}^r A_k^T(t,i)X(t,i)A_k(t,i) \\
&+ \sum_{j=1}^d q_{ij}X(t,j) - [X(t,i)B_0(t,i) + \sum_{k=1}^r A_k^T(t,i)X(t,i)B_k(t,i)] \\
&\qquad \times [R(t,i) + \sum_{k=1}^r B_k^T(t,i)X(t,i)B_k(t,i)]^{-1} \\
&\qquad \times [B_0^T(t,i)X(t,i) + \sum_{k=1}^r B_k^T(t,i)X(t,i)A_k(t,i)] + C_0^T(t,i)C_0(t,i) = 0
\end{aligned} \tag{5.117}
$$

$t \geq 0, i \in \mathcal{D}$ where $R(t,i) = D_0^T(t,i)D_0(t,i)$, $\mathcal{I} = \mathbf{R}_+$.

Remark 5.6.5. It is worth mentioning that under the assumption $0 \in \boldsymbol{\Gamma}^\Sigma$ the SGRDE (5.98) may be rewritten as a system of coupled Riccati differential equation of type (5.117). Indeed, applying Lemma 5.4.5 to (5.98) regarded as a GRDE of type (5.1) with the operator $\Pi(t)$ described in (5.5) and $A(t), B(t)$ given in (5.100) we obtain that $X(\cdot)$ is a solution of SGRDE (5.98) if and only if it is a solution of the following system of Riccati-type equations

$$
\begin{aligned}
&\tfrac{d}{dt}X(t,i) + [A_0(t,i) + B_0(t,i)W(t,i)]^T X(t,i) + X(t,i)[A_0(t,i) + B_0(t,i)W(t,i)] \\
&+ \sum_{k=1}^r [A_k(t,i) + B_k(t,i)W(t,i)]^T X(t,i)[A_k(t,i) + B_k(t,i)W(t,i)] + \sum_{j=1}^d q_{ij}X(t,j) \\
&- (X(t,i)B_0(t,i) + \sum_{k=1}^r (A_k(t,i) + B_k(t,i)W(t,i))^T X(t,i)B_k(t,i)) \\
&\times [R(t,i) + \sum_{k=1}^r B_k^T(t,i)X(t,i)B_k(t,i)]^{-1} [B_0^T(t,i)X(t,i) \\
&+ \sum_{k=1}^r B_k^T(t,i)X(t,i)(A_k(t,i) + B_k(t,i)W(t,i))] \\
&+ M(t,i) - L(t,i)R^{-1}(t,i)L^T(t,i) = 0
\end{aligned} \tag{5.118}
$$

$i \in \mathcal{D}$, where $W(t,i) = -R^{-1}(t,i)L^T(t,i)$. It is clear that the SGRDE (5.118) with $D_0(t,i) = R^{1/2}(t,i)$, $C_0(t,i)$ defined via the factorization $C_0^T(t,i)C_0(t,i) = M(t,i) - L(t,i)R^{-1}(t,i)L^T(t,i)$ and $A_k(t,i) + B_k(t,i)W(t,i)$ replaced by $A_k(t,i)$ is of type (5.117).

The next result will be used in the following developments.

Lemma 5.6.13. *Assume that*

(a) *There exists* $\rho > 0$ *such that* $D_0^T(t,i)D_0(t,i) \geq \rho I_m$ *for all* $t \geq 0, i \in \mathcal{D}$.
(b) *The triple* $(C_0, \mathbf{A}; Q)$ *is detectable.*
(c) *The elements of the matrix* Q *verify* $q_{ij} \geq 0, i \neq j, i, j \in \mathcal{D}$.

Under these assumptions any positive semidefinite and bounded solution of the system (5.117) *is stabilizing.*

Proof. If $(C_0, \mathbf{A}; Q)$ is detectable, then there exists a bounded and continuous function $K : \mathbf{R}_+ \to \mathcal{M}_{np}^d$ such that the operator valued function $\mathcal{L}^K(\cdot)$ generates an exponentially stable evolution where $(\mathcal{L}^K(t)X)(i) = (A_0(t,i) + K(t,i)C_0(t,i))X(i) + X(i)(A_0(t,i) + K(t,i)C_0(t,i))^T + \sum_{k=1}^r A_k(t,i)X(i)A_k^T(t,i) + \sum_{j=1}^d q_{ji}X(j)$. A direct calculation leads to

$$
\begin{aligned}
(\mathcal{L}^K(t)X)(i) &= (A(t,i) + K(t,i)C_0(t,i))X(i) + X(i)(A(t,i) + K(t,i)C_0(t,i))^T \\
&\quad + \sum_{j=1, j\neq i}^d q_{ji}X(j) + \sum_{k=1}^r A_k(t,i)X(i)A_k^T(t,i) := \mathcal{L}_{A+KC_0,\Pi_1^*}(t)[X](i),
\end{aligned}
\tag{5.119}
$$

where $A(t,i) = A_0(t,i) + 1/2q_{ii}I_n$ and $\Pi_1^*(t)$ is the adjoint of the linear operator $X \to \Pi_1(t)[X] : \mathcal{S}_n^d \to \mathcal{S}_n^d$ defined by $\Pi_1(t)[X](i) = \sum_{j=1, j\neq i}^d q_{ij}X(j) + \sum_{k=1}^r A_k(t,i)^T X(i)A_k(t,i)$ which coincides with the element $(1,1)$ of the operator $\Pi(t)$ described in (5.5) in the case $\mathcal{D} = \{1, 2, \ldots, d\}$. Employing (5.119) and the Definition 5.5.1 we may conclude that $(C_0, \mathbf{A}; Q)$ is detectable in the sense of Definition 4.1.2 (b) if and only if the triple $(C_0, \mathbf{A}; \Pi_1)$ is detectable in the sense of Definition 5.5.1. The conclusion follows applying Lemma 5.5.2 in the case of (5.117) regarded as a special case of GRDE (5.1) thus the proof ends. $\square$

Proposition 5.6.14. *Assume that*

(a) *$D_0^T(t,i)D_0(t,i) \geq \rho I_m$ for all $(t,i) \in \mathbf{R}_+ \times \mathcal{D}$*
(b) *The elements of the matrix Q satisfy $q_{ij} \geq 0$ if $i \neq j, i, j \in \mathcal{D}$ and $\sum_{j=1}^d q_{ij} = 0, i \in \mathcal{D}$.*
(c) *$(C_0, A_0, A_1, \ldots, A_r; Q)$ is uniformly observable.*

Then if K is a positive semidefinite and bounded on $\mathbf{R}_+$ solution of system (5.117) we have

 (i) *K is uniform positive;*
(ii) *K is a stabilizing solution.*

Proof. Let K be a positive semidefinite and bounded on $\mathbf{R}_+$ solution of system (5.117). Set

$$
F_K(t,i) = -\mathcal{R}_i^{-1}(t, K(t))\mathcal{P}_i(t, K(t)),
$$

$$
\widetilde{A}_k(t,i) = A_k(t,i) + B_k(t,i)F_K(t,i), \; 0 \leq k \leq r
$$

and $\widetilde{X}(t, t_0)$ be the fundamental matrix solution associated with the linear system

$$
dx(t) = \widetilde{A}_0(t, \eta(t))x(t)dt + \sum_{k=1}^r \widetilde{A}_k(t, \eta(t))x(t)dw_k(t).
$$

We have to prove that $\left(\widetilde{A}_0, \widetilde{A}_1, \ldots \widetilde{A}_r; Q\right)$ is stable.

Let $\tau > 0$ and $\beta > 0$ verifying the inequality in Proposition 4.2.1. Define

$$G(t,i) = E\left[\int_t^{t+\tau} \widetilde{X}^T(s,t)[C_0^T(s,\eta(s))C_0(s,\eta(s)) + F_K^T(s,\eta(s))R(s,\eta(s))\right.$$

$$\left. \times F_K(s,\eta(s))]\widetilde{X}(s,t)ds|\eta(t) = i], t \geq 0, \ i \in \mathcal{D}\right).$$

We shall prove that $\inf\{x^T G(t,i)x; |x| = 1, t \geq 0, i \in \mathcal{D}\} > 0$. Suppose on the contrary that for every $\varepsilon > 0$ there exist $x_\varepsilon \in \mathbf{R}^n, |x_\varepsilon| = 1, t_\varepsilon \geq 0$ and $i_\varepsilon \in \mathcal{D}$ such that $x_\varepsilon^T G(t_\varepsilon, i_\varepsilon)x_\varepsilon < \varepsilon$.

Let $x_\varepsilon(t) = \widetilde{X}(t, t_\varepsilon)x_\varepsilon$ and $u_\varepsilon(t) = F_K(t, \eta(t))x_\varepsilon(t)$. We can write

$$\varepsilon > x_\varepsilon^T G(t_\varepsilon, i_\varepsilon)x_\varepsilon \geq E\left[\int_{t_\varepsilon}^{t_\varepsilon+\tau} u_\varepsilon^T(t)R(t,\eta(t))u_\varepsilon(t)dt|\eta(t_\varepsilon) = i_\varepsilon\right]$$
$$\geq \delta E\left[\int_{t_\varepsilon}^{t_\varepsilon+\tau} |u_\varepsilon(t)|^2 dt|\eta(t_\varepsilon) = i_\varepsilon\right]$$

with some $\delta > 0$. On the other hand, $x_\varepsilon(t) = \Phi(t, t_\varepsilon)x_\varepsilon + \widehat{x}_\varepsilon(t), \quad t \geq t_\varepsilon$ where $\widehat{x}_\varepsilon(t_\varepsilon) = 0$ and

$$d\widehat{x}_\varepsilon(t) = (A_0(t,\eta(t))\widehat{x}_\varepsilon(t) + B_0(t,\eta(t))u_\varepsilon(t))dt$$

$$+ \sum_{k=1}^r [A_k(t,\eta(t))\widehat{x}_\varepsilon(t) + B_k(t,\eta(t))u_\varepsilon(t)]dw_k(t).$$

Hence, by Remark 3.6.1 it exists $\gamma_0 > 0$ such that

$$E\left[|\widehat{x}_\varepsilon(t)|^2|\eta(t_\varepsilon) = i_\varepsilon\right] \leq \gamma_0 E\left[\int_{t_\varepsilon}^{t_\varepsilon+\tau} |u_\varepsilon(t)|^2 dt|\eta(t_\varepsilon) = i_\varepsilon\right] \leq \delta_1\varepsilon.$$

Further, we can write

$$\varepsilon > x_\varepsilon^T G(t_\varepsilon, i_\varepsilon)x_\varepsilon \geq E\left[\int_{t_\varepsilon}^{t_\varepsilon+\tau} |C_0(t,\eta(t))x_\varepsilon(t)|^2 dt|\eta(t_\varepsilon) = i_\varepsilon\right]$$

$$= E\left[\int_{t_\varepsilon}^{t_\varepsilon+\tau} |C_0(t,\eta(t))\Phi(t,t_\varepsilon)x_\varepsilon + C_0(t,\eta(t))\widehat{x}_\varepsilon(t)|^2 dt|\eta(t_\varepsilon) = i_\varepsilon\right]$$

$$\geq \frac{1}{2}E\left[\int_{t_\varepsilon}^{t_\varepsilon+\tau} |C_0(t,\eta(t))\Phi(t,t_\varepsilon)x_\varepsilon|^2 dt|\eta(t_\varepsilon) = i_\varepsilon\right]$$

$$-E\left[\int_{t_\varepsilon}^{t_\varepsilon+\tau} |C_0(t,\eta(t))\widehat{x}_\varepsilon(t)|^2 dt|\eta(t_\varepsilon) = i_\varepsilon\right]$$

$$\geq \frac{1}{2}\beta - \delta_2\varepsilon, \quad \varepsilon > 0.$$

Thus we get a contradiction, because $\beta > 0$. Hence, there exists $\beta_1 > 0$ such that $G(t,i) \geq \beta_1 I_n, \ t \geq 0, i \in \mathcal{D}$. Applying the identity (1.6) to the function $v(t,x,i) = x^T K(t,i)x$ and to the system

$$dx(t) = \widetilde{A}_0(t,\eta(t))x(t)dt + \sum_{k=1}^r \widetilde{A}_k(t,\eta(t))x(t)dw_k(t)$$

and taking into account (5.117) for $K(t,i)$ we get

$$x_0^T E\left[\widetilde{X}^T(t+\tau,t)K(t+\tau,\eta(t+\tau))\widetilde{X}(t+\tau,t)|\eta(t)=i\right]x_0 - x_0^T K(t,i)x_0 =$$
$$-x_0^T G(t,i)x_0, \quad t \geq 0, \ x_0 \in \mathbf{R}^n, \ i \in \mathcal{D}.$$

Therefore

$$\beta_1|x_0|^2 \leq x_0^T K(t,i)x_0 \leq \beta_2|x_0|^2, \quad t \geq 0, i \in \mathcal{D}, x_0 \in \mathbf{R}^n$$

Thus K is a uniform positive function and

$$E\left[\widetilde{X}^T(t+\tau,t)K(t+\tau,\eta(t+\tau))\widetilde{X}(t+\tau,t)|\eta(t)=i\right] \leq \left(1 - \frac{\beta_1}{\beta_2}\right)K(t,i).$$

By virtue of Theorems 3.1.1 and 3.3.8 it follows that $\left(\widetilde{A}_0, \widetilde{A}_1, \ldots \widetilde{A}_r; Q\right)$ is stable and thus the proof is complete. □

Theorem 5.6.15. *Assume that*

(a) *Either the assumptions of Lemma 5.6.13 or the assumptions of Proposition 5.6.14 are fulfilled.*
(b) *The triple $(\mathbf{A}, \mathbf{B}; Q)$ is stabilizable.*
      Then the Riccati-type system (5.117) has a unique positive semidefinite and bounded on $\mathbf{R}_+$ solution. Moreover this solution is stabilizing.

Proof. The proof follows immediately from Theorem 5.6.12, Proposition 5.6.14, Lemma 5.6.13, and Theorem 5.6.5. □

In the particular case when $\mathcal{D} = \{1\}$, the system (5.117) becomes

$$\frac{d}{dt}X(t) + A_0^T(t)X(t) + X(t)A_0(t) + \sum_{k=1}^r A_k^T(t)X(t)A_k(t)$$
$$-[X(t)B_0(t) + \sum_{k=1}^r A_k^T(t)X(t)B_k(t)]$$
$$\times [R(t) + \sum_{k=1}^r B_k^*(t)X(t)B_k(t))]^{-1}$$
$$\times [B_0^T(t)X(t) + \sum_{k=1}^r B_k^T(t)X(t)A_k(t)] + C_0^T(t)C_0(t) = 0$$

$$(5.120)$$

A direct consequence of Theorem 5.6.15 is the following corollary.

Corollary 5.6.16. *Assume that*

(a) *There exists $\rho > 0$ such that $D_0^T(t)D_0(t) \geq \rho I_m$ for all $t \geq 0$.*
(b) *The pair $(\mathbf{A}, \mathbf{B})$ is stabilizable.*
(c) *The pair $(C_0, \mathbf{A})$ is either detectable or uniformly observable.*

Then the Riccati-type equation (5.120) has a unique positive semidefinite and bounded on $\mathbf{R}_+$ solution. Moreover this solution is stabilizing.

Remark 5.6.6. Based on Theorem 5.6.12 one obtains that under the assumption that $(\mathbf{A}, \mathbf{B}; Q)$ is stabilizable, the SGRDE (5.117) has two remarkable positive semidefinite and bounded solutions. We refer to the maximal solution $\widetilde{X}(t)$ and to the minimal solution $\widetilde{X}(t)$, respectively. If additionally $(C_0, \mathbf{A}; Q)$ is either detectable or uniformly observable, then these two solutions coincide, namely $\tilde{X}(t) \equiv \widetilde{X}(t)$. However in the absence of property of detectability and uniform observability $\tilde{X}(t)$ does not always coincide with $\widetilde{X}$. This can be seen in the following numerical example.

Illustrative Example.   Consider $n = 2, d = 1, r = 1, p = 1, m = 1$. In this case (5.117) reduces to:

$$\frac{d}{dt}X(t) + A_0^T(t)X(t) + X(t)A_0(t) + A_1^T(t)X(t)A_1(t)$$

$$- [X(t)B_0(t) + A_1^T(t)X(t)B_1(t)][R(t) + B_1^T(t)X(t)B_1(t)]^{-1}$$

$$\times [B_0^T(t)X(t) + B_1^T(t)X(t)A_1(t)] + C_0^T(t)C_0(t) = 0. \tag{5.121}$$

Choose

$$A_0(t) = \begin{bmatrix} 1 & 0 \\ 0 & 3 \end{bmatrix}, A_1(t) = I_2, B_0(t) = \begin{bmatrix} 2 \\ 1 \end{bmatrix}, B_1(t) = \begin{bmatrix} 0 \\ 0 \end{bmatrix},$$

$$C_0(t) = \begin{bmatrix} 1 & 0 \end{bmatrix}, R(t) = 1.$$

One can see (see Propositions 4.1.5 and 4.3.9) that in the stochastic case the pair $(C_0; (\mathbf{A}_0, \mathbf{A}_1))$ is neither detectable nor observable. The maximal solution of (5.121) is

$$\tilde{X}(t) = \begin{bmatrix} 8 & -21 \\ -21 & 63 \end{bmatrix} > 0$$

and the minimal solution is

$$\widetilde{X}(t) = \begin{bmatrix} 1 & 0 \\ 0 & 0 \end{bmatrix} \geq 0.$$

Indeed, by Theorem 2.7.7 (iv), $\tilde{X}$ is the stabilizing solution of (5.121) and based on Corollary 5.6.7 it coincides with the maximal solution.

On the other hand if $X_\tau(\cdot)$ is the solution of (5.121) with the given final condition $X_\tau(\tau) = 0$, one obtains that

$$X_\tau(t) = \begin{bmatrix} X(t) & 0 \\ 0 & 0 \end{bmatrix}$$

where

$$X(t) = \frac{1 - e^{-5(\tau - t)}}{1 + 4e^{-5(\tau - t)}} \quad \text{for all } t \le \tau.$$

Therefore

$$\lim_{\tau \to \infty} X_\tau(t) = \begin{bmatrix} 1 & 0 \\ 0 & 0 \end{bmatrix} = \widetilde{X}$$

and thus one obtains that $\widetilde{X}$ is the minimal semipositive definite solution of (5.121). Obviously in this case $\check{X} \ne \widetilde{X}$.

5.7   The Filtering Riccati Equation

In this section we focus our attention on the so-called *stochastic generalized filtering Riccati equation* (SGFRE) for stochastic systems. We shall restrict our investigation only to the time-invariant case and $\mathcal{D} = \{1, 2, \ldots, d\}$.

Consider the SGFRE

$$
\begin{aligned}
& A_0(i) Y(i) + Y(i) A_0^T(i) + \sum_{k=1}^{r} A_k(i) Y(i) A_k^T(i) + \sum_{j=1}^{d} q_{ji} Y(j) \\
& - \left(Y(i) C_0^T(i) + \sum_{k=1}^{r} A_k(i) Y(i) C_k^T(i) + \widetilde{L}(i) \right) \\
& \times \left(\widetilde{R}(i) + \sum_{k=1}^{r} C_k(i) Y(i) C_k^T(i) \right)^{-1} \\
& \times \left(Y(i) C_0^T(i) + \sum_{k=1}^{r} A_k(i) Y(i) C_k^T(i) + \widetilde{L}(i) \right)^T + \widetilde{M}(i) = 0
\end{aligned}
\tag{5.122}
$$

with the unknown variables $(Y(1), \ldots, Y(d)) \in \mathcal{S}_n^d$ and $A_k(i) \in \mathbf{R}^{n \times n}$, $C_k(i) \in \mathbf{R}^{p \times n}$, $k = 0, \ldots, r$, $\widetilde{L}(i) \in \mathbf{R}^{n \times p}$, $\widetilde{R}(i) \in \mathcal{S}_p$, $\widetilde{M}(i) \in \mathcal{S}_n$. If $\mathcal{D} = \{1\}$, $A_k(i) = 0$, $C_k(i) = 0$, $k = 1, 2, \ldots, r$ then (5.122) reduces to the well-known Bucy–Kalman [151] filtering algebraic Riccati equation.

The system (5.122) can be rewritten in a compact form as a nonlinear equation in $\mathcal{S}_n^d$ as follows:

$$\mathcal{L} Y - \widetilde{\mathcal{P}}(Y) \widetilde{\mathcal{R}}^{-1}(Y) \widetilde{\mathcal{P}}^*(Y) + \widetilde{M} = 0 \tag{5.123}$$

where $\mathcal{L}$ is the Lyapunov operator defined by the system $(A_0, A_1, \ldots, A_r; Q)$, $\widetilde{\mathcal{P}} : \mathcal{S}_n^d \to \mathcal{M}_{np}^d$ by

$$\widetilde{\mathcal{P}}(Y) = \left(\widetilde{\mathcal{P}}_1(Y), \ldots, \widetilde{\mathcal{P}}_d(Y) \right),$$

$$\widetilde{\mathcal{P}}_i(Y) = Y(i) C_0^T(i) + \sum_{k=1}^{r} A_k(i) Y(i) C_k^T(i) + \widetilde{L}(i), i \in \mathcal{D}$$

$$\widetilde{\mathcal{R}} : \mathcal{S}_n^d \to \mathcal{S}_p \text{ by}$$

$$\widetilde{\mathcal{R}}(Y) = \left(\widetilde{\mathcal{R}}_1(Y), \dots, \widetilde{\mathcal{R}}_d(Y) \right),$$

$$\widetilde{\mathcal{R}}_i(Y) = \widetilde{R}(i) + \sum_{k=1}^{r} C_k(i) Y(i) C_k^T(i), \; i \in \mathcal{D}$$

$$\widetilde{M} = \left(\widetilde{M}(1), \dots, \widetilde{M}(d) \right).$$

Equation (5.123) is defined on a subset of $\mathcal{S}_n^d$ consisting in $Y = (Y(1), \dots, Y(d))$ such that $\det \widetilde{\mathcal{R}}_i(Y) \neq 0$.

The dissipation matrix corresponding to the filtering Riccati equation under investigation is defined as follows

$$\widetilde{\mathcal{N}}(Y) = \left(\widetilde{\mathcal{N}}_1(Y), \dots, \widetilde{\mathcal{N}}_d(Y) \right) \text{ where}$$

$$\widetilde{\mathcal{N}}_i(Y) = \begin{bmatrix} (\mathcal{L}Y)(i) + \widetilde{M}(i) & \widetilde{\mathcal{P}}_i(Y) \\ \widetilde{\mathcal{P}}_i^T(Y) & \widetilde{\mathcal{R}}_i(Y) \end{bmatrix}$$

for all $Y \in \mathcal{S}_n^d, i \in \mathcal{D}$.

Definition 5.7.1. A solution $\widetilde{Y} = \left(\widetilde{Y}(1), \dots, \widetilde{Y}(d) \right)$ of (5.123) is a *stabilizing solution* if the system $\left(A_0 + \widetilde{K}C_0, A_1 + \widetilde{K}C_1, \dots, A_r + \widetilde{K}C_r; Q \right)$ is stable in the sense of Definition 2.7.1, where $\widetilde{K} = \left(\widetilde{K}(1), \dots, \widetilde{K}(d) \right)$,

$$\widetilde{K}(i) = -\widetilde{\mathcal{P}}_i\left(\widetilde{Y} \right) \widetilde{\mathcal{R}}_i^{-1}\left(\widetilde{Y} \right), \; i \in \mathcal{D}. \tag{5.124}$$

Recalling that $\mathbf{A} = (A_0, \dots, A_r)$ and $\mathbf{C} = (C_0, \dots, C_r)$ we prove the following result.

Theorem 5.7.1. *The following are equivalent:*

(i) *$(\mathbf{C}, \mathbf{A}; Q)$ is detectable and it exists $\widehat{Y} = \left(\widehat{Y}(1), \dots, \widehat{Y}(d) \right) \in \mathcal{S}_n^d$ satisfying $\widetilde{\mathcal{N}}\left(\widehat{Y} \right) > 0$;*

(ii) *Equation (5.123) has a stabilizing solution $\widetilde{Y}$ which verifies $\widetilde{\mathcal{R}}_i\left(\widetilde{Y} \right) > 0$.*

Proof. It is easy to see that (5.123) is an equation of type (5.98), associated with the triple $\left(\mathbf{A}^\sharp, \mathbf{C}^\sharp; Q^\sharp \right)$ where $\mathbf{A}^\sharp = \left(A_0^\sharp, \dots, A_r^\sharp \right)$, $\mathbf{C}^\sharp = \left(C_0^\sharp, \dots, C_r^\sharp \right)$ and $Q^\sharp = Q^T, A_k^\sharp = (A_k^T(1), \dots, A_k^T(d)), C_k^\sharp = (C_k^T(1), \dots, C_k^T(d)), k = 0, \dots, r$. From Remark 4.1.4 it follows that $\left(\mathbf{A}^\sharp, \mathbf{C}^\sharp; Q^\sharp \right)$ is stabilizable if and only if $(\mathbf{C}, \mathbf{A}; Q)$ is detectable. The result in the statement follows then from Theorem 5.6.6. $\square$

5.8 Iterative Procedures

In the first part of this section we present an iterative procedure to compute the maximal solution $\tilde{X}(t)$ of (5.102) or equivalently the maximal solution of the SGRDE (5.98). This procedure may also provide a proof of the implication (i) $\to$ (ii) in Theorem 5.6.2.

Lemma 5.8.1. *Assume that the triple* $(\mathbf{A}, \mathbf{B}; Q)$ *is stabilizable. Let* $\tilde{F}_0(t) = (\tilde{F}_0(t, 1),$ $\tilde{F}_0(t, 2), \ldots, \tilde{F}_0(t, d))$ *be a stabilizing feedback gain and let* $X_0(t) = (X_0(t, 1), \ldots,$ $X_0(t, d))$ *be a bounded with bounded derivative solution of the linear differential inequality on* $\mathcal{S}_n^d$:

$$\frac{d}{dt}X_0(t) + \mathcal{L}_{\tilde{F}_0}^*(t)X_0(t) + M_0(t) \leq 0 \qquad (5.125)$$

where $\mathcal{L}_{\tilde{F}_0}^*(t)$ *is the adjoint of the operator defined in (4.2) with* $F(t)$ *replaced by* $\tilde{F}_0(t)$, $M_0(t) = (M_0(t, 1), M_0(t, 2), \ldots, M_0(t, d))$,

$$M_0(t, i) = M(t, i) + \varepsilon I_n + L(t, i)\tilde{F}_0(t, i) + \tilde{F}_0^T(t, i)L^T(t, i) + \tilde{F}_0(t, i)R(t, i)\tilde{F}_0(t, i),$$

$\varepsilon > 0$ *be fixed.*
    Under the considered assumptions we have

$$X_0(t) - \hat{X}(t) \gg 0 \qquad (5.126)$$

for arbitrary $\hat{X}(t) \in \mathbf{\Gamma}^\Sigma$ *which verifies the condition of type (5.106).*

Proof. If $\hat{X}(t) \in \mathbf{\Gamma}^\Sigma$, then applying Lemma 5.1.2 one obtains, via (5.104), that it solves the following linear differential inequality:

$$\frac{d}{dt}\hat{X}(t) + \mathcal{L}^*(t)\hat{X}(t) - \mathcal{P}^*(t, \hat{X}(t))\mathcal{R}^{-1}(t, \hat{X}(t))\mathcal{P}(t, \hat{X}(t)) + M(t) \geq 0, \ t \in \mathbf{R}_+. \qquad (5.127)$$

We define $\hat{M}(t) = (\hat{M}(t, 1), \hat{M}(t, 2), \ldots \hat{M}(t, d))$ by

$$\hat{M}(t) = \frac{d}{dt}\hat{X}(t) + \mathcal{L}^*(t)\hat{X}(t) - \mathcal{P}^*(t, \hat{X}(t))\mathcal{R}^{-1}(t, \hat{X}(t))\mathcal{P}(t, \hat{X}(t)) + M(t), t \in \mathbf{R}_+. \qquad (5.128)$$

Clearly $\hat{M}(t) \geq 0$. By Lemma 5.1.1 we infer that

$$\frac{d}{dt}\hat{X}(t) + \mathcal{L}_{\tilde{F}_0}^*(t)\hat{X}(t) + M(t) + L(t)\tilde{F}_0(t) + \tilde{F}_0^T(t)L^T(t) + \tilde{F}_0^T(t)R(t)\tilde{F}_0(t) - \hat{M}(t)$$
$$- (\hat{F}(t) - \tilde{F}_0(t))^T \mathcal{R}(t, \hat{X}(t))(\hat{F}(t) - \tilde{F}_0(t)) = 0 \qquad (5.129)$$

where $\hat{F}(t) = (\hat{F}(t, 1), \hat{F}(t, 2), \ldots \hat{F}(t, d))$ with

$$\hat{F}(t,i) = -\mathcal{R}_i^{-1}(t,\hat{X}(t))\mathcal{P}_i(t,\hat{X}(t)), \quad t \in \mathcal{I}, \ i \in \mathcal{D}. \tag{5.130}$$

From (5.129) and (5.125) we get:

$$\frac{d}{dt}(X_0(t) - \hat{X}(t)) + \mathcal{L}_{\tilde{F}_0}^*(t)(X_0(t) - \hat{X}(t)) + (\tilde{F}_0(t) - \hat{F}(t))^T \mathcal{R}(t,\hat{X}(t))(\tilde{F}_0(t)$$
$$-\hat{F}(t)) + \varepsilon J^d + \hat{M}(t) \leq 0, \quad t \geq 0.$$

This allows us by Theorem 2.7.5 to conclude that $X_0(t) - \tilde{X}(t) \geq Y(t)$ where $t \to Y(t) = (Y(t,1), Y(t,2), \ldots Y(t,d))$ is the unique bounded solution of the Lyapunov-type equation

$$\frac{d}{dt}Y(t) + \mathcal{L}_{\tilde{F}_0}^*(t)Y(t) + \varepsilon J^d = 0. \tag{5.131}$$

Let $T_0(t,s)$ be the linear evolution operator on $\mathcal{S}_n^d$, defined by the linear differential equation:

$$\frac{d}{dt}S(t) = \mathcal{L}_{\tilde{F}_0}(t)S(t).$$

Since $\tilde{F}_0(t)$ is a stabilizing feedback gain, then there exist positive constants β_0, α_0 such that $\|T_0(t,s)\| \leq \beta_0 e^{-\alpha_0(t-s)}, \forall t \geq s, \ t,s \in \mathcal{I}$. Therefore the unique bounded solution of (5.131) is uniform positive and the proof is complete. $\qquad\square$

Remark 5.8.1. Based on Remark 5.6.4 it follows that if $(A_k(t,i), B_k(t,i))$, $0 \leq k \leq r, i \in \mathcal{D}$ are θ-periodic functions then a stabilizing feedback gain which is θ-periodic function may be chosen. Therefore in the periodic case the inequality (5.125) has a periodic solution with the same period as the coefficients. Moreover if $A_k(t,i) = A_k(i), B_k(t,i) = B_k(i), t \in \mathcal{I}, 0 \leq k \leq r, i \in \mathcal{D}$ we may choose constant solutions of (5.125), $X_0 = (X_0(1), X_0(2) \ldots X_0(d))$. Detailing (5.125) in the time-invariant case, it follows that X_0 may be obtained as a solution of the following LMI system

$$[A_0(i) + B_0(i)\tilde{F}_0(i)]^T X_0(i) + X_0(i)[A_0(i) + B_0(i)\tilde{F}_0(i)]$$
$$+ \sum_{k=1}^r [A_k(i) + B_k(i)\tilde{F}_0(i)]^T X_0(i)[A_k(i) + B_k(i)\tilde{F}_0(i)] + \sum_{j=1}^d q_{ij}X_0(j) \tag{5.132}$$
$$+ M(i) + \varepsilon I_n + L(i)\tilde{F}_0(i) + \tilde{F}_0^T(i)L^T(i) + \tilde{F}_0^T(i)R(i)\tilde{F}_0(i) \leq 0, \ i \in \mathcal{D}.$$

Based on (5.126) we deduce that there exists $\mu_0 > 0$, such that $\mathcal{R}_i(t,X_0(t)) \geq \mu_0 I_m, \ t \in \mathcal{I}, \ i \in \mathcal{D}$. Hence the feedback gain $F_0(t) = (F_0(t,1), \ldots F_0(t,d))$ is well defined by

$$F_0(t,i) = -\mathcal{R}_i^{-1}(t,X_0(t))\mathcal{P}_i(t,X_0(t)), \ i \in \mathcal{D}, \ t \in \mathcal{I}. \tag{5.133}$$

We shall show that $F_0(t)$ is a stabilizing feedback gain for the triple $(\mathbf{A}, \mathbf{B}; Q)$.

To this end we consider $\hat{X}(t) \in \mathbf{\Gamma}^\Sigma$. By direct computation and using (5.127), (5.128) and (5.133) we get

$$\frac{d}{dt}\hat{X}(t) + \mathcal{L}_{F_0}^*(t)\hat{X}(t) + M(t) + L(t)F_0(t) + F_0^T(t)L^T(t) +$$
$$F_0^T(t)R(t)F_0(t) - (\hat{F}(t) - F_0(t))^T \mathcal{R}(t,\hat{X}(t))(\hat{F}(t) - F_0(t)) - \hat{M}(t) = 0. \tag{5.134}$$

Further (5.125) may be rewritten as:

$$\frac{d}{dt}X_0(t) + \mathcal{L}_{F_0}^*(t)X_0(t) + M(t) + L(t)F_0(t) + F_0^T(t)L^T(t) + F_0^T(t)R(t)F_0(t)$$

$$+(F_0(t) - \tilde{F}_0(t))^T \mathcal{R}(t,X_0(t))(F_0(t) - \tilde{F}_0(t)) + \varepsilon J^d \leq 0. \tag{5.135}$$

From (5.134), (5.135), and (5.126) we deduce that $t \to X_0(t) - \hat{X}(t)$ is a bounded and uniform positive solution of the linear differential inequality on $\mathcal{S}_n^d$:

$$\frac{d}{dt}X(t) + \mathcal{L}_{F_0}^*(t)X(t) + \frac{\varepsilon}{2}J^d \ll 0.$$

Using Theorem 2.7.4 (viii) $\to$ (i) we deduce that the system $(A_0 + B_0F_0,\ A_1 + B_1F_0, \ldots, A_r + B_rF_0; Q)$ is stable which shows that $F_0(t)$ is a stabilizing feedback gain. As a consequence we deduce that for each $i \in \mathcal{D}$, the zero state equilibrium of the linear differential equation on $\mathbf{R}^n$

$$\frac{d}{dt}x(t) = \left(A_0(t,i) + \frac{1}{2}q_{ii}I_n + B_0(t,i)F_0(t,i)\right)x(t)$$

is exponentially stable.

Particularly in the time invariant case it follows that the eigenvalues of the matrices $A_0(i) + \frac{1}{2}q_{ii}I_n + B_0(i)F_0(i)$ are located in the half plane $Re\lambda < 0$.

Taking $X_0(t), F_0(t)$ as a first step, we construct iteratively the sequences: $\{X_l(t,i)\}_{l\geq 0}$, $\{F_l(t,i)\}_{l\geq 0}$, $i \in \mathcal{D}$ as follows: $t \to X_{l+1}(t,i)$ is the unique bounded solution of the Lyapunov equation

$$\frac{d}{dt}X_{l+1}(t,i) + \left[\widetilde{A}_0(t,i) + B_0(t,i)F_l(t,i)\right]^T X_{l+1}(t,i)$$
$$+X_{l+1}(t,i)\left[\widetilde{A}_0(t,i) + B_0(t,i)F_l(t,i)\right] + M_{l+1}(t,i) = 0 \tag{5.136}$$

where $M_{l+1}(t) = (M_{l+1}(t,1), \ldots, M_{l+1}(t,d))$ with

$$M_{l+1}(t,i) = M(t,i) + \frac{\varepsilon}{l+2}I_n + L(t,i)F_l(t,i)$$
$$+F_l^T(t,i)L^T(t,i) + F_l^T(t,i)R(t,i)F_l(t,i)$$
$$+\sum_{k=1}^r [A_k(t,i) + B_k(t,i)F_l(t,i)]^T X_l(t,i)[A_k(t,i) + B_k(t,i)F_l(t,i)]$$
$$+\sum_{j\neq i}q_{ij}X_l(t,j) \tag{5.137}$$

$$\tilde{A}_0(t,i) = A_0(t,i) + \frac{1}{2}q_{ii}I_n$$

$$F_{l+1}(t,i) = -(R(t,i) + \sum_{k=1}^r B_k^T(t,i)X_l(t,i)B_k(t,i))^{-1}(B_0^T(t,i)X_{l+1}(t,i)$$
$$+\sum_{k=1}^r B_k^T(t,i)X_l(t,i)A_k(t,i) + L^T(t,i)), \quad l \geq 0,\ i \in \mathcal{D}.$$

Further we show that

(a) $X_l(t,i) - \hat{X}(t,i) \geq \mu_l I_n > 0$ for all integers $l \geq 0$, $i \in \mathcal{D}, l; t \in \mathcal{I}, \hat{X}(t) = (\hat{X}(t,1) \ldots \hat{X}(t,d))$ being an arbitrary bounded function in $\mathbf{\Gamma}^\Sigma$ and μ_l is a positive constant which does not depend upon $\hat{X}(t)$.

(b) The zero state equilibrium of the linear differential equation on $\mathbf{R}^n$

$$\frac{d}{dt}x(t) = [\tilde{A}_0(t,i) + B_0(t,i)F_l(t,i)]x(t)$$

is exponentially stable for each $i \in \mathcal{D}$, $l \geq 0$.

(c) $X_l(t,i) \geq X_{l+1}(t,i), \forall l \geq 0, (t,i) \in \mathcal{I} \times \mathcal{D}$. We remark that the properties (a) and (b) have already been proved for $l = 0$. We shall verify by induction that (a), (b), (c) are fulfilled for every $l \geq 0$.

Let us assume that (a), (b), (c) are fulfilled for the first $l - 1$ terms of the sequences defined by (5.136) and (5.137). By direct computation we obtain that if $\hat{X}(t) \in \mathbf{\Gamma}^\Sigma$ then

$$\frac{d}{dt}\hat{X}(t,i) + (\tilde{A}_0(t,i) + B_0(t,i)F_{l-1}(t,i))^T \hat{X}(t,i)$$

$$+\hat{X}(t,i)(\tilde{A}_0(t,i) + B_0(t,i)F_{l-1}(t,i))$$

$$+\sum_{k=1}^{r} [A_k(t,i) + B_k(t,i)F_{l-1}(t,i)]^T \hat{X}(t,i) [A_k(t,i) + B_k(t,i)F_{l-1}(t,i)]$$

$$+\sum_{j=1,j\neq i}^{d} q_{ij}\hat{X}(t,j) + M(t,i) + L(t,i)F_{l-1}(t,i) + F_{l-1}^T(t,i)L^T(t,i)$$

$$+F_{l-1}^T(t,i)R(t,i)F_{l-1}(t,i) - \hat{M}(t,i)$$

$$-(\hat{F}(t,i) - F_{l-1}(t,i))^T \mathcal{R}_i(t,\hat{X}(t))(\hat{F}(t,i) - F_{l-1}(t,i)) = 0,$$

$\hat{M}(t,i), \hat{F}(t,i)$ being defined in (5.128) and (5.130), respectively.

Using (5.136) with l replaced by $l - 1$ we get

$$\frac{d}{dt}[X_l(t,i) - \hat{X}(t,i)] + [\tilde{A}_0(t,i) + B_0(t,i)F_{l-1}(t,i)]^T [X_l(t,i) - \hat{X}(t,i)]$$
$$+[X_l(t,i) - \hat{X}(t,i)] [\tilde{A}_0(t,i) + B_0(t,i)F_{l-1}(t,i)] + \frac{\varepsilon}{l+1}I_n + \Delta_l(t,i) = 0 \quad (5.138)$$

where we denoted

$$\Delta_l(t,i) = \sum_{k=1}^{r} [A_k(t,i) + B_k(t,i)F_{l-1}(t,i)]^T [X_{l-1}(t,i) - \hat{X}(t,i)]$$

$$\times [A_k(t,i) + B_k(t,i)F_{l-1}(t,i)] + \sum_{j=1,j\neq i}^{d} q_{ij}(X_{l-1}(t,j) - \hat{X}(t,j))$$

$$+\hat{M}(t,i) + (\hat{F}(t,i) - F_{l-1}(t,i))^T \mathcal{R}_i(t,\hat{X}(t))(\hat{F}(t,i) - F_{l-1}(t,i)).$$

Since $X_{l-1}(t,i) - \hat{X}(t,i) \geq \mu_{l-1}I_n$ we get $\Delta_l(t,i) \geq 0$. Taking into account that $\tilde{A}_0(t,i) + B_0(t,i)F_{l-1}(t,i)$ generates an exponentially stable evolution we may conclude that (5.138) has a unique bounded solution which is uniform positive definite. Hence there exists $\mu_l > 0$, such that $X_l(t,i) - \hat{X}(t,i) \geq \mu_l I_n$ and thus (a) is fulfilled. Further we have that $\mathcal{R}_i(t,X_l(t)) \geq \nu_l I_m > 0$.

Using (5.137) we write

$$
\begin{aligned}
&\frac{d}{dt}X_l(t,i) + [\tilde{A}_0(t,i) + B_0(t,i)F_l(t,i)]^T X_l(t,i) + X_l(t,i)[\tilde{A}_0(t,i) + B_0(t,i)F_l(t,i)] \\
&+ \sum_{k=1}^{r}[A_k(t,i) + B_k(t,i)F_l(t,i)]^T X_{l-1}(t,i)[A_k(t,i) + B_k(t,i)F_l(t,i)] \\
&+ \sum_{j=1,j\neq i}^{d} q_{ij}X_{l-1}(t,j) + M(t,i) + \frac{\varepsilon}{l+1}I_n + L(t,i)F_l(t,i) + F_l^T(t,i)L^T(t,i) \\
&+ F_l^T(t,i)R(t,i)F_l(t,i) + [F_l(t,i) - F_{l-1}(t,i)]^T \mathcal{R}_i(t,X_{l-1}(t))[F_l(t,i) - \\
&F_{l-1}(t,i)] = 0.
\end{aligned}
$$

(5.139)

It is easy to see that $t \to \hat{X}(t,i)$ verifies the equation

$$
\begin{aligned}
&\frac{d}{dt}\hat{X}(t,i) + (\tilde{A}_0(t,i) + B_0(t,i)F_l(t,i))^T \hat{X}(t,i) + \hat{X}(t,i)(\tilde{A}_0(t,i) + B_0(t,i)F_l(t,i)) \\
&+ \sum_{k=1}^{r}(A_k(t,i) + B_k(t,i)F_l(t,i))^T \hat{X}(t,i)(A_k(t,i) + B_k(t,i)F_l(t,i)) + \sum_{j=1,j\neq i}^{d} q_{ij}\hat{X}(t,j) \\
&+ M(t,i) + F_l^T(t,i)L^T(t,i) + L(t,i)F_l(t,i) + F_l^T(t,i)R(t,i)F_l(t,i) - \hat{M}(t,i) \\
&- (\hat{F}(t,i) - F_l(t,i))^T \mathcal{R}_i(t,\hat{X}(t))(\hat{F}(t,i) - F_l(t,i)) = 0.
\end{aligned}
$$

Thus we obtain that for each $i \in \mathcal{D}$, $t \to X_l(t,i) - \hat{X}(t,i)$ is a bounded and uniform positive definite solution of the linear differential inequality:

$$
\begin{aligned}
&\frac{d}{dt}Y(t,i) + [\tilde{A}_0(t,i) + B_0(t,i)F_l(t,i)]^T Y(t,i) + Y(t,i)[\tilde{A}_0(t,i) + B_0(t,i)F_l(t,i)] \\
&+ \frac{\varepsilon}{2(l+1)}I_n < 0
\end{aligned}
$$

which allow us to conclude that the zero state equilibrium of the linear differential equation

$$
\frac{d}{dt}x(t) = (\tilde{A}_0(t,i) + B_0(t,i)F_l(t,i))x(t)
$$

(5.140)

is exponentially stable and (b) is fulfilled.

Subtracting (5.136) from (5.139) we get that $t \to X_l(t,i) - X_{l+1}(t,i)$ is a bounded solution of the equation

$$
\begin{aligned}
&\frac{d}{dt}(X_l(t,i) - X_{l+1}(t,i)) + (\tilde{A}_0(t,i) + B_0(t,i)F_l(t,i))^T (X_l(t,i) - X_{l+1}(t,i)) \\
&+ (X_l(t,i) - X_{l+1}(t,i))(\tilde{A}_0(t,i) + B_0(t,i)F_l(t,i)) + \hat{\Delta}_l(t,i) = 0
\end{aligned}
$$

(5.141)

where

$$
\begin{aligned}
\hat{\Delta}_l(t,i) &= \frac{\varepsilon}{(l+1)(l+2)}I_n + [F_l(t,i) - F_{l-1}(t,i)]^T \mathcal{R}_i(t,X_{l-1}(t))[F_l(t,i) - F_{l-1}(t,i)] \\
&+ \sum_{k=1}^{r}[A_k(t,i) + B_k(t,i)F_l(t,i)]^T (X_{l-1}(t,i) - X_l(t,i))[A_k(t,i) + B_k(t,i)F_l(t,i)] \\
&+ \sum_{j=1,j\neq i}^{d} q_{ij}(X_{l-1}(t,j) - X_l(t,j)),
\end{aligned}
$$

for $l \geq 1$ and

$$
\hat{\Delta}_l(t,i) \geq \frac{\varepsilon}{2}I_n + (F_0(t,i) - \tilde{F}_0(t,i))^T \mathcal{R}_i(t,X_0(t))(F_0(t,i) - \tilde{F}_0(t,i))
$$

for $l = 0$.

Since $\hat{\Delta}_0(t,i) \geq 0$ and the zero state equilibrium of (5.140) for $l = 0$ is exponentially stable it follows from (5.141) for $l = 0$ that $X_0(t,i) - X_1(t,i) \geq 0$. Further by induction we obtain that $\hat{\Delta}_l \geq 0$ for $l \geq 1$ which leads to $X_l(t,i) - X_{l+1}(t,i) \geq 0$ and (c) is fulfilled.

From (a) and (c) we conclude that the sequences $\{X_l(t,i)\}_{l \geq 0}, i \in \mathcal{D}$ are convergent. More precisely we have the following theorem.

Theorem 5.8.2. *Assume that*

(a) The system $(\mathbf{A}, \mathbf{B}; Q)$ is stabilizable.
(b) There exists $\hat{X}(t) \in \mathbf{\Gamma}^\Sigma$.

Then for any choice of a stabilizing feedback gain $\tilde{F}_0(t) = (\tilde{F}_0(t,1), \tilde{F}_0(t,2), \ldots \tilde{F}_0(t,d))$, the sequences $\{X_l(t,i)\}_{l \geq 0}, i \in \mathcal{D}$, constructed as solutions of (5.136) (the first terms $X_0(t,i)$ obtained by solving (5.125)) are convergent. If

$$\tilde{X}(t,i) = \lim_{l \to \infty} X_l(t,i), \quad (t,i) \in \mathcal{I} \times \mathcal{D} \tag{5.142}$$

then $\tilde{X}(t) = (\tilde{X}(t,1), \tilde{X}(t,2) \ldots \tilde{X}(t,d))$ is the maximal bounded solution of the system (5.98) verifying (5.106).

Remark 5.8.2. (a) If condition (i) of Theorem 5.6.6 is fulfilled, the solution $\tilde{X}(t)$ provided by (5.142) is just the stabilizing solution of the SGRDE (5.98).
(b) Excepting the first step, when to obtain $X_0(t,i)$ we need to solve a system of linear inequalities of higher dimension, namely (5.125), further to obtain the next terms of the sequences $\{X_l(t,i)\}_{l \geq 1}, i \in \mathcal{D}$, we need to solve a system of d uncoupled Lyapunov equations. We remark that to compute the gains $F_l(t,i)$ in (5.137) we need both the value of $X_l(t,i)$ and the value of $X_{l-1}(t,i)$.
(c) Based on the uniqueness of the bounded solution of a Lyapunov equation, it follows that if the coefficients of the system (5.98) do not depend upon t then the matrices X_l and F_l do not depend upon t. In this case (5.136) and (5.137) become

$$[\tilde{A}_0(i) + B_0(i)F_{l-1}(i)]^T X_l(i) + X_l(i)[\tilde{A}_0(i) + B_0(i)F_{l-1}(i)] + M_l(i) = 0, \, i \in \mathcal{D}$$

$$M_l(i) = M(i) + \tfrac{\varepsilon}{l+1} I_n + L(i)F_{l-1}(i) + F_{l-1}^T(i)L^T(i) + F_{l-1}^T(i)R(i)F_{l-1}(i)$$

$$+ \Sigma_{k=1}^r [A_k(i) + B_k(i)F_{l-1}(i)]^T X_{l-1}(i)[A_k(i) + B_k(i)F_{l-1}(i)] + \Sigma_{j=1, j \neq i}^d q_{ij} X_{l-1}(j), l \geq 1, \tag{5.143}$$

$$\tilde{A}_0(i) = A_0(i) + \tfrac{1}{2} q_{ii} I_n,$$

$$F_l(i) = -[R(i) + \Sigma_{k=1}^r B_k^T(i)X_{l-1}(i)B_k(i)]^{-1}[B_0^T(i)X_l(i) \tag{5.144}$$

$$+ \Sigma_{k=1}^r B_k^T(i)X_{l-1}(i)A_k(i) + L^T(i)], \, l \geq 1$$

while $X_0(i)$ is obtained solving the following system of LMI

$$[A_0(i) + B_0(i)\tilde{F}_0(i)]^T X_0(i) + X_0(i)[A_0(i) + B_0(i)\tilde{F}_0(i)]$$
$$+ \sum_{k=1}^r [A_k(i) + B_k(i)\tilde{F}_0(i)]^T X_0(i)[A_k(i) + B_k(i)\tilde{F}_0(i)] + \sum_{j=1}^d q_{ij} X_0(j)$$
$$+ M(i) + \varepsilon I_n + L(i)\tilde{F}_0(i) + \tilde{F}_0^T(i)L^T(i) + \tilde{F}_0^T(i)R(i)\tilde{F}_0(i) \le 0, \ i \in \mathcal{D} \tag{5.145}$$

and

$$\tilde{F}_0(i) = - \left[R(i) + \sum_{k=1}^r B_k^T(i)X_0(i)B_k(i) \right]^{-1}$$

$$\times \left[B_0^T(i)X_0(i) + \sum_{k=1}^r B_k^T(i)X_0(i)A_k(i) + L^T(i) \right].$$

(d) From the uniqueness of the bounded solution of a Lyapunov equation we also deduce that if the coefficients of the system (5.98) are θ-periodic functions defined on $\mathbf{R}$, then the bounded solution of (5.136) are θ-periodic functions too. Hence it is sufficient to compute the values of $X_l(t,i), F_l(t,i)$ on the interval $[0,\theta]$. At each step l, the initial condition $X_l(0,i)$ is obtained by solving the linear equation

$$X_l(0,i) = \Phi_{l,i}^T(\theta,0)X_l(0,i)\Phi_{l,i}(\theta,0) + \int_0^\theta \Phi_{l,i}^T(s,0)M_l(s,i)\Phi_{l,i}(s,0)ds$$

$\Phi_{l,i}(t,s)$ being the fundamental matrix solution of (5.140). For the first step $X_0(t,i)$ is chosen as a periodic solution of the Lyapunov-type equation on $\mathcal{S}_n$

$$\frac{d}{dt}X_0(t) + \mathcal{L}_{\tilde{F}_0}^*(t)X_0(t) + M_0(t) = 0$$

where $M_0(t) = (M_0(t,1), M_0(t,2), \ldots M_0(t,d))$,

$$M_0(t,i) = M(t,i) + \varepsilon I_n + L(t,i)\tilde{F}_0(t,i) + \tilde{F}_0^T(t,i)L^T(t,i) + \tilde{F}_0^T(t,i)R(t,i)\tilde{F}_0(t,i).$$

If $T_0(t,t_0)$ is the linear evolution operator defined by the linear differential equation on $\mathcal{S}_n^d$:

$$\frac{d}{dt}S(t) = \mathcal{L}_{\tilde{F}_0}(t)S(t) \tag{5.146}$$

then the initial condition $X_0(0) = (X_0(0,1), X_0(0,2), \ldots X_0(0,d))$ is given by

$$X_0(0) = [\tilde{J} - T_0^*(\theta,0)]^{-1} \int_0^\theta T_0^*(s,0)M_0(s)ds$$

where $\tilde{J}$ is the identity operator on $\mathcal{S}_n^d$; $\tilde{J} - T_0^*(\theta,0)$ is invertible due to the exponential stability of the evolution defined by the differential equation (5.146).

In the last part of this section we present a procedure to compute the minimal semipositive solution $\widetilde{\widetilde{X}}(t)$.

First we recall that the minimal solution $\widetilde{\widetilde{X}}(t)$ is obtained as

$$\widetilde{\widetilde{X}}(t) = \lim_{\tau \to \infty} X_\tau(t) \tag{5.147}$$

where $X_\tau(t) = (X_\tau(t,1), X_\tau(t,2), \ldots, X_\tau(t,d))$ is the solution of the system (5.117) with the terminal condition $X_\tau(\tau,i) = 0$, $i \in \mathcal{D}$ (see Theorem 5.6.12).

Let us consider the following systems of Itô differential equations:

$$dx_i(t) = [\tilde{A}_0(t,i)x_i(t) + B_0(t,i)u_i(t)]dt$$

$$+ \sum_{k=1}^{r} [A_k(t,i)x_i(t) + B_k(t,i)u_i(t)]dw_k(t)$$

$$y_i(t) = C_0(t,i)x_i(t), \quad i \in \mathcal{D} \tag{5.148}$$

where

$$\tilde{A}_0(t,i) = A_0(t,i) + \frac{1}{2}q_{ii}I_n.$$

For each $i \in \mathcal{D}$, we consider the Riccati-type differential equation

$$\frac{d}{dt}X_i(t) + \tilde{A}_0^T(t,i)X_i(t) + X_i(t)\tilde{A}_0(t,i) + \sum_{k=1}^r A_k^T(t,i)X_i(t)A_k(t,i)$$
$$-[X_i(t)B_0(t,i) + \sum_{k=1}^r A_k^T(t,i)X_i(t)B_k(t,i)][R(t,i) + \sum_{k=1}^r B_k^T(t,i)X_i(t)B_k(t,i)]^{-1}$$
$$[B_0^T(t,i)X_i(t) + \sum_{k=1}^r B_k^T(t,i)X_i(t)A_k(t,i)] + C_0^T(t,i)C_0(t,i) = 0 \tag{5.149}$$

If for each $i \in \mathcal{D}$, the system (5.148) is stochastically stabilizable and stochastically detectable or stochastically uniformly observable, then invoking Corollary 5.6.16 we obtain that (5.149) has a bounded, stabilizing, and semipositive definite solution $X_i^0(t)$.

Taking $X_i^0(t)$ as a first step, we construct the sequences $\{X_i^l(t)\}_{l \geq 0}$, $i \in \mathcal{D}$ where for each l, $t \to X_i^l(t)$ is the unique bounded semipositive and stabilizing solution of the Riccati differential equation:

$$\frac{d}{dt}X_i^l(t) + \tilde{A}_0^T(t,i)X_i^l(t) + X_i^l(t)\tilde{A}_0(t,i) + \sum_{k=1}^r A_k^T(t,i)X_i^l(t)A_k(t,i)$$
$$-[X_i^l(t)B_0(t,i) + \sum_{k=1}^r A_k^T(t,i)X_i^l(t)B_k(t,i)]$$
$$\times [R(t,i) + \sum_{k=1}^r B_k^T(t,i)X_i^l(t)B_k(t,i)]^{-1} \tag{5.150}$$
$$\times [B_0^T(t,i)X_i^l(t) + \sum_{k=1}^r B_k^T(t,i)X_i^l(t)A_k(t,i)] + \tilde{M}_l(t,i) = 0$$

where $\tilde{M}_l(t,i) = C_0^T(t,i)C_0(t,i) + \sum_{j=1, j \neq i}^d q_{ij}X_j^{l-1}(t)$.

Remark 5.8.3. Clearly, for each fixed $i \in \mathcal{D}$, (5.150) is just the Riccati equation (5.120) associated with the following controlled system with multiplicative white noise:

$$dx_i(t) = \left[\tilde{A}_0(t,i)x_i(t) + B_0(t,i)u(t)\right]dt \tag{5.151}$$

$$+ \sum_{k=1}^{r}[A_k(t,i)x_i(t) + B_k(t,i)u(t)]dw_k(t)$$

$$\tilde{y}_l(t) = \tilde{C}_l(t,i)x_i(t)$$

where

$$\tilde{A}_0(t,i) = A_0(t,i) + \frac{1}{2}q_{ii}I_n,$$

$$\tilde{C}_l(t,i) = \begin{pmatrix} C_0(t,i) \\ \hat{C}_l(t,i) \end{pmatrix}, \hat{C}_l(t,i) = \left[\sum_{j\neq i}q_{ij}X_j^{l-1}(t)\right]^{\frac{1}{2}}.$$

It is easy to see that if the system (5.148) is stochastically detectable, then the system (5.151) is stochastically detectable, and if the system (5.148) is stochastically uniformly observable, then (5.151) is stochastically uniformly observable too.

Proposition 5.8.3. *Assume that for each* $i \in \mathcal{D}$

(a) *The system (5.148) is stochastically stabilizable;*
(b) *The system (5.148) is stochastically detectable or stochastically uniformly observable.*

Under these assumptions we have

(i) $X_i^{l+1}(t) \geq X_i^l(t) \geq 0, \ \forall \ l \geq 0, i \in \mathcal{D}, t \in \mathbf{R}_+;$
(ii) $X_i^l(t) \leq \hat{X}(t,i), \ (t,i) \in \mathbf{R}_+ \times \mathcal{D}, l \geq 0, \ \forall \hat{X}(t) = (\hat{X}(t,1),\ldots,\hat{X}(t,d))$ *semipositive and bounded solution of (5.117).*

Proof. Combining Remark 5.8.3 with Corollary 5.6.16 we deduce that (5.150) has a stabilizing semipositive and bounded solution $X_i^l(t)$, $l \geq 0$, $i \in \mathcal{D}$. By induction we obtain that $\tilde{M}_l(t,i) \geq 0$ which leads to $X_i^l(t) \geq 0$.

For each $l \geq 0$, $i \in \mathcal{D}$ consider the stabilizing feedback gain defined as follows:

$$F_i^l(t) = -\left[R(t,i) + \sum_{k=1}^{r}B_k^T(t,i)X_i^l(t)B_k(t,i)\right]^{-1} \tag{5.152}$$

$$\times \left[B_0^T(t,i)X_i^l(t) + \sum_{k=1}^{r}B_k^T(t,i)X_i^l(t)A_k(t,i)\right].$$

By direct calculation using (5.150) and (5.152) (for l replaced by $l+1$) we obtain

$$\frac{d}{dt}X_i^{l+1}(t) + [\tilde{A}_0(t,i) + B_0(t,i)F_i^{l+1}(t,i)]^TX_i^{l+1} + X_i^{l+1}[\tilde{A}_0(t,i)$$

$$+B_0(t,i)F_i^{l+1}(t)] + \sum_{k=1}^{r}[A_k(t,i) + B_k(t,i)F_i^{l+1}(t)]^TX_i^{l+1}(t)[A_k(t,i)$$

$$+B_k(t,i)F_i^{l+1}(t)] + \tilde{M}_{l+1}(t,i) + (F_i^{l+1}(t))^TR(t,i)F_i^{l+1}(t) = 0$$

$$\frac{d}{dt}X_i^l(t) + [\tilde{A}_0(t,i) + B_0(t,i)F_i^{l+1}(t)]^T X_i^l(t) + X_i^l(t)[\tilde{A}_0(t,i)$$

$$+ B_0(t,i)F_i^{l+1}(t)] + \Sigma_{k=1}^r [A_k(t,i) + B_k(t,i)F_i^{l+1}(t)]^T X_i^l(t)[A_k(t,i)$$

$$+ B_k(t,i)F_i^{l+1}(t)] + \tilde{M}_l(t,i) + (F_i^{l+1}(t))^T R(t,i)F_i^{l+1}(t)$$

$$- (F_i^{l+1}(t) - F_i^l(t))^T (R(t,i) + \Sigma_{k=1}^r B_k^T(t,i)X_i^l(t,i)B_k(t,i))(F_i^{l+1}(t) - F_i^l(t)) = 0$$

which leads to the fact that $t \to X_i^{l+1}(t) - X_i^l(t)$ is the bounded solution of the Lyapunov equation on $\mathcal{S}_n$:

$$\frac{d}{dt}Y(t,i) + [\tilde{A}_0(t,i) + B_0(t,i)F_i^{l+1}(t)]^T Y_i(t) + Y_i(t)[\tilde{A}_0(t,i) + B_0(t,i)F_i^{l+1}(t)]$$

$$+ \Sigma_{k=1}^r [A_k(t,i) + B_k(t,i)F_i^{l+1}(t)]^T Y_i(t)[A_k(t,i) + B_k(t,i)F_i^{l+1}(t)] + \tilde{\Delta}_l(t,i) = 0$$

$$(5.153)$$

where

$$\tilde{\Delta}_l(t,i) = \sum_{j \neq i, j=1}^d q_{ij}[X_j^l(t) - X_j^{l-1}(t)] + (F_i^{l+1}(t) - F_i^l(t))^T [R(t,i)$$

$$+ \sum_{k=1}^r B_k^T(t,i)X_i^l(t)B_k(t,i)](F_i^{l+1}(t) - F_i^l(t)).$$

Let $T_{l+1,i}(t,s)$ be the linear evolution operator on $\mathcal{S}_n$ defined by (5.153) with $\tilde{\Delta}_l(t,i) = 0$.

Since $X_i^{l+1}(t)$ is the stabilizing solution of (5.150) then we have $||T_{l+1,i}(t,s)|| \leq \beta_{l+1,i}e^{-\alpha_{l+1,i}(t-s)}$ for some positive constants $\beta_{l+1,i}, \alpha_{l+1,i}$.

From the uniqueness of the bounded solution of (5.153) we deduce that

$$X_i^{l+1}(t) - X_i^l(t) = \int_t^\infty T_{l+1,i}^*(s,t)\tilde{\Delta}_l(s,i)ds.$$

Since $T_{l+1}^*(s,t)$ is a positive operator on $\mathcal{S}_n$ from the above equality we obtain that

$$(X_i^{l+1}(t) - X_i^l(t)) \geq 0$$

if $\tilde{\Delta}_l(s,i) \geq 0$. This can be checked easily by induction.

For $l = 0$ we have

$$\tilde{\Delta}_0(s,i) = \sum_{j \neq i}^d q_{ij}X_i^0(s) + (F_i^1(s) - F_i^0(s))^T (R(s,i)$$

$$+ \sum_{k=1}^r B_k^T(s,i)X_i^0(t)B_k(s,i))(F_i^1(s) - F_i^0(s)) \geq 0.$$

Thus assertion (i) in the statement is completely proved.

To prove (ii) we recall that

$$X_i^l(t) = \lim_{\tau \to \infty} X_{\tau,i}^l(t) \tag{5.154}$$

(see the proof of Theorem 5.6.12) where $X_{\tau,i}^l(t)$ is the solution of (5.150) with the terminal condition $X_{i,\tau}^l(\tau) = 0$. Let $\hat{X}(t) = (\hat{X}(t,1)\ \hat{X}(t,2)\ ...\hat{X}(t,d))$ be a bounded and semipositive solution of the system (5.117) and let $\hat{F}(t) = (\hat{F}(t,1)\ \hat{F}(t,2)\ ...\hat{F}(t,d))$ be the corresponding feedback gain, i.e. $\hat{F}(t,i) = -\mathcal{R}_i^{-1}(t,\hat{X}(t))\mathcal{P}_i(t,\hat{X}(t))$, $i \in \mathcal{D}, t \geq 0$.

By direct calculation we get

$$\frac{d}{dt}\hat{X}(t,i) + [\tilde{A}_0(t,i) + B_0(t,i)\hat{F}(t,i)]^T\hat{X}(t,i) + \hat{X}(t,i)[\tilde{A}_0(t,i) + B_0(t,i)\hat{F}(t,i)]$$
$$+ \sum_{k=1}^r [A_k(t,i) + B_k(t,i)\hat{F}(t,i)]^T\hat{X}(t,i)[A_k(t,i) + B_k(t,i)\hat{F}(t,i)] + [C_0(t,i)+$$
$$D_0(t,i)\hat{F}(t,i)]^T[C_0(t,i) + D_0(t,i)\hat{F}(t,i)] + \sum_{j=1, j \neq i}^d q_{ij}\hat{X}(t,j) = 0$$

$$\frac{d}{dt}X_{\tau,i}^l(t) + [\tilde{A}_0(t,i) + B_0(t,i)\hat{F}(t,i)]^T X_{\tau,i}^l(t) + X_{\tau,i}^l(t)[\tilde{A}_0(t,i) + B_0(t,i)\hat{F}(t,i)]$$
$$+ \sum_{k=1}^r [A_k(t,i) + B_k(t,i)\hat{F}(t,i)]^T X_{\tau,i}^l(t)[A_k(t,i) + B_k(t,i)\hat{F}(t,i)]$$
$$+ \sum_{j=1, j \neq i}^d q_{ij}X_{\tau,j}^{l-1}(t) - [\hat{F}(t,i) - F_{\tau,i}^l(t)]^T[R(t,i) + \sum_{k=1}^r B_k^T(t,i)X_{\tau,i}^l(t)B_k(t,i)]$$
$$\times [\hat{F}(t,i) - F_{\tau,i}^l(t)] + C_0^T(t,i)C_0(t,i) + \hat{F}^T(t,i)R(t,i)\hat{F}(t,i) = 0$$

where $F_{\tau,i}^l(t)$ is as in (5.152) with $X_i^l(t)$ replaced by $X_{\tau,i}^l(t)$.

We obtain in this way that $t \to \hat{X}(t,i) - X_{\tau,i}^l(t)$ is the solution of the problem

$$\frac{d}{dt}Y_i(t) + \tilde{\mathcal{L}}_i^*(t)Y_i(t) + \hat{\Delta}_l(t,i) = 0 \tag{5.155}$$

$Y_i(\tau) = \hat{X}(\tau,i) \geq 0$, where $\hat{\mathcal{L}}_i^*(t)$ is the adjoint operator of the linear Lyapunov operator on $\mathcal{S}_n$ defined by

$$\hat{\mathcal{L}}_i(t)Y = [\tilde{A}_0(t,i) + B_0(t,i)\hat{F}(t,i)]Y + Y[\tilde{A}_0(t,i) + B_0(t,i)\hat{F}(t,i)]^T$$
$$+ \sum_{k=1}^r [A_k(t,i) + B_k(t,i)\hat{F}(t,i)]Y[A_k(t,i) + B_k(t,i)\hat{F}(t,i)]^T$$

and

$$\hat{\Delta}_l(t,i) = \sum_{j=1, j \neq i}^d q_{ij}(\hat{X}(t,j) - X_{\tau,j}^{l-1}(t)) + (\hat{F}(t,i) - F_{\tau,i}^l(t))^T[R(t,i)$$
$$+ \sum_{k=1}^r B_k^T(t,i)X_{\tau,i}^l(t)B_k(t,i)](\hat{F}(t,i) - F_{\tau,i}^l(t)).$$

If $\hat{T}_i(t,s)$ is the linear evolution operator on $\mathcal{S}_n$ defined by the linear differential equation

$$\frac{d}{dt}Y(t) = \hat{\mathcal{L}}_i(t)Y(t),$$

then from (5.155) we have the representation formula

$$\hat{X}(t,i) - X_{\tau,i}^l(t) = \hat{T}_i^*(\tau,t)\hat{X}(\tau,i) + \int_t^\tau \hat{T}_i^*(s,t)\hat{\Delta}_l(s,i)ds, \ 0 \leq t \leq \tau.$$

Since $\hat{T}_i^*(s,t)$ is a linear positive operator on $\mathcal{S}_n$ then from the above equality we deduce that $\hat{X}(t,i) - X_{\tau,i}^l(t) \geq 0, \forall \ 0 \leq t \leq \tau, i \in \mathcal{D}$ if $\hat{\Delta}_l(s,i) \geq 0$. This last condition may be checked by induction. To this end, we remark that if $l = 0$, we have

$$\hat{\Delta}_0(s,i) = \sum_{j=1,j\neq i}^d q_{ij}\hat{X}(s,j) + (\hat{F}(s,i) - F_{\tau,i}^0(s))^T[R(s,i)$$
$$+ \sum_{k=1}^r B_k^T(s,i)X_{\tau,i}^0(s)B_k(s,i)](\hat{F}(s,i) - F_{\tau,i}^0(s))$$

and it is obvious that $\hat{\Delta}_0(s,i) \geq 0, 0 \leq s \leq \tau < \infty, i \in \mathcal{D}$, which leads to $\hat{X}(t,i) - X_{\tau,i}^0(t) \geq 0$. Further, invoking (5.154) with $l = 0$ we conclude that $\hat{X}(t,i) - X_i^0(t) \geq 0$ and the proof is complete. $\qquad\square$

Theorem 5.8.4. *Assume that:*

(a) $(\mathbf{A},\mathbf{B};Q)$ *is stabilizable.*

(b) For each $i \in \mathcal{D}$, the system (5.148) is either stochastically detectable or stochastically uniformly observable.

Let $\{X_i^l(t)\}_{l\geq 0}, i \in \mathcal{D}$ be the sequences where $X_i^l(t)$ is the unique bounded and stabilizing solution of (5.150). Under the considered assumptions these sequences are convergent and if we define $\widetilde{X}(t,i) = \lim_{l\to\infty} X_i^l(t), (t,i) \in \mathbf{R}_+ \times \mathcal{D}$, then $\widetilde{X}(t) = (\widetilde{X}(t,1) \ldots \widetilde{X}(t,d))$ is the minimal positive semidefinite and bounded solution of the system (5.117).

Proof. If $(\mathbf{A},\mathbf{B};Q)$ is stabilizable, then for each $i \in \mathcal{D}$, the system (5.148) is stochastically stabilizable. Therefore the assumptions of Proposition 5.8.3 are fulfilled and the sequences $\{X_i^l(t)\}_{l\geq 1}, i \in \mathcal{D}$ are well defined and monotonically increasing.

On the other hand if assumption (a) is fulfilled, then applying Theorem 5.6.12 we obtain that the set of semipositive and bounded solutions of the system (5.117) is not empty. From Proposition 5.8.3 (ii) we deduce that the sequences $\{X_i^l(t)\}_{l\geq 1}, i \in \mathcal{D}$ are bounded above. Then the functions $\widetilde{X}(t,i)$ are well defined by $\widetilde{X}(t,i) = \lim_{l\to\infty} X_i^l(t)$. By a standard way (based on Lebesgue Theorem) we obtain that $\widetilde{X}(t) = (\widetilde{X}(t,1) \ldots \widetilde{X}(t,d))$ is a semipositive and bounded solution of (5.117). Applying again Proposition 5.8.3 (ii) we obtain that $\widetilde{X}$ is the minimal semipositive and bounded solution of (5.143) and the proof ends. $\qquad\square$

Remark 5.8.4. (a) In the particular case $A_k(t,i) = 0, B_k(t,i) = 0, k = 1,2,\ldots,r$ and the system is in the time invariant case, the iterative procedure proposed in the previous theorem was used in [1] to compute the stabilizing solution of a system of coupled algebraic Riccati equations associated with a linear system with Markovian jumping.

(b) If for each $i \in \mathcal{D}$ the system (5.148) is stochastically uniformly observable, then the system $(C_0, A_0, \ldots, A_r; Q)$ is uniformly observable (see Proposition 4.2.2

(iii) and in this case the solution $\widetilde{\widetilde{X}}(t)$ obtained in the previous theorem is just the stabilizing, bounded, and positive semidefinite solution of the system (5.117).

(c) At each step $l \geq 0$ the stabilizing solution $X_i^l(t)$ of (5.149) and (5.150), respectively, can be computed using the procedure provided by Theorem 5.8.2.

Numerical Examples

We shall illustrate the above iterative numerical procedures considering the linear time-invariant stochastic system of order $n = 2$, subject to both multiplicative noise and Markovian jumps with $r = 1$ and $\mathcal{D} = \{1,2\}$ having:

$$A_0(1) = \begin{bmatrix} -1 & 0 \\ 1 & -1 \end{bmatrix}, A_0(2) = \begin{bmatrix} -1 & 1 \\ 0 & -1 \end{bmatrix},$$

$$A_1(1) = \begin{bmatrix} -1 & 1 \\ 0 & -2 \end{bmatrix}, A_1(2) = \begin{bmatrix} -2 & 1 \\ 1 & -1 \end{bmatrix},$$

$$B_0(1) = \begin{bmatrix} 1 \\ -1 \end{bmatrix}, B_0(2) = \begin{bmatrix} 1 \\ 1 \end{bmatrix},$$

$$L_0(1) = \begin{bmatrix} 1 \\ -1 \end{bmatrix}, L_0(2) = \begin{bmatrix} 1 \\ 2 \end{bmatrix},$$

$$M_0(1) = \begin{bmatrix} 1 & 1 \\ 1 & 2 \end{bmatrix}, M_0(2) = \begin{bmatrix} 1 & 1 \\ 1 & 4 \end{bmatrix},$$

$$R(1) = 1, R(2) = 2.$$

Our purpose is to solve the SGRDE (5.98) corresponding to the above numerical values using the iterative procedure indicated in the statement of Theorem 5.8.2. Three distinct cases have been considered: the case when the system is subject only to Markov jumps, the case when the system is subject only to multiplicative white noise, and the case when the system is perturbed with both multiplicative white noise and Markovian jumps.

Case a. The Markovian jumping case: $A_1(i) = 0, B_1(i) = 0, i \in \mathcal{D}$. Using Proposition 4.1.3 we determined for the numerical values above:

$$\widetilde{F}_0(1) = [0.5923 \quad -0.7004], \widetilde{F}_0(2) = [-0.0330 \ 0.0653].$$

Then, solving (5.143) we obtained:

$$X_0(1) = 10^3 \begin{bmatrix} 1.5519 & -0.0524 \\ -0.0524 & 1.7776 \end{bmatrix},$$

$$X_0(2) = 10^3 \begin{bmatrix} 1.1139 & 0.2680 \\ 0.2680 & 1.3970 \end{bmatrix}.$$

The solution of (5.98) for this case was determined solving iteratively (5.143). For an imposed level of accuracy $\|X_{l+1}(i) - X_l(i)\| < 10^{-6}$ we obtained after 69 iterations:

$$X(1) = \begin{bmatrix} 30.7868 & 24.3960 \\ 24.3960 & 26.2218 \end{bmatrix}$$

$$X(2) = \begin{bmatrix} 21.5504 & -11.7226 \\ -11.7226 & 19.2254 \end{bmatrix}.$$

Case b. The multiplicative white noise perturbations case: $\mathcal{D}=\{1\}$, $A_i=A_i(1)$; $B_i=B_i(1)$, $i=0,1$.

In this case we obtained the initial values:

$$\tilde{F}_0 = [-0.4094 \quad 0.8482],$$

$$X_0 = \begin{bmatrix} 292.8945 & 163.9337 \\ 163.9337 & 140.9240 \end{bmatrix}$$

and, after 202 iterations, the solution of (5.98):

$$X = \begin{bmatrix} 1.0782 & 1.0307 \\ 1.0307 & 0.5878 \end{bmatrix}.$$

Case c. The case when the system is subject to both Markovian jumps and multiplicative white noise.

In this situation we obtained the initial values:

$$\tilde{F}_0(1) = [-0.3852 \quad 0.8594], \quad \tilde{F}_0(2) = [-0.9000 \quad 0.5763],$$

$$X_0(1) = 10^8 \begin{bmatrix} 5.8005 & -4.5733 \\ -4.5733 & -3.7733 \end{bmatrix}$$

$$X_0(2) = 10^8 \begin{bmatrix} -0.7123 & -0.5110 \\ 0.5110 & -4.8453 \end{bmatrix}.$$

The solution of (5.98) was obtained after 133 iterations solving (5.143); thus, we obtained:

$$X(1) = \begin{bmatrix} 2.1893 & 2.0159 \\ 2.0159 & 2.0998 \end{bmatrix}, X(2) = \begin{bmatrix} 0.7940 & -0.4088 \\ -0.4088 & 3.3714 \end{bmatrix}.$$

5.9   Systems of Generalized Riccati Differential Equations on the Space $\mathcal{S}_n^\infty$

Let $\mathbf{A} = (A_0, A_1, \ldots, A_r), \mathbf{B} = (B_0, B_1, \ldots, B_r)$ be such that $A_k : \mathcal{I} \to \mathcal{M}_n^\infty, B_k : \mathcal{I} \to \mathcal{M}_{nm}^\infty, 0 \leq k \leq r$, are bounded and continuous functions, $\mathcal{I} \subset \mathbf{R}$ being a right unbounded interval. Let $Q = (q_{ij}, i, j \in \mathbf{Z}_+)$ be an infinite matrix which elements q_{ij} satisfy the conditions (2.138) and (2.139). We associate the following system of generalized Riccati differential equations SGRDE on the Banach space $\mathcal{S}_n^\infty$

$$
\begin{aligned}
&\tfrac{d}{dt}X(t,i) + A_0^T(t,i)X(t,i) + X(t,i)A_0(t,i) + \sum_{k=1}^r A_k^T(t,i)X(t,i)A_k(t,i) \\
&+ \sum_{j=0}^\infty q_{ij}X(t,j) - (X(t,i)B_0(t,i) + \sum_{k=1}^r A_k^T(t,i)X(t,i)B_k(t,i) \\
&+ L(t,i))(R(t,i) + \sum_{k=1}^r B_k^T(t,i)X(t,i)B_k(t,i))^{-1}(B_0^T(t,i)X(t,i) \\
&+ \sum_{k=1}^r B_k^T(t,i)X(t,i)A_k(t,i) + L^T(t,i)) + M(t,i) = 0.
\end{aligned}
\tag{5.156}
$$

The SGRDE (5.156) is the special case of GRDE (5.1) for $A(t,i) = A_0(t,i) + \tfrac{1}{2}q_{ii}I_n, B(t,i) = B_0(t,i), (t,i) \in \mathcal{I} \times \mathbf{Z}_+$ and the operator $\Pi(t)$ defined as in (5.5) in the case $\mathcal{D} = \mathbf{Z}_+$. One may check that if the scalars q_{ij} satisfy conditions (2.138) and (2.139) then the operator valued function $\Pi(\cdot)$ satisfies the assumptions $\boldsymbol{\Pi}_1, \boldsymbol{\Pi}_2$ (see the Remarks 5.1.1 and 5.3.4 (d)). Therefore, we may apply the general results derived in Sects. 5.3–5.5 in order to obtain necessary and sufficient conditions for the existence of the bounded and maximal solution, bounded and stabilizing solution or minimal and positive semidefinite solution of SGRDE (5.156). Here, we show how can the concept of stabilizability of a triple (A, B, Π) introduced via Definition 5.3.2 can be restated in terms of the stabilizability of the triple $(\mathbf{A}, \mathbf{B}, Q)$ in the case when Π is described via (5.5) and $\mathcal{D} = \mathbf{Z}_+$.

Definition 5.9.1. We say that the triple $(\mathbf{A}, \mathbf{B}, Q)$ is *stabilizable* if there exists a bounded and continuous function $F : \mathcal{I} \to \mathcal{M}_m^\infty$ such that the operator valued function $\mathfrak{L}_F(\cdot)$ defines an anticausal exponentially stable evolution on $\mathcal{S}_n^\infty$ where $\mathfrak{L}_F(t) : \mathcal{S}_n^\infty \to \mathcal{S}_n^\infty$ is defined by

$$
\begin{aligned}
(\mathfrak{L}_F(t)X)(i) &= [A_0(t,i) + B_0(t,i)F(t,i)]^T X(i) + X(i)[A_0(t,i) + B_0(t,i)F(t,i)] \\
&+ \sum_{k=1}^r [A_k(t,i) + B_k(t,i)F(t,i)]^T X(i)[A_k(t,i) + B_k(t,i)F(t,i)] \\
&+ \sum_{j=0}^\infty q_{ij}X(j), i \in \mathbf{Z}_+,
\end{aligned}
\tag{5.157}
$$

for all $X = \{X(i)\}_{i \in \mathbf{Z}_+} \in \mathcal{S}_n^\infty$.

Remark 5.9.1. (a) Using Definition 2.8.1 and Definition 5.9.1 we deduce that $(\mathbf{A}, \mathbf{B}, Q)$ is stabilizable if and only if there exists a bounded and continuous function $F : \mathcal{I} \to \mathcal{M}_{mn}^\infty$ such that

$$
\|T_F^a(t, t_0)\| \leq \beta e^{\alpha(t - t_0)}, \forall, t \leq t_0, t, t_0 \in \mathcal{I}
\tag{5.158}
$$

for some constants $\alpha > 0, \beta \geq 1$, $T_F^a(t,t_0)$ being the anticausal evolution operator on $\mathcal{S}_n^\infty$ defined by the linear differential equation

$$\frac{d}{dt}Y(t) + \mathfrak{L}_F(t)Y(t) = 0.$$

(b) Based on Corollary 2.8.8 one obtains that if the scalars q_{ij} satisfy the additional condition (2.149) then we may define the stabilizability of the triple (A, B, Q) in terms of exponentially stable evolution defined by the operator valued function $\mathcal{L}_F(\cdot)$ on the Banach space $\ell^1(\mathbf{Z}_+, \mathcal{S}_n)$ where $\mathcal{L}_F(t)$ is defined as in (2.140) with $A_k(t,i)$ replaced by $A_k(t,i) + B_k(t,i)F(t,i)$, $0 \leq k \leq r, (t,i) \in \mathcal{I} \times \mathbf{Z}_+$ for a suitable bounded and continuous function $F : \mathcal{I} \to \mathcal{M}_{mn}^\infty$.

Corollary 5.9.1. *Assume that $A_k(\cdot), B_k(\cdot)$ are bounded and continuous functions and the scalars q_{ij} satisfy the conditions (2.138), (2.139). If $\Pi(t)$ is defined as in (5.5) and $A(t,i) = A_0(t,i) + \frac{1}{2}q_{ii}I_n, B(t,i) = B_0(t,i), (t,i) \in \mathcal{I} \times \mathbf{Z}_+$, then the following are equivalent*

 (i) *The triple (A, B, Π) is stabilizable according to the Definition 5.3.2;*
 (ii) *The triple (A, B, Q) is stabilizable according to the Definition 5.9.1.*

Proof is obvious because, in this special case, the operator valued function $\mathfrak{L}_{A+BF,\Pi_F}(\cdot)$ coincides with the operator valued function $\mathfrak{L}_F(\cdot)$ introduced by (5.157). □

From Corollary 5.9.1 and Remark 5.3.2 one sees that if $A_k(\cdot), B_k(\cdot)$ are periodic functions of period θ, then, without loss of generality in the Definition of stabilizability of the triple (A, B, Q) we may consider only continuous and θ-periodic functions $F : \mathcal{I} \to \mathcal{M}_{mn}^\infty$. Also, if $A_k(t,i) = A_k(i), B_k(t,i) = B_k(i), (t,i) \in \mathcal{I} \times \mathbf{Z}_+, 0 \leq k \leq r$, then in the definition of the triple (A, B, Q) we may involve only constant feedback gains $F \in \mathcal{M}_{mn}^\infty$.

According to the Remark 2.3.7 and Corollary 2.3.9 we deduce that in the time invariant case the triple (A, B, Q) is stabilizable if and only if there exists $F \in \mathcal{M}_{mn}^\infty$ such that the eigenvalues of the linear operator $\mathfrak{L}_F$ be located in the half plane $\mathbf{C}^-$.

Let us assume that the linear stochastic system (4.1) is perturbed by a standard homogeneous Markov process with an infinite countable number of states $(\eta(t), P(t), \mathbf{Z}_+)$. In this case the system (4.1) is stochastically stabilizable if and only if there exists a bounded and continuous function $F : \mathbf{R}_+ \to \mathcal{M}_{mn}^\infty$ such that the closed loop system

$$\begin{aligned}
dx(t) = {}&(A_0(t, \eta(t)) + B_0(t, \eta(t))F(t, \eta(t)))x(t)dt + \\
&\textstyle\sum_{k=1}^r (A_k(t, \eta(t)) + B_k(t, \eta(t))F(t, \eta(t)))x(t)dw_k(t)
\end{aligned} \tag{5.159}$$

is ESMS-C.

Corollary 5.9.2. *Assume that*

(a) *The functions $A_k(\cdot), B_k(\cdot)$ are continuous and bounded.*
(b) *$\mathcal{D} = \mathbf{Z}_+$ and the scalars q_{ij} satisfy the assumptions of Theorem 3.1.4.*

Under these conditions the following are equivalent

 (i) *The system of type (4.1) is stochastic stabilizable;*
(ii) *The triple* $(\mathbf{A}, \mathbf{B}, Q)$ *is stabilizable.*

The results stated in Corollaries 5.9.1 and 5.9.2 allow us to replace the abstract concept of the stabilizability of a triple (A, B, Π) by the stabilizability of the triple $(\mathbf{A}, \mathbf{B}, Q)$ or the stochastic stabilizability of the system (4.1) in the assumptions of the theorems from Sects. 5.3–5.5 in order to deduce conditions for the existence of maximal, stabilizing, minimal global solutions of SGRDE (5.156).

Notes and References

The Riccati equations of stochastic control were generally studied in connection with the linear quadratic problem either for controlled linear stochastic systems with state-dependent noise or for systems with Markov perturbations. For references concerning linear quadratic problems in the stochastic framework, see Chap. 6. Most of the results contained in this chapter were published for the first time in [44, 49]. The iterative procedure to compute the stabilizing solution of SGRDE was also published in [42, 45]. Classes of nonlinear matrix differential equations which contain as particular cases Riccati differential equations arising in control problems for stochastic systems with multiplicative white noise have been studied in [28,29,50,65,67]. Iterative procedures for computation of the stabilizing solution of the algebraic Riccati equations associated with the linear stochastic systems with multiplicative white noise may be found in [73]. Iterative procedures to compute the stabilizing solution of systems of Riccati equations involved in the linear quadratic problem for stochastic systems with Markov parameters can be found, for example, in [1, 69]. Several aspects concerning the algebraic Riccati equations arising in the control of linear stochastic systems may be found in [2, 27] were rich lists of references dealing with symmetric and non-symmetric Riccati equations may be found.

Systems with an infinite number of coupled algebraic Riccati equations arising in connection with linear quadratic problems associated with controlled linear systems perturbed by a standard homogeneous Markov chain with an infinite number countable of states were studied in [8, 63]. An iterative method to compute the minimal solution of a system of coupled algebraic Riccati equations associated with a controlled system perturbed by a standard homogeneous Markov process with a finite number of states may be found in [20].

Chapter 6
Linear Quadratic Optimization Problems for Linear Stochastic Systems

In this chapter as well as in the next chapters one shows how the mathematical results derived in the previous chapters are involved in the design of stabilizing controllers with some imposed performances for a wide class of linear stochastic systems. The design problem of some stabilizing controls minimizing quadratic performance criteria is studied. More precisely, this chapter deals with the so-called *linear quadratic optimization problem* (LQOP). LQOP has received much attention in control literature due to its wide area of applications. The main objective of the theoretical developments presented in the following consists in providing a unified approach to solve LQOP for systems subject both to multiplicative white noise and to Markovian jumping. It will be seen that depending on the class of admissible controls, the corresponding optimal control is obtained either with the stabilizing solution or with the minimal solution of a corresponding system of generalized Riccati differential equations. We also consider the case when the weights matrices are not with definite sign. Such situations may occur in a natural way in economy, ecology, and financial applications. A tracking problem is considered in Sect. 6.4. Throughout this chapter we assume that $\mathcal{D} = \{1, 2, \ldots, d\}$ (for the case $\mathcal{D} = \mathbf{Z}_+$ see Remark 6.3.4).

6.1 Preliminaries

In this section we shall present some auxiliary results which are used in the derivation of the solutions of the optimization problems stated and solved in this chapter.

Let us consider the system subject both to multiplicative white noise and to Markovian jumping, which dynamics is described by the state-space equation:

$$
\begin{aligned}
dx(t) = &\ [A_0(t, \eta(t))x(t) + B_0(t, \eta(t))u(t)]\, dt \\
&+ \Sigma_{k=1}^{r} [A_k(t, \eta(t))x(t) + B_k(t, \eta(t))u(t)]\, dw_k(t)
\end{aligned}
\tag{6.1}
$$

V. Dragan et al., *Mathematical Methods in Robust Control of Linear Stochastic Systems*, 265
DOI 10.1007/978-1-4614-8663-3_6, © Springer Science+Business Media New York 2013

where $t \in \mathbf{R}_+$, with the state vector $x \in \mathbf{R}^n$ and with the control inputs $u \in$ $\mathbf{R}^m$, $\{w(t)\}_{t \in \mathbf{R}}, w(t) = (w_1(t), w_2(t), \ldots, w_r(t))^T$ is a standard Wiener process and $\{\eta(t)\}_{t \geq 0}$ is a standard homogeneous Markov process which satisfy the assumptions stated in Chap. 1. Here we assume that $\eta(t)$ takes values in the finite set $\mathcal{D} = \{1, 2, \ldots, d\}$.

For each quadruple $(t_0, \tau, x_0, i), 0 \leq t_0 < \tau < \infty, x_0 \in \mathbf{R}^n, i \in \mathcal{D}$ we consider the auxiliary cost functions: $J(t_0, \tau, x_0, i, \cdot) : L^2_{\eta,w}([t_0, \tau], \mathbf{R}^m) \to \mathbf{R}$ by

$$J(t_0, \tau, x_0, i; u) = E\left[\int_{t_0}^{\tau} \begin{bmatrix} x^T(t) & u^T(t) \end{bmatrix} \mathcal{Q}(t, \eta(t)) \begin{bmatrix} x(t) \\ u(t) \end{bmatrix} dt \, \Big| \, \eta(t_0) = i \right] \quad (6.2)$$

where

$$\mathcal{Q}(t, i) = \begin{pmatrix} M(t, i) & L(t, i) \\ L^T(t, i) & R(t, i) \end{pmatrix} = \mathcal{Q}^T(t, i)$$

and $x(t) = x_u(t, t_0, x_0)$ is the solution of the system (6.1) corresponding to the input $u(t)$ and having the initial condition (t_0, x_0).

Applying the Itô-type formula (Theorem 1.10.2) we obtain the following result.

Lemma 6.1.1. *If* $t \to K(t, i) : \mathbf{R}_+ \to \mathcal{S}_n, i \in \mathcal{D}$ *are* C^1*-functions, then we have*

$$J(t_0, \tau, x_0, i; u) = x_0^T K(t_0, i) x_0 - E[x^T(\tau) K(\tau, \eta(\tau)) x(\tau) | \eta(t_0) = i]$$

$$+ E\left[\int_{t_0}^{\tau} \begin{bmatrix} x^T(t) & u^T(t) \end{bmatrix} \mathcal{Q}^K(t, \eta(t)) \begin{bmatrix} x(t) \\ u(t) \end{bmatrix} dt \, \Big| \, \eta(t_0) = i \right],$$

for all $0 \leq t_0 < \tau < \infty$, $x_0 \in \mathbf{R}^n$, $i \in \mathcal{D}$, $u \in L^2_{\eta,w}([t_0, \tau], \mathbf{R}^m)$ *where*

$$\mathcal{Q}^K(t, i) = \begin{bmatrix} \mathcal{Q}^K_{11}(t, i) & \mathcal{Q}^K_{12}(t, i) \\ (\mathcal{Q}^K_{12}(t, i))^T & \mathcal{Q}^K_{22}(t, i) \end{bmatrix}$$

with

$$
\begin{aligned}
\mathcal{Q}^K_{11}(t, i) &= \tfrac{d}{dt} K(t, i) + A_0^T(t, i) K(t, i) + K(t, i) A_0(t, i) \\
&\quad + \sum_{k=1}^{r} A_k^T(t, i) K(t, i) A_k(t, i) + \sum_{j=1}^{d} q_{ij} K(t, j) + M(t, i) \\
&= \tfrac{d}{dt} K(t, i) + [\mathcal{L}^*(t) K(t)](i) + M(t, i) \\
\mathcal{Q}^K_{12}(t, i) &= K(t, i) B_0(t, i) + \sum_{k=1}^{r} A_k^T(t, i) K(t, i) B_k(t, i) + L(t, i) \\
&= \mathcal{P}_i^T(t, K(t)) \\
\mathcal{Q}^K_{22}(t, i) &= R(t, i) + \sum_{k=1}^{r} B_k^T(t, i) K(t, i) B_k(t, i) \\
&= \mathcal{R}_i(t, K(t)). \quad \square
\end{aligned}
$$

Corollary 6.1.2. *If $X(t) = (X(t,1), X(t,2) \ldots X(t,d))$ is a solution of the SGRDE of type (5.98) defined on $[t_0, \tau]$, then we have*

$$
\begin{aligned}
J(t_0, \tau, x_0, i; u) &= x_0^T X(t_0, i) x_0 - E[x^T(\tau) X(\tau, \eta(\tau)) x(\tau) | \eta(t_0) = i] \\
&+ E[\int_{t_0}^{\tau} (u(t) - F^X(t, \eta(t)) x(t))^T [R(t, \eta(t)) \\
&+ \sum_{k=1}^{r} B_k^T(t, \eta(t)) X(t, \eta(t)) B_k(t, \eta(t))](u(t) \\
&- F^X(t, \eta(t)) x(t)) dt | \eta(t_0) = i]
\end{aligned}
\tag{6.3}
$$

for all $u \in L_{\eta,w}^2([t_0, \tau], \mathbf{R}^m)$, $x_0 \in \mathbf{R}^n, i \in \mathcal{D}$ where

$$
F^X(t, i) = -\mathcal{R}_i^{-1}(t, X(t)) \mathcal{P}_i(t, X(t))
\tag{6.4}
$$

$(t, i) \in [t_0, \tau] \times \mathcal{D}$ and $x(t) = x_u(t, t_0, x_0)$.

6.2 The Linear Quadratic Optimization Problem for Stochastic Systems: The General Case

6.2.1 The Problem

Let us consider the cost function:

$$
\begin{aligned}
J_1(t_0, x_0, u) &= E \int_{t_0}^{\infty} [x_u^T(t) M(t, \eta(t)) x_u(t) + x_u^T(t) L(t, \eta(t)) u(t) \\
&+ u^T(t) L^T(t, \eta(t)) x_u(t) + u^T(t) R(t, \eta(t)) u(t)] dt
\end{aligned}
\tag{6.5}
$$

where $M(t, i) = M^T(t, i); R(t, i) = R^T(t, i), (t, i) \in \mathbf{R}_+ \times \mathcal{D}$ and $x_u(t)$ denotes the solution of the system (6.1) corresponding to the input $u(.)$ with the initial condition $(t_0, x_0) \in \mathbf{R}_+ \times \mathbf{R}^n$.

The optimization problem treated in this section consists in determining the optimal state-feedback control

$$
u(t) = F(t, \eta(t)) x(t)
\tag{6.6}
$$

which stabilizes (6.1) and minimizes the cost function (6.5). The class of admissible controls for this problem is the set $\tilde{\mathcal{U}}_m(t_0, x_0)$ of stochastic processes $u(t) \in L_{\eta,w}^2([t_0, T], \mathbf{R}^m)$ for all $T > t_0$ with the additional properties that $J_1(t_0, x_0, u)$ exists and it is finite and $\lim_{t \to \infty} E|x_u(t)|^2 = 0$. The fact that $J_1(t_0, x_0, u)$ exists means that there exists

$$
\begin{aligned}
\lim_{T \to \infty} E \int_{t_0}^{T} [x_u^T(t) M(t, \eta(t)) x_u(t) + x_u^T(t) L(t, \eta(t)) u(t) \\
+ u^T(t) L^T(t, \eta(t)) x_u(t) + u^T(t) R(t, \eta(t)) u(t)] dt \in \mathbf{R}.
\end{aligned}
$$

An important feature specific to the systems subject to multiplicative white noise is the one related to the *well-posedness* of the problem. Indeed it will be shown that in contrast to the deterministic case where the matrix

$$\begin{bmatrix} M(t,i) & L(t,i) \\ L^T(t,i) & R(t,i) \end{bmatrix}$$

must be positive semidefinite, in the stochastic case this condition is not necessary. Often in this chapter the optimization problem described by the controlled system (6.1), the cost functional (6.5) and the set of admissible controls $\tilde{\mathcal{U}}_m(t_0, x_0)$ will be called the *first linear quadratic optimization problem* (LQOP1).

6.2.2 The Solution of LQOP1

In the following, we investigate the LQOP described by the cost function (6.5) and the system (6.1). As it is shown in [4, 17], while the cost functions of type (6.20) are always bounded below, the cost function J_1 may have values which approach to $-\infty$. The same thing is expected to happen in the more general case of the systems subject both to multiplicative white noise and to Markovian jumping. In order to clarify this fact, we introduce the notion of the value function of the considered optimization problem. For each $(t_0, x_0) \in \mathbf{R}_+ \times \mathbf{R}^n$ we denote

$$V(t_0, x_0) = \inf_{u \in \tilde{\mathcal{U}}_m(t_0, x_0)} J_1(t_0, x_0, u)$$

the value function associated with the optimization problem LQOP1.

Definition 6.2.1. We say that the optimization problem described by the cost function (6.5) and the system (6.1) is *well-posed* if $-\infty < V(t_0, x_0) < \infty$ for all $(t_0, x_0) \in \mathbf{R}_+ \times \mathbf{R}^n$.

With the notations introduced in the previous chapter we have the following theorem.

Theorem 6.2.1. *Assume that*

 (i) *The system (6.1) is stochastically stabilizable.*
(ii) *The set Γ^Σ defined in (5.104) is not empty.*

Under the above conditions the LQOP described by the cost function (6.5) and the system (6.1) is well-posed. Moreover,

$$V(t_0, x_0) = \sum_{i \in \mathcal{D}} \pi_i(t_0) x_0^T \tilde{X}(t_0, i) x_0 \qquad (6.7)$$

where $\pi_i(t_0) = \mathcal{P}(\eta(t_0) = i)$ and $\tilde{X}(t) = (\tilde{X}(t,1) \ldots \tilde{X}(t,d))$ is the maximal bounded solution of the SGRDE (5.98) which verifies

$$\mathcal{R}_i(t, \tilde{X}(t)) \geq \tilde{\rho} I_m > 0. \tag{6.8}$$

Proof. Let us remark that the assumption (i) implies $\tilde{\tilde{\mathcal{U}}}_m(t_0, x_0) \neq \emptyset$, for all $t_0 \geq 0$ and $x_0 \in \mathbf{R}^n$. Based on Theorem 5.6.2 we deduce that the system (5.98) has a maximal solution $\tilde{X}(t)$ which verifies (6.8). Applying Corollary 6.1.2 for $X(t,i)$ replaced by $\tilde{X}(t,i)$ we get

$$\begin{aligned}
J(t_0, \tau, x_0, i, u) = &\; x_0^T \tilde{X}(t_0, i) x_0 - E[x^T(\tau) \tilde{X}(\tau, \eta(\tau)) x(\tau) | \eta(t_0) = i] \\
&+ E[\int_{t_0}^{\tau} (u(t) - \tilde{F}(t, \eta(t)) x(t))^T \mathcal{R}_{\eta(t)}(t, \tilde{X}(t)) \\
&\times (u(t) - \tilde{F}(t, \eta(t)) x(t)) dt | \eta(t_0) = i]
\end{aligned} \tag{6.9}$$

for all $u \in \tilde{\tilde{\mathcal{U}}}_m(t_0, x_0), t_0 < \tau, x_0 \in \mathbf{R}^n, i \in \mathcal{D}$ where $\tilde{F}(t,i)$ is defined as in (6.4) for $X(t)$ replaced by $\tilde{X}(t)$. Since $\tilde{X}(t)$ is a bounded solution, it follows that there exists $\tilde{c} > 0$ such that $|\tilde{X}(t,i)| \leq \tilde{c}, (\forall)(t,i) \in \mathbf{R}_+ \times \mathcal{D}$. Then from the following inequality

$$|E[x^T(\tau) \tilde{X}(\tau, \eta(\tau)) x(\tau) | \eta(t_0) = i]| \leq \tilde{c} E[|x(\tau)|^2 | \eta(t_0) = i]$$

we obtain

$$\lim_{\tau \to \infty} E[x^T(\tau) \tilde{X}(\tau, \eta(\tau)) x(\tau) | \eta(t_0) = i] = 0.$$

Taking the limit in (6.9) we get

$$\begin{aligned}
J_1(t_0, x_0, u) = &\sum_{i \in \mathcal{D}} \pi_i(t_0) x_0^T \tilde{X}(t_0, i) x_0 + \sum_{i \in \mathcal{D}} \pi_i(t_0) E[\int_{t_0}^{\infty} (u(t) \\
&- \tilde{F}(t, \eta(t)) x(t))^T \mathcal{R}_{\eta(t)}(t, \tilde{X}(t))(u(t) - \tilde{F}(t, \eta(t)) x(t)) dt | \eta(t_0) = i]
\end{aligned} \tag{6.10}$$

for all $u \in \tilde{\tilde{\mathcal{U}}}_m(t_0, x_0), x_0 \in \mathbf{R}^n, t_0 \in \mathbf{R}_+$. Combining (6.10) with (6.8) we obtain that $J_1(t_0, x_0, u) \geq \sum_{i \in \mathcal{D}} \pi_i(t_0) x_0^T \tilde{X}(t_0, i) x_0, \forall u \in \tilde{\tilde{\mathcal{U}}}_m(t_0, x_0)$ which leads to

$$V(t_0, x_0) \geq \sum_{i \in \mathcal{D}} \pi_i(t_0) x_0^T \tilde{X}(t_0, i) x_0.$$

This last inequality shows the well-posedness of the considered optimization problem. It remains to show that (6.7) holds.

To this end let us consider the following perturbed differential equations on $\mathcal{S}_n^d$:

$$\frac{d}{dt} X(t) + \mathcal{L}^*(t) X(t) - \mathcal{P}^*(t, X(t)) \mathcal{R}^{-1}(t, X(t)) \mathcal{P}(t, X(t)) + M(t) + \varepsilon_l J^d = 0 \tag{6.11}$$

where $\{\varepsilon_l\}_{l\geq 0}$ is a monotonically decreasing sequence with $\lim_{l\to\infty}\varepsilon_l = 0$. The SGRDE (6.11) is associated with the quadruple $\Sigma^l = (\mathbf{A},\mathbf{B},Q,Q^l)$ where $\mathbf{A},\mathbf{B},Q$ are as in the case of the SGRDE (5.98) (see Remark 5.6.1) while,

$$Q^l(t,i) = \begin{pmatrix} M(t,i)+\varepsilon_l I_n & L(t,i) \\ L^T(t,i) & R(t,i) \end{pmatrix}.$$

Invoking Lemma 5.1.2 one may show that $\Gamma^\Sigma \subset \tilde{\Gamma}^{\Sigma^l} \subset \tilde{\Gamma}^{\Sigma^{l-1}}$ for all $l \geq 1$. Applying Theorem 5.6.6 (one uses the assumptions (i) and (ii)) we deduce that (6.11) has a bounded and stabilizing solution $X_{\varepsilon_l}(t)$. Since the stabilizing solution of SGRDE (6.11) is just the maximal solution one obtains via Theorem 5.3.6 that the sequence $\{X_{\varepsilon_l}(t)\}_{l\geq 0}$ is convergent and $\lim_{l\to\infty}X_{\varepsilon_l}(t) = \tilde{X}(t)$, where $\tilde{X}(t)$ is the maximal solution of the SGRDE (5.98) which verifies (6.8).

For each $l \geq 0$ we associate the cost function

$$J^{\varepsilon_l}(t_0,x_0,u) = E\left[\int_{t_0}^\infty \{x^T(t)(M(t,\eta(t))+\varepsilon_l I_n)x(t)\right.$$

$$\left. +x^T(t)L(t,\eta(t))u(t)+u^T(t)L^T(t,\eta(t))x(t)+u^T(t)R(t,\eta(t))u(t)\}dt\right]],$$

$u \in \hat{\mathcal{U}}_m(t_0,x_0)$ where $\hat{\mathcal{U}}_m(t_0,x_0)$ consists of all stochastic processes $u \in \tilde{\mathcal{U}}_m(t_0,x_0)$ such that $\lim_{T\to\infty}E\int_{t_0}^T |x_u(t)|^2 dt < \infty$.

Clearly,

$$J^{\varepsilon_l}(t_0,x_0,u) = J_1(t_0,x_0,u) + \varepsilon_l E\left[\int_{t_0}^\infty |x(t)|^2 dt\right] \tag{6.12}$$

for all $u \in \hat{\mathcal{U}}_m(t_0,x_0)$. Reasoning as in the first part of the proof we obtain the analogous of (6.10) for the perturbed cost function $J^{\varepsilon_l}(t_0,x_0,u)$:

$$\begin{aligned} J^{\varepsilon_l}(t_0,x_0,u) = &\sum_{i\in\mathcal{D}}\pi_i(t_0)x_0^T X_{\varepsilon_l}(t_0,i)x_0 + \sum_{i\in\mathcal{D}}\pi_i(t_0)E\left[\int_{t_0}^\infty (u(t)\right. \\ &-F_{\varepsilon_l}(t,\eta(t))x(t))^T \mathcal{R}_{\eta(t)}(t,X_{\varepsilon_l}(t))(u(t)-F_{\varepsilon_l}(t,\eta(t))x(t))dt|\eta(t_0) = i] \end{aligned} \tag{6.13}$$

for all $u \in \hat{\mathcal{U}}_m(t_0,x_0) \subset \tilde{\mathcal{U}}_m(t_0,x_0)$.

Let us consider the control

$$u_{\varepsilon_l}(t) = F_{\varepsilon_l}(t,\eta(t))x_{\varepsilon_l}(t)$$

where

$$F_{\varepsilon_l}(t,i) = -\mathcal{R}_i^{-1}(t,X_{\varepsilon_l}(t))\mathcal{P}_i(t,X_{\varepsilon_l}(t))$$

and $x_{\varepsilon_l}(t)$ is the solution of system (6.1) corresponding to the control $u_{\varepsilon_l}(t)$ and the initial condition $x_{\varepsilon_l}(t_0) = x_0$.

Since $X_{\varepsilon_l}(t)$ is a stabilizing solution of the system (6.11) it follows that $u_{\varepsilon_l} \in \hat{\mathcal{U}}_m(t_0, x_0)$. Hence, from (6.13) with $u(t)$ replaced by $u_{\varepsilon_l}(t)$ we obtain:

$$J^{\varepsilon_l}(t_0, x_0, u_{\varepsilon_l}) = \sum_{i \in \mathcal{D}} \pi_i(t_0) x_0^T X_{\varepsilon_l}(t_0, i) x_0.$$

Therefore:

$$\sum_{i \in \mathcal{D}} \pi_i(t_0) x_0^T X_{\varepsilon_l}(t_0, i) x_0 = J^{\varepsilon_l}(t_0, x_0, u_{\varepsilon_l})$$

$$\geq J_1(t_0, x_0, u_{\varepsilon_l}) \geq V(t_0, x_0) \geq \sum_{i \in \mathcal{D}} \pi_i(t_0) x_0^T \tilde{X}(t_0, i) x_0.$$

Taking the limit for $l \to \infty$, we obtain that (6.7) holds and the proof ends. $\square$

Definition 6.2.2. A pair $(\tilde{x}(t), \tilde{u}(t))$ where $\tilde{u}(t) \in \tilde{\mathcal{U}}_m(t_0, x_0)$ and $\tilde{x}(t) = x_{\tilde{u}}(t, t_0, x_0)$ is the solution of (6.1) corresponding to the input $\tilde{u}(t)$ is called *optimal pair* if $V(t_0, x_0) = J_1(t_0, x_0, \tilde{u})$. In this case the control $\tilde{u}(t)$ is termed the *optimal control*.

Corollary 6.2.2. *Assume that the system (5.98) has a bounded and stabilizing solution,* $\tilde{X}(t) = (\tilde{X}(t, 1) \ldots \tilde{X}(t, d))$ *which verifies (6.8). Set*

$$\tilde{u}(t) = \tilde{F}(t, \eta(t)) \tilde{x}(t), \quad \tilde{F}(t, i) = -\mathcal{R}_i^{-1}(t, \tilde{X}(t)) \mathcal{P}_i(t, \tilde{X}(t))$$

and $\tilde{x}(t)$ is a solution of system (6.1) corresponding to the control $\tilde{u}$ and the initial condition $\tilde{x}(t_0) = x_0$. Under these assumptions $(\tilde{x}(t), \tilde{u}(t))$ is an optimal pair for the optimization problem described by (6.1) and (6.5).

Proof. From Corollary 5.6.7 it follows that the bounded and stabilizing solution of SGRDE (5.98), if it exists, then it is just the maximal bounded solution $\tilde{X}(t)$ which verifies (6.8). Now, the conclusion of this corollary follows in an obvious way, from (6.7), since $\tilde{u} \in \tilde{\mathcal{U}}_m(t_0, x_0)$. $\square$

Theorem 6.2.3. *Assume that the assumptions of Theorem 6.2.1 hold. Then the LQOP described by (6.1) and (6.5) has an optimal pair $(\hat{x}(t), \hat{u}(t))$ for some (t_0, x_0) if and only if*

$$\lim_{t \to \infty} x_0^T \left[T_{\tilde{F}}^*(t, t_0) J^d \right](i) x_0 = 0 \tag{6.14}$$

for all $i \in \mathcal{D}$, where $T_{\tilde{F}}(t, t_0)$ is the linear evolution operator on $\mathcal{S}_n^d$ defined by the linear differential equation

$$\frac{d}{dt} S(t) = \mathcal{L}_{\tilde{F}}(t) S(t), \tag{6.15}$$

$\tilde{F}(t) = (\tilde{F}(t,1) \ldots \tilde{F}(t,d))$ *is associated by (6.4) with the maximal bounded solution of (5.98) which verifies (6.8).*

Proof. Let $(\hat{x}(t), \hat{u}(t))$ be an optimal pair. Using (6.10) we may write

$$V(t_0, x_0) = J_1(t_0, x_0, \hat{u}) = \sum_{i=1}^{d} \pi_i(t_0) x_0^T \tilde{X}(t_0, i) x_0 + E\left[\int_{t_0}^{\infty} (\hat{u}(t) \right.$$
$$\left. -\tilde{F}(t, \eta(t))\hat{x}(t))^T \mathcal{R}_{\eta(t)}(t, \tilde{X}(t))(\hat{u}(t) - \tilde{F}(t, \eta(t))\hat{x}(t))dt\right].$$

Taking into account the value of $V(t_0, x_0)$ given in Theorem 6.2.1, we get

$$E\left[\int_{t_0}^{\infty} (\hat{u}(t) - \tilde{F}(t, \eta(t))\hat{x}(t))^T \mathcal{R}_{\eta(t)}(t, \tilde{X}(t))(\hat{u}(t) - \tilde{F}(t, \eta(t))\hat{x}(t))dt\right] = 0$$

which leads to

$$\hat{u}(t) - \tilde{F}(t, \eta(t))\hat{x}(t) = 0, \qquad a.e.$$

By the uniqueness arguments we deduce that $\hat{x}(t)$ coincides almost surely with the solution $\tilde{x}(t)$ of the problem

$$\begin{aligned} dx(t) &= [A_0(t, \eta(t)) + B_0(t, \eta(t))\tilde{F}(t, \eta(t))]x(t)dt \\ &+ \sum_{k=1}^{r}[A_k(t, \eta(t)) + B_k(t, \eta(t))\tilde{F}(t, \eta(t))]x(t)dw_k(t), \end{aligned} \qquad (6.16)$$

$t \geq t_0, x(t_0) = x_0$.

Hence $\hat{u}(t)$ coincide almost surely with $\tilde{u}(t)$ given by $\tilde{u}(t) = \tilde{F}(t, \eta(t))\tilde{x}(t)$.

Let $\tilde{\Phi}(t, t_0)$ be the fundamental matrix solution of the stochastic differential equation (6.16), hence

$$\tilde{x}(t) = \tilde{\Phi}(t, t_0)x_0.$$

Since the optimal control $\tilde{u}(t) \in \tilde{\tilde{\mathcal{U}}}_m(t_0, x_0)$ it follows that

$$\lim_{t \to \infty} E[|\tilde{\Phi}(t, t_0)x_0|^2 | \eta(t_0) = i] = 0, \ i \in \mathcal{D}.$$

Based on representation formula, given in Theorem 3.1.1 we obtain (6.14). The converse implication follows in the similar way. $\qquad \square$

Corollary 6.2.4. *Suppose that the assumptions of Theorem 6.2.1 are fulfilled. Then the following are equivalent*

(i) *For each $(t_0, x_0) \in \mathbf{R}_+ \times \mathbf{R}^n$ the optimization problem described by (6.1) and (6.5) has an optimal control $u^{(t_0, x_0)}$ that is $V(t_0, x_0) = J_1(t_0, x_0, u^{(t_0, x_0)})$.*

(ii)

$$\lim_{t \to \infty} ||T_{\tilde{F}}(t,t_0)|| = 0, \ \forall \, t_0 \geq 0 \tag{6.17}$$

$T_{\tilde{F}}(t,t_0)$ is the linear evolution operator defined by the differential equation (6.15).

Furthermore if (i) or (ii) holds, then $u^{(t_0,x_0)}(t) = \tilde{F}(t,\eta(t))\tilde{x}(t)$ where $\tilde{x}(t)$ is the solution of (6.16).

Proof. The proof follows immediately taking into account that (6.14) is fulfilled for all $t_0 \geq 0$, $i \in \mathcal{D}, x_0 \in \mathbf{R}^n$ and

$$||T_{\tilde{F}}^*(t,t_0)|| = |T_{\tilde{F}}^*(t,t_0)J^d| = \max_{i \in \mathcal{D}} \sup_{|x_0|=1} \left\{ \left| x_0^T \left[T_{\tilde{F}}^*(t,t_0)J^d \right](i)x_0 \right| \right\}$$

and the norms of the operators $T_{\tilde{F}}^*(t,t_0)$ and $T_{\tilde{F}}(t,t_0)$ are equivalent. $\qquad \square$

Remark 6.2.1. The property of the evolution operator $T_{\tilde{F}}(t,t_0)$ stated in (6.17) shows that the maximal solution $\tilde{X}(t)$ of the system (5.98) has an additional property which consists in the attractivity of the zero solution of the corresponding closed-loop system (6.16), that is

$$\lim_{t \to \infty} E[|\tilde{\Phi}(t,t_0)x_0|^2 | \eta(t_0) = i] = 0, \ i \in \mathcal{D}, \ t_0 \geq 0, \ x_0 \in \mathbf{R}^n.$$

We have to remark that in general, this property is not equivalent to the exponential stability in mean square of the zero solution of the system (6.16); hence, condition (6.17) does not imply that the maximal solution $\tilde{X}(t)$ coincides with the stabilizing solution of the system (5.98).

However, if the coefficients of the system (5.98) are θ-periodic functions, then (6.17) implies that the maximal solution $\tilde{X}(t)$ is just the stabilizing solution of the system (5.98).

This fact is stated in the following theorem.

Theorem 6.2.5. *Assume that the coefficients of the SGRDE (5.98) are θ-periodic functions and the assumptions of Theorem 6.2.1 are fulfilled. Then the following are equivalent*

(i) For all $(t_0,x_0) \in \mathbf{R}_+ \times \mathbf{R}^n$ there exists a control $u^{(t_0,x_0)} \in \tilde{\mathcal{U}}_m(t_0,x_0)$ which verifies

$$V(t_0,x_0) = J_1(t_0,x_0,u^{(t_0,x_0)}).$$

(ii) The system of differential equations (5.98) has a stabilizing and bounded solution $\tilde{X}(t)$ which verifies (6.8).

Proof. From Corollary 6.2.4 we deduce that (i) is equivalent to (6.17). Particularly

$$\lim_{l \to \infty} ||T_{\tilde{F}}(l\theta, 0)|| = 0. \tag{6.18}$$

Based on the identity $T_{\tilde{F}}(t+\theta, t_0+\theta) = T_{\tilde{F}}(t, t_0)$, $\forall t, t_0 \geq 0$ we may show by induction that $T_{\tilde{F}}(l\theta, 0) = (T_{\tilde{F}}(\theta, 0))^l$. Hence (6.18) is equivalent to

$$\lim_{l \to \infty} ||T_{\tilde{F}}(\theta, 0))^l|| = 0. \tag{6.19}$$

Since $T_{\tilde{F}}(\theta, 0) : \mathcal{S}_n^d \to \mathcal{S}_n^d$ is a linear operator acting on a finite dimensional Banach space, we obtain from (6.19) that all eigenvalues of $T_{\tilde{F}}(\theta, 0)$ are located in the inside of the unit disk $|\lambda| < 1$. But $T_{\tilde{F}}(\theta, 0)$ is the monodromy matrix of (6.15) then, applying a well-known result concerning the uniform asymptotic stability of the zero state equilibrium of a linear differential equation with periodic coefficients, (see [74]) we conclude that the zero solution of (6.15) is exponentially stable. This means that the solution $\tilde{X}(t)$ is just the stabilizing solution of the system (5.98) and thus the proof of the implication (i) $\Rightarrow$ (ii) is complete. The implication (ii) $\Rightarrow$ (i) follows from Corollary 6.2.2. $\square$

Corollary 6.2.6. *Assume the following.*

(a) *The system (6.1) and the cost function (6.5) are in the time invariant case;*
(b) $(\mathbf{A}, \mathbf{B}; Q)$ *is stabilizable;*
(c) *The inequality* $\mathcal{L}^* X - \mathcal{P}^*(X)\mathcal{R}^{-1}(X)\mathcal{P}(X) + M \geq 0$ *has a solution*

$$\hat{X} = (\hat{X}(1), \hat{X}(2), \ldots, \hat{X}(d))$$

which verifies the conditions $\mathcal{R}_i(\hat{X}) > 0, i \in \mathcal{D}$. *Then the following are equivalent*

(i) *For all* $x_0 \in \mathbf{R}^n$ *there exists an optimal control* $u^{x_0} \in \tilde{\mathcal{U}}_m(0, x_0)$, *that is* $V(0, x_0) = J_1(0, x_0, u^{x_0})$.
(ii) *The system of algebraic equations (5.103) has a stabilizing solution*

$$\tilde{X} = (\tilde{X}(1), \tilde{X}(2), \ldots, \tilde{X}(d))$$

which verifies $\mathcal{R}_i(\tilde{X}) > 0, \ i \in \mathcal{D}$.
(iii) *The system of linear matrix inequalities*

$$\begin{pmatrix} (\mathcal{L}^* X)(i) + M(i) & \mathcal{P}_i^T(X) \\ \mathcal{P}_i(X) & \mathcal{R}_i(X) \end{pmatrix} > 0, \ i \in \mathcal{D}$$

has solutions in $\mathcal{S}_n^d$. Under these conditions $u^{x_0}(t) = \tilde{F}(\eta(t))\tilde{x}(t)$ where

$$\tilde{F}(i) = -\mathcal{R}_i^{-1}(\tilde{X})\mathcal{P}_i(\tilde{X}), \; i \in \mathcal{D},$$

$\tilde{X}$ being the stabilizing solution of (5.103) and $\tilde{x}(t)$ is a solution of the corresponding closed loop system (6.16).

Proof. (i) $\Leftrightarrow$ (ii) follows from the previous theorem and (ii) $\Leftrightarrow$ (iii) follows from Theorem 5.6.8. $\qquad\qquad\qquad\qquad\qquad\qquad\qquad\qquad\qquad\qquad\qquad\qquad\qquad$ $\square$

6.3 The Linear Quadratic Optimal Regulator for a Stochastic System

6.3.1 The Problem

The second problem treated in the present chapter, known also as a linear quadratic optimal regulator problem (LQORP), requires to find the control of the form (6.6) such that the cost function

$$J_2(t_0, x_0, u) = E \int_{t_0}^{\infty} |y_u(t)|^2 \, dt \qquad (6.20)$$

is minimized in the class $\mathcal{U}(t_0, x_0)$ of all stochastic processes $u \in L_{\eta,w}^2([t_0, T], \mathbf{R}^m)$ for all $T > t_0$, $J_2(t_0, x_0, u) < \infty$, where

$$y_u(t) = y_u(t, t_0, x_0) = C_0(t, \eta(t))x_u(t) + D_0(t, \eta(t))u(t) \qquad (6.21)$$

is an output in $\mathbf{R}^p$. This problem often will be termed the *second linear quadratic optimization problem* (LQOP2).

In order to simplify the expressions involved in the solution of this problem we make the following assumption:

Assumption (A):

(a) There exists $\rho > 0$ such that $D_0^T(t, i)D_0(t, i) \geq \rho I_m$, $\forall \; (t, i) \in \mathbf{R}_+ \times \mathcal{D}$;
(b) $D_0^T(t, i)C_0(t, i) = 0$, $\forall \; (t, i) \in \mathbf{R}_+ \times \mathcal{D}$.

Remark 6.3.1. If the system (6.1) with the output (6.21) verifies the Assumption A (a), then without loss of generality, the Assumption A (b) is fulfilled. Indeed, if A (a) is fulfilled, then by the change of control variables described by

$$u(t) = -[D_0^T(t, \eta(t))D_0(t, \eta(t))]^{-1}D_0^T(t, \eta(t))C_0(t, \eta(t))x(t) + \tilde{u}(t)$$

we may replace the given system (6.1), (6.21) by the following modified system

$$dx(t) = \left[\hat{A}_0(t,\eta(t))x(t) + B_0(t,\eta(t))\tilde{u}(t)\right]dt$$

$$+ \sum_{k=1}^{r}\left[\hat{A}_k(t,\eta(t))x(t) + B_k(t,\eta(t))\tilde{u}(t)\right]dw_k(t)$$

$$y(t) = \hat{C}_0(t,\eta(t))x(t) + D_0(t,\eta(t))\tilde{u}(t)$$

where

$$\hat{A}_k(t,i) = A_k(t,i) - B_k(t,i)R^{-1}(t,i)D_0^T(t,i)C_0(t,i), \quad k = 0,1,\ldots,r,$$
$$\hat{C}_0(t,i) = [I_p - D_0(t,i)R^{-1}(t,i)D_0^T(t,i)]C_0(t,i),$$
$$R(t,i) = D_0^T(t,i)D_0(t,i), \quad (t,i) \in \mathbf{R}_+ \times \mathcal{D}.$$

Clearly, this new system verifies both assumptions A (a) and A (b).

6.3.2   Solution of LQORP

Since the cost functional (6.20) is a particular case of the cost functional (6.5) it follows that the solution of the optimization problem described by the controlled system (6.1), the cost functional (6.20) and the corresponding set of admissible controls $\tilde{\mathcal{U}}_m(t_0,x_0)$ is obtained from the results derived in the previous section. The optimal control of this optimization problem is constructed with the stabilizing solution of SGRDE (5.117).

In this subsection we derive the solution of the optimization problem described by the controlled system (6.1), the cost functional (6.20) and the set of admissible controls $\mathcal{U}(t_0,x_0)$. Let $X(t)$ be a semipositive definite solution of the system (5.117) and let

$$F^X(t) = \left(F^X(t,1)\, F^X(t,2)\, \ldots F^X(t,d)\right)$$

be the corresponding feedback gain defined by (6.4). Set $u^X(t) = F^X(t,\eta(t))x^X(t)$, $t \geq 0$ where $x^X(t)$ is the solution of the system

$$dx(t) = [A_0(t,\eta(t)) + B_0(t,\eta(t))F^X(t,\eta(t))]x(t)dt \\ + \sum_{k=1}^{r}[A_k(t,\eta(t)) + B_k(t,\eta(t))F^X(t,\eta(t))]x(t)dw_k(t), \tag{6.22}$$

$t \geq t_0$, $x(t_0) = x_0$.

Lemma 6.3.1. *For each bounded and positive semidefinite solution $X(t)$ of the system (5.117) the control $u^X(t)$ belongs to $\mathcal{U}(t_0,x_0)$, $t_0 \geq 0$, $x_0 \in \mathbf{R}^n$.*

Proof. Obviously the control $u^X(t) \in L^2_{\eta,w}([t_1,t_2], \mathbf{R}^m)$ for every compact interval $[t_1,t_2] \subset [t_0, \infty)$. Applying Corollary 6.1.2 for

$$M(t,i) = C_0^T(t,i)C_0(t,i),$$

$$L(t,i) = 0,$$

$$R(t,i) = D_0^T(t,i)D_0(t,i), \; u(t) = u^X(t)$$

we obtain

$$E\left[\int_{t_0}^{\tau} |C_0(t,\eta(t))x^X(t) + D_0(t,\eta(t))u^X(t)|^2 dt | \eta(t_0) = i\right] \tag{6.23}$$

$$= x_0^T X(t_0,i)x_0 - E\left[x^T(\tau)X(\tau,\eta(\tau))x(\tau)|\eta(t_0) = i\right],$$

$\forall t_0 < \tau$, $x_0 \in \mathbf{R}^n$, $i \in \mathcal{D}$, $x(t) = x^X(t)$.

Taking into account that $X(t)$ is a positive semidefinite and bounded solution of the system (5.117) it follows that there exists a positive constant c, such that

$$E\left[\int_{t_0}^{\tau} |C_0(t,\eta(t))x(t) + D_0(t,\eta(t))u^X(t)|^2 dt | \eta(t_0) = i\right] \leq x_0^T X(t_0,i)x_0 \leq c|x_0|^2,$$

$(\forall) \; \tau \geq t_0$, $x_0 \in \mathbf{R}^n$, $i \in \mathcal{D}$. Hence

$$E\left[\int_{t_0}^{\infty} |C_0(t,\eta(t))x(t) + D_0(t,\eta(t))u^X(t)|^2 dt | \eta(t_0) = i\right] \leq x_0^T X(t_0,i)x_0$$

which shows that $J_2(t_0,x_0,u^X)$ is well defined and we have

$$J_2(t_0,x_0,u^X) \leq \sum_{j \in \mathcal{D}} \pi_j(t_0)x_0^T X(t_0,j)x_0 \tag{6.24}$$

and the proof is complete. $\square$

Theorem 6.3.2. *Assume that the system* $(\mathbf{A}, \mathbf{B}; Q)$ *is stabilizable. Then the optimization problem LQORP has a solution given by*

$$\tilde{u}(t) = \tilde{\tilde{F}}(t,\eta(t))\tilde{\tilde{x}}(t), \quad t \geq t_0$$

where $\tilde{\tilde{F}}(t,i)$ *is defined as in (6.4) for X replaced by the minimal positive semidefinite and bounded solution* $\tilde{\tilde{X}}(t)$ *of the system (5.117) and* $\tilde{\tilde{x}}$ *is the solution of the problem (6.22) where* $F^X(t,i)$ *is replaced by* $\tilde{\tilde{F}}(t,i)$. *Moreover the optimal value of the cost function is*

$$J_2(t_0,x_0,\tilde{u}) = \sum_{i=1}^{d} \pi_i(t_0)x_0^T \tilde{\tilde{X}}(t_0,i)x_0.$$

Proof. Let $X_\tau(t) = (X_\tau(t,1) \ \ldots \ X_\tau(t,d))$ be the solution of the SGRDE (5.117) which verifies the terminal condition $X_\tau(\tau,i) = 0$.

Based on Corollary 5.2.3 and Theorem 5.6.12 it follows that the solution $X_\tau(t)$ is defined for all $t \in [0,\tau]$ and

$$\lim_{\tau \to \infty} X_\tau(t) = \tilde{\tilde{X}}(t).$$

Applying Corollary 6.1.2 for $X(t,i)$ replaced by $X_\tau(t,i)$ we obtain

$$E\left[\int_{t_0}^{\tau} |C_0(t,\eta(t))x(t) + D_0(t,\eta(t))u(t)|^2 dt | \eta(t_0) = i\right] = x_0^T X_\tau(t_0,i)x_0$$

$$+ E\left[\int_{t_0}^{\tau} [u(t) - F_\tau(t,\eta(t))x(t)]^T \mathcal{R}_{\eta(t)}(t,X_\tau(t))[u(t) - F_\tau(t,\eta(t))x(t)]dt | \eta(t_0) = i\right],$$

$$(6.25)$$

$\forall\, u \in L^2_{\eta,w}([t_0,\tau], \mathbf{R}^m)$.

Hence

$$E\left[\int_{t_0}^{\tau} |y_u(t)|^2 dt | \eta(t_0) = i\right] \geq x_0^T X_\tau(t_0,i)x_0 \qquad (6.26)$$

and equality is possible if $u(t) = F_\tau(t,\eta(t))x_\tau(t)$, $t \in [t_0,\tau]$, $x_\tau(t)$ being the solution of the problem (6.22) for $F^X(t,i)$ replaced by $F_\tau(t,i) = -\mathcal{R}_i^{-1}(t,X_\tau(t))\mathcal{P}_i(t,X_\tau(t))$. From (6.26) for $u(t) = \tilde{\tilde{u}}(t)$ we obtain easily that

$$J_2(t_0,x_0,\tilde{\tilde{u}}) \geq \sum_{i \in \mathcal{D}} \pi_i(t_0)x_0^T \tilde{\tilde{X}}(t_0,i)x_0. \qquad (6.27)$$

Combining (6.24) with (6.27) we get

$$J_2(t_0,x_0,\tilde{\tilde{u}}) = \sum_{i \in \mathcal{D}} \pi_i(t_0)x_0^T \tilde{\tilde{X}}(t_0,i)x_0.$$

Let $u \in \mathcal{U}(t_0,x_0)$ be arbitrary. Applying (6.25) to the restriction of u to the interval $[t_0,\tau]$ and taking the limit for $\tau \to \infty$ we obtain:

$$E\left[\int_{t_0}^{\infty} |y_u(t)|^2 dt | \eta(t_0) = i\right] = x_0^T \tilde{\tilde{X}}(t_0,i)x_0$$

$$+ E\left[\int_{t_0}^{\infty} (u(t) - \tilde{\tilde{F}}(t,\eta(t))x(t))^T \mathcal{R}_{\eta(t)}(t,\tilde{\tilde{X}}(t))(u(t) - \tilde{\tilde{F}}(t,\eta(t))x(t))dt | \eta(t_0) = i\right]$$

which leads to

$$J_2(t,x_0,u) = \sum_{i \in \mathcal{D}} \pi_i(t_0)x_0^T \tilde{\tilde{X}}(t_0,i)x_0$$

$$+ \sum_{i \in \mathcal{D}} \pi_i(t_0)E\left[\int_{t_0}^{\infty} (u(t) - \tilde{\tilde{F}}(t,\eta(t))x(t))^T \mathcal{R}_{\eta(t)}(t,\tilde{\tilde{X}}(t))(u(t)\right.$$

$$\left. - \tilde{\tilde{F}}(t,\eta(t))x(t))dt | \eta(t_0) = i\right]$$

$\forall\, u \in \mathcal{U}(t_0,x_0)$ which completes the proof.   □

Remark 6.3.2. From (6.23) and (6.26) for $u(t) = \tilde{\tilde{u}}(t)$ we obtain

$$\lim_{\tau \to \infty} E\left[\tilde{x}^T(\tau)\tilde{X}(\tau, \eta(\tau))\tilde{x}(\tau) | \eta(t_0) = i\right] = 0$$

which is the single information concerning the behavior of the optimal trajectory of the system for $t \to \infty$.

Theorem 6.3.3. *Assume that the assumptions in Theorem 5.6.15 are fulfilled. Under these conditions the solutions of the optimization problems LQOP1 and LQORP described by the cost function (6.20) and the controlled system (6.1) coincide and they are given by*

$$\tilde{u}(t) = \tilde{F}(t, \eta(t))\tilde{x}(t) \tag{6.28}$$

where $\tilde{F}(t, i)$ is defined as in (6.4) with $X(t)$ replaced by the stabilizing and bounded solution $\tilde{X}(t)$ of the system (5.117) and $\tilde{x}(t)$ is the solution of the problem (6.22) with $F^X(t, i)$ replaced by $\tilde{F}(t, i)$. Moreover the optimal value of the cost function is given by

$$J_2(t_0, x_0, \tilde{u}) = \sum_{i \in \mathcal{D}} \pi_i(t_0) x_0^T \tilde{X}(t_0, i) x_0.$$

Proof. Under the considered assumptions the SGRDE (5.117) has a unique bounded and positive semidefinite solution and that solution is a stabilizing one. Therefore the control $\tilde{u}(t)$ given by (6.28) coincides with $\tilde{\tilde{u}}(t)$ and hence the conclusion of the theorem follows immediately. $\square$

Remark 6.3.3. Since $\tilde{\tilde{\mathcal{U}}}_m(t_0, x_0) \subseteq \mathcal{U}(t_0, x_0)$ it follows that

$$J_2(t_0, x_0, \check{u}) = \min_{u \in \tilde{\mathcal{U}}_m(t_0, x_0)} J_2(t_0, x_0, u) \geq \min_{u \in \mathcal{U}(t_0, x_0)} J_2(t_0, x_0, u)$$
$$= J_2\left(t_0, x_0, \tilde{\tilde{u}}\right). \tag{6.29}$$

On the other hand, from Theorem 6.3.3 and Corollary 6.2.2 it follows that if the system (6.1) is stochastic stabilizable and the system

$$dx(t) = A_0(t, \eta(t))x(t)\,dt + \sum_{k=1}^r A_k(t, \eta(t))x(t)\,dw_k(t)$$
$$y(t) = C_0(t, \eta(t))x(t)$$

is either stochastic detectable or stochastic uniformly observable, then in (6.29) we have equality and additionally, $\tilde{u} = \tilde{\tilde{u}}$ (a.s.).

The next illustrative example shows that in the absence of the properties of detectability and observability in (6.29) the equality does not always take place.

Illustrative example. Consider the system (6.1) in the particular case $n = 2, r = 1, d = 1, m = 1$. In this case the system becomes:

$$dx(t) = (A_0 x(t) + B_0 u(t)) dt + (A_1 x(t) + B_1 u(t)) dw_1(t) \qquad (6.30)$$

$$x = \begin{bmatrix} x_1 \\ x_2 \end{bmatrix} \in \mathbf{R}^2, \, u(t) \in \mathbf{R}$$

and the coefficient matrices are those from the numerical example at the end of Sect. 5.6. The cost functional is

$$J_2(0, x_0, u) = E\left[\int_0^\infty \left(x_1^2(t) + u^2(t)\right) dt\right]. \qquad (6.31)$$

From Corollary 6.2.2 one obtains that the solution of the optimization problem described by the system (6.30), the cost functional (6.31), and the set of admissible controls $\tilde{\mathcal{U}}_1(0, x_0)$ is constructed with the stabilizing solution of the SGRAE (5.103) and the optimal value is given by:

$$J_2(0, x_0, \tilde{u}) = \begin{bmatrix} x_{10} & x_{20} \end{bmatrix} \begin{bmatrix} 8 & -21 \\ -21 & 63 \end{bmatrix} \begin{bmatrix} x_{10} \\ x_{20} \end{bmatrix} \qquad (6.32)$$

where $x_0 = \begin{bmatrix} x_{10} & x_{20} \end{bmatrix}^T$. On the other hand from Theorem 6.3.2 it follows that the solution of the optimization problem described by the system (6.30), the cost function (6.31), and the set of admissible controls $\mathcal{U}(0, x_0)$ is constructed with the minimal solution of the SGRAE (5.103). The optimal value is:

$$J_2(0, x_0, \tilde{\tilde{u}}) = \begin{bmatrix} x_{10} & x_{20} \end{bmatrix} \begin{bmatrix} 1 & 0 \\ 0 & 0 \end{bmatrix} \begin{bmatrix} x_{10} \\ x_{20} \end{bmatrix}. \qquad (6.33)$$

From (6.32) and (6.33) one sees that $J_2(0, x_0, \tilde{u}) \neq J_2(0, x_0, \tilde{\tilde{u}})$.

Remark 6.3.4. Using the results of the previous chapter (Sect. 5.9) and the proofs of Theorems 6.2.1, 6.3.2, and 6.3.3, we may conclude that the results stated in these theorems are valid in the case $\mathcal{D} = \mathbf{Z}_+$.

6.4 A Tracking Problem

Consider the stochastic system (6.1) with the output (6.21) together with assumptions A (a) and A (b) stated in Sect. 6.2. Then if $t \to r(t) = (r(t, 1), r(t, 2), \ldots, r(t, d))$: $\mathbf{R}_+ \to (\mathbf{R}^p)^d$ is a continuous and bounded function, the *tracking problem* consists in finding a control $\hat{u}(\cdot) \in \tilde{\mathcal{U}}_m(t_0, x_0)$ which minimizes the cost function

$$\hat{J}(u) = \varlimsup_{T \to \infty} \frac{1}{T - t_0} E\left[\int_{t_0}^T |y_u(t, t_0, x_0) - r(t, \eta(t))|^2 dt\right] \qquad (6.34)$$

in the class of all stochastic processes $\tilde{\mathcal{U}}_m(t_0,x_0)$, where $\tilde{\mathcal{U}}_m(t_0,x_0)$ is the set of all stochastic processes $u : [t_0,\infty) \times \Omega \to \mathbf{R}^m$ with the properties: $u \in L^2_{\eta,w}([t_0,T],\mathbf{R}^m)$ for all $T > t_0$ and $\sup E |x_u(t,t_0,x_0)|^2 < \infty, t \geq t_0$.

For each $(t_0,\tau,x_0,i) \in \mathbf{R}_+ \times \mathbf{R}_+ \times \mathbf{R}^n \times \mathcal{D}$ with $0 \leq t_0 < \tau$ we consider the auxiliary cost functions

$$\mathcal{W}(t_0,\tau,x_0,i,u) = E\left[\int_{t_0}^{\tau} |y_u(t,t_0,x_0) - r(t,\eta(t))|^2 dt | \eta(t_0) = i\right],$$

for all $u \in L^2_{\eta,w}([t_0,\tau],\mathbf{R}^m)$. Based on Itô-type formula given in Theorem 1.10.2 we obtain the following lemma.

Lemma 6.4.1. *Let* $t \to K(t,i) : \mathbf{R}_+ \to \mathcal{S}_n, t \to g(t,i) : \mathbf{R}_+ \to \mathbf{R}^n, t \to h(t,i) : \mathbf{R}_+ \to \mathbf{R}, i \in \mathcal{D}$ *be* C^1-*functions, and let*

$$v(t,x,i) = x^T K(t,i)x + 2g^T(t,i)x + h(t,i).$$

Then

$$\mathcal{W}(t_0,\tau,x_0,i,u) = v(t_0,x_0,i) - E\left[v(\tau,x(\tau),\eta(\tau)) | \eta(t_0) = i\right]$$

$$+E\left[\int_{t_0}^{\tau} \{(x^T(t) \quad u^T(t))\mathcal{Q}^K(t,\eta(t))\begin{pmatrix} x(t) \\ u(t) \end{pmatrix} + 2\left[\frac{\partial}{\partial t}g^T(t,\eta(t))\right.\right.$$

$$+g^T(t,\eta(t))A_0(t,\eta(t)) + \textstyle\sum_{j=1}^d q_{\eta(t)j}g^T(t,j) - r^T(t,\eta(t))C_0(t,\eta(t))\Big]x(t)$$

$$+2\left[g^T(t,\eta(t))B_0(t,\eta(t)) - r^T(t,\eta(t))D_0(t,\eta(t))\right]u(t)$$

$$+r^T(t,\eta(t))r(t,\eta(t)) + \tfrac{\partial}{\partial t}h(t,\eta(t)) + \textstyle\sum_{j=1}^d q_{\eta(t)j}h(t,j)\}dt | \eta(t_0) = i\bigg]$$

for all $t_0, 0 \leq t_0 < \tau, x_0 \in \mathbf{R}^n, i \in \mathcal{D}, u \in L^2_{\eta,w}([t_0,\tau],\mathbf{R}^m)$ *where* $x(t) = x_u(t,t_0,x_0), \mathcal{Q}^K(t,i)$ *being as in Lemma 6.1.1 with* $M(t,i) = C_0^T(t,i)C_0(t,i), L(t,i) = 0, R(t,i) = D_0^T(t,i)D_0(t,i).$

Let $\tilde{X}(t)$ be the stabilizing and bounded solution of the SGRDE (5.117). Set $\tilde{F}(t) = (\tilde{F}(t,1), \tilde{F}(t,2) \ldots \tilde{F}(t,d)), \tilde{F}(t,i) = -\mathcal{R}_i^{-1}(t,\tilde{X}(t))\mathcal{P}_i(t,\tilde{X}(t))$ be the stabilizing feedback gain. This means that the zero state equilibrium of the corresponding closed-loop system (6.16) is ESMS-C. Then by Corollary 3.2.10, the zero solution of the differential equation with Markovian jumping

$$\frac{d}{dt}x(t) = \left[A_0(t,\eta(t)) + B_0(t,\eta(t))\tilde{F}(t,\eta(t))\right]x(t)$$

is ESMS-C.

Now, applying Theorem 3.2.8 we deduce that the zero state equilibrium of the linear differential equation on $(\mathbf{R}^n)^d$:

$$\frac{d}{dt}y_i(t) = \left[A_0(t,i) + B_0(t,i)\tilde{F}(t,i)\right]y_i(t) + \sum_{j=1}^d q_{ji}y_j(t), i \in \mathcal{D}$$

is exponentially stable.

Let $\tilde{g}(t) = (\tilde{g}(t,1)\ \tilde{g}(t,2)\ \ldots \tilde{g}(t,d))$ (see Corollary 3.2.9) be the unique bounded solution on $\mathbf{R}_+$ of the affine differential equations

$$\frac{d}{dt}y_i(t) + [A_0(t,i) + B_0(t,i)\tilde{F}(t,i)]^T y_i(t) + \sum_{j=1}^d q_{ij} y_j(t) \tag{6.35}$$
$$- [C_0(t,i) + D_0(t,i)\tilde{F}(t,i)]^T r(t,i) = 0$$

$i \in \mathcal{D}$. From the previous lemma we have the following corollary.

Corollary 6.4.2. *Assume that the system (5.117) has a bounded and stabilizing solution $\tilde{X}(t)$. Take $\tilde{g}(t)$ the unique bounded solution of (6.35) and $h(t,i)$ be arbitrary C^1-function as in the previous lemma. If $v(t,x,i) = x^T\tilde{X}(t,i)x + 2\tilde{g}^T(t,i)x + h(t,i)$, we have*

$$\mathcal{W}(t_0,\tau,x_0,i,u) = v(t_0,x_0,i) - E[v(\tau,x(\tau),\eta(\tau))|\eta(t_0) = i]$$
$$+E\left[\int_{t_0}^\tau \left\{ [u(t) - \tilde{F}(t,\eta(t))x(t)]^T \mathcal{R}_{\eta(t)}(t,\tilde{X}(t)) [u(t) - \tilde{F}(t,\eta(t))x(t)] \right.\right.$$
$$+2\left[\tilde{g}^T(t,\eta(t))B_0(t,\eta(t)) - r^T(t,\eta(t))D_0(t,\eta(t))\right][u(t) - \tilde{F}(t,\eta(t))x(t)] \tag{6.36}$$
$$\left.\left. + \frac{\partial}{\partial t}h(t,\eta(t)) + \sum_{j=1}^d q_{\eta(t)j}h(t,j) + r^T(t,\eta(t))r(t,\eta(t))\}dt\bigg|\eta(t_0) = i\right], \right.$$

for all t_0, $0 \le t_0 < \tau$, $x_0 \in \mathbf{R}^n$, $i \in \mathcal{D}$, $u \in L_{\eta,w}^2([t_0,\tau],\mathbf{R}^m)$, $x(t) = x_u(t,t_0,x_0)$.

Remark 6.4.1. If $\tilde{X}$ is a bounded and stabilizing solution of the SGRDE (5.117), then we may write

$$\frac{d}{dt}\tilde{X}(t) + \mathcal{L}_{\tilde{F}}^*(t)\tilde{X}(t) + [C_0(t) + D_0(t)\tilde{F}(t)]^T [C_0(t) + D_0(t)\tilde{F}(t)] = 0$$

which shows that the stabilizing and bounded solution of the system (5.117) if it exists is always positive semidefinite. Therefore, the condition $\mathcal{R}_i(t,\tilde{X}(t)) \ge \rho I_m > 0$ is fulfilled.

For each $\tau > 0$ set $h_\tau(t) = (h_\tau(t,1)\ \ldots h_\tau(t,d))^T$, the solution of the system of affine differential equations

$$\frac{d}{dt}h(t) + Qh(t) + \tilde{m}(t) = 0$$

with the terminal condition $h_\tau(\tau) = 0$, where

$$\tilde{m}(t) = (\tilde{m}_1(t)\ \tilde{m}_2(t)\ \ldots\ \tilde{m}_d(t))^T$$
$$\tilde{m}_j(t) = r^T(t,j)r(t,j) - [\tilde{g}^T(t,j)B_0(t,j) - r^T(t,j)D_0(t,j)]\mathcal{R}_j^{-1}(t,\tilde{X}(t)) \tag{6.37}$$
$$\times [B_0^T(t,j)\tilde{g}(t,j) - D_0^T(t,j)r(t,j)],$$

$i \in \mathcal{D}, t \ge 0, Q = \{q_{ij}\}_{i,j\in\mathcal{D}}$. Let $v_\tau(t,x,i)$ be defined by

$$v_\tau(t,x,i) = x^T\tilde{X}(t,i)x + 2\tilde{g}^T(t,i)x + h_\tau(t,i).$$

From Corollary 6.4.2 we get:

$$
\begin{aligned}
\mathcal{W}(t_0,\tau,x_0,i,u) &= v_\tau(t_0,x_0,i) - E[v_\tau(\tau,x(\tau),\eta(\tau))|\eta(t_0)=i] \\
&\quad + E\left[\int_{t_0}^\tau (u(t) - \tilde{F}(t,\eta(t))x(t) - \psi(t,\eta(t))^T \mathcal{R}_{\eta(t)}(t,\tilde{X}(t))(u(t)\right. \\
&\quad \left. - \tilde{F}(t,\eta(t))x(t) - \psi(t,\eta(t))dt|\eta(t_0)=i\right]
\end{aligned}
\tag{6.38}
$$

for all t_0, $0 \le t_0 < \tau$, $x_0 \in \mathbf{R}^n$, $i \in \mathcal{D}$, $u \in L^2_{\eta,w}([t_0,\tau],\mathbf{R}^m)$, $x(t) = x_u(t,t_0,x_0)$ where

$$
\psi(t,i) = -\mathcal{R}_i^{-1}(t,\tilde{X}(t))[B_0^T(t,i)\tilde{g}(t,i) - D_0^T(t,i)r(t,i)].
\tag{6.39}
$$

Now we are able to prove the main result of this section.

Theorem 6.4.3. *Assume that the system of differential equations (5.117) has a bounded and stabilizing solution $\tilde{X}(t)$. Let $\tilde{g}(t) = (\tilde{g}(t,1),\ \tilde{g}(t,2),\ \dots \tilde{g}(t,d))$ be the unique bounded on $\mathbf{R}_+$ solution of (6.35) and $\psi(t,i)$ defined by (6.39). Under these conditions we have*

$$
\min_{u \in \tilde{\mathcal{U}}_m(t_0,x_0)} \hat{J}(u) = \hat{J}(\bar{u}) = \varlimsup_{T \to \infty} \frac{1}{T}\int_0^T \sum_{i=1}^d \sum_{j=1}^d \pi_i(t_0)\tilde{p}_{ij}\tilde{m}_j(t)dt,
$$

for all $t_0 \ge 0$, $x_0 \in \mathbf{R}^n$, where $\bar{u}(t) = \tilde{F}(t,\eta(t))\bar{x}(t) + \psi(t,\eta(t))$, $\bar{x}(t)$ being the solution of the problem

$$
\begin{aligned}
dx(t) &= \left[(A_0(t,\eta(t)) + B_0(t,\eta(t))\tilde{F}(t,\eta(t)))x(t) + B_0(t,\eta(t))\psi(t,\eta(t))\right]dt \\
&\quad + \sum_{k=1}^r \left[(A_k(t,\eta(t)) + B_k(t,\eta(t))\tilde{F}(t,\eta(t)))x(t)\right. \\
&\quad \left. + B_k(t,\eta(t))\psi(t,\eta(t))\right]dw_k(t)
\end{aligned}
\tag{6.40}
$$

$t \ge t_0$, $\bar{x}(t_0) = x_0$ and

$$
\tilde{P} = (\tilde{p}_{ij})_{i,j \in \mathcal{D}} = \lim_{t \to \infty} P(t) = \lim_{t \to \infty} e^{Qt}.
$$

Proof. Applying Theorem 3.6.1 to the system (6.40) we deduce that $\sup_{t \ge t_0} E|\bar{x}(t)|^2 < \infty$ and therefore $\bar{u}(t)$ belongs to $\tilde{\mathcal{U}}_m(t_0,x_0)$. It is easy to see that for each $u \in \tilde{\mathcal{U}}_m(t_0,x_0)$ we have

$$
\hat{J}(u) = \limsup_{T \to \infty} \frac{1}{T-t_0} \sum_{i=1}^d \pi_i(t_0)\mathcal{W}(t_0,T,x_0,i,u).
$$

Then from (6.38) we have for $u \in \tilde{\mathcal{U}}_m(t_0,x_0)$

$$
\hat{J}(u) \ge \limsup_{T \to \infty} \frac{1}{T-t_0} \sum_{i=1}^d \pi_i(t_0)\left\{v_T(t_0,x_0,i) - E[v_T(T,x(T),\eta(T))|\eta(t_0)=i]\right\}
$$

$$
= \limsup_{T \to \infty} \frac{1}{T-t_0} \sum_{i=1}^d \pi_i(t_0)h_T(t_0,i) = \hat{J}(\bar{u}).
$$

But

$$h_T(t) = \int_t^T e^{Q(s-t)} \tilde{m}(s)\,ds = \int_t^T P(s-t)\tilde{m}(s)\,ds.$$

Therefore

$$h_T(t_0) = \int_{t_0}^T \left[P(s-t_0) - \tilde{P}\right]\tilde{m}(s)\,ds + \int_{t_0}^T \tilde{P}\tilde{m}(s)\,ds.$$

Since $\lim_{t\to\infty} P(t) = \tilde{P}$ and $\tilde{m}(t)$ is a continuous and bounded function we have

$$\lim_{T\to\infty} \frac{1}{T-t_0} \int_{t_0}^T \left(P(s-t_0) - \tilde{P}\right)\tilde{m}(s)\,ds = 0.$$

Hence

$$\lim_{T\to\infty} \sup \frac{1}{T-t_0} \sum_{i=1}^d \pi_i(t_0) h_T(t_0,i)$$

$$= \overline{\lim}_{T\to\infty} \frac{1}{T-t_0} \int_{t_0}^T \sum_{i=1}^d \sum_{j=1}^d \pi_i(t_0) \tilde{p}_{ij}\tilde{m}_j(t)\,dt$$

$$= \overline{\lim}_{T\to\infty} \frac{1}{T} \int_0^T \sum_{i=1}^d \sum_{j=1}^d \pi_i(t_0) \tilde{p}_{ij}\tilde{m}_j(t)\,dt.$$

The last equality follows since $\sum_{i=1}^d \sum_{j=1}^d \pi_i(t_0) \tilde{p}_{ij}\tilde{m}_j(t)$ is a bounded function on $\mathbf{R}_+$. Thus the proof ends.   □

Remark 6.4.2. Concerning the feasibility aspects of the control $\bar{u}(t) = \tilde{F}(t,\eta(t))$ $\bar{x}(t) + \psi(t,\eta(t))$ which is the solution of the above tracking problem, we distinguish two important situations:

(a) If the system (6.1), (6.21) is in the time invariant case and the signal $r(t)$ satisfies $r(t,i) = r(i)$, $(t,i) \in \mathbf{R}_+ \times \mathcal{D}$, then the stabilizing solution of the system (5.117) is constant and solves the system of algebraic equations. This solution may be computed applying the iterative procedure described in Sect. 5.8. By uniqueness arguments it follows that the bounded solution of the system (6.35) is constant and it solves the system of linear equations

$$\left[A_0(i) + B_0(i)\tilde{F}(i)\right]^T \tilde{g}(i) + \sum_{j=1}^d q_{ij}\tilde{g}(j) - \left[C_0(i) + D_0(i)\tilde{F}(i)\right]^T r(i) = 0, \ i \in \mathcal{D}.$$

(b) If the coefficients of the system (6.1), (6.21) are θ-periodic functions, then the stabilizing solution of the system (5.117) is a θ-periodic function, and it can be computed with the iterative procedure given in Sect. 5.8. From the uniqueness arguments the bounded solution of the system (6.35) is a θ-periodic function and its initial conditions can be obtained by solving a linear system of algebraic equations.

(c) Under the assumptions of Theorem 6.4.3 it follows that the optimal value of the tracking problem does not depend upon x_0.

Notes and References

The results presented in this chapter are mainly based on the papers [44, 45, 117]. The linear quadratic problem in the stochastic case has been investigated starting with [151]. For stochastic linear systems with multiplicative noise we mention [3, 4, 11, 17, 76, 77, 82–84, 98, 111, 126, 127, 147, 149, 150] and for infinite dimensional case we cite [23–25, 123, 139]. In the case of stochastic systems subject to Markovian perturbations, the linear quadratic problem has been addressed in [86, 87, 109, 112].

Chapter 7
Stochastic H_2 Optimal Control

In this chapter the problem of H_2-control of a continuous-time linear system subject to Markovian jumping and independent multiplicative and additive white noise perturbations is considered. Several kinds of H_2 type of performance criteria (often called H_2-norms) are introduced and characterized via solutions of some suitable linear equations on the spaces of symmetric matrices. The purpose of such performance criteria is to provide a measure of the effect of additive white noise perturbation over an output of the controlled system. Different aspects specific to systems affected by Markov processes are emphasized. Firstly, the problem of optimization of H_2-norms is solved under the assumption that full state vector is available for measurements. One shows that among all stabilizing controllers of arbitrary dimension, the best performance is achieved by a zero order controller. The corresponding feedback gain of the optimal controller is constructed based on the stabilizing solution of a system of generalized Riccati equations. Secondly, the H_2 optimization problem is solved under the assumption that only an output is available for measurements. The state-space realization of the H_2-optimal controller coincides with the stochastic version of the well-known Kalman–Bucy filter. In the construction of the optimal controller the stabilizing solutions of two systems of coupled Riccati equations are involved. Finally a problem of the H_2 filtering in the case of stochastic systems affected by multiplicative and additive white noise and Markovian switching is solved. Throughout this chapter we assume that $\mathcal{D} = \{1, 2, \ldots, d\}$ and the controlled systems are in the time invariant setting.

7.1 Stochastic H_2 Norms

Consider the linear stochastic system **G** described by

$$
\begin{aligned}
dx(t) &= A_0(\eta(t))x(t)\,dt + \textstyle\sum_{k=1}^{r} A_k(\eta(t))x(t)\,dw_k(t) \\
&\quad + B_v(\eta(t))\,dv(t) \\
z(t) &= C(\eta(t))x(t)
\end{aligned}
\tag{7.1}
$$

with $x \in \mathbf{R}^n$, $z \in \mathbf{R}^p$, $A_k(i) \in \mathbf{R}^{n \times n}$, $k = 0, \ldots, r$, $B_\nu(i) \in \mathbf{R}^{n \times m_\nu}$, $C(i) \in \mathbf{R}^{p \times n}$, $i \in \mathcal{D}$, $w_k(t)$, $t \geq 0$ is a scalar Wiener process and $\nu(t)$, $t \geq 0$ is an m_ν-dimensional Wiener process.

As in the previous chapters, $w(t) = (w_1(t), \ldots, w_r(t))^T$ and $\eta(t)$ are standard Wiener process and Markov process, respectively, with the properties in Sect. 1.8; $\nu(t)$, $t \geq 0$ is an m_ν-dimensional standard Wiener process independent of the pair $(w(t), \eta(t))$, $t \geq 0$. Throughout this chapter, $\mathcal{F}_t$, $\mathcal{G}_t$, $\mathcal{H}_t$ are the σ-algebras defined in Chap. 1 related to the processes $w(t)$ and $\eta(t)$ and $\widehat{\mathcal{H}}_t$ is the smallest σ-algebra containing $\mathcal{H}_t$ and the σ-algebra generated by $\nu(s)$, $0 \leq s \leq t$.

Denoting by $\Phi(t,s)$ the fundamental matrix solution of the system

$$dx(t) = A_0(\eta(t))x(t)\,dt + \sum_{k=1}^{r} A_k(\eta(t))x(t)\,dw_k(t), \tag{7.2}$$

according to (1.29) the solutions of (7.1) have the following representation

$$x(t) = \Phi(t,0)x_0 + \Phi(t,0)\int_0^t \Phi^{-1}(s,0)B_\nu(\eta(s))\,d\nu(s). \tag{7.3}$$

Particularly, the solution of (7.1) with zero initial conditions is:

$$x_0(t) = \Phi(t,0)\int_0^t \Phi^{-1}(s,0)B_\nu(\eta(s))\,d\nu(s). \tag{7.4}$$

We prove

Lemma 7.1.1. *For each* $\tau > 0$ *and* $j \in \mathcal{D}$ *we have*

$$E\left[x_0(\tau)x_0^T(\tau)\chi_{\eta(\tau)=j}\right] = E\left[\int_0^\tau \Phi(\tau,s)B_\nu(\eta(s))B_\nu^T(\eta(s))\Phi^T(\tau,s)\chi_{\eta(\tau)=j}ds\right]$$

$$\tag{7.5}$$

Proof. Set

$$\Psi(s) = \Phi^{-1}(s,0)B_\nu(\eta(s)). \tag{7.6}$$

It is obvious that the components of Ψ belong to $L^{2p}_{\eta,w}[0,\tau]$ for all integer $p \geq 1$ and in particular for $p = 2$.

We show that

$$E\left[\Phi(\tau,0)\int_0^\tau \Psi(t)\,d\nu(t)\left(\Phi(\tau,0)\int_0^\tau \Psi(t)\,d\nu(t)\right)^T \chi_{\eta(\tau)=j}\right]$$

$$= E\int_0^\tau \Phi(\tau,0)\Psi(t)\Psi^T(t)\Phi^T(\tau,0)\chi_{\eta(\tau)=j}dt. \tag{7.7}$$

To this end we prove (7.7) for the case when the elements of Ψ are step functions in $L^4_{\eta,w}[0,\tau]$. Indeed, let

$$\Psi(t) = \sum_{i=0}^{k-1} \Psi(t_i)\chi_{[t_i,t_{i+1})}, \ 0 = t_0 < t_1 < \ldots < t_{k-1} < t_k = \tau,$$

$\Psi(t_i)$ being $\mathcal{H}_{t_i}$ measurable, $0 \leq i \leq k$, $E|\Psi(t_i)|^4 < \infty$. We have

$$E\left[\Phi(\tau,0)\int_0^\tau \Psi(t)\,dv(t)\left(\Phi(\tau,0)\int_0^\tau \Psi(t)\,dv(t)\right)^T \chi_{\eta(\tau)=j} \mid \mathcal{H}_\tau\right]$$

$$= E\left[\Phi(\tau,0)\sum_{i,l}\Psi(t_i)\left(v(t_{i+1})-v(t_i)\right)\left(v(t_{l+1})-v(t_l)\right)^T \Psi^T(t_l)\right.$$

$$\left.\times\Phi^T(\tau,0)\chi_{\eta(\tau)=j} \mid \mathcal{H}_\tau\right] \tag{7.8}$$

$$= \Phi(\tau,0)\sum_{i,l}\Psi(t_i)E\left[\left(v(t_{i+1})-v(t_i)\right)\left(v(t_{l+1})-v(t_l)\right)^T \mid \mathcal{H}_\tau\right]$$

$$\times\Psi^T(t_l)\Phi^T(\tau,0)\chi_{\eta(\tau)=j}$$

$$= \Phi(\tau,0)\left(\sum_{i=0}^{k-1}\Psi(t_i)\Psi^T(t_i)(t_{i+1}-t_i)\right)\Phi^T(\tau,0)\chi_{\eta(\tau)=j}.$$

The last equality above has been obtained by taking into account that the σ-algebra generated by $\{v(t)-v(s), t,s \in [0,\tau]\}$ is independent of $\mathcal{H}_\tau$ and therefore

$$E\left[\left(v(t_{i+1})-v(t_i)\right)\left(v(t_{l+1})-v(t_l)\right)^T \mid \mathcal{H}_\tau\right]$$

$$= E\left[\left(v(t_{i+1})-v(t_i)\right)\left(v(t_{l+1})-v(t_l)\right)^T\right] = \delta_{i,l}(t_{i+1}-t_i)I_{m_v},$$

where $\delta_{i,l}$ are the Kronecker coefficients. Hence by taking expectation in (7.8) one concludes that (7.5) holds if the elements of Ψ are step functions in $L^4_{\eta,w}([0,\tau])$. Now, based on Remark 1.9.2 take a sequence $\{\Psi_k(t)\}_{k=0,1,\ldots}$ of step functions in $L^4_{\eta,w}([0,\tau])$ such that

$$\lim_{k\to\infty} E\int_0^\tau |\Psi_k(t) - \Psi(t)|^4\,dt = 0. \tag{7.9}$$

Writing (7.7) for each Ψ_k one obtains

$$E\left[\left(\Phi(\tau,0)\int_0^\tau \Psi_k(t)\,dv(t)\right)\left(\Phi(\tau,0)\int_0^\tau \Psi_k(t)\,dv(t)\right)^T \chi_{\eta(\tau)=j}\right]$$
$$= E\int_0^\tau \Phi(\tau,0)\Psi_k(t)\Psi_k^T(t)\Phi^T(\tau,0)\chi_{\eta(\tau)=j}\,dt. \tag{7.10}$$

Using Theorem 1.9.3 and (7.9) above it follows that

$$\lim_{k \to \infty} E\left[\left(\Phi(\tau,0)\int_0^\tau \Psi_k(t)\,dv(t)\right)\left(\Phi(\tau,0)\int_0^\tau \Psi_k(t)\,dv(t)\right)^T \chi_{\eta(\tau)=j}\right]$$

$$= E\left[\left(\Phi(\tau,0)\int_0^\tau \Psi(t)\,dv(t)\right)\left(\Phi(\tau,0)\int_0^\tau \Psi(t)\,dv(t)\right)^T \chi_{\eta(\tau)=j}\right]$$

and

$$\lim_{k \to \infty} E\int_0^\tau \Phi(\tau,0)\Psi_k(t)\Psi_k^T(t)\Phi^T(\tau,0)\chi_{\eta(\tau)=j}dt$$

$$= E\int_0^\tau \Phi(\tau,0)\Psi(t)\Psi^T(t)\Phi^T(\tau,0)\chi_{\eta(\tau)=j}dt.$$

Combining the last two equalities with (7.10) one obtains (7.7). By replacing $\Psi(t)$ in (7.7) with (7.6), (7.5) directly follows because $\Phi(\tau,0)\Phi^{-1}(s,0) = \Phi(\tau,s)$ a.s. and thus the proof ends.   □

Remark 7.1.1. If we consider the particular case when $A_k(i) = 0, 1 \le k \le r, i \in \mathcal{D}$ the proof of the above lemma does not become simpler. This is due to the fact that in the representation formula (7.4) we cannot write

$$x_0(\tau) = \int_0^\tau \Phi(\tau,s)B_v(\eta(s))\,dv(s) \tag{7.11}$$

because the expression under integral is random and it is measurable with respect to $\mathcal{H}_\tau$. On the other hand, the integral in (7.11) is well defined if the function under integral is measurable with respect to

$$\widehat{\mathcal{H}}_s = \mathcal{H}_s \vee \sigma(v(t), 0 \le t \le s)$$

for all $s < \tau$.

Let us introduce the following notations

$$\pi_i(t) = \mathcal{P}\{\eta(t) = i\}, \tag{7.12}$$

$$\widetilde{P} = \lim_{t \to \infty} P(t) \quad \text{with elements } \tilde{p}_{ij} \tag{7.13}$$

$$\pi_i = \mathcal{P}(\eta(0) = i) = \pi_i(0) \tag{7.14}$$

$$\pi_{i\infty} = \sum_{j=1}^d \pi_j \tilde{p}_{ji}. \tag{7.15}$$

It is obvious that

$$\pi_i(t) = \sum_{j=1}^d \pi_j p_{ji}(t).$$

Hence

$$\lim_{t \to \infty} \pi_i(t) = \pi_{i\infty}.$$

Set

$$\widehat{B}_v(s,i) = \pi_i(s) B_v(i) B_v^T(i) \tag{7.16}$$

$$\widehat{B}_v(i) = \pi_{i\infty} B_v(i) B_v^T(i). \tag{7.17}$$

It is clear that

$$\lim_{s \to \infty} \widehat{B}_v(s,i) = \widehat{B}_v(i) \text{ for all } i \in \mathcal{D} \tag{7.18}$$

With these notations we prove the following lemma.

Lemma 7.1.2. *With $x_0(t)$ defined by (7.4) we have*

$$E\left[x_0(\tau) x_0^T(\tau) \chi_{\eta(\tau)=j}\right] = \int_0^\tau \left(e^{\mathcal{L}(\tau-s)} \widehat{B}_v(s)\right)(j) \, ds$$

where $\widehat{B}_v(s) = \left(\widehat{B}_v(s,1), \ldots, \widehat{B}_v(s,d)\right)$ with $\widehat{B}_v(s,i)$ given by (7.16) and $\mathcal{L}$ is the Lyapunov operator defined by the system $(A_0, A_1, \ldots, A_r; Q)$.

Proof. Based on Lemma 7.1.1 we may write successively:

$$E\left[x_0(\tau) x_0^T(\tau) \chi_{\eta(\tau)=j}\right]$$

$$= \int_0^\tau E\left[\Phi(\tau,s) B_v(\eta(s)) B_v^T(\eta(s)) \Phi^T(\tau,s) \chi_{\eta(\tau)=j}\right] ds$$

$$= \int_0^\tau \sum_{i=1}^d \pi_i(s) E\left[\Phi(\tau,s) B_v(\eta(s)) B_v^T(\eta(s)) \Phi^T(\tau,s) \chi_{\eta(\tau)=j} \mid \eta(s) = i\right] ds$$

$$= \int_0^\tau \sum_{i=1}^d E\left[\Phi(\tau,s) \widehat{B}_v(\eta(s)) \Phi^T(\tau,s) \chi_{\eta(\tau)=j} \mid \eta(s) = i\right] ds$$

$$= \int_0^\tau \left(T(\tau,s) \widehat{B}_v(s)\right)(j) \, ds.$$

For the last equality above we used the representation formula (3.8) of the evolution operator $T(t,s)$. The conclusion follows because in time-invariant case $T(t,s) = e^{\mathcal{L}(t-s)}$ (see Remark 2.6.3). □

Lemma 7.1.3. *Assume that the system $(A_0, A_1, \ldots, A_r; Q)$ is stable. Then we have*

$$\lim_{\tau \to \infty} E\left[x_0(\tau) x_0^T(\tau) \chi_{\eta(\tau)=j}\right] = \widehat{P}_c(j) \tag{7.19}$$

where $\widehat{P}_c = \left(\widehat{P}_c(1), \ldots, \widehat{P}_c(d) \right)$ is the unique positive semidefinite solution of the Lyapunov like equation $\mathcal{L}P + \widehat{B}_v = 0$ with $\widehat{B}_v = \left(\widehat{B}_v(1), \ldots, \widehat{B}_v(d) \right)$, $\widehat{B}_v$ being defined by (7.17).

Proof. Based on Lemma 7.1.2 we have

$$E\left[x_0(\tau) x_0^T(\tau) \chi_{\eta(\tau)=j} \right] = \int_0^\tau \left(e^{\mathcal{L}(\tau-s)} \widehat{B}_v(s) \right)(j)\,ds$$

$$= \int_0^\tau \left(e^{\mathcal{L}(\tau-s)} \left(\widehat{B}_v(s) - \widehat{B}_v \right) \right)(j)\,ds$$

$$+ \int_0^\tau \left(e^{\mathcal{L}(\tau-s)} \widehat{B}_v \right)(j)\,ds.$$

By a simple change of integration variable we get

$$E\left[x_0(\tau) x_0^T(\tau) \chi_{\eta(\tau)=j} \right] = \int_0^\tau \left(e^{\mathcal{L}(\tau-s)} \left(\widehat{B}_v(s) - \widehat{B}_v \right) \right)(j)\,ds$$

$$+ \int_0^\tau \left(e^{\mathcal{L}s} \widehat{B}_v \right)(j)\,ds. \qquad (7.20)$$

Since the system $(A_0, A_1, \ldots, A_r; Q)$ is stable, there exist $\beta \geq 1$, $\alpha > 0$ such that $\left\| e^{\mathcal{L}s} \right\| \leq \beta e^{-\alpha s}$ for all $s \geq 0$. Further we have:

$$\left| \int_0^\tau \left(e^{\mathcal{L}(\tau-s)} \left(\widehat{B}_v(s) - \widehat{B}_v \right) \right)(j)\,ds \right| \leq \left| \int_0^\tau \left(e^{\mathcal{L}(\tau-s)} \left(\widehat{B}_v(s) - \widehat{B}_v \right) \right)ds \right|$$

$$\leq \beta \int_0^\tau e^{-\alpha(\tau-s)} \left| \widehat{B}_v(s) - \widehat{B}_v \right|\,ds.$$

Taking $\tau \to \infty$ one obtains from (7.18) by standard arguments

$$\lim_{\tau\to\infty} \beta \int_0^\tau e^{-\alpha(\tau-s)} \left| \widehat{B}_v(s) - \widehat{B}_v \right|\,ds = 0$$

which leads to

$$\lim_{\tau\to\infty} \int_0^\tau \left(e^{\mathcal{L}(\tau-s)} \left(\widehat{B}_v(s) - \widehat{B}_v \right) \right)(j)\,ds = 0.$$

Hence from (7.20) we get

$$\lim_{\tau\to\infty} E\left[x_0(\tau) x_0^T(\tau) \chi_{\eta(\tau)=j} \right] = \int_0^\infty \left(e^{\mathcal{L}s} \widehat{B}_v \right)(j)\,ds = \widehat{P}_c(j).$$

The last equality follows from the proof of Theorem 2.7.7 combined with Theorem 2.3.7 (iii). Thus the proof is complete. $\qquad \square$

Remark 7.1.2. From the representation formulae (7.3) and (7.4) together with Lemma 7.1.3 it follows that if the system $(A_0, A_1, \ldots, A_r; Q)$ is stable then

$$\lim_{t \to \infty} E\left[x(t) x^T(t) \chi_{\eta(t)=j}\right] = \lim_{t \to \infty} E\left[x_0(t) x_0^T(t) \chi_{\eta(t)=j}\right] = \widehat{P}_c(j)$$

for all $j \in \mathcal{D}$ and for any solution $x(t)$ of the system (7.1).

Theorem 7.1.4. *Assume that the system $(A_0, A_1, \ldots, A_r; Q)$ is stable. Then*

$$\lim_{t \to \infty} E|z(t)|^2 = \sum_{j=1}^{d} Tr\left(C(j) \widehat{P}_c(j) C^T(j)\right)$$

$$= \sum_{j=1}^{d} \pi_{j\infty} Tr\left(B_v^T(j) \widehat{P}_o(j) B_v(j)\right)$$

where $\widehat{P}_o = \left(\widehat{P}_o(1), \ldots, \widehat{P}_o(d)\right)$ is the unique positive semidefinite solution of the equation

$$\mathcal{L}^* P_o + \widetilde{C} = 0$$

with $\widetilde{C} = \left(\widetilde{C}(1), \ldots, \widetilde{C}(d)\right), \widetilde{C}(j) = C^T(j) C(j), j \in \mathcal{D}.$

Proof. First we shall prove the result in the statement for $z_0 = C(\eta(t)) x_0(t)$. To this end we have

$$\lim_{t \to \infty} E|z_0(t)|^2 = \lim_{t \to \infty} Tr E\left[z_0(t) z_0^T(t)\right]$$

$$= \lim_{t \to \infty} Tr E\left[C(\eta(t)) x_0(t) x_0^T(t) C^T(\eta(t))\right]$$

$$= \lim_{t \to \infty} \sum_{j=1}^{d} Tr E\left[C(\eta(t)) x_0(t) x_0^T(t) \chi_{\eta(t)=j} C^T(\eta(t))\right]$$

$$= \lim_{t \to \infty} \sum_{j=1}^{d} Tr C(j) E\left[x_0(t) x_0^T(t) \chi_{\eta(t)=j}\right] C^T(j).$$

Then based on Lemma 7.1.3 we get

$$\lim_{t \to \infty} E|z_0(t)|^2 = \sum_{j=1}^{d} Tr\left(C(j) \widehat{P}_c(j) C^T(j)\right). \tag{7.21}$$

Taking into account the definition of the inner product in $\mathcal{S}_n^d$ (see (2.16)) and the representation formulae of $\widehat{P}_c$ and $\widehat{P}_o$, we have

$$\sum_{j=1}^{d} Tr\left(C(j)\widehat{P}_c(j)C^T(j)\right) = \sum_{j=1}^{d} Tr\left(\widehat{P}_c(j)C^T(j)C(j)\right)$$

$$= \left\langle \widehat{P}_c, \widetilde{C} \right\rangle = \int_0^\infty \left\langle e^{\mathcal{L}t}\widehat{B}_v, \widetilde{C} \right\rangle dt$$

$$= \int_0^\infty \left\langle \widehat{B}_v, e^{\mathcal{L}^*t}\widetilde{C} \right\rangle dt = \left\langle \widehat{B}_v, \widehat{P}_o \right\rangle$$

$$= \sum_{j=1}^{d} Tr\left(\widehat{B}_v(j)\widehat{P}_o(j)\right)$$

$$= \sum_{j=1}^{d} \pi_{j\infty} Tr\left(B_v^T(j)\widehat{P}_o(j)B_v(j)\right).$$

Finally we remark that based on the representation formula (7.3) it follows that for any output $z(t)$ we have

$$\lim_{t\to\infty} E\,|z(t)|^2 = \lim_{t\to\infty} E\,|z_0(t)|^2$$

and the proof is complete. $\square$

For the system $\mathbf{G}$ defined by (7.1), under the assumption of Theorem 7.1.4 we introduce the following norm.

Definition 7.1.1. We call the H_2 *norm* of the system (7.1)

$$\|\mathbf{G}\|_2 = \left[\lim_{t\to\infty} E\,|z(t)|^2\right]^{\frac{1}{2}}. \tag{7.22}$$

Remark 7.1.3. The result in Theorem 7.1.4 shows that the right-hand side of (7.22) is well defined and a characterization of the H_2 norm can be given in terms of the controllability and observability Gramians $\widehat{P}_c$ and $\widehat{P}_o$ respectively, which extends to the case of stochastic systems of type (7.1) the well-known results from the deterministic setting.

Further we prove the following theorem.

Theorem 7.1.5. *Under the assumption of Theorem 7.1.4 we have*

$$\lim_{T\to\infty} \frac{1}{T}E\left[\int_0^T |z(s)|^2\,ds \mid \eta(0)=i\right] = \sum_{j=1}^{d} Tr\left(B_v^T(j)\widehat{P}_o(j)B_v(j)\right)\tilde{p}_{ij}. \tag{7.23}$$

Proof. Applying the Itô-type formula (Theorem 1.10.2) for the system (7.1) and for the function $v(x,i) = x^T\widehat{P}_o(i)x$, $x \in \mathbf{R}^n$, $i \in \mathcal{D}$ one obtains

$$E\left[\int_0^T |z(s)|^2 \, ds \mid \eta(0) = i\right] \tag{7.24}$$

$$= E\left[\int_0^T Tr\left(B_v^T(\eta(s))\widehat{P}_o(\eta(s))B_v(\eta(s))\right) ds \mid \eta(0) = i\right]$$

$$+ x_0^T \widehat{P}_o(i) x_0 - E\left[x^T(T)\widehat{P}_o(\eta(T))x(T) \mid \eta(0) = i\right].$$

But

$$E\left[\int_0^T Tr\left(B_v^T(\eta(s))\widehat{P}_o(\eta(s))B_v(\eta(s))\right) ds \mid \eta(0) = i\right]$$

$$= E\left[\int_0^T \sum_{j=1}^d Tr\left(B_v^T(j)\widehat{P}_o(j)B_v(j)\chi_{\eta(s)=j}\right) ds \right. \tag{7.25}$$

$$\left. \mid \eta(0) = i\right]$$

$$= \sum_{j=1}^d Tr\left(B_v^T(j)\widehat{P}_o(j)B_v(j)\right) \int_0^T E\left[\chi_{\eta(s)=j} \mid \eta(0) = i\right] ds$$

$$= \sum_{j=1}^d Tr\left(B_v^T(j)\widehat{P}_o(j)B_v(j)\right) \int_0^T p_{ij}(s) \, ds.$$

Since $\lim_{s\to\infty} p_{ij}(s) = \tilde{p}_{ij}$ we obtain from (7.25) that

$$\lim_{T\to\infty} \frac{1}{T} E\left[\int_0^T Tr\left(B_v^T(\eta(s))\widehat{P}_o(\eta(s))B_v(\eta(s))\right) ds \mid \eta(0) = i\right]$$

$$= \sum_{j=1}^d Tr\left(B_v^T(j)\widehat{P}_o(j)B_v(j)\right) \lim_{T\to\infty} \frac{1}{T} \int_0^T p_{ij}(s) \, ds \tag{7.26}$$

$$= \sum_{j=1}^d Tr\left(B_v^T(j)\widehat{P}_o(j)B_v(j)\right) \tilde{p}_{ij}.$$

Based on Lemma 7.1.3 it follows that

$$\lim_{T\to\infty} \frac{1}{T} \left\{ x_0^T \widehat{P}_o(i) x_0 - E\left[x^T(T)^T \widehat{P}_o(\eta(T))x(T) \mid \eta(0) = i\right]\right\}$$

$$= \lim_{T\to\infty} \frac{1}{T} \left\{ x_0^T \widehat{P}_o(i) x_0 - \sum_{j=1}^d Tr\left(\widehat{P}_o(j)\right) \right. \tag{7.27}$$

$$\times E\left[x(T)x^T(T)\chi_{\eta(T)=j} \mid \eta(0) = i\right]\right\}$$

$$= 0.$$

Finally from (7.24) combined with (7.26) and (7.27) we get (7.23) and the proof ends.   □

Evidently that the next result holds.

Corollary 7.1.6. *Under the assumptions of Theorem 7.1.4 the following hold:*

$$\lim_{T \to \infty} \frac{1}{T} \int_0^T |z(t)|^2 \, dt = \lim_{T \to \infty} E |z(T)|^2 = \|\mathbf{G}\|_2^2.$$ □

Theorem 7.1.7. *Assume that the system $(A_0, A_1, \ldots, A_r; Q)$ is stable. Then*

$$\lim_{T \to \infty} \frac{1}{T} \sum_{i=1}^d E \left[\int_0^T |z(s)|^2 \, ds \mid \eta(0) = i \right]$$

$$= \sum_{j=1}^d \delta_j Tr \left(B_v^T(j) \widehat{P}_o(j) B_v(j) \right)$$

$$= \sum_{j=1}^d Tr \left(C(j) \widetilde{P}_c(j) C^T(j) \right),$$

where

$$\delta_j = \sum_{i=1}^d \tilde{p}_{ij}$$

and

$$\widetilde{P}_c = \left(\widetilde{P}_c(1), \ldots, \widetilde{P}_c(d) \right)$$

is the unique positive semidefinite solution of the equation $\mathcal{L}P + \widehat{M} = 0$, with $\widehat{M}(i) = \delta_i B_v(i) B_v^T(i)$.

Proof. From Theorem 7.1.5 we have

$$\lim_{T \to \infty} \frac{1}{T} \sum_{i=1}^d \left[\int_0^T |z(s)|^2 \, ds \mid \eta(0) = i \right]$$

$$= \sum_{i,j=1}^d Tr \left(B_v^T(j) \widehat{P}_o(j) B_v(j) \right) \tilde{p}_{ij}$$

$$= \sum_{j=1}^d \delta_j Tr \left(B_v^T(j) \widehat{P}_o(j) B_v(j) \right)$$

$$= \left\langle \widehat{P}_o, \widehat{M} \right\rangle = \int_0^\infty \left\langle e^{\mathcal{L}^* t} \widetilde{C}, \widehat{M} \right\rangle dt$$

$$= \int_0^\infty \left\langle \widetilde{C}, e^{\mathcal{L} t} \widehat{M} \right\rangle dt = \left\langle \widetilde{C}, \widetilde{P}_c \right\rangle$$

$$= \sum_{j=1}^{d} Tr\left(C(j)\tilde{P}_c(j)C^T(j)\right)$$

and hence the proof is complete. □

Using the result in the above theorem, one can introduce a new norm for the system **G** given by Theorem 7.1.7.

Definition 7.1.2. If the zero-solution of the system (7.1) in the absence of the additive noise $v(t)$ is ESMS, then we define:

$$|||\mathbf{G}|||_2^2 = \lim_{T\to\infty} \frac{1}{T} \sum_{i=1}^{d} E\left[\int_0^T |z(s)|^2 ds \mid \eta(0) = i\right].$$

Remark 7.1.4. (a) Based on the results in Theorems 7.1.4 and 7.1.7 it follows that while $\|\mathbf{G}\|_2$ depends on the initial distribution $\pi = (\pi_1, \ldots, \pi_d)$ of the process $\eta(t)$, the norm $|||\mathbf{G}|||_2$ does not depend on the initial distribution of $\eta(t)$.
(b) In the particular case when the system (7.1) is subject only to white noise perturbations, the two norms defined above coincides. The difference between them is due to the Markov jump perturbations.
(c) It is obvious that

$$\|\mathbf{G}\|_2 \le |||\mathbf{G}|||_2.$$

7.2 Stochastic H_2 Optimal Control: The Case of Perfect State Measurements

In this section we shall state and solve the design problem of a stabilizing controller that minimizes the H_2 norm of a controlled system whose states are accessible for measurements.

Consider the system **G** described by

$$dx(t) = [A_0(\eta(t))x(t) + B_0(\eta(t))u(t)]dt$$

$$+ \sum_{k=1}^{r} [A_k(\eta(t))x(t) + B_k(\eta(t))u(t)]dw_k(t) \qquad (7.28)$$

$$+ B_v(\eta(t))dv(t)$$

$$z(t) = C(\eta(t))x(t) + D(\eta(t))u(t)$$

where $x \in \mathbf{R}^n$ is the state vector, $u \in \mathbf{R}^m$ denotes the vector of control variables, $z \in \mathbf{R}^p$ is the regulated output, and $A_k(i), B_k(i), 0 \le k \le r, C(i), D(i), B_v(i), i \in \mathcal{D}$ are constant matrices of appropriate dimensions with real elements. The stochastic processes $\{w(t)\}_{t \geq 0} = (w_1(t), \ldots, w_r(t))^T$, $\{\eta(t)\}_{t \geq 0}$ and $\{v(t)\}_{t \geq 0}$ have the properties stated in the preceding section.

Consider the following family of controllers $\mathbf{G}_c$ described by

$$\dot{x}_c(t) = A_c(\eta(t)) x_c(t) + B_c(\eta(t)) u_c(t) \tag{7.29}$$

$$y_c(t) = C_c(\eta(t)) x_c(t) + D_c(\eta(t)) u_c(t)$$

where $x_c \in \mathbf{R}^{n_c}, u_c \in \mathbf{R}^n, y_c \in \mathbf{R}^m$. Let us remark that the controller $\mathbf{G}_c$ of form (7.29) is completely determined by the set of parameters $(n_c, A_c(i), B_c(i), C_c(i), D_c(i), i \in \mathcal{D})$ where $n_c \geq 0$ denotes the controller order. In the particular case $n_c = 0$ the controller (7.29) reduces to

$$y_c(t) = D_c(\eta(t)) u_c(t)$$

which shows that the zero order (state-feedback) controllers are included in the set of controllers (7.29).

The resulting system $\mathbf{G}_{cl}$ obtained by coupling a controller of the form (7.29) to the system (7.28) by taking $u_c(t) = x(t)$ and $u(t) = y_c(t)$ is

$$dx_{cl}(t) = A_{0cl}(\eta(t)) x_{cl}(t) dt + \sum_{k=1}^{r} A_{kcl}(\eta(t)) x_{cl}(t) dw_k(t)$$

$$+ B_{vcl}(\eta(t)) dv(t) \tag{7.30}$$

$$y_{cl}(t) = C_{cl}(\eta(t)) x_{cl}(t)$$

where

$$x_{cl} = \begin{bmatrix} x \\ x_{cl} \end{bmatrix};$$

$$A_{0cl}(i) = \begin{bmatrix} A_0(i) + B_0(i) D_c(i) & B_0(i) C_c(i) \\ B_c(i) & A_c(i) \end{bmatrix};$$

$$A_{kcl}(i) = \begin{bmatrix} A_k(i) + B_k(i) D_c(i) & B_k(i) C_c(i) \\ 0 & 0 \end{bmatrix};$$

$$B_{vcl}(i) = \begin{bmatrix} B_v(i) \\ 0 \end{bmatrix};$$

$$C_{cl}(i) = [C(i) + D(i) D_c(i) \quad D(i) C_c(i)].$$

Definition 7.2.1. A controller $\mathbf{G}_c$ of form (7.29) is called *stabilizing* for the system (7.28) if the zero solution of the closed-loop system (7.30) (in the absence of the noise v) is ESMS.

By $\mathcal{K}_s(\mathbf{G})$ we denote the set of all stabilizing controllers $\mathbf{G}_c$ of the form (7.29). Then two optimization problems will be formulated and solved in the following:

(OP1) Find a stabilizing controller of the form (7.29) minimizing $\|\mathbf{G}_{cl}\|_2$.
(OP2) Find a stabilizing controller of the form (7.29) minimizing $\||\mathbf{G}_{cl}\||_2$.

For the sake of simplicity we shall unify the notations writing $\|\cdot\|_{2,\ell}$, $\ell = 1,2$ where $\|\cdot\|_{2,1}$ stands for $\|\cdot\|_2$ and $\|\cdot\|_{2,2}$ stands for $\||\cdot\||_2$. Thus from Theorems 7.1.4 and 7.1.7 we have

$$\|\mathbf{G}_{cl}\|_{2,\ell}^2 = \sum_{i=1}^{d} \varepsilon_i Tr\left(B_{vcl}^T(i)\,\widehat{P}_{ocl}(i)\,B_{vcl}(i)\right) \tag{7.31}$$

where

$$\varepsilon_i = \pi_{i\infty} \quad \text{for} \quad \ell = 1 \quad \text{and} \tag{7.32}$$

$$\varepsilon_i = \delta_i \quad \text{for} \quad \ell = 2$$

and $\widehat{P}_{ocl}(i) = \left(\widehat{P}_{ocl}(1),\ldots,\widehat{P}_{ocl}(d)\right)$ is the unique positive semidefinite solution of the Lyapunov-type equation on $\mathcal{S}_{n+n_c}^d$, with n_c denoting the order of the controller:

$$\begin{aligned}
&A_{0cl}^T(i)\,\widehat{P}_{ocl}(i) + \widehat{P}_{ocl}(i)\,A_{0cl}(i) + \Sigma_{k=1}^{r}\,A_{kcl}^T(i)\,\widehat{P}_{ocl}(i)\,A_{kcl}(i)\\
&+ \Sigma_{j=1}^{d}\,q_{ij}\widehat{P}_{ocl}(j) + C_{cl}^T(i)\,C_{cl}(i) = 0,\ i \in \mathcal{D}.
\end{aligned} \tag{7.33}$$

One can associate with the system (7.28) the following *stochastic generalized Riccati algebraic equations* (SGRAE)

$$\begin{aligned}
&A_0^T(i)\,X(i) + X(i)\,A_0(i) + \Sigma_{k=1}^{r}\,A_k^T(i)\,X(i)\,A_k(i)\\
&+ \Sigma_{j=1}^{d}\,q_{ij}X(j) - \left[X(i)\,B_0(i) + \Sigma_{k=1}^{r}\,A_k^T(i)\,X(i)\,B_k(i) + C^T(i)\,D(i)\right]\\
&\times \left[D^T(i)\,D(i) + \Sigma_{k=1}^{r}\,B_k^T(i)\,X(i)\,B_k(i)\right]^{-1}\\
&\times \left[B_0^T(i)\,X(i) + \Sigma_{k=1}^{r}\,B_k^T(i)\,X(i)\,A_k(i) + D^T(i)\,C(i)\right] + C^T(i)\,C(i) = 0,
\end{aligned} \tag{7.34}$$

$i \in \mathcal{D}$, which can be written in a compact form as

$$\mathcal{L}^* X - \mathcal{P}^T(X)\,\mathcal{R}^{-1}(X)\,\mathcal{P}(X) + \widetilde{C} = 0$$

where $\mathcal{L}$ is the Lyapunov operator defined by the system $(A_0, A_1, \ldots, A_r; Q)$ and

$$\mathcal{P}(X) = (\mathcal{P}_1(X), \ldots, \mathcal{P}_d(X))$$

with

$$\mathcal{P}_i(X) = B_0^T(i)\,X(i) + \sum_{k=1}^{r} B_k^T(i)\,X(i)\,A_k(i) + D^T(i)\,C(i)$$

and

$$\mathcal{R}(X) = (\mathcal{R}_1(X),\ldots,\mathcal{R}_d(X))$$

with

$$\mathcal{R}_i(X) = D^T(i)D(i) + \sum_{k=1}^{r} B_k^T(i)X(i)B_k(i).$$

Denote by

$$\Lambda^\Sigma(X) = \left(\Lambda_1^\Sigma(X),\ldots,\Lambda_d^\Sigma(X)\right) \in \mathcal{S}_{n+m}^d$$

the *generalized dissipation matrix*, where

$$\Lambda_i^\Sigma(X) = \begin{bmatrix} (\mathcal{L}^*X)(i) + \widetilde{C}(i) & \mathcal{P}_i^T(X) \\ \mathcal{P}_i(X) & \mathcal{R}_i(X) \end{bmatrix}.$$

Assume that the following conditions are fulfilled

H1. The system $(\mathbf{A},\mathbf{B};Q)$ is stabilizable, where as usual, $\mathbf{A} = (A_0,A_1,\ldots,A_r)$, $\mathbf{B} = (B_0,B_1,\ldots,B_r)$;

H2. It exists $\widehat{X} = \left(\widehat{X}(1),\ldots,\widehat{X}(d)\right)$ such that $\Lambda^\Sigma\left(\widehat{X}\right) > 0$.

Applying Theorem 5.6.9 we deduce that the SGRAE (7.34) has a stabilizing solution $\widetilde{X}$. Defining now the gains

$$\widetilde{F}(i) = -\mathcal{R}_i^{-1}\left(\widetilde{X}\right)\mathcal{P}_i\left(\widetilde{X}\right), \; i \in \mathcal{D} \tag{7.35}$$

it results that the control

$$u = \widetilde{F}(\eta(t))x(t)$$

stabilizes the system (7.28) in the absence of the additive noise $v(t)$.

The corresponding closed-loop system $\widetilde{\mathbf{G}}_{cl}$ is

$$dx_{cl}(t) = \left[A_0(\eta(t)) + B_0(\eta(t))\widetilde{F}(\eta(t))\right]x(t)\,dt$$

$$+ \sum_{k=1}^{r}\left[A_k(\eta(t)) + B_k(\eta(t))\widetilde{F}(\eta(t))\right]x(t)\,dw_k(t)$$

$$+ B_v(\eta(t))\,dv(t) \tag{7.36}$$

$$z(t) = \left[C(\eta(t)) + D(\eta(t))\widetilde{F}(\eta(t))\right]x(t)$$

Then the following result is valid.

Proposition 7.2.1. *Under the assumptions* **H1** *and* **H2** *we have*

$$\left\|\widetilde{\mathbf{G}}_{cl}\right\|_{2,\ell}^2 = \sum_{j=1}^{d} \varepsilon_j Tr\left(B_v^T(j)\widetilde{X}(j)B_v(j)\right).$$

Proof. By direct algebraic manipulations (see also Lemma 5.1.1 with $W(i) = \widetilde{F}(i)$) we obtain that the SGRAE (7.34) satisfied by $\widetilde{X}$ can be written in a Lyapunov form as follows:

$$\left[A_0(i) + B_0(i)\widetilde{F}(i)\right]^T \widetilde{X}(i) + \widetilde{X}(i)\left[A_0(i) + B_0(i)\widetilde{F}(i)\right]$$
$$+ \Sigma_{k=1}^{r}\left[A_k(i) + B_k(i)\widetilde{F}(i)\right]^T \widetilde{X}(i)\left[A_k(i) + B_k(i)\widetilde{F}(i)\right]$$
$$+ \Sigma_{j=1}^{d} q_{ij}\widetilde{X}(j) + \left[C(i) + D(i)\widetilde{F}(i)\right]^T \left[C(i) + D(i)\widetilde{F}(i)\right] = 0$$

which shows that the observability Gramian $\widehat{P}_{ocl}$ associated with the closed-loop system (7.36) coincides with the stabilizing solution $\widetilde{X}$ of the SGRAE (7.34). The conclusion in the statement follows from Theorems 7.1.4 and 7.1.7. $\square$

The main result of this section is the following theorem.

Theorem 7.2.2. *Assume that* **H1** *and* **H2** *are fulfilled. Under these conditions we have*

$$\min_{\mathbf{G}_c \in \mathcal{K}_s(\mathbf{G})} \|\mathbf{G}_{cl}\|_{2,\ell} = \left[\sum_{j=1}^{d} \varepsilon_j Tr\left(B_v^T(j)\widetilde{X}(j)B_v(j)\right)\right]^{\frac{1}{2}}$$

and the optimal control is

$$u(t) = \widetilde{F}(\eta(t))x(t)$$

where $\widetilde{X}$ is the stabilizing solution of SGRAE (7.34), $\widetilde{F} = \left(\widetilde{F}(1), \ldots, \widetilde{F}(d)\right)$ is the stabilizing feedback gain defined by (7.35) and ε_i are defined in (7.32).

Proof. Let $\mathbf{G}_c \in \mathcal{K}_s(\mathbf{G})$ and $\mathbf{G}_{cl}$ be the corresponding closed-loop system and $\widehat{P}_{ocl}(i)$ denotes the observability Gramian. Let

$$\begin{bmatrix} U_{11}(i) & U_{12}(i) \\ U_{12}^T(i) & U_{22}(i) \end{bmatrix}$$

be a partition of $\widehat{P}_{ocl}(i)$ conformably with the partition of the state matrix of the resulting system. Partitioning (7.33) according to the partition of $\widehat{P}_{ocl}(i)$ we get:

$$(A_0(i) + B_0(i)D_c(i))^T U_{11}(i) + B_c^T(i)U_{12}^T(i)$$
$$+ U_{11}(i)(A_0(i) + B_0(i)D_c(i)) + U_{12}(i)B_c(i)$$
$$+ \Sigma_{k=1}^{r}(A_k(i) + B_k(i)D_c(i))^T U_{11}(i)(A_k(i) + B_k(i)D_c(i)) \qquad (7.37)$$
$$+ \Sigma_{j=1}^{d} q_{ij}U_{11}(j) + (C(i) + D(i)D_c(i))^T (C(i) + D(i)D_c(i)) = 0$$

$$(A_0(i) + B_0(i)D_c(i))^T U_{12}(i) + B_c^T(i)U_{22}(i) + U_{11}(i)B_0(i)C_c(i)$$
$$+U_{12}(i)A_c(i) + \sum_{k=1}^r (A_k(i) + B_k(i)D_c(i))^T U_{11}(i)B_k(i)C_c(i) \qquad (7.38)$$
$$+\sum_{j=1}^d q_{ij}U_{12}(j) + (C(i) + D(i)D_c(i))^T D(i)C_c(i) = 0$$

$$C_c^T(i)B_0^T(i)U_{12}(i) + A_c^T(i)U_{22}(i) + U_{12}^T B_0(i)C_c(i)$$
$$+U_{22}(i)A_c(i) + \sum_{k=1}^r C_c^T(i)B_k^T(i)U_{11}(i)B_k(i)C_c(i) \qquad (7.39)$$
$$+\sum_{j=1}^d q_{ij}U_{22}(j) + C_c^T(i)D^T(i)D(i)C_c(i) = 0.$$

Using Lemma 5.1.1 with $W(i) = D_c(i)$, SGRAE (7.34) satisfied by the stabilizing solution $\widetilde{X}$ can be written as follows:

$$(A_0(i) + B_0(i)D_c(i))^T \widetilde{X}(i) + \widetilde{X}(i)(A_0(i) + B_0(i)D_c(i))$$
$$+\sum_{k=1}^r (A_k(i) + B_k(i)D_c(i))^T \widetilde{X}(i)(A_k(i) + B_k(i)D_c(i))$$
$$+\sum_{j=1}^d q_{ij}\widetilde{X}(j) + (C(i) + D(i)D_c(i))^T (C(i) + D(i)D_c(i)) \qquad (7.40)$$
$$- \left(D_c(i) - \widetilde{F}(i)\right)^T \mathcal{R}_i\left(\widetilde{X}\right)\left(D_c(i) - \widetilde{F}(i)\right) = 0.$$

Denoting by

$$\widetilde{U}_{11}(i) = U_{11}(i) - \widetilde{X}(i)$$

and subtracting (7.40) from (7.37) one easily obtains that the triplets $\left(\widetilde{U}_{11}(i), U_{12}(i),\right.$ $\left. U_{22}(i)\right)$ solve the following system of algebraic equations:

$$(A_0(i) + B_0(i)D_c(i))^T \widetilde{U}_{11}(i) + \widetilde{U}_{11}(i)(A_0(i) + B_0(i)D_c(i))$$
$$+B_c^T(i)U_{12}^T(i) + U_{12}(i)B_c(i) + \sum_{k=1}^r (A_k(i) + B_k(i)D_c(i))^T$$
$$\times \widetilde{U}_{11}(i)(A_k(i) + B_k(i)D_c(i)) + \sum_{j=1}^d q_{ij}\widetilde{U}_{11}(j) \qquad (7.41)$$
$$+ \left(D_c(i) - \widetilde{F}(i)\right)^T \mathcal{R}_i\left(\widetilde{X}\right)\left(D_c(i) - \widetilde{F}(i)\right) = 0$$

$$(A_0(i) + B_0(i)D_c(i))^T U_{12}(i) + B_c^T(i)U_{22}(i) + \widetilde{U}_{11}(i)B_0(i)C_c(i)$$
$$+U_{12}(i)A_c(i) + \sum_{k=1}^r (A_k(i) + B_k(i)D_c(i))^T \widetilde{U}_{11}(i)B_k(i)C_c(i) \qquad (7.42)$$
$$+\sum_{j=1}^d q_{ij}U_{12}(j) + \left(D_c(i) - \widetilde{F}(i)\right)^T \mathcal{R}_i\left(\widetilde{X}\right)C_c(i) = 0$$

$$C_c^T(i)B_0^T(i)U_{12}(i) + A_c^T(i)U_{22}(i) + U_{12}^T(i)B_0(i)C_c(i)$$
$$+U_{22}(i)A_c(i) + \sum_{k=1}^r C_c^T(i)B_k^T(i)\widetilde{U}_{11}(i)B_k(i)C_c(i) \qquad (7.43)$$
$$+\sum_{j=1}^d q_{ij}U_{22}(j) + C_c^T(i)\mathcal{R}_i\left(\widetilde{X}\right)C_c(i) = 0.$$

Setting

$$\widetilde{U}(i) = \begin{bmatrix} \widetilde{U}_{11}(i) & U_{12}(i) \\ U_{12}^T(i) & U_{22}(i) \end{bmatrix},$$

(7.41)–(7.43) can be written in a compact form as follows:

$$A_{0cl}^T(i)\widetilde{U}(i)+\widetilde{U}(i)A_{0cl}(i)+\sum_{k=1}^r A_{kcl}^T(i)\widetilde{U}(i)A_{kcl}(i)$$
$$+\sum_{j=1}^d q_{ij}\widetilde{U}(i)+\Theta^T(i)\mathcal{R}_i\left(\widetilde{X}\right)\Theta(i)=0,$$

where

$$\Theta(i)=\left[D_c(i)-\widetilde{F}(i)\quad C_c(i)\right].$$

Since the system $(A_{0cl},A_{1cl},\ldots,A_{rcl};Q)$ is stable it follows that $\widetilde{U}(i)\geq 0$. Further we have

$$\|\mathbf{G}_{cl}\|_{2,\ell}^2=\sum_{i=1}^d \varepsilon_i Tr\left(B_{vcl}^T(i)\widehat{P}_{0cl}(i)B_{vcl}(i)\right)$$

$$=\sum_{i=1}^d \varepsilon_i Tr\left(B_v^T(i)\widetilde{X}(i)B_v(i)\right)+\sum_{i=1}^d \varepsilon_i Tr\left(B_{vcl}^T(i)\widetilde{U}(i)B_{vcl}(i)\right),$$

Since $\widetilde{U}(i)$ is positive semidefinite it follows that

$$\|\mathbf{G}_{cl}\|_{2,\ell}^2\geq\sum_{i=1}^d \varepsilon_i Tr\left(B_v^T(i)\widetilde{X}(i)B_v(i)\right)$$

for all stabilizing controllers $\mathbf{G}_c$. Using Proposition 7.2.1 the conclusion in the statement immediately follows. □

Remark 7.2.1. From Theorem 7.2.2 it follows that both optimization problems *(OP1)* and *(OP2)* have the same optimal solution given by the controllers with the set of parameters $n_c=0, A_c(i)=0, B_c(i)=0, C_c(i)=0, D_c(i)=\widetilde{F}(i), i\in\mathcal{D}$.

The theoretical results derived in this section are illustrated by the following numerical example.

Consider the stochastic linear system subject both to Markovian jumps and to multiplicative noise of form (7.28) with $n=2, \mathcal{D}=\{1,2\}$ and $r=1$, where:

$$A_0(1)=\begin{bmatrix}-1&0\\1&-1\end{bmatrix},A_0(2)=\begin{bmatrix}-1&1\\0&-1\end{bmatrix},$$

$$A_1(1)=\begin{bmatrix}-1&1\\0&-2\end{bmatrix},A_1(2)=\begin{bmatrix}-2&1\\1&-1\end{bmatrix},$$

$$B_0(1)=\begin{bmatrix}1\\-1\end{bmatrix},B_0(2)=\begin{bmatrix}1\\1\end{bmatrix},$$

$$B_1(1) = \begin{bmatrix} -1 \\ 1 \end{bmatrix}, B_1(2) = \begin{bmatrix} -2 \\ 1 \end{bmatrix},$$

$$B_v(1) = \begin{bmatrix} 1 \\ -2 \end{bmatrix}; B_v(2) = \begin{bmatrix} 1 \\ 3 \end{bmatrix},$$

$$C(1) = [1 \ \ 3], \ C(2) = [2 \ -1];$$

$$D(1) = 1; \ \ D(2) = 3,$$

$$Q = \begin{bmatrix} -1 & 1 \\ 1 & -1 \end{bmatrix},$$

and the initial distribution $(0.5 \ 0.5)$. Applying the iterative algorithm presented in Sect. 5.8 for a precision of 10^{-6}, after 205 iterations the following solution has been obtained:

$$F(1) = [-0.2863 \quad -1.5672];$$

$$F(2) = [-0.8547 \quad 0.2353],$$

providing the optimal H_2 norm of the resulting system which equals 4.4028.

7.3 Stochastic H_2 Optimal Control: The Output Feedback Control

Consider the system **G** described by:

$$dx(t) = [A_0(\eta(t))x(t) + B_0(\eta(t))u(t)]dt$$

$$+ \sum_{k=1}^{r} [A_k(\eta(t))x(t) + B_k(\eta(t))u(t)]dw_k(t)$$

$$+ B_v(\eta(t))dv(t) \tag{7.44}$$

$$dy(t) = C_0(\eta(t))x(t)dt + \sum_{k=1}^{r} C_k(\eta(t))x(t)dw_k(t)$$

$$+ D_v(\eta(t))dv(t)$$

$$z(t) = C(\eta(t))x(t) + D(\eta(t))u(t)$$

where $x \in \mathbf{R}^n$ denotes the state, $u \in \mathbf{R}^m$ is the control variable, $y \in \mathbf{R}^p$ is the measured output, and $z \in \mathbf{R}^s$ denotes the regulated output; $\eta(t), w(t), v(t), t \geq 0$ are stochastic processes with the properties given in the previous sections.

Associate with the system (7.44) the following class of controllers $\mathbf{G}_c$ of the form

$$
\begin{aligned}
dx_c(t) &= A_c(\eta(t))x_c(t)dt + \sum_{k=1}^{r} A_{kc}(\eta(t))x_c(t)dw_k(t) \\
&\quad + B_c(\eta(t))dy(t) \\
u(t) &= C_c(\eta(t))x_c(t).
\end{aligned} \tag{7.45}
$$

By coupling $\mathbf{G}_c$ to $\mathbf{G}$ one obtains the resulting system $\mathbf{G}_{cl}$ with the state equations

$$
dx_{cl}(t) = A_{0cl}(\eta(t))x_{cl}(t)dt + \sum_{k=1}^{r} A_{kcl}(\eta(t))x_{cl}(t)dw_k(t)
$$

$$
+ B_{vcl}(\eta(t))dv(t) \tag{7.46}
$$

$$
z(t) = C_{cl}(\eta(t))x_{cl}(t)
$$

where

$$
x_{cl} = \begin{bmatrix} x \\ x_c \end{bmatrix},
$$

$$
A_{0cl}(i) = \begin{bmatrix} A_0(i) & B_0(i)C_c(i) \\ B_c(i)C_0(i) & A_c(i) \end{bmatrix},
$$

$$
A_{kcl}(i) = \begin{bmatrix} A_k(i) & B_k(i)C_c(i) \\ B_c(i)C_k(i) & A_{kc}(i) \end{bmatrix}, k = 1,\ldots,r
$$

$$
B_{vcl}(i) = \begin{bmatrix} B_v(i) \\ B_c(i)D_v(i) \end{bmatrix},
$$

$$
C_{cl}(i) = [C(i) \quad D(i)C_c(i)], \ i \in \mathcal{D}.
$$

Definition 7.3.1. The controller $\mathbf{G}_c$ is said *stabilizing controller* of $\mathbf{G}$ if the zero-solution of the closed-loop system (7.46) in the absence of the white noise $v(t)$ is ESMS. The set of all stabilizing controllers will be denoted by $\mathcal{K}(\mathbf{G})$.

A controller in $\mathcal{K}(\mathbf{G})$ is determined by the set of the following parameters: $n_c \geq 1, A_c(i) \in \mathbf{R}^{n_c \times n_c}, B_c(i) \in \mathbf{R}^{n_c \times p}, C_c(i) \in \mathbf{R}^{m \times n_c}$. The controller order n_c is not a priori fixed. For a stabilizing controller $\mathbf{G}_c$ we consider the norms $\|\mathbf{G}_{cl}\|_2$ and $\||\mathbf{G}_{cl}|\|_2$ corresponding to the closed-loop system. Then two optimization problems will be formulated and solved in the following:

(OP1') Find a stabilizing controller of type (7.45) minimizing $\|\mathbf{G}_{cl}\|_2$.
(OP2') Find a stabilizing controller of type (7.45) minimizing $\||\mathbf{G}_{cl}|\|_2$.

It is expected that the solutions of the two problems formulated above be different. In the particular case when the whole state vector is available for measurements the solutions of (OP1') and (OP2') coincide and they are given by a stabilizing state feedback.

Consider the associated SGRAE

$$
\begin{aligned}
&A_0\,(i)\,Y\,(i)+Y\,(i)\,A_0^T\,(i)+\textstyle\sum_{k=1}^r A_k\,(i)\,Y\,(i)\,A_k^T\,(i)+\textstyle\sum_{j=1}^d q_{ji}Y\,(j)\\
&-\left[Y\,(i)\,C_0^T\,(i)+\textstyle\sum_{k=1}^r A_k\,(i)\,Y\,(i)\,C_k^T\,(i)+\varepsilon_i B_v\,(i)\,D_v^T\,(i)\right]\\
&\times\left[\varepsilon_i D_v\,(i)\,D_v^T\,(i)+\textstyle\sum_{k=1}^r C_k\,(i)\,Y\,(i)\,C_k^T\,(i)\right]^{-1}\\
&\times\left[C_0\,(i)\,Y\,(i)+\textstyle\sum_{k=1}^r C_k\,(i)\,Y\,(i)\,A_k^T\,(i)+\varepsilon_i D_v\,(i)\,B_v^T\,(i)\right]\\
&+\varepsilon_i B_v\,(i)\,B_v^T\,(i)=0,\ \ i\in\mathcal{D}
\end{aligned}
\tag{7.47}
$$

where ε_i have been introduced in the previous section. Recall that

$$
\widetilde{Y}=\left(\widetilde{Y}(1),\ldots,\widetilde{Y}(d)\right)\in\mathcal{S}_n^d
$$

is a stabilizing solution of (7.47) if the system

$$
\left(A_0+\widetilde{K}C_0,A_1+\widetilde{K}C_1,\ldots,A_r+\widetilde{K}C_r;Q\right)
$$

is stable, where

$$
\begin{aligned}
\widetilde{K}(i)=&-\left[\widetilde{Y}(i)\,C_0^T\,(i)+\textstyle\sum_{k=1}^r A_k\,(i)\,\widetilde{Y}(i)\,C_k^T\,(i)+\varepsilon_i B_v\,(i)\,D_v^T\,(i)\right]\\
&\times\left[\varepsilon_i D_v\,(i)\,D_v^T\,(i)+\textstyle\sum_{k=1}^r C_k\,(i)\,Y\,(i)\,C_k^T\,(i)\right]^{-1},\ i\in\mathcal{D}.
\end{aligned}
\tag{7.48}
$$

A necessary and sufficient condition which guarantees the existence of the stabilizing solution of (7.47) is provided by Theorem 5.7.1. To this end we introduce the corresponding generalized dissipation matrix

$$
\widetilde{\mathcal{N}}(Y)=\left(\widetilde{\mathcal{N}}_1(Y),\ldots,\widetilde{\mathcal{N}}_d(Y)\right)
$$

where

$$
\widetilde{\mathcal{N}}_i(Y)=\begin{bmatrix}(\mathcal{L}Y)(i)+\varepsilon_i B_v\,(i)\,B_v^T\,(i) & \widetilde{\mathcal{P}}_i(Y)\\[4pt]\widetilde{\mathcal{P}}_i^T(Y) & \widetilde{\mathcal{R}}_i(Y)\end{bmatrix}
\tag{7.49}
$$

with

$$
\widetilde{\mathcal{P}}_i(Y)=Y\,(i)\,C_0^T\,(i)+\sum_{k=1}^r A_k\,(i)\,Y\,(i)\,C_k^T\,(i)+\varepsilon_i B_v\,(i)\,D_v^T\,(i)
$$

and

$$
\widetilde{\mathcal{R}}_i(Y)=\varepsilon_i D_v\,(i)\,D_v^T\,(i)+\sum_{k=1}^r C_k\,(i)\,Y\,(i)\,C_k^T\,(i),\ i\in\mathcal{D}
$$

for all $Y=(Y(1),\ldots,Y(d))\in\mathcal{S}_n^d$. From Theorem 5.7.1 it follows that the SGRAE (7.47) has a stabilizing solution if and only if the triplet $(\mathbf{C},\mathbf{A};Q)$ is detectable and there exists $\widehat{Y} \in \mathcal{S}_n^d$ such that $\widetilde{\mathcal{N}}\left(\widehat{Y}\right) > 0$. Further, if $\mathbf{G}_{cl}$ is the closed-loop system obtained by coupling a stabilizing controller of the set $\mathcal{K}(\mathbf{G})$ to the system (7.44), then according to Theorems 7.1.4 and 7.1.7 we have

$$\|G_{cl}\|_{2,\ell}^2 = \sum_{i=1}^d \varepsilon_i Tr\left(B_{vcl}^T(i)\,\widehat{P}_{ocl}(i)\,B_{vcl}(i)\right) \tag{7.50}$$

where

$$\widehat{P}_{ocl} = \left(\widehat{P}_{ocl}(1),\ldots,\widehat{P}_{ocl}(d)\right)$$

is the observability Gramian of the closed-loop system and it verifies the Lyapunov-type system:

$$\begin{aligned} A_{0cl}^T(i)\,\widehat{P}_{ocl}(i) + \widehat{P}_{ocl}(i)\,A_{0cl}(i) + \sum_{k=1}^r A_{kcl}^T(i)\,\widehat{P}_{ocl}(i)\,A_{kcl}(i) \\ + \sum_{j=1}^d q_{ji}\widehat{P}_{ocl}(j) + C_{cl}^T(i)C_{cl}(i) = 0. \end{aligned} \tag{7.51}$$

Since $(A_{0cl}, A_{1cl},\ldots,A_{rcl};Q)$ is stable, the system (7.51) has a unique positive semidefinite solution $\widehat{P}_{ocl}(i)$.

Let $\widetilde{X} = \left(\widetilde{X}(1),\ldots,\widetilde{X}(d)\right)$ be the stabilizing solution of SGRAE (7.34). Denote by

$$U(i) = \widehat{P}_{ocl}(i) - \begin{bmatrix} \widetilde{X}(i) & 0 \\ 0 & 0 \end{bmatrix}, i \in \mathcal{D}.$$

By direct calculation one obtains as in the proof of Theorem 7.2.2 that

$$U = (U(1),\ldots,U(d)) \in \mathcal{S}_{n+n_c}^d$$

is the solution of the Lyapunov-type equation

$$\begin{aligned} A_{0cl}^T(i)\,U(i) + U(i)\,A_{0cl}(i) + \sum_{k=1}^r A_{kcl}^T(i)\,U(i)\,A_{kcl}(i) \\ + \sum_{j=1}^d q_{ji}U(j) + \widehat{C}_{cl}^T(i)\,\widehat{C}_{cl}(i) = 0, \ i \in \mathcal{D} \end{aligned} \tag{7.52}$$

where

$$\widehat{C}_{cl}(i) = \left[-\Pi(i)\widetilde{F}(i) \quad \Pi(i)C_c(i)\right]$$

with

$$\Pi(i) = \left(D^T(i)D(i) + \sum_{k=1}^r B_k^T(i)\,\widetilde{X}(i)\,B_k(i)\right)^{\frac{1}{2}}.$$

Since the system $(A_{0cl}, A_{1cl}, \ldots, A_{rcl}; Q)$ is stable it follows that the unique solution of (7.52) is positive semidefinite. As in the proof of Theorem 7.2.2, the equality (7.50) can be written as

$$\|\mathbf{G}_{cl}\|_{2,\ell}^2 = \sum_{i=1}^d \varepsilon_i Tr\left(B_v^T(i)\widetilde{X}(i)B_v(i)\right) \tag{7.53}$$

$$+ \sum_{i=1}^d \varepsilon_i Tr\left(B_{vcl}^T(i)U(i)B_{vcl}(i)\right).$$

On the other hand since

$$U = (U(1), \ldots, U(d))$$

is the observability Gramian associated with the triplet

$$\left(\widehat{C}_{cl}, (A_{0cl}, \ldots, A_{rcl}); Q\right),$$

then according to the results in Theorems 7.1.4 and 7.1.7 we get

$$\sum_{i=1}^d \varepsilon_i Tr\left(B_{vcl}^T(i)U(i)B_{vcl}(i)\right) = \sum_{i=1}^d Tr\left(\widehat{C}_{cl}(i)\widehat{P}_{ccl}(i)\widehat{C}_{cl}^T(i)\right) \tag{7.54}$$

where

$$\widehat{P}_{ccl} = \left(\widehat{P}_{ccl}(1), \ldots, \widehat{P}_{ccl}(d)\right)$$

is the unique solution of the Lyapunov equation on $\mathcal{S}_{n+n_c}^d$:

$$A_{0cl}(i)\widehat{P}_{ccl}(i) + \widehat{P}_{ccl}(i)A_{0cl}^T(i) + \sum_{k=1}^r A_{kcl}(i)\widehat{P}_{ccl}(i)A_{kcl}^T(i) \atop + \sum_{j=1}^d q_{ji}\widehat{P}_{ccl}(j) + \varepsilon_i B_{vcl}(i)B_{vcl}^T(i) = 0. \tag{7.55}$$

From (7.53) and (7.54) one obtains

$$\|\mathbf{G}_{cl}\|_{2,\ell}^2 = \sum_{i=1}^d \varepsilon_i Tr\left(B_v^T(i)\widetilde{X}(i)B_v(i)\right) \tag{7.56}$$

$$+ \sum_{i=1}^d Tr\left(\widehat{C}_{cl}(i)\widehat{P}_{ccl}(i)\widehat{C}_{cl}^T(i)\right).$$

Let

$$\widetilde{Y} = \left(\widetilde{Y}(1), \ldots, \widetilde{Y}(d)\right)$$

be the stabilizing solution of the SGRAE (7.47) and define:

$$V(i) = \widehat{P}_{ccl}(i) - \begin{bmatrix} \widetilde{Y}(i) & 0 \\ 0 & 0 \end{bmatrix}.$$

Let

$$\begin{bmatrix} Y_{11}(i) & Y_{12}(i) \\ Y_{12}^T(i) & Y_{22}(i) \end{bmatrix}$$

be the partition of $\widehat{P}_{ccl}(i)$ according to the partition of the state matrix of the closed-loop system. It is easy to see that (7.55) can be partitioned as follows:

$$\begin{aligned}
& A_0(i)Y_{11}(i) + Y_{11}(i)A_0^T(i) + B_0(i)C_c(i)Y_{12}^T(i) \\
& + Y_{12}(i)C_c^T(i)B_0^T(i) + \sum_{k=1}^r \left(A_k(i)Y_{11}(i)A_k^T(i) + B_k(i)C_c(i)Y_{12}^T(i)A_k^T(i) \right. \\
& \left. + A_k(i)Y_{12}(i)C_c^T(i)B_k^T(i) + B_k(i)C_c(i)Y_{22}(i)C_c^T(i)B_k^T(i) \right) \\
& + \sum_{j=1}^d q_{ji}Y_{11}(j) + \varepsilon_i B_v(i)B_v^T(i) = 0
\end{aligned}$$

$$\begin{aligned}
& A_0(i)Y_{12}(i) + B_0(i)C_c(i)Y_{22}(i) + Y_{11}(i)C_0^T(i)B_c^T(i) + Y_{12}(i)A_c^T(i) \\
& + \sum_{k=1}^r \left(A_k(i)Y_{11}(i)C_k^T(i)B_c^T(i) + B_k(i)C_c(i)Y_{12}^T(i)C_k^T(i)B_c^T(i) \right. \\
& \left. + A_k(i)Y_{12}(i)A_{kc}^T(i) + B_k(i)C_c(i)Y_{22}(i)A_{kc}^T(i) \right) \\
& + \sum_{j=1}^d q_{ji}Y_{12}(j) + \varepsilon_i B_v(i)D_v^T(i)B_c^T(i) = 0
\end{aligned} \qquad (7.57)$$

$$\begin{aligned}
& B_c(i)C_0(i)Y_{12}(i) + A_c(i)Y_{22}(i) + Y_{12}^T(i)C_0^T(i)B_c^T(i) \\
& + Y_{22}(i)A_c^T(i) + \sum_{k=1}^r \left(B_c(i)C_k(i)Y_{11}(i)C_k^T(i)B_c^T(i) + \right. \\
& A_{kc}(i)Y_{12}^T(i)C_k^T(i)B_c^T(i) + B_c(i)C_k(i)Y_{12}(i)A_{kc}^T(i) + A_{kc}(i)Y_{22}(i)A_{kc}^T(i) \big) \\
& + \sum_{j=1}^d q_{ji}Y_{22}(j) + \varepsilon_i B_c(i)D_v(i)D_v^T(i)B_c^T(i) = 0.
\end{aligned}$$

By direct calculations based on (7.57) and (7.47) we deduce that $V = (V(1), \ldots, V(d))$ is a solution of the following Lyapunov-type equation on $\mathcal{S}_{n+n_c}^d$:

$$\begin{aligned}
& A_{0cl}(i)V(i) + V(i)A_{0cl}^T(i) + \sum_{k=1}^r A_{kcl}(i)V(i)A_{kcl}^T(i) \\
& + \sum_{j=1}^d q_{ji}V(j) + \widehat{B}_{vcl}(i)\widehat{B}_{vcl}^T(i) = 0, \ i \in \mathcal{D}
\end{aligned} \qquad (7.58)$$

where

$$\widehat{B}_{vcl}(i) = \begin{bmatrix} -\widetilde{K}(i) \\ B_c(i) \end{bmatrix} \widehat{\Pi}(i)$$

with

$$\widehat{\Pi}(i) = \left(\varepsilon_i D_v(i)D_v^T(i) + \sum_{k=1}^r C_k(i)\widetilde{Y}(i)C_k^T(i) \right)^{\frac{1}{2}}, \ i \in \mathcal{D}.$$

Since the system

$$(A_{0cl}, \ldots, A_{rcl}; Q)$$

is stable, (7.58) has a unique solution $V(i) \geq 0$. Furthermore (7.56) can be rewritten in the form

$$\|G_{cl}\|_{2,\ell}^2 = \sum_{i=1}^{d} \varepsilon_i Tr\left(B_v^T(i) \tilde{X}(i) B_v(i) \right)$$

$$+ \sum_{i=1}^{d} Tr\left(\Pi(i) \tilde{F}(i) \tilde{Y}(i) \tilde{F}^T(i) \Pi(i) \right) \tag{7.59}$$

$$+ \sum_{i=1}^{d} Tr\left(\hat{C}_{cl}(i) V(i) \hat{C}_{cl}^T(i) \right).$$

Now we are able to prove the main result of this section.

Theorem 7.3.1. *Assume that*

(i) *The triplet* $(\mathbf{A}, \mathbf{B}; Q)$ *is stabilizable and* $(\mathbf{C}, \mathbf{A}; Q)$ *is detectable.*
(ii) *There exists* $\hat{X} \in \mathcal{S}_n^d$ *verifying*

$$\Lambda^{\Sigma}\left(\hat{X} \right) > 0$$

   where Λ^{Σ} *denotes the generalized dissipation matrix associated with SGRAE (7.34).*
(iii) *There exists* $\hat{Y} \in \mathcal{S}_n^d$ *verifying*

$$\tilde{\mathcal{N}}\left(\hat{Y} \right) > 0$$

where $\tilde{\mathcal{N}}$ *is defined by (7.49).*
   Under the above conditions we have

$$\min_{G_c \in \mathcal{K}(\mathbf{G})} \|G_{cl}\|_{2,\ell}^2 = \sum_{i=1}^{d} \varepsilon_i Tr\left(B_v^T(i) \tilde{X}(i) B_v(i) \right)$$

$$+ \sum_{i=1}^{d} Tr\left(\Pi(i) \tilde{F}(i) \tilde{Y}(i) \tilde{F}^T(i) \Pi(i) \right)$$

and this minimum is attained by the optimal controller

$$dx_c(t) = \tilde{A}_{0c}(\eta(t)) x_c(t) dt$$

$$+ \sum_{k=1}^{r} \tilde{A}_{kc}(\eta(t)) x_c(t) dw_k(t) \tag{7.60}$$

$$+\widetilde{B}_c\left(\eta\left(t\right)\right)dy\left(t\right)$$

$$u\left(t\right)=\widetilde{C}_c\left(\eta\left(t\right)\right)x_c\left(t\right)$$

with

$$\widetilde{A}_{kc}\left(i\right)=A_k\left(i\right)+\widetilde{K}\left(i\right)C_k\left(i\right)+B_k\left(i\right)\widetilde{F}\left(i\right),\, k=0,\ldots,r$$
$$\widetilde{B}_c\left(i\right)=-\widetilde{K}\left(i\right)$$
$$\widetilde{C}_c\left(i\right)=\widetilde{F}\left(i\right),\, i\in\mathcal{D},$$

where $\widetilde{K}\left(i\right)$ and $\widetilde{F}\left(i\right)$ are defined by (7.48) and (7.35), respectively.

Proof. From (7.59) and from the positivity of the solution V of (7.58) it follows that

$$\|\mathbf{G}_{cl}\|_{2,\ell}^2\geq\sum_{i=1}^{d}\varepsilon_i Tr\left(B_v^T\left(i\right)\widetilde{X}\left(i\right)B_v\left(i\right)\right)\qquad(7.61)$$

$$+\sum_{i=1}^{d}Tr\left(\Pi\left(i\right)\widetilde{F}\left(i\right)\widetilde{Y}\left(i\right)\widetilde{F}^T\left(i\right)\Pi\left(i\right)\right)$$

for all stabilizing controllers $\mathbf{G}_c\in\mathcal{K}\left(\mathbf{G}\right)$. We show now that the controller given by (7.60) belongs to the class of stabilizing controllers $\mathcal{K}\left(\mathbf{G}\right)$ and for this controller (7.61) becomes equality. The closed-loop system corresponding to the controller (7.60) is

$$dx\left(t\right)=\left(A_0\left(\eta\left(t\right)\right)x\left(t\right)+B_0\left(\eta\left(t\right)\right)\widetilde{F}\left(\eta\left(t\right)\right)x_c\left(t\right)\right)dt$$

$$+\sum_{k=1}^{r}\left(A_k\left(\eta\left(t\right)\right)x\left(t\right)+B_k\left(\eta\left(t\right)\right)\widetilde{F}\left(\eta\left(t\right)\right)x_c\left(t\right)\right)dw_k\left(t\right)$$

$$+B_v\left(\eta\left(t\right)\right)dv\left(t\right)$$

$$dx_c\left(t\right)=\left(-\widetilde{K}\left(\eta\left(t\right)\right)C_0\left(\eta\left(t\right)\right)x\left(t\right)\right.\qquad(7.62)$$

$$+\left(A_0\left(\eta\left(t\right)\right)+B_0\left(\eta\left(t\right)\right)\widetilde{F}\left(\eta\left(t\right)\right)+\widetilde{K}\left(\eta\left(t\right)\right)C_0\left(\eta\left(t\right)\right)\right)x_c\left(t\right)\right)dt$$

$$+\sum_{k=1}^{r}\left(-\widetilde{K}\left(\eta\left(t\right)\right)C_k\left(\eta\left(t\right)\right)x\left(t\right)\right.$$

$$+\left.\left(A_k\left(\eta\left(t\right)\right)+B_k\left(\eta\left(t\right)\right)\widetilde{F}\left(\eta\left(t\right)\right)+\widetilde{K}\left(\eta\left(t\right)\right)C_k\left(\eta\left(t\right)\right)\right)x_c\left(t\right)\right)$$

$$\times dw_k\left(t\right)-\widetilde{K}\left(\eta\left(t\right)\right)D_v\left(\eta\left(t\right)\right)dv\left(t\right)$$

$$z\left(t\right)=C\left(\eta\left(t\right)\right)x\left(t\right)+D\left(\eta\left(t\right)\right)\widetilde{F}\left(\eta\left(t\right)\right)x_c\left(t\right).$$

If $\left[x^T(t) \ x_c^T(t)\right]^T$ is a solution of (7.62) in the absence of the additive noise $v(t)$, we define:

$$\xi(t) = x(t) - x_c(t), t \geq 0.$$

Then by direct computations it follows that the stochastic process $\left[x^T(t) \ \xi^T(t)\right]^T$ verifies the system

$$
\begin{aligned}
dx(t) = & \left(\left(A_0(\eta(t)) + B_0(\eta(t)) \widetilde{F}(\eta(t)) \right) x(t) \right. \\
& \left. - B_0(\eta(t)) \widetilde{F}(\eta(t)) \xi(t) \right) dt \\
& + \sum_{k=1}^{r} \left(\left(A_k(\eta(t)) + B_k(\eta(t)) \widetilde{F}(\eta(t)) \right) x(t) \right. \\
& \left. - B_k(\eta(t)) \widetilde{F}(\eta(t)) \xi(t) \right) dw_k(t) \\
d\xi(t) = & \left(A_0(\eta(t)) + \widetilde{K}(\eta(t)) C_0(\eta(t)) \right) \xi(t) dt \\
& + \sum_{k=1}^{r} \left(A_k(\eta(t)) + \widetilde{K}(\eta(t)) C_k(\eta(t)) \right) \xi(t) dw_k(t).
\end{aligned}
\tag{7.63}
$$

Since $\widetilde{Y}$ is the stabilizing solution of the SGRAE (7.47), from the second equation of (7.63) one obtains

$$E\left[|\xi(t)|^2 \mid \eta(0) = i \right] \leq \beta e^{-\alpha t} |\xi(0)|^2, t \geq 0, i \in \mathcal{D} \tag{7.64}$$

for some $\alpha > 0$ and $\beta \geq 1$. Further the first equation of (7.63) can be rewritten as follows:

$$
\begin{aligned}
dx(t) = & \left(\left(A_0(\eta(t)) + B_0(\eta(t)) \widetilde{F}(\eta(t)) \right) x(t) + f_0(t) \right) dt \\
& + \sum_{k=1}^{r} \left(\left(A_k(\eta(t)) + B_k(\eta(t)) \widetilde{F}(\eta(t)) \right) x(t) + f_k(t) \right) dw_k(t)
\end{aligned}
$$

with

$$f_k(t) = -B_k(\eta(t)) \widetilde{F}(\eta(t)) \xi(t), t \geq 0, k = 0, 1, \ldots, r.$$

Applying Theorem 3.6.1 part (i), one deduces that there exist $\tilde{\beta} \geq 1$ and $\tilde{\alpha} > 0$ such that

$$E\left[|x(t)|^2 \mid \eta(0) = i \right] \leq \tilde{\beta} e^{-\tilde{\alpha} t} \left(|x(0)|^2 + |\xi(0)|^2 \right). \tag{7.65}$$

From (7.64) and (7.65) we get

$$E\left[|x_c(t)|^2 \mid \eta(0) = i\right] \leq \hat{\beta} e^{-\hat{\alpha} t}\left(|x(0)|^2 + |\xi(0)|^2\right)$$

where $\hat{\alpha} = \min(\alpha, \tilde{\alpha})$, $\hat{\beta} = \max\left(\beta, \tilde{\beta}\right)$. Therefore we may conclude that the controller (7.60) is a stabilizing controller. On the other hand we may write with this controller:

$$\sum_{i=1}^{d} Tr\left(\tilde{C}_{cl}(i) V(i) \tilde{C}_{cl}^T(i)\right) = \sum_{i=1}^{d} Tr\left(\Pi(i)\tilde{F}(i)\right. \tag{7.66}$$

$$\left. \times \left(V_{11}(i) - V_{12}(i) - V_{12}^T(i) + V_{22}(i)\right) \tilde{F}^T(i)\Pi(i)\right),$$

where

$$\begin{bmatrix} V_{11}(i) & V_{12}(i) \\ V_{12}^T(i) & V_{22}(i) \end{bmatrix}$$

is the partition of the solution $V(i)$ of (7.58) corresponding to the controller (7.60). Partitioning (7.58) we obtain the following system:

$$A_0(i) V_{11}(i) + B_0(i) \tilde{F}(i) V_{12}^T(i) + V_{11}(i) A_0^T(i) + V_{12}(i) \tilde{F}^T(i) B_0^T(i)$$
$$+ \Sigma_{k=1}^{r}\left(A_k(i) V_{11}(i) A_k^T(i) + B_k(i) \tilde{F}(i) V_{12}^T(i) A_k^T(i)\right.$$
$$\left. + A_k(i) V_{12}(i) \tilde{F}^T(i) B_k^T(i) + B_k(i) \tilde{F}(i) V_{22}(i) \tilde{F}^T(i) B_k^T(i)\right)$$
$$+ \Sigma_{j=1}^{d} q_{ji} V_{11}(j) + \tilde{K}(i) \hat{\Pi}^2(i) \tilde{K}^T(i) = 0$$

$$A_0(i) V_{12}(i) + B_0(i) \tilde{F}(i) V_{22}(i) - V_{11}(i) C_0^T(i) \tilde{K}^T(i) + V_{12}(i) \tilde{A}_{0c}^T(i)$$
$$+ \Sigma_{k=1}^{r}\left(-A_k(i) V_{11}(i) C_k^T(i) \tilde{K}^T(i) - B_k(i) \tilde{F}(i) V_{12}^T(i) C_k^T(i) \tilde{K}^T(i)\right. \tag{7.67}$$
$$\left. + A_k(i) V_{12}(i) \tilde{A}_{kc}^T(i) + B_k(i) \tilde{F}(i) V_{22}(i) \tilde{A}_{kc}^T(i)\right) + \Sigma_{j=1}^{d} q_{ji} V_{12}(j)$$
$$+ \tilde{K}(i) \hat{\Pi}^2(i) \tilde{K}^T(i) = 0$$

$$-\tilde{K}(i) C_0(i) V_{12}(i) + \tilde{A}_{0c}(i) V_{22}(i) - V_{12}^T(i) C_0^T 9I0\tilde{K}^T(i) + V_{22}(i) \tilde{A}_{0c}^T(i)$$
$$+ \Sigma_{k=1}^{r}\left(\tilde{K}(i) C_k(i) V_{11}(i) C_k^T(i) \tilde{K}^T(i) - \tilde{A}_{kc}(i) V_{12}^T(i) C_k^T(i) \tilde{K}^T(i)\right.$$
$$\left. -\tilde{K}(i) C_k(i) V_{12}(i) \tilde{A}_{kc}^T(i) + \tilde{A}_{kc}(i) V_{22}(i) \tilde{A}_{kc}^T(i)\right) + \Sigma_{j=1}^{d} q_{ji} V_{22}(j)$$
$$+ \tilde{K}(i) \hat{\Pi}^2(i) \tilde{K}^T(i) = 0$$

By summing the first and the third equations (7.67) and by subtracting then the second equation of (7.67) and its transposed one obtains that

$$W(i) = V_{11}(i) - V_{12}(i) - V_{12}^T(i) + V_{22}(i)$$

solves the equation

$$
\left(A_0\left(i\right)+\tilde{K}\left(i\right)C_0\left(i\right)\right)W\left(i\right)+W\left(i\right)\left(A_0\left(i\right)+\tilde{K}\left(i\right)C_0\left(i\right)\right)^T
$$
$$
+\Sigma_{k=1}^r\left(A_k\left(i\right)+\tilde{K}\left(i\right)C_k\left(i\right)\right)W\left(i\right)\left(A_k\left(i\right)+\tilde{K}\left(i\right)C_k\left(i\right)\right)^T
$$
$$
+\Sigma_{j=1}^d q_{ji}W\left(j\right)=0.
$$

Since the system

$$
\left(A_0+\tilde{K}C_0,A_1+\tilde{K}C_1,\ldots,A_r+\tilde{K}C_r;Q\right)
$$

is stable, the above equation has a unique solution from which we deduce that $W\left(i\right)=0,\ i\in\mathcal{D}$. Based on (7.65) this shows that

$$
\sum_{i=1}^d Tr\left(\tilde{C}_{cl}\left(i\right)V\left(i\right)\tilde{C}_{cl}^T\left(i\right)\right)=0
$$

and therefore

$$
\left\|\tilde{\mathbf{G}}_{cl}\right\|_{2,\ell}^2=\sum_{i=1}^d\varepsilon_iTr\left(B_v^T\left(i\right)\tilde{X}\left(i\right)B_v\left(i\right)\right)
$$

$$
+\sum_{i=1}^d Tr\left(\Pi(i)\tilde{F}\left(i\right)\tilde{Y}\left(i\right)\tilde{F}^T\left(i\right)\Pi(i)\right),
$$

where $\tilde{\mathbf{G}}_{cl}$ is the closed-loop system corresponding to the controller (7.60) and thus the proof is complete. $\square$

Remark 7.3.1. In the particular case when $\mathcal{D}=\{1\}$, $A_k=0$, $B_k=0$, $C_k=0$, $k=1,2,\ldots,r$ the controller (7.60) reduces to the well-known Kalman–Bucy filter which solves the classic H_2 optimization problem. Therefore it is natural that in the general framework considered here, the solution of the H_2 optimization problem has a similar form with the Kalman–Bucy filter. Unfortunately in the general case when the nominal plant is corrupted with multiplicative white noise the solution of the H_2 optimization problem is a stochastic system with multiplicative noise, which fact leads to implementation difficulties. This fact suggests to consider an H_2 optimization problem in the class of controllers with $A_{kc}\left(i\right)=0,\ k=1,\ldots,r$, which still remains an open problem.

At the end of this section we focused our attention on the strictly Markovian case, namely $d>1$, $A_k\left(i\right)=0$, $B_k\left(i\right)=0$, $C_k\left(i\right)=0$, $A_{kc}\left(i\right)=0,\ 1\leq k\leq r,\ i\in\mathcal{D}$. Therefore the controlled system is in this case:

$$
dx\left(t\right)=\left(A_0\left(\eta\left(t\right)\right)x\left(t\right)+B_0\left(\eta\left(t\right)\right)u\left(t\right)\right)dt+B_v\left(\eta\left(t\right)\right)dv\left(t\right)
$$
$$
dy\left(t\right)=C_0\left(\eta\left(t\right)\right)x\left(t\right)dt+D_v\left(\eta\left(t\right)\right)dv\left(t\right)   \tag{7.68}
$$
$$
z\left(t\right)=C\left(\eta\left(t\right)\right)x\left(t\right)+D\left(\eta\left(t\right)\right)u\left(t\right)
$$

In this particular case Theorem 7.3.1 leads to the following corollary:

Corollary 7.3.2. *Assume that:*

(i) *The triplet $(A_0, B_0; Q)$ is stabilizable and $(C_0, A_0; Q)$ is detectable.*

(ii) *It exists $\widehat{X} = \left(\widehat{X}(1), \ldots, \widehat{X}(d)\right) \in \mathcal{S}_n^d$ satisfying the LMI:*

$$\Lambda_i^{\Sigma} := \begin{bmatrix} \mathcal{L}^*\widehat{X}(i) + C^T(i)C(i) & \widehat{X}(i)B_0(i) + C^T(i)D(i) \\ B_0^T(i)\widehat{X}(i) + D^T(i)C(i) & D^T(i)D(i) \end{bmatrix} > 0$$

where

$$\mathcal{L}^*\widehat{X}(i) = A_0^T(i)\widehat{X}(i) + \widehat{X}(i)A_0(i) + \sum_{j=1}^{d} q_{ij}\widehat{X}(j).$$

(iii) *It exists $\widehat{Y} = \left(\widehat{Y}(1), \ldots, \widehat{Y}(d)\right) \in \mathcal{S}_n^d$ satisfying the LMI:*

$$\widetilde{\mathcal{N}}_i := \begin{bmatrix} \mathcal{L}\left(\widehat{Y}\right)(i) + \varepsilon_i B_v(i)B_v^T(i) & \widehat{Y}(i)C_0^T(i) + \varepsilon_i B_v(i)D_v^T(i) \\ C_0(i)\widehat{Y}(i) + \varepsilon_i D_v(i)B_v^T(i) & \varepsilon_i D_v(i)D_v^T(i) \end{bmatrix} > 0$$

where ε_i are either $\pi_{i\infty}$ or δ_i introduced in Sect. 7.2. Then the controller

$$dx_c(t) = \widetilde{A}_c(\eta(t))x_c(t)\,dt + \widetilde{B}_c(\eta(t))\,dy(t) \qquad (7.69)$$
$$u(t) = \widetilde{C}_c(\eta(t))x_c(t)$$

with

$$\widetilde{A}_c(i) = A_0(i) + B_0(i)\widetilde{F}(i) + \widetilde{K}(i)C_0(i)$$
$$\widetilde{B}_c(i) = -\widetilde{K}(i)$$
$$\widetilde{C}_c(i) = \widetilde{F}(i)$$

stabilizes the system (7.68) and

$$\left\|\widetilde{\mathbf{G}}_{cl}\right\|_{2,\ell}^2 = \Sigma_{i=1}^{d}\,\varepsilon_i Tr\left(B_v^T(i)\widetilde{X}(i)B_v(i)\right)$$
$$+ \Sigma_{i=1}^{d}\,Tr\left(\left(D^T(i)D(i)\right)^{\frac{1}{2}}\widetilde{F}(i)\widetilde{Y}(i)\widetilde{F}^T(i)\left(D^T(i)D(i)\right)^{\frac{1}{2}}\right)$$
$$= \min_{\mathbf{G}_c \in \mathcal{K}(\mathbf{G})}\|\mathbf{G}_{cl}\|_{2,\ell}^2$$

where $\widetilde{\mathbf{G}}_{cl}$ is the closed-loop system obtained by coupling the controller (7.69) to the system (7.68), $\widetilde{X}$ and $\widetilde{Y}$ are the stabilizing solutions of the Riccati-type equations

$$A_0^T(i)X(i) + X(i)A_0(i) + \sum_{j=1}^d q_{ij}X(j) - (X(i)B_0(i) + C^T(i)D(i))$$
$$\times (D^T(i)D(i))^{-1} (B_0^T(i)X(i) + D^T(i)C(i)) + C^T(i)C(i) = 0$$

and

$$A_0(i)Y(i) + Y(i)A_0^T(i) + \sum_{j=1}^d q_{ji}Y(j) - (Y(i)C_0^T(i) + \varepsilon_i B_v(i)D_v^T(i))$$
$$\times (\varepsilon_i D_v(i)D_v^T(i))^{-1} (C_0(i)Y(i) + \varepsilon_i D_v(i)B_v^T(i)) + \varepsilon_i B_v(i)B_v^T(i) = 0,$$

and $\widetilde{F}$ and $\widetilde{K}$ are given by

$$\widetilde{F}(i) = -(D^T(i)D(i))^{-1} (B_0^T(i)\widetilde{X}(i) + D^T(i)C(i))$$

$$\widetilde{K}(i) = -(\widetilde{Y}(i)C_0^T(i) + \varepsilon_i B_v(i)D_v^T(i)) (\varepsilon_i D_v(i)D_v^T(i))^{-1}. \qquad \square$$

In order to illustrate the above results we shall present a numerical example. The helicopter dynamics is considered having the state-space equations

$$\dot{x}(t) = A(\eta)x(t) + B(\eta(t))u(t) + Ew(t)$$
$$z(t) = C_1 x(t) + D_1 u(t)$$
$$y(t) = C_2 x(t) + D_2 w(t)$$

where $\eta(t)$ indicates the airspeed and the state variables are the horizontal velocity x_1, the vertical velocity x_2, the pitch rate x_3, and the pitch angle x_4. The matrices in the above state-space representation have the form (see [30]):

$$A(i) = \begin{bmatrix} -0.0366 & 0.0271 & 0.0188 & -0.4555 \\ 0.0482 & -1.01 & 0.0024 & -4.0208 \\ 0.1002 & a_{32}(i) & -0.707 & a_{34}(i) \\ 0 & 0 & 1 & 0 \end{bmatrix};$$

$$B(i) = \begin{bmatrix} 0.4422 & 0.1761 \\ b_{21}(i) & -7.5922 \\ -5.5200 & 4.4900 \\ 0 & 0 \end{bmatrix};$$

$$E = [I_{4\times4} \quad 0_{4\times1}]; \quad C_1 = \begin{bmatrix} I_{4\times4} \\ 0_{2\times4} \end{bmatrix}; \quad D_1 = \begin{bmatrix} 0_{4\times2} \\ I_{2\times2} \end{bmatrix};$$

$$C = [0 \quad 1 \quad 0 \quad 0]; \quad D_2 = [0 \quad 0 \quad 0 \quad 0 \quad 1], i = 1,2,3$$

Table 7.1 Parameters as function of speed

Airspeed (knots)	a_{32}	a_{34}	b_{21}
135	0.3681	1.4200	3.5446
60	0.0664	0.1198	0.9775
170	0.5047	2.5460	5.1120

Table 7.2 Optimal H_2 norms

Q	Optimal H_2 norms computed by the method in the present paper
Q_1	$\|\mathbf{G}_{cl}\|_{2,1} = 4.6735$; $\|\mathbf{G}_{cl}\|_{2,2} = 8.0988$
Q_2	$\|\mathbf{G}_{cl}\|_{2,1} = 4.5196$; $\|\mathbf{G}_{cl}\|_{2,2} = 7.8264$
Q_3	$\|\mathbf{G}_{cl}\|_{2,1} = 4.8113$; $\|\mathbf{G}_{cl}\|_{2,2} = 8.3333$

where $a_{32}(\cdot)$, $a_{34}(\cdot)$ and $b_{21}(\cdot)$ are given in Table 7.1 as function of the airspeed. The behavior of $\eta(t)$ is modeled as a Markov chain with three states corresponding to the three values of the airspeed: 135, 60, and 170 knots.

The following three different transition matrices have been considered:

$$
Q_1 = \begin{bmatrix} -0.0907 & 0.0671 & 0.0236 \\ 0.0671 & -0.0671 & 0 \\ 0.0236 & 0 & -0.0236 \end{bmatrix};
$$

$$
Q_2 = \begin{bmatrix} -0.0171 & 0.0007 & 0.0164 \\ 0.0013 & -0.0013 & 0 \\ 0.0986 & 0 & -0.0986 \end{bmatrix};
$$

$$
Q_3 = \begin{bmatrix} -0.0450 & 0.0002 & 0.0448 \\ 0.0171 & -0.0171 & 0 \\ 0.0894 & 0 & -0.0894 \end{bmatrix}.
$$

The initial assumed distributions are (0.333 0.333 0.333), (0.6 0.3 0.1), and (0.6 0.1 0.3), respectively. The optimal H_2 corresponding norms obtained using the method described in this section are presented in Table 7.2.

Here only the optimal H_2 controller for the case $Q = Q_1$ is given. Its realization is the following:

$$
A_c(1) = \begin{bmatrix} -0.4431 & 0.3328 & 0.4106 & 0.0327 \\ -3.4133 & -10.3798 & 4.8501 & 6.3131 \\ 5.3252 & 5.2657 & -6.8663 & -9.4439 \\ 0 & 1.7630 & 1 & 0 \end{bmatrix}, B_c(1) = \begin{bmatrix} -0.1509 \\ 3.0100 \\ -1.1841 \\ -1.7630 \end{bmatrix},
$$

$$
C_c(1) = \begin{bmatrix} -0.9282 & 0.0139 & 0.9616 & 1.3881 \\ 0.0226 & 0.8442 & -0.1896 & -0.7131 \end{bmatrix},
$$

$$A_c(2) = \begin{bmatrix} -0.4133 & 0.4164 & 0.3727 & -0.0675 \\ -2.0379 & -9.7852 & 3.6641 & 4.2692 \\ 5.8528 & 3.3426 & -7.5378 & -10.9517 \\ 0 & 1.3828 & 1 & 0 \end{bmatrix}, B_c(2) = \begin{bmatrix} -0.1727 \\ 2.6160 \\ -0.4174 \\ -1.3828 \end{bmatrix},$$

$$C_c(2) = \begin{bmatrix} -0.9144 & 0.1586 & 0.9440 & 1.2483 \\ 0.1570 & 0.8317 & -0.3607 & -0.9312 \end{bmatrix},$$

$$A_c(3) = \begin{bmatrix} -0.4517 & 0.2545 & 0.4437 & 0.1318 \\ -4.3958 & -11.1936 & 5.5719 & 7.2984 \\ 5.0354 & 6.8942 & -6.4680 & -7.9318 \\ 0 & 2.2062 & 1 & 0 \end{bmatrix}, B_c(3) = \begin{bmatrix} -0.1030 \\ 3.4319 \\ -2.2534 \\ -2.2062 \end{bmatrix},$$

$$C_c(3) = \begin{bmatrix} -0.9240 & -0.0573 & 0.9882 & 1.5154 \\ -0.0368 & 0.8507 & -0.0682 & -0.4705 \end{bmatrix}.$$

Let us finally remark that no ill-conditioned computations occurred when the iterative procedure described in this section has been applied.

7.4   A Kalman Filtering Problem for Stochastic Systems with State-Dependent Noise and Markovian Jumps

As mentioned in the introductory part of this chapter, the H_2 norm optimization problem is strongly related to the Kalman filtering. In this section we shall formulate and solve a Kalman-type filtering problem for stochastic systems with state-dependent noise and Markovian jumping.

7.4.1   Problem Formulation and Solution

Consider the following ESMS stochastic system subject both to multiplicative white noise and to jump Markovian parameters:

$$\begin{aligned} dx(t) &= A(\eta(t))x(t)dt + D(\eta(t))x(t)d\xi(t) \\ &\quad + B(\eta(t))d\beta(t) \\ dy(t) &= C(\eta(t))x(t)dt + G(\eta(t))x(t)dv(t) + d\mu(t) \end{aligned} \tag{7.70}$$

where $x \in \mathbf{R}^n$ denotes the state vector, $y \in \mathbf{R}^p$ is the measurement, $\eta(t)_{t \geq 0}$ is a standard homogeneous Markov process on a given probability space $(\Omega, \mathcal{F}, P)$ with the state space $\mathcal{D} = \{1, 2, 3 \ldots, N\}$ and the transition probability matrix $P(t) = e^{Qt}$ and ξ, β, v, μ are zero-mean independent Wiener processes. The elements of the infinitesimal generator Q satisfy $q_{ij} \geq 0$ if $i \neq j$ and $\sum_{j=1}^{N} q_{ij} = 0, 1 \leq i \leq N$ (see Theorem 1.8.2 (vii)).

Given the ESMS stochastic system (7.70), the problem analyzed in the following consists in determining the Luenberger observer-type stable filter of form

$$d\hat{x}(t) = A(\eta(t))\hat{x}(t)dt + L(\eta(t))[dy(t) - C(\eta(t))\hat{x}(t)dt] \qquad (7.71)$$

such that the H_2 norm of the mapping

$$\begin{bmatrix} \beta(t) \\ \mu(t) \end{bmatrix} \rightarrow e(t) := x(t) - \hat{x}(t) \qquad (7.72)$$

is minimized.

Remark 7.4.1. The observer (7.71) does not include multiplicative white noise. This structure could provide worse results than in the case of an observer with state-dependent noises but this is difficult to implement because the state-dependent noises cannot be directly measured.

Before determining the solution of this filtering problem we state the following useful result concerning the monotonicity of the stabilizing solution of the Riccati equation with respect to its free term (see Theorem 5.3.6 and [2]).

Proposition 7.4.1. *Consider the following systems of coupled Riccati-type equations*

$$A(i)X(i) + X(i)A^T(i) - X(i)M(i)X(i) + D(i)X(i)D^T(i)$$
$$+ \sum_{j=1}^{N} q_{ij}X(j) + P(i) = 0$$

and

$$A(i)\tilde{X}(i) + \tilde{X}(i)A^T(i) - \tilde{X}(i)M(i)\tilde{X}(i) + D(i)\tilde{X}(i)D^T(i)$$
$$+ \sum_{j=1}^{N} q_{ij}\tilde{X}(j) + R(i) = 0$$

where $M(i) \geq 0$ and $R(i) \geq P(i) \geq 0$, $i = 1,\ldots,N$. If $X(i) \geq 0$ and $\tilde{X}(i) \geq 0$ denote the stabilizing solutions of the above Riccati-type systems and $R(i) \geq P(i)$ then $\tilde{X}(i) \geq X(i)$, $i = 1,\ldots,N$.

The solution of the above problem is given by the following theorem.

Theorem 7.4.2. *The optimal gains $L(i)$, $i = 1,\ldots,N$ of the stochastic filter with Markovian jumps (7.71) for which the H_2 norm of the mapping (7.72) is minimized are given by*

$$L(i) = X(i)C^T(i)K(i) \qquad (7.73)$$

where $X(i), i = 1, \ldots, N$ is the stabilizing solution of the coupled system of Riccati-
type equations

$$A(i)X(i) + X(i)A^T(i) - X(i)C^T(i)K^{-1}(i)C(i)X(i)$$
$$+ \sum_{j=1}^{N} q_{ij}X(j) + D(i)Y(i)D^T(i) + \pi_{i\infty}B(i)B^T(i) = 0, \qquad (7.74)$$

$Y(i), i = 1, \ldots, N$ is the solution of the system of Lyapunov-type equations

$$A(i)Y(i) + Y(i)A^T(i) + D(i)Y(i)D^T(i) +$$
$$\sum_{j=1}^{N} q_{ij}Y(j) + \pi_{i\infty}B(i)B^T(i) = 0, \qquad (7.75)$$

$$K(i) := \pi_{i\infty}I + G(i)Y(i)G^T(i), i = 1, \ldots, N, \qquad (7.76)$$

with $\pi_{i\infty}$ defined in (7.15)

Proof. Coupling the systems (7.70) and (7.71) one obtains

$$dx = A(\eta)xdt + D(\eta)xd\xi + B(\eta)d\beta$$
$$d\hat{x} = A(\eta)\hat{x}dt + L(\eta)C(\eta)xdt$$
$$+ L(\eta)G(\eta)xdv + L(\eta)d\mu.$$

Subtracting the above equations one obtains the following equivalent system

$$\begin{bmatrix} de \\ dx \end{bmatrix} = \begin{bmatrix} A(\eta) - L(\eta)C(\eta) & 0 \\ 0 & A(\eta) \end{bmatrix} \begin{bmatrix} e \\ x \end{bmatrix}$$
$$+ \begin{bmatrix} 0 & D(\eta) \\ 0 & D(\eta) \end{bmatrix} \begin{bmatrix} e \\ x \end{bmatrix} d\xi + \begin{bmatrix} 0 & -L(\eta)G(\eta) \\ 0 & 0 \end{bmatrix} \qquad (7.77)$$
$$\times \begin{bmatrix} e \\ x \end{bmatrix} dv + \begin{bmatrix} B(\eta) & -L(\eta) \\ B(\eta) & 0 \end{bmatrix} \begin{bmatrix} d\beta \\ d\mu \end{bmatrix}.$$

Using Theorem 7.1.4 it follows that the H_2 norm of the above stochastic system
with the output $z(t) = e(t)$ equals $\sum_{j=1}^{N} Tr\mathcal{C}(j)\mathcal{P}_c(j)\mathcal{C}^T(j)$ where $\mathcal{C}(j) := [I \ 0]$, $j = 1, \ldots, N$, $\mathcal{P}$ denoting the solution of the Lyapunov-type equation $\mathcal{L}P + \hat{B}_v = 0$ (see
Lemma 7.1.3) written for the system (7.77). Using the definition (2.124) of the
operator $\mathcal{L}$, this Lyapunov-type equation becomes

$$\mathcal{A}(i)\mathcal{P}_c(i) + \mathcal{P}_c(i)\mathcal{A}^T(i) + \mathcal{D}_1(i)\mathcal{P}(i)\mathcal{D}_1(i)^T + \mathcal{D}_2(i)\mathcal{P}(i)\mathcal{D}_2(i)^T$$
$$+ \sum_{j=1}^{N} q_{ij}\mathcal{P}_c(j) + \pi_{i\infty}\mathcal{B}(i)\mathcal{B}^T(i) = 0, i = 1, \ldots, N \qquad (7.78)$$

where we denoted

$$\mathcal{A}(i) = \begin{bmatrix} A(i) - L(i)C(i) & 0 \\ 0 & A(i) \end{bmatrix}, \mathcal{B}(i) = \begin{bmatrix} B(i) & -L(i) \\ B(i) & 0 \end{bmatrix},$$
$$\mathcal{D}_1(i) = \begin{bmatrix} 0 & D(i) \\ 0 & D(i) \end{bmatrix}, \mathcal{D}_2(i) = \begin{bmatrix} B(i) & -L(i) \\ B(i) & 0 \end{bmatrix}.$$

Partitioning $\mathcal{P}(i)$ as

$$\mathcal{P}_c(i) = \begin{bmatrix} X(i) & Z(i) \\ Z^T(i) & Y(i) \end{bmatrix} \geq 0$$

direct computations gives that the block $(1,1)$ of (7.77) is

$$\begin{aligned}
&A(i)X(i) + X(i)A^T(i) - X(i)C^T(i)K^{-1}(i)C(i)X(i) \\
&+ \sum_{j=1}^{N} q_{ij}X(j) + D(i)Y(i)D^T(i) + \pi_{i\infty}B(i)B^T(i) \\
&+ \left[L(i) - X(i)C^T(i)K^{-1}(i) \right] K(i) \\
&\times \left[L(i) - X(i)C^T(i)K^{-1}(i) \right]^T = 0, \ i = 1,\ldots,N
\end{aligned} \tag{7.79}$$

and the block $(2,2)$ coincides with (7.75) in the statement. Then applying Proposition 7.4.1 it follows that for $L(i)$, $i = 1,\ldots,N$ given by (7.74) the solutions $X(i)$ of the system (7.79) are minimal and thus, since the trace function of positive semidefinite matrices is increasing, one concludes the H_2 norm of the system (7.77) is minimized.   □

7.4.2   A Numerical Example

We consider a numerical example that demonstrates the advantages of applying the estimator of the present paper to a tracking problem where the position x_1 and the velocity x_2 of an evasive guided-vehicle have to be filtered, utilizing their noisy measurements. The acceleration x_3 is related to its command via a first order lag with a time constant of $1/5$ s. The acceleration command is generated applying a proportional-navigation (PN) guidance law to a virtual target which is assumed at all times. The system is given by:

$$dx_1 = x_2 dt$$

$$dx_2 = x_3 dt$$

$$dx_3 = -5x_3 dt - 5 \times 3[(x_1 + x_2 t_{go})/t_{go}^2]dt + \sqrt{\bar{q}}d\beta + 2x_1 d\xi$$

where $t_{go} = 10$ s and $\bar{q} = 10^{-4}$. The measurements are given by

$$dy = \begin{bmatrix} x_1 \\ x_2 \end{bmatrix} dt + \bar{R}^{1/2}d\mu$$

where $\bar{R} = \begin{bmatrix} 10 & 0 \\ 0 & 0.1 \end{bmatrix}$. We note that in dx_3, the PN term $u_{PN} := [-3(x_1 + x_2 t_{go})/t_{go}^2] dt$
drives x_1 to zero, while the state-multiplicative noise term $2x_1 d\xi$ produces evasive
maneuvers that eventually decay to zero. The above system is readily described
by (7.70) where

$$A = \begin{bmatrix} 0 & 1 & 0 \\ 0 & 0 & 1 \\ -15/t_{go} & -15/t_{go^2} & -5 \end{bmatrix}, B = \sqrt{\bar{q}} \begin{bmatrix} 0 \\ 0 \\ 5 \end{bmatrix}$$

$$C = \begin{bmatrix} 1 & 0 & 0 \\ 0 & 1 & 0 \end{bmatrix}, G = \begin{bmatrix} 0 & 0 & 0 \\ 0 & 0 & 0 \end{bmatrix}$$

and

$$D = \begin{bmatrix} 0 & 0 & 0 \\ 0 & 0 & 0 \\ 2 & 0 & 0 \end{bmatrix}$$

Also due to non-unit intensity in the measurement noise, we replace in (7.70) $d\mu$
by $\sqrt{\bar{R}} d\mu$.

Two filters will be compared: the *Kalman filter* (KF) and the *stochastic filter* (SF)
of the present paper.

The KF is derived by solving the Riccati equations

$$\dot{P} = A\bar{P} + \bar{P}A^T - \bar{P}C^T \bar{R}^{-1} C\bar{P} + BB^T, P(0) = 10 \cdot I_3$$

where its gains were computed using $K(k) = \bar{P}C^T \bar{R}^{-1}$ resulting in the steady-state
values of

$$K = \begin{bmatrix} 0.0002 & 0.0035 \\ 0.0000 & 0.0647 \end{bmatrix}.$$

The SF gains were found by applying a transient version of Theorem 7.4.2 where
the zeros at the right-hand sides of (7.74) and (7.75) are replaced by $\dot{X}(i)$ and $\dot{Y}(i)$,
respectively, and where $\pi_{i\infty}$ was replaced by $\pi_{i,t}$ and where the term $\pi_{i\infty}I$ in (7.76)
was replaced by $\pi_{i,t}\bar{R}$ where $\pi_{i,t} := \sum_{j=1}^{N} \pi_j p_{ji}$. Note that $\pi_i = \mathcal{P}(\eta(0) = i)$ denoting
the initial distribution where $p_{ij}(t)$ are the elements of the matrix $P(t)$ depicted in
Fig. 7.1.

The filters were compared using time-domain simulation, the results of which
are in Figs. 7.2–7.5 where KF and SF estimated states are compared with the
corresponding true values (Figs. 7.2 and 7.3) and where the errors are compared

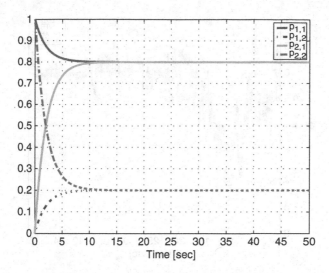

Fig. 7.1 Transition probabilities

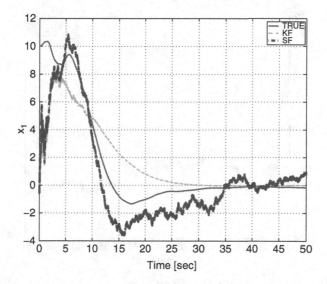

Fig. 7.2 True and estimated position with KF and SF

with each other (Figs. 7.4 and 7.5). The results clearly indicate that the SF of the
present paper outperforms, in the estimation bias sense, the standard KF which
ignores both multiplicative noise and Markov transitions of the mode. The SF
provides somewhat more noisy estimates of the states, a phenomenon which is
mainly observed in the velocity estimates. The lower bias in the estimations by the
SF over the KF is the result of closer prediction by the SF of the true covariances
of the state-estimation errors. To statistically establish the merits of the SF over the

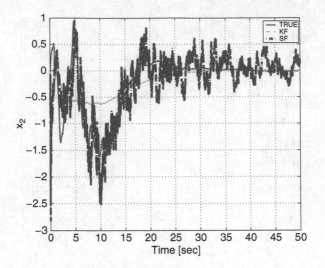

Fig. 7.3 True and estimated velocity with KF and SF

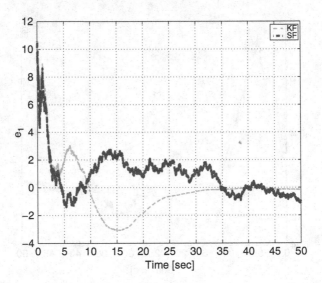

Fig. 7.4 Position estimation error with KF and SF

KF, a Monte-Carlo simulation study of 20 runs has been performed. The results of this study are depicted in Fig. 7.6 (position estimation error) and Fig. 7.7 (velocity estimation error) which show the smaller standard deviations of the estimation error computed on the 20 runs ensemble as a function of time obtained using the SF than with the KF.

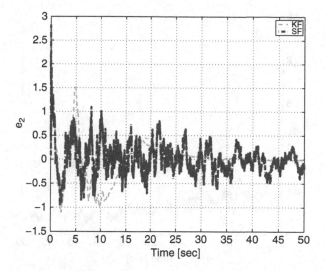

Fig. 7.5 Velocity estimation error with KF and SF

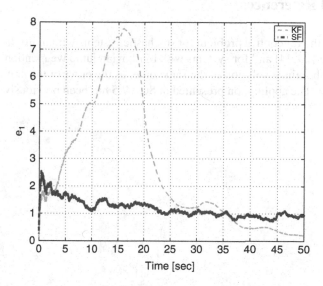

Fig. 7.6 Position estimation error with KF and SF—1σ from 20 Monte-Carlo runs

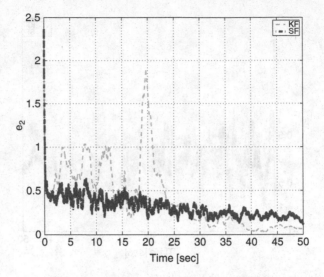

Fig. 7.7 Velocity estimation error with KF and SF—1σ from 20 Monte-Carlo runs

Notes and References

As concerns the H_2 control problem for stochastic systems with multiplicative white noise we cite [24,34] and for systems with Markovian jump we mention [19,30,43, 48] where suboptimal solution of the same order as the order of the nominal system is considered. The application presented in Sect. 7.5 has been previously considered in [136].

Chapter 8
Stochastic Version of Bounded Real Lemma and Applications

The main goal of this chapter is to investigate the robustness properties of a stable linear stochastic system with respect to various classes of uncertainties.

A crucial role in determining a lower bound of robustness radius will be played by the norm of a linear bounded operator associated with the given plant. This operator will be called *input–output operator* and it will be introduced in Sect. 8.1. In the next section a stochastic version of the so-called *Bounded Real Lemma* will be proved. This result provides an estimation of the norm of the input–output operator in terms of feasibility of some linear matrix inequalities (LMI) or in terms of existence of the stabilizing solutions of a generalized algebraic Riccati-type equation.

Further the stochastic version of the so-called *Small Gain Theorem* (SGT) will be proved. This result will be used to derive a lower bound of robustness with respect to linear structural uncertainties. Then we shall investigate the stability robustness with respect to a wide class of nonlinear uncertainties.

As in the previous chapters a unitary approach will be used for systems subject both to multiplicative white noise disturbances and to Markovian switching. Throughout this chapter we assume that the Markov process takes values in the finite set $\mathcal{D} = \{1, 2, \ldots, d\}$. In order to simplify the developments in this chapter we restrict our attention to the systems which are in time-invariant case.

8.1 Input–Output Operators

Consider the linear system described by:

$$dx(t) = [A_0(\eta(t))x(t) + B_0(\eta(t))u(t)]dt$$

$$+ \sum_{k=1}^{r} [A_k(\eta(t))x(t) + B_k(\eta(t))u(t)]dw_k(t) \qquad (8.1)$$

$$y(t) = C(\eta(t))x(t) + D(\eta(t))u(t)$$

V. Dragan et al., *Mathematical Methods in Robust Control of Linear Stochastic Systems*, DOI 10.1007/978-1-4614-8663-3_8, © Springer Science+Business Media New York 2013

with the state $x(t) \in \mathbf{R}^n$, the input $u(t) \in \mathbf{R}^m$ and the output $y(t) \in \mathbf{R}^p$. $A_k(i), B_k(i), k = 0, 1, \ldots, r, C(i), D(i), i \in \mathcal{D}$ are constant matrices of appropriate dimensions. The stochastic processes $\eta(t), t \geq 0, w(t) = (w_1(t), \ldots, w_r(t))_{t \geq 0}^T$ have the properties given in Chap. 1. If $u(t), t \geq 0$ is a stochastic process having the components in $L_{\eta,w}^2[0, \infty), x_u(t), t \geq 0$ stands for the solution of (8.1) with the initial condition $x_u(0) = 0$. According to the results derived in Chap. 1, Sect. 1.12, the components of the process $x_u(t), t \geq 0$ are in $L_{\eta,w}^2[0, \tau], \forall \tau > 0$. Moreover, if the system $(A_0, \ldots, A_r; Q)$ is stable, then based on Theorem 3.6.1, with $f_k(t) = B_k(\eta(t))u(t)$ it follows that $x_u(\cdot)$ is in $L_{\eta,w}^2([0, \infty), \mathbf{R}^n)$. On the other hand, by uniqueness arguments one easily obtains that the map $u \mapsto x_u(\cdot)$ is linear. Therefore, if the system $(A_0, \ldots, A_r; Q)$ is stable, we may consider the operator $\mathcal{T}$ defined on the space of stochastic processes $L_{\eta,w}^2([0, \infty), \mathbf{R}^m)$ with values in $L_{\eta,w}^2([0, \infty), \mathbf{R}^p)$, as follows:

$$(\mathcal{T}u)(t) = y_u(t),$$

where

$$y_u(t) = C(\eta(t))x_u(t) + D(\eta(t))u(t). \tag{8.2}$$

From Theorem 1.6.1 it follows that $L_{\eta,w}^2([0, \infty), \mathbf{R}^\ell)$ is a closed subspace of the Hilbert space $L^2([0, \infty), \mathbf{R}^\ell)$. Therefore $L_{\eta,w}^2([0, \infty), \mathbf{R}^\ell)$ is a real Hilbert space with the usual inner product

$$\langle u, v \rangle = E \int_0^\infty u^T(t)v(t)\,dt = \int_0^\infty E u^T(t)v(t)\,dt.$$

The norm induced by this inner product will be denoted by $\|\cdot\|$. Obviously

$$\|z\| = \left(E \int_0^\infty |z(t)|^2\,dt \right)^{\frac{1}{2}} \tag{8.3}$$

$$= \left(\sum_{j=1}^d \pi_j E \left[\int_0^\infty |z(t)|^2\,dt \mid \eta(0) = j \right] \right)^{\frac{1}{2}}$$

for all $z \in L_{\eta,w}^2([0, \infty), \mathbf{R}^\ell)$, where $\pi_i = \mathcal{P}\{\eta(0) = i\}$. Invoking again Theorem 3.6.1, immediately follows that it exists $c > 0$ not depending on u such that

$$\|x_u\|^2 = \sum_{j=1}^d \pi_j E \left[\int_0^\infty |x_u(t)|^2\,dt \mid \eta(0) = j \right]$$

$$\leq c \sum_{j=1}^d \pi_j E \left[\int_0^\infty |u(t)|^2\,dt \mid \eta(0) = j \right] = c \|u\|^2.$$

This allows us to conclude that the operator $\mathcal{T}$ defined by (8.2) is linear and bounded. The operator $\mathcal{T}$ introduced above will be termed the *input–output operator* associated with the system (8.1), and the system (8.1) will be a state-space realization of the operator $\mathcal{T}$. As in the deterministic case the state-space realization of the input–output operator is not unique. The set of operators

$$\mathcal{T} : L^2_{\eta,w}([0,\infty),\mathbf{R}^m) \to L^2_{\eta,w}([0,\infty),\mathbf{R}^p)$$

which admit state-space realizations is a subspace of the Banach space

$$\mathbf{B}\left(L^2_{\eta,w}([0,\infty),\mathbf{R}^m),L^2_{\eta,w}([0,\infty),\mathbf{R}^p)\right).$$

Indeed, one can easily check that if

$$\mathcal{T}_\ell : L^2_{\eta,w}([0,\infty),\mathbf{R}^m) \to L^2_{\eta,w}([0,\infty),\mathbf{R}^p), \ell = 1,2$$

have the state-space realization

$$dx_\ell(t) = [A_{0\ell}(\eta(t))x_\ell(t) + B_{0\ell}(\eta(t))u(t)]dt \tag{8.4}$$

$$+ \sum_{k=1}^{r}[A_{k\ell}(\eta(t))x_\ell(t) + B_{k\ell}(\eta(t))u(t)]dw_k(t)$$

$$y_\ell(t) = C_\ell(\eta(t))x_\ell(t) + D_\ell(\eta(t))u(t), \ell = 1,2$$

then the operators $\alpha_1\mathcal{T}_1 + \alpha_2\mathcal{T}_2$ will have the state-space realization of form (8.1) with:

$$A_k(i) = \begin{bmatrix} A_{k1}(i) & 0 \\ 0 & A_{k2}(i) \end{bmatrix},$$

$$B_k(i) = \begin{bmatrix} B_{k1}(i) \\ B_{k2}(i) \end{bmatrix},$$

$$C(i) = [\alpha_1 C_1(i) \quad \alpha_2 C_2(i)],$$

$$D(i) = \alpha_1 D_1(i) + \alpha_2 D_2(i) \quad \text{and}$$

$$x = \begin{bmatrix} x_1 \\ x_2 \end{bmatrix}.$$

Remark 8.1.1. For every $\tau > 0$, the system (8.1) defines a linear operator

$$\mathcal{T}_\tau : L^2_{\eta,w}([0,\tau],\mathbf{R}^m) \to L^2_{\eta,w}([0,\tau],\mathbf{R}^p)$$

by $y = \mathcal{T}_\tau u$ with

$$y(t) = C(\eta(t))x_u(t) + D(\eta(t))u(t), t \in [0,\tau], \forall u \in L^2_{\eta,w}([0,\tau],\mathbf{R}^m), t \in [0,\tau].$$

Based on Remark 3.6.1, one immediately deduces that $\mathcal{T}_\tau$ is a bounded operator. One expects that the norm $\|\mathcal{T}_\tau\|$ depends on τ. Moreover, for any $0 < \tau_1 < \tau_2$, we have

$$\|\mathcal{T}_{\tau_1}\| \le \|\mathcal{T}_{\tau_2}\|.$$

If the system $(A_0, A_1, \ldots, A_r; Q)$ is stable, then

$$\|\mathcal{T}\| = \sup_{\tau > 0} \|\mathcal{T}_\tau\|.$$

The last assertion in the above remark is also true if the linear operator $\mathcal{T}$ defined by (8.2) on the space $L^2_{\eta,w}([0,\infty), \mathbf{R}^m)$ is a bounded operator with values in the space $L^2_{\eta,w}([0,\infty), \mathbf{R}^p)$.

Concerning the product and the inversion of the input–output operators we have the following proposition.

Proposition 8.1.1. *(i) If*

$$\mathcal{T}^1_\tau : L^2_{\eta,w}([0,\tau], \mathbf{R}^m) \to L^2_{\eta,w}([0,\tau], \mathbf{R}^p),$$

$$\mathcal{T}^2_\tau : L^2_{\eta,w}([0,\tau], \mathbf{R}^{m_1}) \to L^2_{\eta,w}([0,\tau], \mathbf{R}^m),$$

have the state-space realizations as in (8.4) with $A_{k\ell}(i) \in \mathbf{R}^{n_\ell \times n_\ell}, B_{k1}(i) \in \mathbf{R}^{n_1 \times m}, B_{k2}(i) \in \mathbf{R}^{n_2 \times m_1}, 0 \le k \le r, C_1(i) \in \mathbf{R}^{p \times n_1}, C_2(i) \in \mathbf{R}^{m \times n_2}, D_1(i) \in \mathbf{R}^{p \times m}, D_2(i) \in \mathbf{R}^{m \times m_1}, i \in \mathcal{D}$ then the product

$$\mathcal{T}^1_\tau \mathcal{T}^2_\tau : L^2_{\eta,w}([0,\tau], \mathbf{R}^{m_1}) \to L^2_{\eta,w}([0,\tau], \mathbf{R}^p)$$

has the state-space realization of the form (8.1) where

$$A_k(i) = \begin{bmatrix} A_{k1}(i) & B_{k1}(i) C_2(i) \\ 0 & A_{k2}(i) \end{bmatrix},$$

$$B_k(i) = \begin{bmatrix} B_{k1}(i) D_2(i) \\ B_{k2}(i) \end{bmatrix}, \ 0 \le k \le r,$$

$$C(i) = [C_1(i) \ \ D_1(i) C_2(i)],$$

$$D(i) = D_1(i) D_2(i), \ i \in \mathcal{D};$$

(ii) Assume that in (8.1) we have $p = m$ and $\det D(i) \ne 0$, $i \in \mathcal{D}$. Then for every $\tau > 0$, the input–output operator $\mathcal{T}_\tau : L^2_{\eta,w}([0,\tau], \mathbf{R}^m) \to L^2_{\eta,w}([0,\tau], \mathbf{R}^m)$ is invertible and its inverse $\mathcal{T}_\tau^{-1}$ has the state-space realization:

$$d\xi(t) = \left[\tilde{A}_0(\eta(t)) \xi(t) + \tilde{B}_0(\eta(t)) y(t)\right] dt$$
$$+ \sum_{k=1}^r \left[\tilde{A}_k(\eta(t)) \xi(t) + \tilde{B}_k(\eta(t)) y(t)\right] dw_k(t) \qquad (8.5)$$
$$u(t) = \tilde{C}(\eta(t)) \xi(t) + \tilde{D}(\eta(t)) y(t),$$

where

$$\begin{aligned}
\tilde{A}_k(i) &= A_k(i) - B_k(i)D^{-1}(i)C(i) \\
\tilde{B}_k(i) &= B_k(i)D^{-1}(i), 0 \le k \le r; \\
\tilde{C}(i) &= -D^{-1}(i)C(i) \\
\tilde{D}(i) &= D^{-1}(i), i \in D.
\end{aligned}$$

Moreover if the systems $(A_0, A_1, \ldots, A_r; Q)$ *and* $(\tilde{A}_0, \tilde{A}_1, \ldots, \tilde{A}_r; Q)$ *are stable, then the input–output operator* $\mathcal{T}$ *associated with (8.1) is invertible and its inverse* $\mathcal{T}^{-1}$ *has the realization given by (8.5).*

Proof. Part (i) of the statement immediately follows by the uniqueness of the solution $x_u(\cdot)$ of the linear system (8.1).

(ii) Denote by $\hat{\mathcal{T}}_\tau$ the input–output operator defined by (8.5) on $[0, \tau]$. Applying the result of part (a) one easily checks that

$$\mathcal{T}_\tau \hat{\mathcal{T}}_\tau = I_{L^2_{\eta,w}([0,\tau],\mathbf{R}^m)} = \hat{\mathcal{T}}_\tau \mathcal{T}_\tau$$

where $I_{L^2_{\eta,w}([0,\tau],\mathbf{R}^m)}$ is the identity operator on $L^2_{\eta,w}([0,\tau], \mathbf{R}^m)$. The last assertion follows in the same way as above. $\square$

In the following we shall proof a result that will play an important role in the proof of the Bounded Real Lemma in the next section. For each continuous function $F : [0, \tau] \to \mathcal{M}^d_{m,n}$, $F(t) = (F(t,1), \ldots, F(t,d))$, consider the following Lyapunov-type equation on $\mathcal{S}^d_n$:

$$\begin{aligned}
&\tfrac{d}{dt}K(t,i) + (A_0(i) + B_0(i)F(t,i))^T + K(t,i)(A_0(i) + B_0(i)F(t,i)) \\
&+ \textstyle\sum_{k=1}^r (A_k(i) + B_k(i)F(t,i))^T K(t,i)(A_k(i) + B_k(i)F(t,i)) \\
&+ \textstyle\sum_{j=1}^d q_{ij}K(t,j) + (C(i) + D(i)F(t,i))^T (C(i) + D(i)F(t,i)) \\
&- \gamma^2 F^T(t,i)F(t,i) = 0, \; i \in \mathcal{D}.
\end{aligned} \tag{8.6}$$

For each $\gamma > 0$, denote by

$$K_\gamma(t) = (K_\gamma(t,1), \ldots, K_\gamma(t,d))$$

the solution of (8.6) verifying the condition $K_\gamma(\tau, i) = 0$, $i \in \mathcal{D}$.

Lemma 8.1.2. *Assume that for a fixed* $\tau > 0$ *we have* $\|\mathcal{T}_\tau\| < \gamma$. *Then for all* ε_0 *such that* $0 < \varepsilon_0^2 < \gamma^2 - \|\mathcal{T}_\tau\|^2$ *we have*

$$\gamma^2 I_m - D^T(i)D(i) - \sum_{k=1}^r B_k^T(i)K_\gamma(t,i)B_k(i) \ge \varepsilon_0^2 I_m \tag{8.7}$$

for all $t \in [0, \tau]$, $i \in \mathcal{D}$.

Proof. Denoting

$$\Gamma_\gamma(t,i) = \gamma^2 I_m - D^T(i)D(i) - \sum_{k=1}^{r} B_k^T(i)K_\gamma(t,i)B_k(i),$$

(8.7) can be written as $\Gamma_\gamma(t,i) \geq \varepsilon_0^2 I_m$, $\forall t \in [0,\tau]$, $i \in \mathcal{D}$. The proof has then two stages.

Stage 1. We firstly proof that for each γ satisfying the condition $\gamma > \|\mathcal{T}_\tau\|$ we have

$$\Gamma_\gamma(t,i) \geq 0, \ \forall t \in (0,\tau), i \in \mathcal{D}. \tag{8.8}$$

If (8.8) does not hold, then it follows that there exists $t_0 \in (0,\tau)$, $i_0 \in \mathcal{D}$, $u_0 \in \mathbf{R}^m$ with $|u_0| = 1$ such that $u_0^T\Gamma_\gamma(t_0,i_0)u_0 < 0$. Since the function $t \to u_0^T\Gamma_\gamma(t,i_0)u_0$ is continuous it follows that there exist $\delta_0 > 0$, $v > 0$ such that

$$u_0^T\Gamma_\gamma(t,i_0)u_0 < -v < 0, \ \forall t \in [t_0, t_0 + \delta_0] \tag{8.9}$$

with $t_0 + \delta_0 < \tau$. Let $\delta \in (0, \delta_0)$ be arbitrary but fixed and define the stochastic process

$$v_\delta(t) = \begin{cases} 0 \text{ if } t \notin [t_0, t_0 + \delta] \\ u_0 \chi_{\eta(t)=i_0} \text{ if } t \in [t_0, t_0 + \delta] \end{cases}.$$

It is obvious that $v_\delta \in L^2_{\eta,w}([0,\tau],\mathbf{R}^m)$. Let $x_\delta(t), t \in [0,\tau]$ be the solution of the following problem with initial conditions:

$$\begin{aligned} dx(t) &= \{[A_0(\eta(t)) + B_0(\eta(t))F(t,\eta(t))]x(t) + B_0(\eta(t))v_\delta(t)\}dt \\ &+ \sum_{k=1}^{r}\{[A_k(\eta(t)) + B_k(\eta(t))F(t,\eta(t))]x(t) \\ &+ B_k(\eta(t))v_\delta(t)\}dw_k(t), \ t \in [0,\tau], \ x_\delta(0) = 0. \end{aligned} \tag{8.10}$$

Define $u_\delta(t) = v_\delta(t) + F(t,\eta(t))x_\delta(t)$, $t \in [0,\tau]$. Since $u_\delta(t) \in L^2_{\eta,w}([0,\tau],\mathbf{R}^m)$, from (8.10) one deduces that

$$x_{u_\delta}(t) = x_\delta(t), \ t \in [0,\tau].$$

Let $y_\delta = \mathcal{T}_\tau u_\delta$, therefore

$$y_\delta(t) = C(\eta(t))x_\delta(t) + D(\eta(t))u_\delta(t), t \in [0,\tau].$$

By direct computations taking into account the definition of $u_\delta(t)$ we obtain that:

$$
\begin{aligned}
|y_\delta(t)|^2 - \gamma^2 |u_\delta(t)|^2 &= x_\delta^T(t) \Big[(C(\eta(t)) + D(\eta(t)) F(t,\eta(t)))^T \\
&\quad \times (C(\eta(t)) + D(\eta(t)) F(t,\eta(t))) \\
&\quad - \gamma^2 F^T(t,\eta(t)) F(t,\eta(t)) \Big] x_\delta(t) + 2 x_\delta^T(t) \\
&\quad \times \Big[(C(\eta(t)) + D(\eta(t)) F(t,\eta(t)))^T D(\eta(t)) \\
&\quad - \gamma^2 F^T(t,\eta(t)) \Big] v_\delta(t) + v_\delta^T(t) \\
&\quad \times \Big[D^T(\eta(t)) D(\eta(t)) - \gamma^2 I_m \Big] v_\delta(t).
\end{aligned}
\tag{8.11}
$$

Using the Itô-type formula for the function

$$
v(t,x,i) = x^T K_\gamma(t,i) x
$$

and for the process $x_\delta(t)$, $t \in [0,\tau]$, based on (8.6) and (8.11) one obtains that

$$
\begin{aligned}
&E\left[\int_0^\tau \left(|y_\delta(t)|^2 - \gamma^2 |u_\delta|^2 \right) dt \mid \eta(0) = i \right] \\
&= E\left[\int_0^\tau \left\{ 2 x_\delta^T(t) \mathcal{P}_\gamma(t,\eta(t)) v_\delta(t) - v_\delta^T(t) \Gamma_\gamma(t,\eta(t)) v_\delta(t) \right\} dt \mid \eta(0) = i \right]
\end{aligned}
$$

for all $i \in \mathcal{D}$, where $\mathcal{P}_\gamma(t,i)$ is defined as

$$
\begin{aligned}
\mathcal{P}_\gamma(t,i) &= K_\gamma(t,i) B_0(i) + \sum_{k=1}^r (A_k(i) + B_k(i) F(t,i))^T K_\gamma(t,i) B_k(i) \\
&\quad + (C(i) + D(i) F(t,i))^T D(i) - \gamma^2 F^T(t,i).
\end{aligned}
$$

Taking into account the definition of v_δ we further can write

$$
\begin{aligned}
&E\left[\int_0^\tau \left(|y_\delta(t)|^2 - \gamma^2 |u_\delta|^2 \right) dt \mid \eta(0) = i \right] \\
&= E\left[\int_{t_0}^{t_0+\delta} \left\{ 2 x_\delta^T(t) \mathcal{P}_\gamma(t,\eta(t)) u_0 - u_0^T \Gamma_\gamma(t,\eta(t)) u_0 \right\} \chi_{\eta(t)=i_0} dt \mid \eta(0) = i \right] \\
&= \sum_{j=1}^d E\left[\int_{t_0}^{t_0+\delta} \left\{ 2 x_\delta^T(t) \mathcal{P}_\gamma(t,j) u_0 - u_0^T \Gamma_\gamma(t,j) u_0 \right\} \chi_{\eta(t)=j} \chi_{\eta(t)=i_0} dt \mid \eta(0) = i \right].
\end{aligned}
$$

Since $\chi_{\eta(t)=i} \chi_{\eta(t)=i_0} = 0$ for $i \neq i_0$ and $\chi_{\eta(t)=i} \chi_{\eta(t)=i_0} = \chi_{\eta(t)=i_0}$ for $i = i_0$, we obtain:

$$
\begin{aligned}
&E\left[\int_0^\tau \left(|y_\delta(t)|^2 - \gamma^2 |u_\delta|^2 \right) dt \mid \eta(0) = i \right] \\
&= E\left[\int_{t_0}^{t_0+\delta} \left\{ 2 x_\delta^T(t) \mathcal{P}_\gamma(t,i_0) u_0 - u_0^T \Gamma_\gamma(t,i_0) u_0 \right\} \chi_{\eta(t)=i_0} dt \mid \eta(0) = i \right],
\end{aligned}
\tag{8.12}
$$

$i \in \mathcal{D}$. Based on (8.9) one immediately obtains that:

$$E\left[\int_0^\tau \left(|y_\delta(t)|^2 - \gamma^2 |u_\delta|^2\right) dt \mid \eta(0) = i\right] \geq$$
$$-2E\left[\int_{t_0}^{t_0+\delta} |x_\delta^T(t)\mathcal{P}_\gamma(t,i_0)u_0| \chi_{\eta(t)=i_0} dt \mid \eta(0) = i\right]$$
$$+\nu E\left[\int_{t_0}^{t_0+\delta} \chi_{\eta(t)=i_0} dt \mid \eta(0) = i\right],$$

and therefore

$$E\left[\int_0^\tau \left(|y_\delta(t)|^2 - \gamma^2 |u_\delta|^2\right) dt \mid \eta(0) = i\right] \geq$$
$$-2E\left[\int_{t_0}^{t_0+\delta} |x_\delta^T(t)\mathcal{P}_\gamma(t,i_0)u_0| \chi_{\eta(t)=i_0} dt \mid \eta(0) = i\right] \tag{8.13}$$
$$+\nu \int_{t_0}^{t_0+\delta} p_{i,i_0}(t) dt, \ i \in \mathcal{D}.$$

Based on Remark 3.6.1 one deduces that there exists $c_1 > 0$ depending on τ such that

$$\sup_{0 \leq t \leq \tau} E\left[|x_\delta(t)|^2 \mid \eta(0) = i\right] \leq c_1 E\left[\int_0^\tau |v_\delta(t)|^2 dt \mid \eta(0) = i\right]$$

$$\leq c_1 \delta. \tag{8.14}$$

On the other hand we have

$$E\left[\int_0^\tau |x_\delta^T(t)\mathcal{P}_\gamma(t,i_0)u_0| \chi_{\eta(t)=i_0} dt \mid \eta(0) = i\right]$$
$$\leq \int_0^\tau \left(E\left[|x_\delta(t)|^2 dt \mid \eta(0) = i\right]\right)^{\frac{1}{2}} |\mathcal{P}_\gamma(t,i_0)| dt.$$

Hence, using (8.14) we obtain

$$2E\left[\int_{t_0}^{t_0+\delta} |x_\delta^T(t)\mathcal{P}_\gamma(t,i_0)u_0| \chi_{\eta(t)=i_0} dt \mid \eta(0) = i\right] \leq c_2 \delta\sqrt{\delta} \tag{8.15}$$

where $c_2 > 0$ is a constant depending on τ. Then we have

$$E\int_0^\tau \left(|y_\delta(t)|^2 - \gamma^2 |u_\delta|^2\right) dt$$

$$= \sum_{i=1}^d \pi_i E\left[\int_0^\tau \left(|y_\delta(t)|^2 - \gamma^2 |u_\delta|^2\right) dt \mid \eta(0) = i\right] \tag{8.16}$$

$$\geq \int_{t_0}^{t_0+\delta} h(t) dt - c_2 \delta\sqrt{\delta}$$

where we denoted

$$h(t) = v \sum_{i=1}^{d} \pi_i p_{i,i_0}(t).$$

Since $p_{i_0,i_0}(t)$ is a continuous function it follows that there exists $\delta \in (0, \delta_0)$ such that

$$p_{i_0,i_0}(t) \geq \frac{1}{2} p_{i_0,i_0}(t_0) > 0 \quad \forall t_0 \leq t \leq t_0 + \delta.$$

Then for $\delta > 0$ small enough (8.16) becomes

$$\|y_\delta\|^2 - \gamma^2 \|u_\delta\|^2 = E \int_0^\tau \left(|y_\delta(t)|^2 - \gamma^2 |u_\delta|^2 \right) dt$$

$$\geq \frac{1}{2} \delta v \pi_{i_0} p_{i_0,i_0}(t_0) - c_2 \delta \sqrt{\delta} > 0.$$

This contradicts the assumption in the statement $\|\mathcal{T}_\tau\| < \gamma$. It follows then that (8.8) is accomplished for $t \in (0, \tau)$. From the continuity with respect to t it follows that (8.8) is accomplished for $t \in [0, \tau]$.

Stage 2. Let ε_0 be such that

$$0 < \varepsilon_0^2 < \gamma^2 - \|\mathcal{T}_\tau\|^2.$$

Then for $\hat{\gamma} = (\gamma^2 - \varepsilon_0^2)^{\frac{1}{2}}$ it is obvious that $\|\mathcal{T}_\tau\| < \hat{\gamma}$. According to *Stage 1* we have

$$\Gamma_{\hat{\gamma}}(t,i) \geq 0, \ t \in [0, \tau], \ i \in \mathcal{D}.$$

This leads to

$$\gamma^2 I_m - D^T(i) D(i) - \sum_{k=1}^{r} B_k^T(i) K_{\hat{\gamma}}(t,i) B_k(i) \geq \varepsilon_0^2 I_m. \tag{8.17}$$

On the other hand one can immediately check that

$$\frac{d}{dt} \left[K_{\hat{\gamma}}(t,i) - K_\gamma(t,i) \right]$$
$$+ \left[A_0(i) + B_0(i) F(t,i) \right]^T \left[K_{\hat{\gamma}}(t,i) - K_\gamma(t,i) \right]$$
$$+ \left[K_{\hat{\gamma}}(t,i) - K_\gamma(t,i) \right] \left[A_0(i) + B_0(i) F(t,i) \right]$$
$$+ \sum_{k=1}^{r} \left[A_k(i) + B_k(i) F(t,i) \right]^T \left[K_{\hat{\gamma}}(t,i) - K_\gamma(t,i) \right]$$
$$\times \left[A_k(i) + B_k(i) F(t,i) \right] + \sum_{j=1}^{d} q_{ij} \left[K_{\hat{\gamma}}(t,i) - K_\gamma(t,i) \right]$$
$$+ \varepsilon_0^2 F^T(t,i) F(t,i) = 0,$$

from which it follows that $K_{\hat{\gamma}}(t,i) - K_{\gamma}(t,i) \geq 0$. Therefore from (8.17) we deduce that

$$\gamma^2 I_m - D^T(i)D(i) - \sum_{i=1}^{r} B_k^T(i)K_{\gamma}(t,i)B_k(i) \geq \varepsilon_0 I_m$$

and hence the proof is complete.   □

Corollary 8.1.3. *If there exists $\tau > 0$ such that $\|\mathcal{T}_{\tau}\| < \gamma$, then $D^T(i)D(i) < \gamma^2 I_m$, $i \in \mathcal{D}$.*

Remark 8.1.2. If the system $(A_0, A_1, \ldots, A_r; Q)$ is stable and if $\|\mathcal{T}\| < \gamma$, then $\|\mathcal{T}_{\tau}\| < \gamma$ for all $\tau > 0$.

8.2   Stochastic Version of the Bounded Real Lemma

In the present section we shall derive necessary and sufficient conditions under which the norm of the input–output operator is less than a prescribed level of attenuation γ. These conditions extend at the stochastic systems of form (8.1) the well-known conditions given by the Bounded Real Lemma in the deterministic framework. The results proved in this section include as particular cases the results separately proved for stochastic systems with multiplicative white noise and for systems with Markovian jumps, respectively.

Consider the following system of generalized Riccati algebraic equations:

$$\begin{aligned}
&A_0^T(i)X(i) + X(i)A_0(i) + \sum_{k=1}^{r} A_k^T(i)X(i)A_k(i) + \sum_{j=1}^{d} q_{ij}X(j) \\
&+ \left(X(i)B_0(i) + \sum_{k=1}^{r} A_k^T(i)X(i)B_k(i) + C^T(i)D(i)\right) \\
&\times \left(\gamma^2 I_m - D^T(i)D(i) - \sum_{k=1}^{r} B_k^T(i)X(i)B_k(i)\right)^{-1} \\
&\times \left(B_0^T(i)X(i) + \sum_{k=1}^{r} B_k^T(i)X(i)A_k(i) + D^T(i)C(i)\right) \\
&+ C^T(i)C(i) = 0, \ i \in \mathcal{D}.
\end{aligned} \tag{8.18}$$

One can notice that in the particular case when $A_k(i) = 0$, $B_k(i) = 0$, $1 \leq k \leq r$, $\mathcal{D} = \{1\}$, the SGRAE (8.18) reduces to the well-known algebraic Riccati equation used in the deterministic framework in order to determine the H_∞ norm of a linear system. With the notations introduced in Sect. 5.6, the SGRAE (8.18) can be written as the following nonlinear equation on $\mathcal{S}_n^d$:

$$\mathcal{L}^* X - \mathcal{P}^T(X)\mathcal{R}^{-1}(X)\mathcal{P}(X) + C^T C = 0, \tag{8.19}$$

$\mathcal{L} : \mathcal{S}_n^d \to \mathcal{S}_n^d$ being the Lyapunov-type operator defined by the system $(A_0, A_1, \ldots, A_r; Q)$,

$$\mathcal{P}(X) = (\mathcal{P}_1(X), \ldots, \mathcal{P}_d(X))$$

with

$$\mathcal{P}_i(X) = B_0^T(i)X(i) + \sum_{k=1}^{r} B_k^T(i)X(i)A_k(i) + D^T(i)C(i),$$

$$\mathcal{R}(X) = (\mathcal{R}_1(X), \ldots, \mathcal{R}_d(X))$$

where

$$\mathcal{R}_i(X) = -\gamma^2 I_m + D^T(i)D(i) + \sum_{k=1}^{r} B_k^T(i)X(i)B_k(i), \; i \in \mathcal{D}$$

and $X = (X(1), \ldots, X(d))$. We shall also use the following differential equations on $\mathcal{S}_n^d$:

$$\frac{d}{dt}X(t) + \mathcal{L}^*X(t) - \mathcal{P}^T(X)\mathcal{R}^{-1}(X)\mathcal{P}(X) + C^TC = 0 \qquad (8.20)$$

$$\frac{d}{dt}K(t) = \mathcal{L}^*K(t) - \mathcal{P}^T(K(t))\mathcal{R}^{-1}(K(t))\mathcal{P}(K(t)) + C^TC. \qquad (8.21)$$

Remark 8.2.1. (i) Both the algebraic equation (8.19) and the differential equations (8.20) and also (8.21) are defined on the subset $\mathcal{U}_\gamma \subset \mathcal{S}_n^d$ with the elements $X = (X(1), \ldots, X(d))$ for which $det\,\mathcal{R}_i(X) \neq 0$, $i \in \mathcal{D}$. From Corollary 8.1.3 it follows that if exists $\tau > 0$ such that $\|\mathcal{T}_\tau\| < \gamma$ then the null element $(0, 0, \ldots, 0) \in \mathcal{S}_n^d$ is in $\mathcal{U}_\gamma$.

(ii) A C^1-function $X : [0, \tau] \to \mathcal{U}_\gamma$ is a solution of (8.20) if and only if $K : [0, \tau] \to \mathcal{U}_\gamma$ defined as $K(t) = X(\tau - t)$ is a solution of (8.21).

For every $\tau > 0$, $x_0 \in \mathbf{R}^n$, $\gamma > 0$, $i \in \mathcal{D}$, consider the following cost functions:

$$\mathcal{H}_\gamma(\tau, x_0, i, \cdot) : L_{\eta,w}^2([0, \tau]; \mathbf{R}^m) \to \mathbf{R}$$

$$\mathcal{H}_\gamma(\tau, x_0, \cdot) : L_{\eta,w}^2([0, \tau]; \mathbf{R}^m) \to \mathbf{R}$$

defined by

$$\mathcal{H}_\gamma(\tau, x_0, i, u) = E\left[\int_0^\tau \left(|y_u(t, x_0)|^2 - \gamma^2 |u(t)|^2 \right) dt \mid \eta(0) = i \right]$$

and

$$\mathcal{H}_\gamma(\tau, x_0, u) = E \int_0^\tau \left(|y_u(t, x_0)|^2 - \gamma^2 |u(t)|^2 \right) dt$$

where

$$y_u(t, x_0) = C(\eta(t))x_u(t, x_0) + D(\eta(t))u(t), \; t \in [0, \tau],$$

$x_u(t,x_0)$ being the solution of the system (8.1) determined by the input $u(t)$ and the initial condition $x_u(0,x_0) = x_0$. It is obvious that

$$\mathcal{H}_\gamma(\tau,x_0,u) = \sum_{i=1}^{d} \pi_i \mathcal{H}_\gamma(\tau,x_0,i,u).$$

From Corollary 6.1.2 and from Remark 8.2.1 (ii) one directly obtains:

Lemma 8.2.1. *If* $X : [0,\tau] \to \mathcal{S}_n^d$, $X(t) = (X(t,1),\ldots,X(t,d))$ *is a solution of* (8.20) *and* $K(t) = X(\tau - t)$, *then*

$$
\begin{aligned}
\mathcal{H}_\gamma(\tau,x_0,i,u) &= x_0^T X(0,i) x_0 - E\left[x^T(\tau) X(\tau,\eta(\tau)) x(\tau) \mid \eta(0) = i\right] \\
&\quad - E\left[\int_0^\tau \left(u(t) - F^X(t,\eta(t)) x(t)\right)^T \right. \\
&\quad \times \left[\gamma^2 I_m - D^T(\eta(t)) D(\eta(t)) - \sum_{k=1}^r B_k^T(\eta(t)) X(t,\eta(t)) B_k(\eta(t))\right] \\
&\quad \left. \times \left(u(t) - F^X(t,\eta(t)) x(t)\right) dt \mid \eta(0) = i\right] \\
&= x_0^T K(\tau,i) x_0 - E\left[x^T(\tau) K(0,\eta(t)) x(\tau) \mid \eta(0) = i\right] \\
&\quad - E\left[\int_0^\tau \left(u(t) - F^K(t,\eta(t)) x(t)\right)^T \right. \\
&\quad \times \left[\gamma^2 I_m - D^T(\eta(t)) D(\eta(t)) - \sum_{k=1}^r B_k^T(\eta(t)) K(\tau - t,\eta(t)) B_k(\eta(t))\right] \\
&\quad \left. \times \left(u(t) - F^K(t,\eta(t)) x(t)\right) dt \mid \eta(0) = i\right],
\end{aligned}
$$

for all $x_0 \in \mathbf{R}^n$, $i \in \mathcal{D}$, $u \in L^2_{\eta,w}([0,\tau],\mathbf{R}^m)$, $x(t) = x_u(t,x_0)$

$$
\begin{aligned}
F^X(t,i) &= -\left(R(i) + \sum_{k=1}^r B_k^T(i) X(t,i) B_k(i)\right)^{-1} \\
&\quad \times \left(B_0^T(i) X(t,i) + \sum_{k=1}^r B_k^T(i) X(t,i) A_k(i) + D^T(i) C(i)\right) \\
F^K(t,i) &= -\left(R(i) + \sum_{k=1}^r B_k^T(i) K(\tau - t,i) B_k(i)\right)^{-1} \\
&\quad \times \left(B_0^T(i) K(\tau - t,i) + \sum_{k=1}^r B_k^T(i) K(\tau - t,i) A_k(i) + D^T(i) C(i)\right),
\end{aligned}
$$

where $R(i) = -\gamma^2 I_m + D^T(i) D(i)$.

We prove now the following useful result.

Lemma 8.2.2. *Assume that the system* $(A_0,A_1,\ldots,A_r;Q)$ *is stable and* $\|\mathcal{T}\| < \gamma$. *In these conditions it exists a constant* $\rho > 0$ *such that*

$$\mathcal{H}_\gamma(\tau,x_0,i,u) \leq \rho |x_0|^2 \ \forall \tau > 0, \ x_0 \in \mathbf{R}^n, \ u \in L^2_{\eta,w}([0,\tau],\mathbf{R}^m).$$

Proof. Let $x_u(t,x_0)$ be the solution of the system (8.1) corresponding to the arbitrary control $u \in L^2_{\eta,w}([0,\tau],\mathbf{R}^m)$. Then one can write

$$x_u(t,x_0) = x_0(t,x_0) + x_u(t,0),$$

where $x_0(t,x_0)$ is the solution of the system (8.1) for $u = 0$ satisfying $x_0(0,x_0) = x_0$. Therefore $x_0(t,x_0) = \Phi(t,0) x_0$. As in the preceding section the process $x_u(t) = x_u(t,0)$ is the solution of the system (8.1) satisfying the initial condition $x_u(0,0) = 0$. Denoting

$$y_0(t,x_0) = C(\eta(t))x_0(t,x_0) \text{ and}$$
$$y_u(t) = C(\eta(t))x_u(t) + D(t)u(t),$$

one obtains that

$$y_u(t,x_0) = y_0(t,x_0) + y_u(t). \tag{8.22}$$

Since the system $(A_0,A_1,\ldots,A_r;Q)$ is stable it exists $\rho_1 > 0$ not depending on x_0, such that

$$E\left[\int_0^\infty |y_0(t,x_0)|^2 \, dt\right] \leq \rho_1^2 |x_0|^2, \ \forall \, x_0 \in \mathbf{R}^n. \tag{8.23}$$

On the other hand from the inequalities

$$\|\mathcal{T}_\tau\| \leq \|\mathcal{T}\| < \gamma$$

it follows that it exists $v > 0$ not depending on $u(t)$ such that

$$E\int_0^\tau \left(|y_u(t)|^2 - \gamma^2 |u(t)|^2\right) dt \leq -v^2 E \int_0^\tau |u(t)|^2 \, dt, \tag{8.24}$$

$\forall \, u \in L^2_{\eta,w}([0,\tau],\mathbf{R}^m)$. Using the decomposition (8.22) of y_u one obtains that

$$\mathcal{H}_\gamma(\tau,x_0,u) = E\int_0^\tau |y_0(t,x_0)|^2 \, dt + 2E\int_0^\tau y_0^T(t,x_0)y_u(t) \, dt$$
$$+ E\int_0^\tau \left(|y_u(t)|^2 - \gamma^2 |u(t)|^2\right) dt.$$

Taking into account (8.23) and (8.24) one immediately obtains:

$$\mathcal{H}_\gamma(\tau,x_0,u) \leq \rho_1^2 |x_0|^2 + 2\rho_1 \gamma |x_0| \|u\| - v^2 \|u\|^2, \tag{8.25}$$

$\forall \, u \in L^2_{\eta,w}([0,\tau],\mathbf{R}^m)$, where $\|u\| = \left(E\int_0^\tau |u(t)|^2 \, dt\right)^{\frac{1}{2}}$. Since the right-hand side of (8.25) is a second degree polynomial with respect to $\|u\|$ one immediately deduces that

$$\mathcal{H}_\gamma(\tau,x_0,u) \leq \rho^2 |x_0|^2 \tag{8.26}$$

where $\rho = \rho_1 v^{-1}\sqrt{\gamma^2 + v^2}$ and therefore the proof is complete. $\qquad\square$

We shall denote in the following by $X_\tau(t) = (X_\tau(t,1),\ldots,X_\tau(t,d))$ the solution of (8.20) satisfying the condition $X_\tau(\tau,i) = 0, \ i \in \mathcal{D}$. Let $\mathcal{I}_\tau(\gamma) \subset [0,\tau]$ the maximal interval on which the solution $X_\tau(\cdot)$ is defined. From Remark 8.2.1 (i) it follows that if $\|\mathcal{T}_\tau\| < \gamma$ then $\mathcal{I}_\tau(\gamma)$ is nonempty. Then from Lemma 8.1.2 one obtains the following lemma.

Lemma 8.2.3. *If* $\sup_{\tau > 0} \|\mathcal{T}_\tau\| < \gamma$ *then*

$$\gamma^2 I_m - D^T(i) D(i) - \sum_{k=1}^{r} B_k^T(i) X_\tau(t,i) B_k(i) \geq \varepsilon_0^2 I_m, \, t \in \mathcal{I}_\tau(\gamma), \tag{8.27}$$

$i \in \mathcal{D}, \tau > 0$, *where* $\varepsilon_0 > 0$ *does not depend upon* τ.

Proof. Let $\varepsilon_0 > 0$ be such that $\varepsilon_0^2 < \gamma^2 - \sup_{\tau > 0} \|\mathcal{T}_\tau\|^2$. Let $\tau > 0$ and $t_1 \in \mathcal{I}_\gamma(\tau)$, $t_1 < \tau$. Obviously $[t_1, \tau] \subset \mathcal{I}_\gamma(\tau)$. Denote

$$F_\tau(t,i) = \left(\gamma^2 I_m - D^T(i) D(i) - \sum_{k=1}^{r} B_k^T(i) X_\tau(t,i) B_k(i) \right)^{-1} \tag{8.28}$$

$$\times \left(B_0^T(i) X_\tau(t,i) + \sum_{k=1}^{r} B_k^T(i) X_\tau(t,i) A_k(i) + D^T(i) C(i) \right),$$

$t \in [t_1, \tau]$, $i \in \mathcal{D}$. With Lemma 5.1.1 one immediately obtains that (8.20) satisfied by $X_\tau(\cdot)$ can be written in a Lyapunov form on $\mathcal{S}_n^d$ as follows:

$$\begin{aligned}
&\tfrac{d}{dt} X_\tau(t,i) + [A_0(i) + B_0(i) F_\tau(t,i)]^T X_\tau(t,i) \\
&+ X_\tau(t,i) [A_0(i) + B_0(i) F_\tau(t,i)] \\
&+ \sum_{k=1}^{r} [A_k(i) + B_k(i) F_\tau(t,i)]^T X_\tau(t,i) [A_k(i) + B_k(i) F_\tau(t,i)] \\
&+ \sum_{j=1}^{d} q_{ij} X_\tau(t,j) - \gamma^2 F_\tau^T(t,i) F_\tau(t,i) \\
&+ [C(i) + D(i) F_\tau(t,i)]^T [C(i) + D(i) F_\tau(t,i)] = 0,
\end{aligned} \tag{8.29}$$

$t \in [t_1, \tau]$, $i \in \mathcal{D}$.

Let $F : [0, \tau] \to \mathcal{M}_{mn}^d$ defined as:

$$F(t) = (F(t,1), \ldots, F(t,d)), \tag{8.30}$$

$$F(t,i) = \begin{cases} F_\tau(t,i), \, t \in [t_1, \tau] \\ F_\tau(t_1,i), \, t \in [0, t_1], i \in \mathcal{D} \end{cases}$$

and let $X(t) = (X(t,1), \ldots, X(t,d))$ with $X(\tau) = 0$ be the solution of (8.6) corresponding to the feedback $F(\cdot)$ defined as in (8.30). Then from (8.29) and (8.30) it follows that $X(t) = X_\tau(t)$, $t \in [t_1, \tau]$. Applying Lemma 8.1.2 one obtains that (8.27) is true for all $t \subset [t_1, \tau]$ and the proof is complete. $\qquad\square$

In the following we shall denote by $K^0(t) = \left(K^0(t,1), \ldots, K^0(t,d) \right)$ the solution of (8.21) satisfying the initial condition $K^0(0,i) = 0$, $i \in \mathcal{D}$. We also denote $[0, t_f)$ the maximal interval on which this solution is defined. The next lemma summarizes some properties of the solution $K^0(t)$.

Lemma 8.2.4. *Assume that the system* $(A_0, A_1, \ldots, A_r; Q)$ *is stable and* $\|\mathcal{T}\| < \gamma$. *Then the solution* $K^0(t)$ *of (8.21) has the following properties:*

(i)

$$\gamma^2 I_m - D^T(i) D(i) - \sum_{k=1}^{r} B_k^T(i) K^0(t, i) B_k(i) \geq \varepsilon_0^2 I_m$$

$\forall t \in [0, t_f), \ \varepsilon_0$ *independent of* t;

(ii)

$$x_0^T K^0(\tau, i) x_0 = \mathcal{H}_\gamma(\tau, x_0, i, \tilde{u}_\tau) \geq \mathcal{H}_\gamma(\tau, x_0, i, u)$$

$\forall \tau \in (0, t_f), \ x_0 \in \mathbf{R}^n, i \in \mathcal{D}, u \in L^2_{\eta, w}([0, \tau], \mathbf{R}^m),$ *where* $\tilde{u}_\tau(t) = F_\tau(t, \eta(t)) \tilde{x}_\tau(t)$ *and*

$$F_\tau(t, i) = \left(\gamma^2 I_m - D^T(i) D(i) - \sum_{k=1}^{r} B_k^T(i) K^0(\tau - t, i) B_k(i) \right)^{-1} \cdot$$

$$\times \left(B_0^T(i) K^0(\tau - t, i) + \sum_{k=1}^{r} B_k^T(i) K^0(\tau - t, i) A_k(i) + D^T(i) C(i) \right)$$

and $\tilde{x}_\tau(t), \ t \in [0, \tau]$ *is the solution of the equation*

$$d\tilde{x}(t) = [A_0(\eta(t)) + B_0(\eta(t)) F_\tau(t, \eta(t))] \tilde{x}(t) \, dt$$

$$+ \sum_{k=1}^{r} [A_k(\eta(t)) + B_k(\eta(t)) F_\tau(t, \eta(t))] \tilde{x}(t) \, dw_k(t)$$

with the initial condition $\tilde{x}_0(0) = x_0$;

(iii) *It exists* $\tilde{\rho} > 0$ *not depending on* τ *such that*

$$0 \leq K^0(\tau, i) \leq \tilde{\rho} I_n, \ \forall \tau \in [0, t_f), \ i \in \mathcal{D};$$

(iv)

$$K^0(\tau_1, i) \leq K^0(\tau_2, i), \ \forall 0 \leq \tau_1 < \tau_2 < t_f.$$

Proof. (i) Let $\tau \in (0, t_f)$ be arbitrary but fixed and denote

$$X_\tau(t) = (X_\tau(t, 1), \ldots, X_\tau(t, d))$$

defined by

$$X_\tau(t) = K^0(\tau - t, i), \ t \in [0, \tau], \ i \in \mathcal{D}.$$

Then $X_\tau(t)$ is the solution of (8.20) with the final condition $X_\tau(\tau) = 0$. Based on Lemma 8.2.3 and Remark 8.1.1 one obtains

$$\gamma^2 I_m - D^T(i) D(i) - \sum_{k=1}^r B_k^T(i) X_\tau(t,i) B_k(i) \geq \varepsilon_0^2 I_m, \quad t \in [0,\tau]. \qquad (8.31)$$

Since ε_0 does not depend on τ, based on (8.31) and on the definition of X_τ the proof of part (i) is complete.

(ii) Applying Lemma 8.2.1 for $K^0(t) = X_\tau(\tau - t)$ one obtains:

$$\mathcal{H}_\gamma(\tau, x_0, i, u) = x_0^T K^0(\tau, i) x_0 - E\left[\int_0^\tau (u(t) - F_\tau(t, \eta(t) x_u(t, x_0))^T \right. \qquad (8.32)$$

$$\times \left(\gamma^2 I_m - D^T(\eta(t)) D(\eta(t)) - \sum_{k=1}^r B_k^T(\eta(t)) X_\tau(t, \eta(t)) B_k(\eta(t))\right)$$

$$\left. \times (u(t) - F_\tau(t, \eta(t) x_u(t, x_0)) dt \mid \eta(0) = i\right],$$

$\forall x_0 \in \mathbf{R}^n, i \in \mathcal{D}, u \in L^2_{\eta,w}([0,\tau], \mathbf{R}^m)$. From (8.32) and (i) immediately follows that

$$\mathcal{H}_\gamma(\tau, x_0, i, u) \leq x_0^T K^0(\tau, i) x_0 \qquad (8.33)$$

and for

$$u(t) = F_\tau(t, \eta(t)) x_u(t, x_0) = F_\tau(t, \eta(t)) \tilde{x}(t)$$

the inequality (8.33) becomes equality.

(iii) From (8.33) one immediately deduces that

$$0 \leq \mathcal{H}_\gamma(\tau, x_0, i, 0) \leq x_0^T K^0(\tau, i) x_0. \qquad (8.34)$$

On the other hand for every $i \in \mathcal{D}$ one can write:

$$\pi_i x_0^T K^0(\tau, i) x_0 \leq \sum_{j=1}^d \pi_j \mathcal{H}_\gamma(\tau, x_0, j, \tilde{u}) = \mathcal{H}_\gamma(\tau, x_0, \tilde{u}).$$

From Lemma 8.2.2 we have

$$\mathcal{H}_\gamma(\tau, x_0, \tilde{u}) \leq \rho^2 |x_0|^2. \qquad (8.35)$$

Then from (8.34) and (8.35) it follows that (iii) is satisfied for

$$\tilde{\rho} = \max_{i \in \mathcal{D}} \frac{\rho^2}{\pi_i}.$$

(iv) Let $0 < \tau_1 < \tau_2 < t_f$ and consider the stochastic process $u_{\tau_2}, t \in [0, \tau_2]$, as follows

$$u_{\tau_2}(t) = \begin{cases} \tilde{u}_{\tau_1}(t), \, t \in [0, \tau_1] \\ 0, \, t \in (\tau_1, \tau_2] \end{cases}.$$

It is obvious that $u_{\tau_2} \in L^2_{\eta,w}([0, \tau_2], \mathbf{R}^m)$. Let $x_{\tau_2}(t), \, t \in [0, \tau_2]$ be the solution of the system (8.1) determined by the input variable $u_{\tau_2}(t)$ and by the initial condition $x_{\tau_2}(0) = x_0$. One can easily check that $x_{\tau_2}(t) = \tilde{x}_{\tau_1}(t)$ for $t \in [0, \tau_1]$ and

$$\mathcal{H}_\gamma(\tau_1, x, i, \tilde{u}_{\tau_1}) \le \mathcal{H}_\gamma(\tau_2, x_0, i, u_{\tau_2}).$$

Invoking again the maximality properties in (ii) one obtains:

$$x_0^T K^0(\tau_1, i) x_0 = \mathcal{H}_\gamma(\tau_1, x_0, i, \tilde{u}_{\tau_1}) \le \mathcal{H}_\gamma(\tau_2, x_0, i, \tilde{u}_{\tau_2})$$

$$\le x_0^T K^0(\tau_2, i) x_0, \quad \forall x_0 \in \mathbf{R}^n, \, i \in \mathcal{D},$$

and therefore the proof is complete. □

Remark 8.2.2. From (i) and (iii) in Lemma 8.2.4 it follows that the solution $K^0(\cdot)$ is defined on $[0, \infty)$, i.e. $t_f = \infty$.

Consider the following subsets of $\mathcal{S}_n^d$:

$$\Pi = \{ X = (X(1), \dots, X(d)) \in \mathcal{S}_n^d \mid \\ \mathcal{L}^* X - \mathcal{P}^T(X) \mathcal{R}^{-1}(X) \mathcal{P}(X) + C^T C \le 0, \, \mathcal{R}(X) < 0 \} \tag{8.36}$$

and

$$\tilde{\Pi} = \{ X = (X(1), \dots, X(d)) \in \mathcal{S}_n^d \mid \\ \mathcal{L}^* X - \mathcal{P}^T(X) \mathcal{R}^{-1}(X) \mathcal{P}(X) + C^T C < 0, \, \mathcal{R}(X) < 0 \}. \tag{8.37}$$

Remark 8.2.3. (i) $\tilde{\Pi} \subset \Pi$.
(ii) If the system $(A_0, A_1, \dots, A_r; Q)$ is stable, then $\Pi \subset \mathcal{S}_{n+}^d$.
(iii) Let us introduce the generalized dissipation matrix

$$\mathcal{N}(X) = (\mathcal{N}_1(X, \gamma), \dots, \mathcal{N}_d(X, \gamma))$$

associated with the system (8.1) and with the scalar γ, as follows

$$\mathcal{N}_i(X,\gamma) = \begin{bmatrix} \mathcal{N}_{11}^i(X,\gamma) & \mathcal{N}_{12}^i(X,\gamma) \\ (\mathcal{N}_{12}^i)^T(X,\gamma) & \mathcal{N}_{22}^i(X,\gamma) \end{bmatrix}$$

where

$$\mathcal{N}_{11}^i(X,\gamma) = A_0^T(i)X(i) + X(i)A_0(i) + \sum_{k=1}^r A_k^T(i)X(i)A_k(i)$$

$$+ \sum_{j=1}^d q_{ij}X(j) + C^T(i)C(i)$$

$$\mathcal{N}_{12}^i(X,\gamma) = X(i)B_0(i) + \sum_{k=1}^r A_k^T(i)X(i)B_k(i) + C^T(i)D(i) = \mathcal{P}_i^T(X)$$

$$\mathcal{N}_{22}^i(X,\gamma) = -\gamma^2 I_m + D^T(i)D(i) + \sum_{k=1}^r B_k^T(i)X(i)B_k(i) = \mathcal{R}_i(X).$$

It is easy to check that

$$\Pi = \left\{ X \in \mathcal{S}_n^d \mid \mathcal{N}(X) \le 0, \mathcal{R}(X) < 0 \right\}$$

and

$$\tilde{\Pi} = \left\{ X \in \mathcal{S}_n^d \mid \mathcal{N}(X) < 0 \right\}.$$

From the above inequalities one easily deduces that both Π and $\tilde{\Pi}$ are convex sets. The set Π includes the solutions of (8.19) for which the condition $\mathcal{R}(X) < 0$ is accomplished.

Proposition 8.2.5. *Assume that the system $(A_0, \ldots, A_r; Q)$ is stable and $\Pi \ne \varnothing$. Then for all*

$$\hat{X} = (\hat{X}(1), \ldots, \hat{X}(d)) \in \Pi$$

we have

$$K^0(t) \le \hat{X}, \, \forall t \in [0, t_f),$$

K^0 *denoting the solution of (8.21) verifying the initial condition $K^0(0) = 0$.*

Proof. Under the above assumptions, by Remark 8.2.3 (ii) it follows that it exists $X \ge 0$ with $\mathcal{R}(X) < 0$. Therefore $\gamma^2 I_m - D^T(i)D(i) > 0$, $i \in \mathcal{D}$. Thus we may conclude that the solution $K^0(t)$ is defined on an interval $[0, \tau]$, $\tau > 0$. Let $\hat{X} = (\hat{X}(1), \ldots, \hat{X}(d)) \in \Pi$ be arbitrary but fixed. Define

$$\hat{M} = (\hat{M}(1), \ldots, \hat{M}(d))$$

by

$$\hat{M} = -\mathcal{L}^* \hat{X} + \mathcal{P}^T(\hat{X}) \mathcal{R}^{-1}(\hat{X}) \mathcal{P}(\hat{X}) - C^T C.$$

From the definition of $\hat{M}$ it follows that $\hat{X}$ verifies the algebraic equation

$$\mathcal{L}^* \hat{X} - \mathcal{P}^T(\hat{X}) \mathcal{R}^{-1}(\hat{X}) \mathcal{P}(\hat{X}) + C^T C + \hat{M} = 0. \tag{8.38}$$

Let $\tau \in (0, t_f)$ and let $X_\tau(t) = (X_\tau(t, 1), \ldots, X_\tau(t, d))$ defined as

$$X_\tau(t, i) = K^0(\tau - t, i), \, t \in [0, \tau], \, i \in \mathcal{D}.$$

Thus one deduces that $X_\tau(\cdot)$ is the solution of (8.20) satisfying the terminal condition $X_\tau(\tau) = 0$. Define

$$F_\tau(t) = (F_\tau(t, 1), \ldots, F_\tau(t, d)),$$
$$F_\tau(t, i) = -\mathcal{R}_i^{-1}(X_\tau(t)) \mathcal{P}_i(X_\tau(t)), \, i \in \mathcal{D}, \, t \in [0, \tau].$$

By direct computations (see also Lemma 5.1.1) one obtains that $\hat{X}$ verifying (8.38) is also a solution of the equation parameterized with respect to t

$$\mathcal{L}_{F_\tau}^*(t) X - \gamma^2 F_\tau^T(t) F_\tau(t) + (C + DF_\tau(t))^T (C + DF_\tau(t))$$
$$+ \hat{M} - (F_\tau(t) - \hat{F})^T \mathcal{R}(\hat{X}) (F_\tau(t) - \hat{F}) = 0, \, t \in [0, \tau], \tag{8.39}$$

where $\mathcal{L}_{F_\tau}(t)$ denotes as usual, the Lyapunov-type operator defined by the system $(A_0 + B_0 F_\tau, \ldots, A_r + B_r F_\tau; Q)$ and

$$\hat{F} = (\hat{F}(1), \ldots, \hat{F}(d)),$$
$$\hat{F}(i) = -\mathcal{R}_i^{-1}(\hat{X}) \mathcal{P}_i(\hat{X}), \, i \in \mathcal{D}.$$

On the other hand, based on (8.29) one obtains that (8.20) verified by $X_\tau(\cdot)$ can be rewritten as:

$$\frac{d}{dt} X_\tau(t) + \mathcal{L}_{F_\tau}^*(t) X_\tau(t) - \gamma^2 F_\tau^T(t) F_\tau(t)$$
$$+ (C + DF_\tau(t))^T (C + DF_\tau(t)) = 0. \tag{8.40}$$

Let $Y(t) = \hat{X} - X_\tau(t)$, $t \in [0, \tau]$. From (8.39) and (8.40) one obtains that

$$\frac{d}{dt}Y(t) + \mathcal{L}_{F_\tau}^*(t)Y(t) + \overline{M}(t) = 0 \qquad (8.41)$$

where

$$\overline{M}(t) = -\left(F_\tau(t) - \hat{F}\right)^T \mathcal{R}\left(\hat{X}\right)\left(F_\tau(t) - \hat{F}\right) + \hat{M}$$

and it immediately follows that $\overline{M}(t) \geq 0$. Based on Remark 8.2.3 (ii) it follows that $Y(\tau) = \hat{X} \geq 0$. According to constant variation formula, we have

$$Y(t) = T_\tau^*(\tau, t)Y(\tau) + \int_t^\tau T_\tau^*(s, t)\overline{M}(s)\,ds, \; t \in [0, \tau] \qquad (8.42)$$

where $T_\tau(t, s)$ is the linear operator of evolution on $\mathcal{S}_n^d$ defined by the differential equation

$$\frac{dY}{dt} = \mathcal{L}_{F_\tau}(t)Y(t).$$

Since $T_\tau^*(s, t)$ is a positive operator on $\mathcal{S}_n^d$ for any $s \geq t$ from (8.42) it follows that $Y(t) \geq 0$ for all $t \in [0, \tau]$, which leads to $X_\tau(t) \leq \hat{X}$, $t \in [0, \tau]$, or equivalently,

$$K^0(t) \leq \hat{X}, \; \forall\, t \in [0, \tau]. \qquad (8.43)$$

Since τ has been arbitrarily chosen in $[0, t_f)$ it follows that (8.43) is verified for any $t \in [0, t_f)$. $\square$

Before proving the main result of this section we remind the following known result from the theory of differential equations.

Lemma 8.2.6. *Let $F : \mathcal{X} \to \mathcal{X}$ be a continuous function defined on the Banach space $\mathcal{X}$. If $\xi : [0, \infty) \to \mathcal{X}$ is a solution of the differential equation*

$$\dot{\xi}(t) = F(\xi(t))$$

with the property $\lim_{t \to \infty} \xi(t) = \hat{\xi} \in \mathcal{X}$, then $F\left(\hat{\xi}\right) = 0$.

Proof. Let $\varphi : \mathcal{X} \to \mathbf{R}$ be a linear and continuous functional. Then $t \to \varphi(\xi(t))$ satisfies:

$$\frac{d}{dt}\varphi(\xi(t)) = \varphi(F(\xi(t)))$$

and

$$\lim_{t \to \infty} \varphi(\xi(t)) = \varphi\left(\hat{\xi}\right).$$

Since

$$\varphi\left(\xi\left(t\right)\right) - \varphi\left(\xi\left(t_0\right)\right) = \int_{t_0}^{t} \varphi\left(F\left(\xi\left(s\right)\right)\right) ds$$

it follows that

$$\lim_{t \to \infty} \int_{t_0}^{t} \varphi\left(F\left(\xi\left(s\right)\right)\right) ds = \varphi\left(\hat{\xi}\right) - \varphi\left(\xi\left(t_0\right)\right) \in \mathbf{R}.$$

Then the integral $\int_{t_0}^{\infty} \varphi\left(F\left(\xi\left(s\right)\right)\right) ds$ is convergent. On the other hand

$$\lim_{t \to \infty} \varphi\left(F\left(\xi\left(t\right)\right)\right) = \varphi\left(F\left(\hat{\xi}\right)\right).$$

From the convergence of the above integral it follows that $\varphi\left(F\left(\hat{\xi}\right)\right) = 0$. Since φ is an arbitrary linear and continuous functional we deduce that $F\left(\hat{\xi}\right) = 0$ and hence the proof is complete. □

The main result of this section is the following theorem.

Theorem 8.2.7 (Bounded Real Lemma). *The following assertions are equivalent:*

(i) *The system $(A_0, A_1, \ldots, A_r; Q)$ is stable and $\|\mathcal{T}\| < \gamma$;*
(ii) *It exists $\hat{X} = \left(\hat{X}\left(1\right), \ldots, \hat{X}\left(d\right)\right) \in \mathcal{S}_n^d$, $\hat{X}\left(i\right) > 0$ satisfying the following LMI on $\mathcal{S}_{n+m}^d$:*

$$\mathcal{N}\left(\hat{X}, \gamma\right) < 0,$$

$\mathcal{N}\left(X, \gamma\right)$ *denoting the generalized dissipation matrix associated with the system (8.1) and with the parameter γ.*
(iii) *The SGRAE (8.18) has a stabilizing solution $\tilde{X} = \left(\tilde{X}\left(1\right), \ldots, \tilde{X}\left(d\right)\right)$ satisfying $\tilde{X}\left(i\right) \geq 0$ and*

$$\gamma^2 I_m - D^T\left(i\right) D\left(i\right) - \sum_{k=1}^{r} B_k^T\left(i\right) \tilde{X}\left(i\right) B_k\left(i\right) > 0, \ i \in \mathcal{D}. \tag{8.44}$$

Proof. (i) $\Rightarrow$ (ii). For every $\delta > 0$ consider the linear and bounded operator:

$$\mathcal{T}_\delta : L_{\eta,w}^2\left([0,\infty), \mathbf{R}^m\right) \to L_{\eta,w}^2\left([0,\infty), \mathbf{R}^{n+p}\right)$$

defined by

$$\mathcal{T}_\delta u = y_{u,\delta}$$

where

$$y_{u,\delta}(t) = \begin{bmatrix} C(\eta(t)) \\ \delta I_n \end{bmatrix} x_u(t) + \begin{bmatrix} D(\eta(t)) \\ 0 \end{bmatrix} u(t)$$

where $x_u(t)$ is the solution of the system (8.1) with the initial condition $x_u(0) = 0$. Then

$$E \int_0^\infty |y_{u,\delta}(t)|^2 dt = E \int_0^\infty |y_u(t)|^2 dt + \delta^2 E \int_0^\infty |x_u(t)|^2 dt.$$

Applying Theorem 3.6.1 (ii), one deduces that there exists $c > 0$ not depending on u such that

$$E \int_0^\infty |y_{u,\delta}(t)|^2 dt \le E \int_0^\infty |y_u(t)|^2 dt + \delta^2 c E \int_0^\infty |u(t)|^2 dt$$

$$\le \left(\|\mathcal{T}\|^2 + \delta^2 c \right) E \int_0^\infty |u(t)|^2 dt$$

$\forall u \in L^2_{\eta,w}([0,\infty), \mathbf{R}^m)$. Hence we obtain that $\|\mathcal{T}_\delta\|^2 \le \|\mathcal{T}\|^2 + \delta^2 c$. Therefore it exists $\delta_0 > 0$ such that

$$\sup_{0 \le \delta < \delta_0} \|\mathcal{T}_\delta\| < \gamma. \tag{8.45}$$

For $0 < \delta < \delta_0$ let us denote by $K_\delta^0(t)$ the solution of the differential equation

$$\frac{d}{dt} K(t) = \mathcal{L}^* K(t) - \mathcal{P}^T (K(t)) \mathcal{R}^{-1} (K(t)) \mathcal{P}(K(t)) \tag{8.46}$$

$$+ C^T C + \delta^2 J^d$$

satisfying the initial condition $K_\delta^0(0) = 0$. Since the system $(A_0, A_1, \ldots, A_r; Q)$ is stable and $\|\mathcal{T}_\delta\| < \gamma$ it follows that one can apply Lemma 8.2.4 and Remark 8.2.2 to the solution $K_\delta^0(t)$, $\delta \in (0, \delta_0]$. Therefore it exists $\tilde{\rho} > 0$ such that

$$0 \le K_\delta^0(t,i) \le \tilde{\rho} I_n,\ t \ge 0, i \in \mathcal{D} \tag{8.47}$$

$$K_\delta^0(\tau_1, i) \le K_\delta^0(\tau_2, i),\ \forall\, 0 \le \tau_1 < \tau_2 \tag{8.48}$$

$$\gamma^2 I_m - D^T(i) D(i) - \sum_{k=1}^r B_k^T(i) K_\delta^0(t,i) B_k(i) \ge \varepsilon_0^2 I_m, \tag{8.49}$$

where $\varepsilon_0 > 0$.

From (8.47) and (8.48) it also follows that

$$\tilde{K}_\delta = \left(\tilde{K}_\delta\left(1\right),\ldots,\tilde{K}_\delta\left(d\right)\right)$$

with

$$\tilde{K}_\delta\left(i\right) = \lim_{t\to\infty} K_\delta^0\left(t,i\right),\ i \in \mathcal{D} \tag{8.50}$$

is well defined.

From Lemma 8.2.6 it follows that $\tilde{K}_\delta$ is a stationary solution of the differential equation (8.46) and hence it satisfies:

$$\mathcal{L}^*\tilde{K}_\delta - \mathcal{P}^T\left(\tilde{K}_\delta\right)\mathcal{R}^{-1}\left(\tilde{K}_\delta\right)\mathcal{P}\left(\tilde{K}_\delta\right) + C^TC + \delta^2 J^d = 0.$$

Using (8.49) one also obtains that $\tilde{K}_\delta$ defined by (8.50) verifies:

$$\mathcal{R}\left(\tilde{K}_\delta\right) \le -\varepsilon_0^2 I_m, i \in \mathcal{D}. \tag{8.51}$$

Since $(A_0, A_1, \ldots, A_r; Q)$ is stable, one easily obtains the following representation:

$$\tilde{K}_\delta = \int_0^\infty e^{\mathcal{L}^*s}\left[C^TC + \delta^2 J^d - \mathcal{P}^T\left(\tilde{K}_\delta\right)\mathcal{R}^{-1}\left(\tilde{K}_\delta\right)\mathcal{P}\left(\tilde{K}_\delta\right)\right] ds.$$

Taking into account the positivity of the operator $e^{\mathcal{L}^*s}$ and the inequality (8.51) it follows that

$$\tilde{K}_\delta \ge \delta^2 \int_0^\infty e^{\mathcal{L}^*s} J^d ds. \tag{8.52}$$

From Theorem 2.6.1 (ii) it follows that there exists $v > 0$ such that $e^{\mathcal{L}^*s} J^d \ge e^{-vs} J^d$. Therefore (8.52) reduces to $\tilde{K}_\delta \ge \frac{\delta^2}{v} J^d > 0$. Finally, notice that for all $\delta \in (0, \delta_0)$, $\tilde{K}_\delta$ verifies:

$$\mathcal{L}^*\tilde{K}_\delta - \mathcal{P}^T\left(\tilde{K}_\delta\right)\mathcal{R}^{-1}\left(\tilde{K}_\delta\right)\mathcal{P}\left(\tilde{K}_\delta\right) + C^TC < 0.$$

This shows, together with (8.51) and Lemma 5.1.2, that $\tilde{K}_\delta$ satisfies $\mathcal{N}\left(\tilde{K}_\delta\right) < 0$ and therefore (ii) is true.

(ii) $\Rightarrow$ (iii). From Remark 8.2.3 (iii) it follows that $\hat{X} \in \tilde{\Pi}$. Therefore $\mathcal{L}^*\hat{X} < 0$ and $\hat{X} > 0$. By using Theorem 2.7.7 (iv) $\to$ (i) one concludes that $(A_0, \ldots, A_r; Q)$ is stable. Hence $(\mathbf{A}, \mathbf{B}; Q)$ is stabilizable. Now, by virtue of Theorem 5.6.9 where

$$M\left(i\right) = -C^T\left(i\right)C\left(i\right),$$
$$L\left(i\right) = -C^T\left(i\right)D\left(i\right),$$
$$R\left(i\right) = \gamma^2 I_m - D^T\left(i\right)D\left(i\right), i \in \mathcal{D},$$

we may conclude that the SGRAE (8.18) has a stabilizing solution $\tilde{X} = (\tilde{X}(1),\ldots,\tilde{X}(d))$ verifying $\mathcal{R}(\tilde{X}) < 0$. It remains only to show that $\tilde{X} \geq 0$. Indeed, since the system $(A_0,\ldots,A_r;Q)$ is stable and $\mathcal{R}(\tilde{X}) < 0$, from (8.19) for $\tilde{X}$ and Theorem 2.7.5 (iii) and (iv) it follows directly that $\tilde{X} \geq 0$.

(iii) $\Rightarrow$ (i). Assume that the SGRAE (8.18) has a stabilizing solution $\tilde{X} \geq 0$ verifying (8.44). To prove that the system $(A_0,\ldots,A_r;Q)$ is stable we write the SGRAE (8.18) verified by $\tilde{X}$ in the equivalent form

$$\mathcal{L}^*\tilde{X} + \overline{C}^T\overline{C} = 0 \tag{8.53}$$

where

$$\overline{C} = (\overline{C}(1),\ldots,\overline{C}(d)), \ \overline{C}(i) = \begin{bmatrix} \overline{C}_1(i) \\ \overline{C}_2(i) \end{bmatrix}$$

with

$$\overline{C}_1(i) = \left(\gamma^2 I_m - D^T(i)D(i) - \sum_{k=1}^r B_k^T(i)\tilde{X}(i)B_k(i)\right)^{-\frac{1}{2}}$$

$$\times \left(B_0^T(i)\tilde{X}(i) + \sum_{k=1}^r B_k^T(i)\tilde{X}(i)A_k(i) + D^T(i)C(i)\right),$$

$$\overline{C}_2(i) = C(i), \ i \in \mathcal{D}.$$

Further take

$$H_k = (H_k(1),\ldots,H_k(d)), \ k = 0,1,\ldots,r$$

where

$$H_k(i) = \left[B_k(i)\left(-\mathcal{R}_i(\tilde{X})\right)^{-\frac{1}{2}} \ 0 \right], \ i \in \mathcal{D}.$$

With the above notations one obtains that

$$\left(A_0 + H_0\overline{C},\ldots,A_r + H_r\overline{C};Q\right) = \left(A_0 + B_0\tilde{F},\ldots,A_r + B_r\tilde{F};Q\right)$$

is stable. If $x(t), t \geq 0$ is an arbitrary solution of the equation

$$dx(t) = A_0(\eta(t))x(t)\,dt + \sum_{k=1}^r A_k(\eta(t))x(t)\,dw_k(t)$$

then we can write

$$dx(t) = \left[(A_0(\eta(t)) + H_0(\eta(t))\overline{C}(\eta(t)))x(t) + f_0(t)\right]dt \qquad (8.54)$$

$$+ \sum_{k=1}^{r} \left[(A_k(\eta(t)) + H_k(\eta(t))\overline{C}(\eta(t)))x(t) + f_k(t)\right]dw_k(t).$$

Based on a similar reasoning as in the proof of Lemma 5.6.13 (see also Theorem 4.1.7) one deduces that the null solution of (8.54) is ESMS. It remains to prove that $\|\mathcal{T}\| < \gamma$. Applying the Itô-type formula for the function $x^T \tilde{X}(i)x$ and to the system (8.1) one obtains that:

$$E \int_0^\infty \left\{ |y_u(t)|^2 - \gamma^2 |u(t)|^2 \, dt \right\} \qquad (8.55)$$

$$= -E \int_0^\infty \left| (-\mathcal{R}_{\eta(t)}(\tilde{X}))^{\frac{1}{2}} (u(t) - \tilde{F}(\eta(t))x_u(t)) \right|^2 dt$$

for any $u \in L^2_{\eta,w}([0,\infty), \mathbf{R}^m)$, $x_u(t), t \geq 0$ denoting the solution of (8.1) with the input $u(t), t \geq 0$ and with zero initial conditions. The equality (8.55) can be rewritten as follows:

$$\|\mathcal{T}u\|^2 - \gamma^2 \|u\|^2 = -\|g_u\|^2, \qquad (8.56)$$

where

$$g_u(t) = \left(\gamma^2 I_m - D^T(i)D(i) - \sum_{k=1}^{r} B_k^T(i)\tilde{X}(i)B_k(i) \right)^{\frac{1}{2}} \qquad (8.57)$$

$$\times (u(t) - \tilde{F}(\eta(t))x_u(t)).$$

From (8.56) it follows that

$$\|\mathcal{T}\| \leq \gamma. \qquad (8.58)$$

It remains to prove that the equality cannot take place in (8.58). Indeed, if $\|\mathcal{T}\| = \gamma$ there exists a sequence of stochastic processes $u_l, l \geq 0$, $\{u_l\} \subset L^2_{\eta,w}([0,\infty), \mathbf{R}^n)$ with

$$\|u_l\| = 1, \ \forall l \geq 0 \qquad (8.59)$$

and

$$\lim_{l \to \infty} \|\mathcal{T}u_l\| = \gamma. \qquad (8.60)$$

Let $x_l(t)$, $t \geq 0$, be the solution of the system (8.1) determined by the input $u_l(t)$ and having the initial conditions $x_l(0) = 0, l \geq 0$. We also denote by $g_l(t)$ the process defined by (8.57) in which u has been replaced by u_l. Using (8.59), the equality (8.56) becomes:

$$\|\mathcal{T}u_l\|^2 - \gamma^2 = -\|g_l\|^2.$$

Then taking into account (8.60) one obtains:

$$\lim_{l \to \infty} \|g_l\| = 0. \tag{8.61}$$

Further, from (8.57) and (8.44) it results that:

$$\lim_{l \to \infty} \|\tilde{g}_l\| = 0, \tag{8.62}$$

where we denoted $\tilde{g}_l(t) = u_l(t) - \tilde{F}(\eta(t))x_l(t)$, $t \geq 0$. The differential equation verified by x_l can be rewritten as:

$$
\begin{aligned}
dx_l(t) = &\left\{ \left[A_0(\eta(t)) + B_0(\eta(t))\tilde{F}(\eta(t)) \right] x_l(t) \right. \\
&\left. + B_0(\eta(t))\tilde{g}_l(t) \right\} dt \\
&+ \sum_{k=1}^{r} \left\{ \left[A_k(\eta(t)) + B_k(\eta(t))\tilde{F}(\eta(t)) \right] x_l(t) \right. \\
&\left. + B_k(\eta(t))\tilde{g}_l(t) \right\} dw_k(t).
\end{aligned}
$$

Since the system $\left(A_0 + B_0\tilde{F}, \ldots, A_r + B_r\tilde{F}; Q \right)$ is stable, combining the result in Theorem 3.6.1 and (8.62) we obtain that $\lim_{l \to \infty} \|x_l\| = 0$ and then, using again (8.62), it immediately follows that $\lim_{l \to \infty} \|u_l\| = 0$ which contradicts (8.59) and thus the proof is complete. □

Remark 8.2.4. (i) From the above theorem it follows that

$$\|\mathcal{T}\| = \inf\left\{ \gamma > 0, \text{ for which it exists } X = (X(1), \ldots, X(d)) \in \mathcal{S}_n^d, \right.$$

$$X > 0 \text{ such that } \mathcal{N}_i(X) < 0 \Big\}$$

$$= \inf\{ \gamma > 0, \text{ SGRAE (8.18) has the stabilizing solution}$$

$$\tilde{X} = \left(\tilde{X}(1), \ldots, \tilde{X}(d) \right) \text{ satisfying } \tilde{X}(i) \geq 0, \mathcal{R}_i(\tilde{X}) < 0, i \in \mathcal{D} \}.$$

(ii) Let us notice that in contrast to the H_2 norms associated with a stochastic linear system that can be directly computed by the results in Theorems 7.1.4 and 7.1.5, the norm of the input–output operator associated with a stochastic linear system cannot be directly computed. This norm can be estimated using a γ-procedure as in the deterministic case.

(iii) From the numerical point of view, the equivalence (i) $\Leftrightarrow$ (ii) is more effective to compute $\|\mathcal{T}\|$ because for every γ it reduces to test the feasibility of an LMI system. The equivalence (i) $\Leftrightarrow$ (iii) of Theorem 8.2.7 is useful to develop mixed H_2/H_∞ procedures for robust stabilization.

(iv) In the particular case when it exists $r_1 \geq 1$, such that $A_k(i) = 0$, $r_1 \leq k \leq r$ and $B_k(i) = 0$, $0 \leq k \leq r_1 - 1$, $C^T(i)D(i) = 0$, $i \in \mathcal{D}$, SGRAE (8.18) reduces to the following Lyapunov-type equation:

$$A_0^T(i)X(i) + X(i)A_0(i) + \sum_{k=1}^{r_1-1} A_k^T(i)X(i)A_k(i) \\ + \sum_{j=1}^{d} q_{ij}X(j) + C^T(i)C(i) = 0, \ i \in \mathcal{D}. \tag{8.63}$$

By convention, if $r_1 = 1$, the first sum in (8.63) is missing. If the system $(A_0,\ldots,A_{r_1-1};Q)$ is stable, then (8.63) has a unique solution $\tilde{X} = (\tilde{X}(1),\ldots,\tilde{X}(d)) \geq 0$. Moreover, if (i) in Theorem 8.2.7 is fulfilled, then the solution of (8.63) satisfies the condition:

$$D^T(i)D(i) + \sum_{k=r_1}^{r} B_k^T(i)\tilde{X}(i)B_k(i) < \gamma^2 I_m, \ i \in \mathcal{D}.$$

Remark 8.2.5. $L^2_{\eta,w}([0,\infty),\mathbf{R}^m)$ can be organized also as a real Hilbert space taking the inner product

$$(u,v) = \sum_{i=1}^{d} E\left[\int_0^\infty u^T(t)v(t)\,dt \mid \eta(0) = i\right].$$

The corresponding induced norm will be denoted by $||| \cdot |||$.

Proposition 8.2.8. *Suppose that* $(A_0,\ldots,A_r;Q)$ *is stable. Then* $\|\mathcal{T}\| = |||\mathcal{T}|||$.

Proof. It is easy to see that all preceding results and remarks hold if the norm $\|\cdot\|$ is replaced by $||| \cdot |||$. In this case the performance index $H_\gamma(\tau,x_0,u)$ is replaced by $\sum_{i=1}^{d} H_\gamma(\tau,x_0,i,u)$. Therefore, taking into account Remark 8.2.4 (i), we have:

$$|||\mathcal{T}||| = \inf\{\gamma > 0, \text{ SGRAE (8.18) has a stabilizing solution} \\ \tilde{X} \geq 0 \text{ with } R_i(\tilde{X}) < 0, \ i \in \mathcal{D}\}.$$

Hence $|||\mathcal{T}||| = \|\mathcal{T}\|$ and thus the proof is complete. $\square$

From Theorem 8.2.7 and Remark 8.2.4 (iv) one immediately obtains the following corollary.

Corollary 8.2.9. *Consider the system*

$$dx(t) = A_0(\eta(t))x(t)dt + \sum_{k=1}^{r_1-1} A_k(\eta(t))x(t)dw_k(t)$$

$$+ \sum_{k=r_1}^{r} B_k(\eta(t))u(t)dw_k(t) \qquad (8.64)$$

$$y(t) = C(\eta(t))x(t) + D(\eta(t))u(t)$$

with $C^T(i)D(i) = 0$, $i \in \mathcal{D}$. Assume that the system $(A_0,\ldots,A_{r_1-1};Q)$ is stable and denote by

$$\mathcal{T} : L_{\eta,w}^2([0,\infty);\mathbf{R}^m) \to L_{\eta,w}^2([0,\infty);\mathbf{R}^p)$$

the input–output operator associated with the system (8.64). Then

$$\|\mathcal{T}\| = \max_{i\in\mathcal{D}} \sqrt{\lambda_{\max}(i)}$$

where $\lambda_{\max}(i)$ is the largest eigenvalue of the matrix

$$D^T(i)D(i) + \sum_{k=r_1}^{r} B_k^T(i)\tilde{X}(i)B_k(i), \; i \in \mathcal{D},$$

$\tilde{X} = (\tilde{X}(1),\ldots,\tilde{X}(d))$ being the unique solution of (8.63).

Proposition 8.2.10. *Let $\mathbf{D} : L_{\eta,w}^2([0,\infty);\mathbf{R}^m) \to L_{\eta,w}^2([0,\infty);\mathbf{R}^p)$ be the linear bounded operator defined by $(\mathbf{D}u)(t) = D(\eta(t))u(t)$, $u \in L_{\eta,w}^2([0,\infty);\mathbf{R}^m)$. Then*

$$\|\mathbf{D}\| = |D| = \max\{|D(i)|, i \in \mathcal{D}\}.$$

Proof. Since $D^T(i)D(i) \le |D|^2 I_m$, we have

$$\|\mathbf{D}u\|^2 = E \int_0^\infty u^T(t)D^T(\eta(t))D(\eta(t))u(t)dt$$

$$\le |D|^2 E \int_0^\infty |u(t)|^2 dt = |D|^2 \|u\|^2$$

for every $u \in L_{\eta,w}^2([0,\infty);\mathbf{R}^m)$. Hence $\|\mathbf{D}\| \le |D|$.
    Further let $i \in \mathcal{D}$, $u \in \mathbf{R}^m$ arbitrary but fixed. Take

$$\hat{u}(t) = \begin{cases} u\chi_{\eta(t)=i} & \text{if } t \in [0,1] \\ 0 & \text{if } t > 1 \end{cases}.$$

Obviously $\hat{u} \in L_\eta^2([0,\infty); \mathbf{R}^m)$ and therefore $\hat{u} \in L_{\eta,w}^2([0,\infty); \mathbf{R}^m)$. The inequality

$$\|\mathbf{D}\hat{u}\|^2 \le \|\mathbf{D}\|^2 \|\hat{u}\|^2$$

becomes

$$\int_0^1 E\left(|D(\eta(t))u|^2 E\chi_{\eta(t)=i}\right) dt \le \|\mathbf{D}\|^2 \int_0^1 |u|^2 E\chi_{\eta(t)=i} dt.$$

Therefore

$$\int_0^1 |D(i)u|^2 E\chi_{\eta(t)=i} dt \le \|\mathbf{D}\|^2 |u|^2 \int_0^1 E\chi_{\eta(t)=i} dt.$$

But

$$\int_0^1 E\chi_{\eta(t)=i} dt = \int_0^1 \sum_{j=1}^d \pi_j E\left[\chi_{\eta(t)=i} \mid \eta(0) = j\right] dt$$

$$= \sum_{j=1}^d \int_0^1 \pi_j p_{ji}(t) dt \ge \int_0^1 \pi_i p_{ii}(t) dt > 0.$$

Thus we may conclude that

$$|D(i)u| \le \|D\| |u|$$

which leads to $|D| \le \|D\|$ and thus the proof is complete. $\qquad\square$

Remark 8.2.6. (i) Evidently if $u \in L_\eta^2([0,\infty); \mathbf{R}^m)$ then $\mathbf{D}u \in L_\eta^2([0,\infty); \mathbf{R}^p)$. The proof in Proposition 8.2.10 shows that $\|\hat{\mathbf{D}}\| = |D| = \|\mathbf{D}\|$ where $\hat{\mathbf{D}}$ is the restriction of the operator $\mathbf{D}$ to the subspace $L_\eta^2([0,\infty); \mathbf{R}^m) \subset L_{\eta,w}^2([0,\infty); \mathbf{R}^m)$.
(ii) The conclusion of Proposition 8.2.10 can be obtained directly from Corollary 8.2.9. Indeed, if we take $C(i) = 0$, $i \in \mathcal{D}$ it follows that $\tilde{X}(i) = 0$, $i \in \mathcal{D}$ and therefore $\|\mathbf{D}\|^2 = \max_{i \in \mathcal{D}} |D(i)|^2$.

The following result allows us to increase the number of relations of equivalence in Theorem 8.2.7 and it is useful in some applications:

Proposition 8.2.11. *Let $\mathcal{N}(X) = (\mathcal{N}_1(X), \ldots, \mathcal{N}_d(X))$ be the generalized dissipation matrix associated with the matrices $A_k(i), B_k(i), C(i), D(i)$ and with the scalar $\gamma > 0$. Then the following assertions are equivalent:*

(i) There exists $X = (X(1), \ldots, X(d)) \in \mathcal{S}_n^d$, $X > 0$ such that $\mathcal{N}_i(X) < 0$, $\forall i \in \mathcal{D}$.
(ii) There exists $Y = (Y(1), \ldots, Y(d)) \in \mathcal{S}_n^d$, $Y > 0$ such that

$$\begin{bmatrix} \mathcal{W}_{0,0}(Y,i) & \mathcal{W}_{0,1}(Y,i) & \cdots & \mathcal{W}_{0,r}(Y,i) & \mathcal{W}_{0,r+1}(Y,i) & \mathcal{W}_{0,r+2}(Y,i) \\ \mathcal{W}_{0,1}^T(Y,i) & \mathcal{W}_{1,1}(Y,i) & \cdots & \mathcal{W}_{1,r}(Y,i) & \mathcal{W}_{1,r+1}(Y,i) & \mathcal{W}_{1,r+2}(Y,i) \\ \vdots & \vdots & \ddots \vdots & \vdots & \vdots \\ \mathcal{W}_{0,r}^T(Y,i) & \mathcal{W}_{1,r}^T(Y,i) & \cdots & \mathcal{W}_{r,r}(Y,i) & \mathcal{W}_{r,r+1}(Y,i) & \mathcal{W}_{r,r+2}(Y,i) \\ \mathcal{W}_{0,r+1}^T(Y,i) & \mathcal{W}_{1,r+1}^T(Y,i) & \cdots & \mathcal{W}_{r,r+1}^T(Y,i) & \mathcal{W}_{r+1,r+1}(Y,i) & \mathcal{W}_{r+1,r+2}(Y,i) \\ \mathcal{W}_{0,r+2}^T(Y,i) & \mathcal{W}_{1,r+2}^T(Y,i) & \cdots & \mathcal{W}_{r,r+2}^T(Y,i) & \mathcal{W}_{r+1,r+2}^T(Y,i) & \mathcal{W}_{r+2,r+2}(Y,i) \end{bmatrix} < 0,$$

$$(8.65)$$

$i \in \mathcal{D}$, where

$$
\begin{aligned}
\mathcal{W}_{0,0}(Y,i) &= \left(A_0(i) + \tfrac{1}{2}q_{ii}I_n\right)Y(i) \\
&\quad + Y(i)\left(A_0(i) + \tfrac{1}{2}q_{ii}I_n\right)^T + B_0(i)B_0^T(i), \\
\mathcal{W}_{0,k}(Y,i) &= Y(i)A_k^T(i) + B_0(i)B_k^T(i),\ k = 1,\ldots,r, \\
\mathcal{W}_{0,r+1}(Y,i) &= Y(i)C^T(i) + B_0(i)D^T(i), \\
\mathcal{W}_{0,r+2}(Y,i) &= \left(\sqrt{q_{i1}}Y(i),\ldots,\sqrt{q_{i,i-1}}Y(i)\ \ \sqrt{q_{i,i+1}}Y(i),\ldots,\ \sqrt{q_{id}}Y(i)\right), \\
\mathcal{W}_{l,k}(Y,i) &= B_l(i)B_k^T(i),\ 1 \le l,k \le r,\ l \ne k, \\
\mathcal{W}_{l,l}(Y,i) &= B_l(i)B_l^T(i) - Y(i),\ 1 \le l \le r, \\
\mathcal{W}_{l,r+1}(Y,i) &= B_l(i)D^T(i),\ 1 \le l \le r, \\
\mathcal{W}_{r+1,r+1}(Y,i) &= D(i)D^T(i) - \gamma^2 I_p, \\
\mathcal{W}_{l,r+2}(Y,i) &= 0,\ 1 \le l \le r+1, \\
\mathcal{W}_{r+2,r+2}(Y,i) &= diag\left(-Y(1),\ldots,-Y(i-1)\ \ -Y(i+1),\ldots,-Y(d)\right).
\end{aligned}
$$

Proof. It is easy to see that the existence of $X = (X(1),\ldots,X(d)) > 0$ such that $\mathcal{N}_i(X) < 0$ is equivalent with the existence of $X = (X(1),\ldots,X(d)) > 0$ such that

$$\begin{bmatrix} \mathcal{V}_{11}(X,i) & \mathcal{V}_{12}(X,i) & \mathcal{V}_{13}(X,i) \\ \mathcal{V}_{12}^T(X,i) & \mathcal{V}_{22}(X,i) & \mathcal{V}_{23}(X,i) \\ \mathcal{V}_{13}^T(X,i) & \mathcal{V}_{23}^T(X,i) & \mathcal{V}_{33}(X,i) \end{bmatrix} < 0 \qquad (8.66)$$

where $\mathcal{V}_{11}(X,i)$ is a $(n+m) \times (n+m)$ matrix given by

$$\mathcal{V}_{11}(X,i) = \begin{bmatrix} A_0^T(i)X(i) + X(i)A_0(i) + \sum_{j=1}^d q_{ij}X(j) & X(i)B_0(i) \\ B_0^T(i)X(i) & -\gamma^2 I_m \end{bmatrix},$$

$\mathcal{V}_{12}(X,i)$ is a $(n+m) \times (r \cdot n)$ matrix

$$\mathcal{V}_{12}(X,i) = \begin{bmatrix} A_1^T(i)X(i) & \ldots & A_r^T(i)X(i) \\ B_1^T(i)X(i) & \ldots & B_r^T(i)X(i) \end{bmatrix},$$

$\mathcal{V}_{13}(X,i)$ is a $(n+m) \times p$ matrix defined by

$$\mathcal{V}_{13}(X,i) = \begin{bmatrix} C^T(i) \\ D^T(i) \end{bmatrix},$$

$\mathcal{V}_{22}(X,i)$ is a $(n \cdot r) \times (n \cdot r)$ matrix

$$\mathcal{V}_{22}(X,i) = diag(-X(i),\ldots,-X(i)),$$

$\mathcal{V}_{23}(X,i)$ is a $(n \cdot r) \times p$ matrix given by

$$\mathcal{V}_{23}(X,i) = 0,$$

and

$$\mathcal{V}_{33}(X,i) = -I_p.$$

Let us introduce

$$\Psi(i) = diag\left(X^{-1}(i) \quad I_m \quad -\mathcal{V}_{22}^{-1}(X,i) \quad I_p\right).$$

It is obvious that $\Psi(i) = \Psi^T(i) > 0$. Pre and post multiplication of (8.66) by $\Psi(i)$ one obtains that it exists $X = (X(1),\ldots,X(d)) > 0$, such that

$$\begin{bmatrix} \tilde{\mathcal{V}}_{11}(X,i) & B_0(i) & \tilde{\mathcal{V}}_{13}(X,i) & X^{-1}(i)C^T(i) \\ B_0^T(i) & -\gamma^2 I_m & \tilde{\mathcal{V}}_{23}(X,i) & D^T(i) \\ \tilde{\mathcal{V}}_{13}^T(X,i) & \tilde{\mathcal{V}}_{23}^T(X,i) & \tilde{\mathcal{V}}_{33}(X,i) & \tilde{\mathcal{V}}_{34}(X,i) \\ C(i)X^{-1}(i) & D(i) & \tilde{\mathcal{V}}_{34}^T(X,i) & -I_p \end{bmatrix} < 0 \qquad (8.67)$$

where

$$\tilde{\mathcal{V}}_{11}(X,i) = A_0(i)X^{-1}(i) + X^{-1}(i)A_0^T(i) + \sum_{j=1}^{d} q_{ij}X^{-1}(i)X(j)X^{-1}(i),$$

$$\tilde{\mathcal{V}}_{13}(X,i) = \left[X^{-1}(i)A_1^T(i) \;\ldots.\; X^{-1}(i)A_r^T(i)\right],$$

$$\tilde{\mathcal{V}}_{23}(X,i) = \left[B_1^T(i) \;\ldots\; B_r^T(i)\right],$$

$\tilde{\mathcal{V}}_{33}(X,i)$ is a $rn \times rn$ matrix defined by

$$\tilde{\mathcal{V}}_{33}(X,i) = diag\left(-X^{-1}(i),\ldots,-X^{-1}(i)\right),$$

and $\tilde{\mathcal{V}}_{34}(X,i)$ is an $rn \times p$ matrix, $\tilde{\mathcal{V}}_{34}(X,i) = 0$. Denoting $Z(i) = X^{-1}(i)$ one immediately obtains that (8.67) is equivalent with the existence of $Z = (Z(1),\ldots,Z(d)) > 0$ satisfying

$$
\begin{bmatrix}
\hat{\mathcal{V}}_{11}(Z,i) & \hat{\mathcal{V}}_{12}(Z,i) & Z(i)C^T(i) & \hat{\mathcal{V}}_{14}(Z,i) & B_0(i) \\
\hat{\mathcal{V}}_{12}^T(Z,i) & \hat{\mathcal{V}}_{22}(Z,i) & \hat{\mathcal{V}}_{23}(Z,i) & \hat{\mathcal{V}}_{24}(Z,i) & \hat{\mathcal{V}}_{25}(Z,i) \\
C(i)Z(i) & \hat{\mathcal{V}}_{23}^T(Z,i) & -I_p & \hat{\mathcal{V}}_{34}(Z,i) & D(i) \\
\hat{\mathcal{V}}_{14}^T(Z,i) & \hat{\mathcal{V}}_{24}^T(Z,i) & \hat{\mathcal{V}}_{34}^T(Z,i) & \hat{\mathcal{V}}_{44}(Z,i) & \hat{\mathcal{V}}_{45}(Z,i) \\
B_0^T(i) & \hat{\mathcal{V}}_{25}^T(Z,i) & D^T(i) & \hat{\mathcal{V}}_{45}^T(Z,i) & -\gamma^2 I_m
\end{bmatrix} < 0, \qquad (8.68)
$$

where

$$
\hat{\mathcal{V}}_{11}(Z,i) = \left(A_0(i) + \frac{1}{2}q_{ii}I_n\right)Z(i) + Z(i)\left(A_0(i) + \frac{1}{2}q_{ii}I_n\right)^T,
$$

$$
\hat{\mathcal{V}}_{12}(Z,i) = \left[Z(i)A_1^T(i) \ \dots \ Z(i)A_r^T(i)\right],
$$

$$
\hat{\mathcal{V}}_{14}(Z,i) = \left[\sqrt{q_{i1}}Z(i) \ \dots \ \sqrt{q_{i,i-1}}Z(i) \ \sqrt{q_{i,i+1}}Z(i) \ \dots \ \sqrt{q_{id}}Z(i)\right]
$$

is an $n \times (d-1)n$ matrix,

$$
\hat{\mathcal{V}}_{22}(Z,i) = diag\left(-Z(i) \ \dots \ -Z(i)\right)
$$

has the dimensions $rn \times rn$, $\hat{\mathcal{V}}_{23}(Z,i) = 0$ is an $nr \times p$ matrix, $\hat{\mathcal{V}}_{24}(Z,i) = 0$ is an $nr \times (d-1)n$ matrix,

$$
\hat{\mathcal{V}}_{25}(Z,i) = \begin{bmatrix} B_1(i) \\ \vdots \\ B_r(i) \end{bmatrix}
$$

is an $nr \times m$ matrix, $\hat{\mathcal{V}}_{34}(Z,i) = 0$ has the dimensions $p \times (d-1)n$,

$$
\hat{\mathcal{V}}_{44}(Z,i) = diag\left(-Z(1) \ \dots \ -Z(i-1) \ -Z(i+1) \ \dots \ -Z(d)\right)
$$

is a $(d-1)n \times (d-1)n$ matrix and $\hat{\mathcal{V}}_{45}(Z,i) = 0$ has the dimensions $(r-1)n \times m$.

Taking the Schur complement of the block $-\gamma^2 I_m$ of (8.68) it follows that this condition is accomplished if and only if there exists $Z = (Z(1),\dots,Z(d)) > 0$ such that

$$
\begin{bmatrix}
\hat{\mathcal{W}}_{11}(Z,i) & \hat{\mathcal{W}}_{12}(Z,i) & Z(i)C^T(i)+\gamma^{-2}B_0(i)D^T(i) & \hat{\mathcal{W}}_{14}(Z,i) \\
\hat{\mathcal{W}}_{12}^T(Z,i) & \hat{\mathcal{W}}_{22}(Z,i) & \hat{\mathcal{W}}_{23}(Z,i) & \hat{\mathcal{W}}_{24}(Z,i) \\
C(i)Z(i)+\gamma^{-2}D(i)B_0^T(i) & \hat{\mathcal{W}}_{23}^T(Z,i) & -I_p+\gamma^{-2}D(i)D^T(i) & \hat{\mathcal{W}}_{34}(Z,i) \\
\hat{\mathcal{W}}_{14}^T(Z,i) & \hat{\mathcal{W}}_{24}^T(Z,i) & \hat{\mathcal{W}}_{34}^T(Z,i) & \hat{\mathcal{W}}_{44}(Z,i)
\end{bmatrix} < 0,
$$

$$
(8.69)
$$

where

$$\hat{W}_{11}(Z,i) = \left(A_0(i) + \frac{1}{2}q_{ii}I_n\right)Z(i) + Z(i)\left(A_0(i) + \frac{1}{2}q_{ii}I_n\right)^T$$
$$+ \gamma^{-2}B_0(i)B_0^T(i),$$
$$\hat{W}_{12}(Z,i) = \left[Z(i)A_1^T(i) + \gamma^{-2}B_1(i)B_1^T(i) \quad \dots \quad Z(i)A_r^T(i) + \gamma^{-2}B_r(i)B_r^T(i)\right],$$
$$\hat{W}_{14}(Z,i) = \hat{V}_{14}(Z,i), \quad \hat{W}_{22}(Z,i) = \hat{V}_{22}(Z,i),$$
$$\hat{W}_{23}(Z,i) = \hat{V}_{23}(Z,i), \quad \hat{W}_{24}(Z,i) = \hat{V}_{24}(Z,i),$$
$$\hat{W}_{34}(Z,i) = \hat{V}_{34}(Z,i), \quad \hat{W}_{44}(Z,i) = \hat{V}_{44}(Z,i).$$

Consider the $(n(r+d)+p) \times (n(r+d)+p)$ matrix

$$\Gamma = diag\left(\gamma I_n, \gamma I_{rn}, \gamma I_p, \gamma I_{n(d-1)}\right).$$

By pre and post multiplication of (8.69) with Γ and denoting $Y(i) = \gamma^2 Z(i), i \in \mathcal{D}$ one obtains (8.65) and therefore the proof is complete.   □

At the end of this section we consider the particular cases when the system (8.1) is subject either only to Markov perturbations or to white noise multiplicative perturbations.

Assume that in (8.1) we have $A_k(i) = 0, B_k(i) = 0, k = 1,\dots,r, i \in \mathcal{D}$. Then (8.1) becomes

$$\dot{x}(t) = A_0(\eta(t))x(t) + B_0(\eta(t))u(t) \tag{8.70}$$

$$y(t) = C(\eta(t))x(t) + D(\eta(t))u(t).$$

The generalized dissipation matrix is in this case

$$\mathcal{N}(X) = (\mathcal{N}_1(X),\dots,\mathcal{N}_d(X))$$

with

$$\mathcal{N}_i(X) = \begin{bmatrix} A_0^T(i)X(i) + X(i)A_0(i) \\ + \sum_{j=1}^d q_{ij}X(j) + C^T(i)C(i) & X(i)B_0(i) + C^T(i)D(i) \\ B_0^T(i)X(i) + D^T(i)C(i) & -\gamma^2 I_m + D^T(i)D(i) \end{bmatrix} \tag{8.71}$$

for any $X = (X(1),\dots,X(d)) \in \mathcal{S}_n^d, i \in \mathcal{D}$. The SGRAE (8.18) becomes in this case

$$A_0^T(i)X(i) + X(i)A_0(i) + \sum_{j=1}^d q_{ij}X(j) + \left[X(i)B_0(i) + C^T(i)D(i)\right]$$
$$\times \left[\gamma^2 I_m - D^T(i)D(i)\right]^{-1}\left[B_0^T(i)X(i) + D^T(i)C(i)\right] + C^T(i)C(i) = 0, \tag{8.72}$$

$i \in \mathcal{D}$. Combining Theorem 8.2.7 and Proposition 8.2.11 one directly obtains the Bounded Real Lemma in the case of systems subject to Markov perturbations.

Corollary 8.2.12. *For the system (8.70) and for a $\gamma > 0$ the following assertions are equivalent:*

(i) *The pair $(A_0; Q)$ is stable and the input–output operator $\mathcal{T}$ defined by the system (8.70) satisfies*

$$\|\mathcal{T}\| < \gamma.$$

(ii) *It exists $X = (X(1), \ldots, X(d)) > 0$ such that $\mathcal{N}_i(X, \gamma) < 0, \forall i \in \mathcal{D}$.*

(iii) *$\gamma^2 I_m - D^T(i) D(i) > 0$ and the SGRAE (8.72) has a stabilizing solution $\tilde{X} = (\tilde{X}(1), \ldots, \tilde{X}(d)) \geq 0$.*

(iv) *There exists $Y = (Y(1), \ldots, Y(d)) > 0$ satisfying the following system of LMI:*

$$\begin{bmatrix} \mathcal{W}_{0,0}(Y,i) & \mathcal{W}_{0,r+1}(Y,i) & \mathcal{W}_{0,r+2}(Y,i) \\ \mathcal{W}_{0,r+1}^T(Y,i) & \mathcal{W}_{r+1,r+1}(Y,i) & \mathcal{W}_{r+1,r+2}(Y,i) \\ \mathcal{W}_{0,r+2}^T(Y,i) & \mathcal{W}_{r+1,r+2}^T(Y,i) & \mathcal{W}_{r+2,r+2}(Y,i) \end{bmatrix} < 0, \ i \in \mathcal{D},$$

where $\mathcal{W}_{ij}(Y,i)$ are the same as in (8.65).

In the following we assume that $\mathcal{D} = \{1\}$, $q_{11} = 0$ and $r \geq 1$. In this case the system (8.1) becomes

$$dx(t) = [A_0 x(t) + B_0 u(t)] dt + \sum_{k=1}^r [A_k x(t) + B_k u(t)] dw_k(t)$$

$$y(t) = Cx(t) + Du(t). \tag{8.73}$$

Then the generalized dissipation matrix is:

$$\mathcal{N}(X)$$

$$= \begin{bmatrix} A_0^T X + X A_0 + \sum_{k=1}^r A_k^T X A_k + C^T C & X B_0 + \sum_{k=1}^r A_k^T X B_k + C^T D \\ B_0^T X + \sum_{k=1}^r B_k^T X A_k + D^T C & -\gamma^2 I_m + D^T D + \sum_{k=1}^r B_k^T X B_k \end{bmatrix}$$

for any $X \in \mathcal{S}_n$. The SGRAE (8.18) becomes in this case:

$$A_0^T X + X A_0 + \sum_{k=1}^r A_k^T X A_k + \left[X B_0 + \sum_{k=1}^r A_k^T X B_k + C^T D\right]$$
$$\times \left[\gamma^2 I_m - D^T D - \sum_{k=1}^r B_k^T X B_k\right]^{-1} \left[B_0^T X + \sum_{k=1}^r B_k^T X A_k + D^T C\right] \tag{8.74}$$
$$+ C^T C = 0.$$

Applying again Theorem 8.2.7 and Proposition 8.2.11 one directly obtains the Bounded Real Lemma for systems subject only to multiplicative white noise perturbations.

Corollary 8.2.13. *For the system (8.73) and for a $\gamma > 0$, the following are equivalent:*

(i) *The system* $(A_0, \ldots, A_r)$ *is stable and the input–output operator* $\mathcal{T}$ *associated with the system* (8.73) *satisfies the condition* $\|\mathcal{T}\| < \gamma$.

(ii) *There exists a matrix* $\hat{X} > 0$ *satisfying* $\mathcal{N}(\hat{X}) < 0$;

(iii) *The SGRAE* (8.74) *has a positive semidefinite stabilizing solution* $\tilde{X}$ *satisfying* $\gamma^2 I_m - D^T D - \sum_{k=1}^r B_k^T X B_k > 0.$

(iv) *There exists* $Y > 0, Y \in \mathcal{S}_n^d$ *verifying the following LMI:*

$$
\begin{bmatrix}
A_0 Y + Y A_0^T + B_0 B_0^T & Y A_1^T + B_0 B_1^T & \cdots & Y A_r^T + B_0 B_r^T & Y C_0^T + B_0 D^T \\
A_1 Y + B_1 B_0^T & -Y + B_1 B_1^T & \cdots & B_1 B_r^T & B_1 D^T \\
\vdots & \vdots & \ddots & \vdots & \vdots \\
A_r Y + B_r B_0^T & B_r B_1^T & \cdots & -Y + B_r B_r^T & B_r D^T \\
C Y + D^T B_0 & D B_1^T & \cdots & D B_r^T & -\gamma^2 I_p + D D^T
\end{bmatrix} < 0.
$$

Remark 8.2.7. It is easy to see that in the case $\mathcal{D} = \{1\}, A_k = 0, B_k = 0, k = 1, \ldots, r$ the results stated in Corollaries 8.2.12 and 8.2.13 reduce to the well-known version of the Bounded Real Lemma of the deterministic case.

8.3 Robust Stability with Respect to Linear Structured Uncertainties

At the beginning of this section we shall prove the stochastic version of the so-called SGT. As it is known from the deterministic framework this is a powerful tool in analyzing the robust stabilization with respect to different classes of linear perturbations.

8.3.1 Small Gain Theorem

We prove first the following theorem.

Theorem 8.3.1. *Assume that*

(a) *The system* $(A_0, \ldots, A_r; Q)$ *is stable.*

(b) *The system* (8.1) *has the same number of inputs and outputs.*

(c) *The input–output operator* $\mathcal{T}$ *defined by the system* (8.1) *satisfies the condition* $\|\mathcal{T}\| < 1$.

Then we have:

(i) *The matrices* $I_m \pm D(i), i \in \mathcal{D}$ *are invertible.*

(ii) *The system* $(\bar{A}_0, \ldots, \bar{A}_r; Q)$ *is stable, where*

$$
\bar{A}_k(i) = A_k(i) \pm B_k(i)(I_m \mp D(i))^{-1} C(i), k = 0, 1, \ldots, r;
$$

Proof. (i) Using Corollary 8.1.3 and Remark 8.1.2 for the case $\gamma = 1$ one obtains that $I_m - D^T(i)D(i) > 0$, $i \in \mathcal{D}$. It follows that all eigenvalues of the matrices $D(i)$, $i \in \mathcal{D}$ are inside of the unit circle and therefore $\det(I_m \pm D(i)) \neq 0$, which shows that $I_m \pm D(i)$, $i \in \mathcal{D}$ are invertible.

(ii) From the implication (i) $\Rightarrow$ (ii) of Theorem 8.2.7 for $\gamma = 1$ we deduce that there exists $\hat{X} = (\hat{X}(1), \ldots, \hat{X}(d)) > 0$ satisfying

$$\mathcal{N}_i(\hat{X}, 1) < 0, i \in \mathcal{D}. \tag{8.75}$$

Using the Schur complement of the block $(2,2)$ one obtains that (8.75) is equivalent with the condition

$$\mathcal{L}^*\hat{X} - \mathcal{P}^T(\hat{X})\mathcal{R}^{-1}(\hat{X})\mathcal{P}(\hat{X}) + C^T C + \hat{M} = 0 \tag{8.76}$$

$$\mathcal{R}(\hat{X}) < 0$$

for a certain $\hat{M} > 0$, $\hat{M} = (\hat{M}(1), \ldots, \hat{M}(d)) \in \mathcal{S}_n^d$. By direct computations (see Lemma 5.1.1) one obtains that (8.76) can be rewritten as:

$$\begin{aligned}
&\mathcal{L}_G^*\hat{X} - G^T G + (C + DG)^T(C + DG) \\
&- (G - \hat{F})^T \mathcal{R}(\hat{X})(G - \hat{F}) + \hat{M} = 0,
\end{aligned} \tag{8.77}$$

where

$$G = (G(1), \ldots, G(d)), \; G(i) = \pm(I_m \mp D(i))^{-1}C(i),$$

$$\hat{F} = (\hat{F}(1), \ldots, \hat{F}(d)), \; \hat{F}(i) = -\mathcal{R}_i^{-1}(\hat{X})\mathcal{P}_i(\hat{X}), \; i \in \mathcal{D}.$$

Then one obtains:

$$(C(i) + D(i)G(i))^T(C(i) + D(i)G(i)) - G^T(i)G(i)$$

$$= C^T(i)\left[I_m \pm (I_m \mp D^T(i))^{-1}D^T(i)\right]\left[I_m \pm D(i)(I_m \mp D(i))^{-1}\right]C(i)$$

$$- G^T(i)G(i)$$

$$= C^T(i)(I_m \mp D^T(i))^{-1}(I_m \mp D(i))^{-1}C(i) - G^T(i)G(i)$$

$$= G^T(i)G(i) - G^T(i)G(i) = 0.$$

Thus it follows that (8.77) reduces to:

$$\mathcal{L}_G^*\hat{X} - (G - \hat{F})^T \mathcal{R}(\hat{X})(G - \hat{F}) + \hat{M} = 0.$$

Since $\hat{M} - (G - \hat{F})^T \mathcal{R}(\hat{X})(G - \hat{F}) > 0$ and $\hat{X} > 0$, one obtains via Theorem 2.7.7 one obtains that the system $(A_0 + B_0 G, \ldots, A_r + B_r G; Q)$ is stable. But $A_k(i) + B_k(i)G(i) = \bar{A}_k(i)$ and thus the proof is complete. $\square$

Theorem 8.3.2 (The First Small Gain Theorem). *Assume that the assumptions in Theorem 8.3.1 hold. Then the operators*

$$I \mp \mathcal{T} : L^2_{\eta,w}\{[0,\infty), \mathbf{R}^m\} \to L^2_{\eta,w}\{[0,\infty), \mathbf{R}^m\}$$

are invertible and the operators

$$(I \mp \mathcal{T})^{-1} : L^2_{\eta,w}\{[0,\infty), \mathbf{R}^m\} \to L^2_{\eta,w}\{[0,\infty), \mathbf{R}^m\}$$

have the following state-space realization:

$$dx(t) = \left[\overline{A}_0(\eta(t)) x(t) + \overline{B}_0(\eta(t)) y(t) \right] dt \tag{8.78}$$

$$+ \sum_{k=1}^{r} \left[\overline{A}_k(\eta(t)) x(t) + \overline{B}_k(\eta(t)) y(t) \right] dw_k(t)$$

$$u(t) = \overline{C}(\eta(t)) x(t) + \overline{D}(\eta(t)) y(t),$$

$\overline{A}(i)$ *being defined as in Theorem 8.3.1,* $\overline{B}_k(i) = B_k(i)(I_m \mp D(i))^{-1}$, $\overline{C}_k(i) = \pm (I_m \mp D(i))^{-1} C(i)$, $\overline{D}(i) = (I_m \mp D(i))^{-1}$, $0 \le k \le r, i \in \mathcal{D}$.

The proof immediately follows using Theorem 8.3.1 and Proposition 8.1.1 (ii).

$\square$

Remark 8.3.1. If $\|\mathcal{T}\| < 1$, then the invertibility of the operators $I \mp \mathcal{T}$ can be also obtained by a well-known result from the theory of linear and bounded operators on a Banach space. Theorem 8.3.2 additionally shows that the operators $(I \mp \mathcal{T})^{-1}$ have realizations in the state-space.

Consider the following systems

$$dx_1(t) = [A_{01}(\eta(t)) x_1(t) + B_{01}(\eta(t)) u_1(t)] dt$$

$$+ \sum_{k=1}^{r} [A_{k1}(\eta(t)) x_1(t) + B_{k1}(\eta(t)) u_1(t)] dw_k(t) \tag{8.79}$$

$$y_1(t) = C_1(\eta(t)) x_1(t),$$

$$dx_2(t) = [A_{02}(\eta(t)) x_2(t) + B_{02}(\eta(t)) u_2(t)] dt$$

$$+ \sum_{k=1}^{r} [A_{k2}(\eta(t)) x_2(t) + B_{k2}(\eta(t)) u_2(t)] dw_k(t) \tag{8.80}$$

$$y_2(t) = C_2(\eta(t)) x_2(t) + D_2(\eta(t)) u_2(t),$$

with the states $x_l \in \mathbf{R}^{n_l}$, $l = 1,2$, the output variables $y_1 \in \mathbf{R}^p$, $y_2 \in \mathbf{R}^m$, the inputs $u_1 \in \mathbf{R}^m$, $u_2 \in \mathbf{R}^p$. When coupling (8.79) and (8.80) by taking $u_2 = y_1$ and $u_1 = y_2$ one obtains the following resulting system:

$$d\xi(t) = A_{0cl}(\eta(t))\,\xi(t)\,dt + \sum_{k=1}^{r} A_{kcl}(\eta(t))\,\xi(t)\,dw_k(t) \qquad (8.81)$$

where

$$A_{kcl}(i) = \begin{bmatrix} A_{k1}(i) + B_{k1}(i)D_2(i)C_1(i) & B_{k1}(i)C_2(i) \\ B_{k2}(i)C_1(i) & A_{k2}(i) \end{bmatrix}, \; k = 0,1,\ldots,r.$$

Then another consequence of Theorem 8.3.1 is as follows.

Theorem 8.3.3 (The Second Small Gain Theorem). *Assume that the following assumptions hold:*

(i) The systems $(A_{0l},\ldots,A_{rl};Q)$, $l = 1,2$ are stable.
(ii) $\|\mathcal{T}_1\| < \gamma$, $\|\mathcal{T}_2\| < \gamma^{-1}$ for a certain $\gamma > 0$, where

$$\mathcal{T}_1 : L^2_{\eta,w}\{[0,\infty),\mathbf{R}^m\} \to L^2_{\eta,w}\{[0,\infty),\mathbf{R}^p\},$$

$$\mathcal{T}_2 : L^2_{\eta,w}\{[0,\infty),\mathbf{R}^p\} \to L^2_{\eta,w}\{[0,\infty),\mathbf{R}^m\}$$

are the input–output operators defined by the systems (8.79) and (8.80), respectively.

In these conditions the zero solution of the system (8.81) is ESMS.

Proof. From Proposition 8.1.1 one deduces that a state-space realization of the operator $\mathcal{T}_1\mathcal{T}_2$ is:

$$dx(t) = [A_0(\eta(t))x(t) + B_0(\eta(t))u(t)]\,dt$$

$$+ \sum_{k=1}^{r} [A_k(\eta(t))x(t) + B_k(\eta(t))u(t)]\,dw_k(t) \qquad (8.82)$$

$$y(t) = C(\eta(t))x(t),$$

where $A_k(\cdot)$, $B_k(\cdot)$ are defined as in Proposition 8.1.1 and

$$C(i) = [C_1(i) \quad 0], \; x = \begin{bmatrix} x_1 \\ x_2 \end{bmatrix}.$$

It is easy to see that

$$A_{kcl}(i) = A_k(i) + B_k(i)C(i) = \overline{A}_k(i), \; k = 0,\ldots,r, \; i \in \mathcal{D},$$

$\overline{A}_k(i)$ being the ones in Theorem 8.3.1 with $D(i) = 0$. The conclusion in the statement follows applying Theorem 8.3.1 to the system (8.82). We show now that the assumptions in this theorem are fulfilled. Thus, from assumption (i) in the statement and from the triangular structure of the matrices $A_k(i)$, using Theorem 3.6.1 one deduces that the zero solution of the system (8.82) for $u(t) = 0$ is ESMS. From assumption (ii) we have $\|\mathcal{T}_1\mathcal{T}_2\| \leq \|\mathcal{T}_1\| \|\mathcal{T}_2\| < 1$ and hence the proof is complete. $\square$

Remark 8.3.2. Without important changes, the result in Theorem 8.3.3 also remains valid in the case when the output equation of (8.79) has the form:

$$y_1(t) = C_1(\eta(t))x_1(t) + D_1(\eta(t))u_1(t).$$

From assumption (ii) of Theorem 8.3.3 immediately follows that $I_m - D_1(i)D_2(i)$ is invertible for all $i \in \mathcal{D}$. The coefficients of the closed-loop system will be changed accordingly. We shall not detail them since they will be not used in the following developments.

An interesting case is the one when in the system (8.79) we have $n_1 > 0$ and in (8.80) $n_2 = 0$. In this situation the resulting system obtained by coupling (8.79) with (8.80) reduces to

$$dx_1(t) = [A_{01}(\eta(t)) + B_{01}(\eta(t))D_2(\eta(t))C_1(\eta(t))]x_1(t)\,dt \qquad (8.83)$$

$$+ \sum_{k=1}^{r} [A_{k1}(\eta(t)) + B_{k1}(\eta(t))D_2(\eta(t))C_1(\eta(t))]x_1(t)\,dw_k(t).$$

The input–output operator $\mathcal{T}_2$ associated with the system (8.80) becomes

$$(\mathcal{T}_2u_2)(t) = D_2(\eta(t))u_2(t),\ t \geq 0\ \forall u_2 \in L^2_{\eta,w}([0,\infty),\mathbf{R}^p).$$

From Proposition 8.2.10 it follows that $\|\mathcal{T}_2\| = |D| = \max\{|D(i)|,\ i \in \mathcal{D}\}$. Consider the system

$$dx(t) = [A_0(\eta(t))x(t) + B_0(\eta(t))u(t)]\,dt$$

$$+ \sum_{k=1}^{r} [A_k(\eta(t))x(t) + B_k(\eta(t))u(t)]\,dw_k(t) \qquad (8.84)$$

$$y(t) = C(\eta(t))x(t).$$

Then we have

Corollary 8.3.4. *Assume that*

(i) The system $(A_0,\ldots,A_r;Q)$ is stable.
(ii) $\|\mathcal{T}\| < \gamma$ and $|D| < \gamma^{-1}$ where

$$\mathcal{T}: L^2_{\eta,w}([0,\infty),\mathbf{R}^m) \to L^2_{\eta,w}([0,\infty),\mathbf{R}^p)$$

denotes the input–output operator associated with the system (8.84) and $D = (D(1), \ldots, D(d)) \in \mathcal{M}_{mp}^d$.

Then the zero solution of the system

$$dx(t) = [A_0(\eta(t)) + B_0(\eta(t))D(\eta(t))C(\eta(t))]x(t)\,dt$$
$$+ \sum_{k=1}^{r} [A_k(\eta(t)) + B_k(\eta(t))D(\eta(t))C(\eta(t))]x(t)\,dw_k(t)$$

is ESMS.

8.3.2 Robust Stability with Respect to Linear Parametric Uncertainties

It is a known fact that the exponential stability of a solution of a linear deterministic system is not essentially influenced when the coefficients of the equation describing the system are subject to "small perturbations." Taking into account the equivalence between the exponential stability in mean square of a zero solution of a stochastic differential equation and the exponential stability of the zero solution of a Lyapunov-type linear differential equation, one expects that the exponential stability in mean square not to be affected by the small perturbations of the coefficients in the given equation. When analyzing the robustness of the solution of a system of stochastic differential equations we refer to the preservation of the stability property when the system is subject to coefficients variations not necessarily small. Such variations or uncertainties are due to the inaccurate knowledge of the system coefficients or to some simplifications of the mathematical model. One must take into account that a controller designed for the simplified model will be used for the real system subject to perturbations.

In the present section the robust stability with respect to a class of linear uncertainties will be investigated. Consider the linear system described by:

$$dx(t) = [A_0(\eta(t)) + B_0(\eta(t))\Delta(\eta(t))C(\eta(t))]x(t)\,dt$$
$$+ \sum_{k=1}^{r} [A_k(\eta(t) + B_k(\eta(t))\Delta(\eta(t))C(\eta(t))]x(t)\,dw_k(t) \tag{8.85}$$

where $A_k(i) \in \mathbf{R}^{n \times n}, 0 \le k \le r, B_k(i) \in \mathbf{R}^{n \times m}, 0 \le k \le r, C(i) \in \mathbf{R}^{p \times n}, i \in \mathcal{D}$ are assumed known and $\Delta(i) \in \mathbf{R}^{n \times p}$ are unknown matrices. Thus the system (8.85) is the perturbed system of the nominal one:

$$dx(t) = A_0(\eta(t))x(t)\,dt + \sum_{k=1}^{r} A_k(\eta(t)x(t)\,dw_k(t) \tag{8.86}$$

and the matrices $B_k(i), C(i)$ determine the structure of the uncertainties. If the zero solution of the nominal system (8.86) is ESMS, we shall analyze if the zero solution of the perturbed system (8.85) remains ESMS for $\Delta(i) \neq 0$. This is a primary formulation of the robust stability with respect to structured linear uncertainties for a stochastic system. For a more precise formulation we shall introduce a norm in the set of uncertainties. If $\Delta = (\Delta(1), \ldots, \Delta(d)) \in \mathcal{M}_{mp}^d$, one defines

$$|\Delta| = \max\{|\Delta(i)|, i \in \mathcal{D}\} = \max_{i \in \mathcal{D}} \sqrt{\lambda_{\max}(i)}$$

where $\lambda_{\max}(i)$ is the largest eigenvalue of the matrix $\Delta^T(i)\Delta(i)$.

As a measure of the stability robustness we introduce the *stability radius* with respect to linear structured uncertainties.

Definition 8.3.1. The *stability radius of the pair* $(A_0, \ldots, A_r; Q)$ with respect to the structure of linear uncertainties described by $(B_0, \ldots, B_r; C)$ is the number

$$\rho_L(\mathbf{A}, Q \mid \mathbf{B}, C) = \inf\{\rho > 0 \mid \exists \ \Delta = (\Delta(1), \ldots, \Delta(d)) \in \mathcal{M}_{mp}^d$$
with $|\Delta| \leq \rho$ for which the zero solution of the corresponding
system of type (8.85) is not ESMS}.

The result stated in Corollary 8.3.4 allows us to obtain a lower bound of the stability radius defined above. To this end, let us introduce the fictitious system

$$dx(t) = [A_0(\eta(t))x(t) + B_0(\eta(t))u(t)]dt$$

$$+ \sum_{k=1}^{r} [A_k(\eta(t)x(t) + B_k(\eta(t))u(t)]dw_k(t) \qquad (8.87)$$

$$y(t) = C(\eta(t)x(t))$$

with the known matrices of the perturbed system (8.85).

Corollary 8.3.5. *Assume that the zero solution of the nominal system (8.86) is ESMS. Let*

$$\mathcal{T} : L_{\eta,w}^2([0,\infty), \mathbf{R}^m) \to L_{\eta,w}^2([0,\infty), \mathbf{R}^p)$$

be the input–output operator associated with the fictitious system (8.87). Then

$$\rho_L(\mathbf{A}, Q \mid \mathbf{B}, C) \geq \|\mathcal{T}\|^{-1}. \qquad (8.88)$$

Proof. Let $\rho < \|\mathcal{T}\|^{-1}$ be an arbitrarily fixed number. We show that for any $\Delta \in \mathcal{M}_{mp}^d$ with $|\Delta| < \rho$ the zero solution of the perturbed system (8.85) is ESMS. Let Δ be such that $|\Delta| < \rho < \|\mathcal{T}\|^{-1}$. Denoting $\gamma = \rho^{-1}$, we have $\|\mathcal{T}\| < \gamma$ and $|\Delta| < \gamma^{-1}$. Applying the result of Corollary 8.3.4 one deduces that the zero solution of the system (8.85) is ESMS for the considered perturbation Δ. Therefore $\rho_L(\mathbf{A}, Q \mid \mathbf{B}, C) \geq \rho$. Since ρ is arbitrary it follows that (8.88) holds and thus the proof is complete. $\square$

At the end of this subsection we shall show certain structures of the linear uncertainty frequently used in the literature can be embedded in the general form of the system (8.85).

Consider first the perturbed system

$$dx(t) = \left[A_0(\eta(t)) + \hat{B}_0(\eta(t)) \Delta_0(\eta(t)) C(\eta(t)) \right] x(t) \, dt \qquad (8.89)$$

$$+ \sum_{k=1}^{r} \left[A_k(\eta(t) + \hat{B}_k(\eta(t)) \Delta_k(\eta(t)) C(\eta(t)) \right] x(t) \, dw_k(t)$$

where $A_k(i) \in \mathbf{R}^{n \times n}$, $\hat{B}_k(i) \in \mathbf{R}^{n \times m_k}$, $0 \leq k \leq r$, $i \in \mathcal{D}$ are known and $\Delta_k(i) \in \mathbf{R}^{m_k \times p}$, $0 \leq k \leq r$, $i \in \mathcal{D}$ are assumed unknown. In order to show that the system (8.89) is in fact a particular case of the system (8.85) we define $B_k(i) \in \mathbf{R}^{n \times m}$, $m = \sum_{k=0}^{r} m_k$ as follows

$$B_0(i) = \begin{bmatrix} \hat{B}_0(i) & 0 & \cdots & 0 \end{bmatrix}$$

$$B_k(i) = \begin{bmatrix} 0 & 0 & \cdots \hat{B}_k(i) \cdots 0 \end{bmatrix} \qquad (8.90)$$

$$1 \leq k \leq r, i \in \mathcal{D}, \Delta(i) = \begin{bmatrix} \Delta_0(i) \\ \vdots \\ \Delta_r(i) \end{bmatrix}.$$

With these notations the system (8.89) can be rewritten in the equivalent form (8.85). Further we have

$$|\Delta(i)|^2 = \lambda_{\max} \left[\Delta^T(i) \Delta(i) \right] = \lambda_{\max} \left[\sum_{k=0}^{r} \Delta_k^T(i) \Delta_k(i) \right].$$

Another interesting structure of perturbations is the situation when

$$dx(t) = \left[A_0(\eta(t)) + \hat{B}_0(\eta(t)) \Delta_0(\eta(t)) \hat{C}_0(\eta(t)) \right] x(t) \, dt \qquad (8.91)$$

$$+ \sum_{k=1}^{r} \left[A_k(\eta(t) + \hat{B}_k(\eta(t)) \Delta_k(\eta(t)) \hat{C}_k(\eta(t)) \right] x(t) \, dw_k(t)$$

where $A_k(i) \in \mathbf{R}^{n \times n}$, $\hat{B}_k(i) \in \mathbf{R}^{n \times m_k}$, $\hat{C}_k(i) \in \mathbf{R}^{p_k \times n}$, are assumed known and $\Delta_k(i) \in \mathbf{R}^{m_k \times p_k}$, $0 \leq k \leq r, i \in \mathcal{D}$ are unknown matrices describing the modeling uncertainties. Define $B_k(i) \in \mathbf{R}^{n \times m}$, $m = \sum_{k=0}^{r} m_k$ as in (8.90),

$$C(i) \in \mathbf{R}^{p \times n}, \ p = \sum_{k=0}^{r} p_k, \ C(i) = \begin{bmatrix} \hat{C}_0(i) \\ \vdots \\ \hat{C}_r(i) \end{bmatrix},$$

and $\Delta(i) = diag(\Delta_0(i), \cdots, \Delta_r(i))$.

With these notations the system (8.91) can be written in (8.85) form. Obviously we have

$$|\Delta(i)|^2 = \lambda_{\max}\left[\Delta^T(i)\Delta(i)\right] = \max_{0 \le k \le r} \lambda_{\max}\left[\Delta_k^T(i)\Delta_k(i)\right]$$

$$= \max_{0 \le k \le r} |\Delta_k(i)|^2.$$

8.3.3 Robust Stability with Respect to a Class of Nonlinear Uncertainties

In this subsection we shall consider the case when a stochastic linear system is subject to a class of nonlinear uncertainties. We shall also define the stability radius and we shall provide an estimation of its lower bound.

Consider the system:

$$
\begin{aligned}
dx(t) &= [A_0(\eta(t))x(t) + B_0(\eta(t))\Delta(t,y(t),\eta(t))]dt \\
&\quad + \textstyle\sum_{k=1}^{r} [A_k(\eta(t)x(t) + B_k(\eta(t))\Delta(t,y(t),\eta(t))]dw_k(t) \quad (8.92) \\
y(t) &= C(\eta(t))x(t)
\end{aligned}
$$

where $A_k(i) \in \mathbf{R}^{n \times n}$, $B_k(i) \in \mathbf{R}^{n \times m}$, $0 \le k \le r$, $C(i) \in \mathbf{R}^{p \times n}$ are assumed known and $\Delta : \mathbf{R}_+ \times \mathbf{R}^p \times \mathcal{D} \to \mathbf{R}^m$ are functions with the following properties:

(i) For any $i \in \mathcal{D}$, $(t,y) \to \Delta(t,y,i)$ is a continuous function on $\mathbf{R}_+ \times \mathbf{R}^p$ and $\Delta(t,0,i) = 0$ for all $t \ge 0$.

(ii) For every $\tau > 0$ there exists $\nu(\tau) > 0$ such that

$$|\Delta(t,y_1,i) - \Delta(t,y_2,i)| \le \nu(\tau)|y_1 - y_2|$$

for all $t \in [0,\tau]$, $y_1, y_2 \in \mathbf{R}^p$, $i \in \mathcal{D}$.

(iii) There exists $\delta > 0$ such that $|\Delta(t,y,i| \le \delta|y|$, $\forall (t,y,i) \in \mathbf{R}_+ \times \mathbf{R}^p \times \mathcal{D}$.

In this section we shall denote by Δ the set of all functions $\Delta : \mathbf{R}_+ \times \mathbf{R}^p \times \mathcal{D} \to \mathbf{R}^m$ satisfying the above conditions. Let us notice that both constants $\nu(\tau)$ and δ in (ii) and in (iii) depend on the function $\Delta(\cdot,\cdot) \in \Delta$.

For every Δ in $\mathbf{\Delta}$ denote

$$\|\Delta\| = \sup \left\{ \frac{|\Delta(t,y,i)|}{|y|} ; t \geq 0, \, y \neq 0, \, i \in \mathcal{D} \right\}. \tag{8.93}$$

Let $\mathcal{X}_{t_0}$ the set of all random n-dimensional $\mathcal{H}_{t_0}$-measurable vectors ξ which additionally satisfy $E\,|\xi|^2 < \infty$. It is obvious that $\mathbf{R}^n \subset \mathcal{X}_{t_0} \, \forall \, t_0 \geq 0$. For every $t_0 \geq 0$, $\xi \in \mathcal{X}_{t_0}$ and $\Delta \in \mathbf{\Delta}$ denote by $x_\Delta(t,t_0,\xi)$ the solution of the perturbed system (8.92) satisfying the initial condition $x_\Delta(t_0,t_0,\xi) = \xi$. Applying Theorem 1.11.1, one deduces that $x_\Delta(\cdot,t_0,\xi) \in L_{\eta,w}^2([t_0,T],\mathbf{R}^n)$ for every $T > t_0$. Moreover, if $E\,|\xi|^{2b} < \infty$, $b \geq 1$, then

$$\sup_{t_0 \leq t \leq T} \left\{ E\left[|x_\Delta(t,t_0,\xi)|^{2b} \mid \eta(t_0) = i\right] \right\} \leq K\left(1 + E\left[|\xi|^{2b} \mid \eta(t_0) = i\right]\right)$$

where K depends on T and on $T - t_0$.

Definition 8.3.2. The zero solution of the perturbed system (8.92) is called *exponentially stable in mean square with conditioning (ESMS-C)* if there exist $\alpha > 0$ and $\beta \geq 1$ such that

$$E\left[|x_\Delta(t,t_0,x_0)|^2 \mid \eta(t_0) = i\right] \leq \beta e^{-\alpha(t-t_0)} |x_0|^2$$

for any $t \geq t_0 \geq 0$, $x_0 \in \mathbf{R}^n$, $i \in \mathcal{D}$.

The constants α, β of the above definition may depend on the perturbation $\Delta \in \mathbf{\Delta}$, but they do not depend on t, t_0, x_0.

In order to characterize the robustness of the nominal system (8.86) with respect to the nonlinear perturbations $\Delta \in \mathbf{\Delta}$ we introduce the following definition.

Definition 8.3.3. The *robustness radius* with respect to the nonlinear stochastic uncertainties which structure is determined by $\mathbf{B} = (B_0, \ldots, B_r)$ and C, is

$$\rho_{NL}(\mathbf{A},\mathcal{Q} \mid \mathbf{B},C) = \inf \{\rho > 0 \mid \exists \, \Delta \in \mathbf{\Delta} \text{ with } \|\Delta\| \leq \rho$$
$$\text{for which the zero solution of the system (8.92)}$$
$$\text{is not ESMS-C}\}.$$

Remark 8.3.3. Since the class of uncertainties Δ also includes the functions $\Delta(t,y,i) = \Delta(i)y$ modeling the linear uncertainties considered in the previous section, it is easy to check that

$$\rho_{NL}(\mathbf{A},\mathcal{Q} \mid \mathbf{B},C) \leq \rho_L(\mathbf{A},\mathcal{Q} \mid \mathbf{B},C).$$

In order to prove the main result of this section, two additional results are required.

Lemma 8.3.6. *Let $\varphi : \mathbf{R}^n \times \Omega \to \mathbf{R}_+$ measurable with respect to $\mathcal{B}(\mathbf{R}^n) \otimes \mathcal{R}_t$ and $g : \Omega \to \mathbf{R}^n$ measurable with respect to $\mathcal{H}_t, t \geq 0$ being fixed, where $\mathcal{R}_t$ and $\mathcal{H}_t$ are defined in Chap. 1. Let*

$$h(x,i) = E[\varphi(x,\cdot) \mid \eta(t) = i] \quad \forall x \in \mathbf{R}^n, \, i \in \mathcal{D}, and$$
$$\hat{\varphi}(\omega) = \varphi(g(\omega),\omega).$$

If $\hat{\varphi}(\cdot)$ and $\varphi(x,\cdot)$ are integrable, then

$$h(g(\omega),\eta(t,\omega)) = E[\hat{\varphi} \mid \mathcal{H}_t](\omega) \quad a.s. \tag{8.94}$$

Proof. We prove first (8.94) for the case when $\varphi(x,\omega) = \varphi_1(x)\varphi_2(\omega)$, with $\varphi_1(x) \geq 0$ measurable with respect to $\mathcal{B}(\mathbf{R}^n)$ and bounded while $\varphi_2(\cdot) \geq 0$, is $\mathcal{R}_t$-measurable and bounded. From Theorem 1.10.1 one obtains

$$E[\varphi_2 \mid \mathcal{H}_t] = E[\varphi_2 \mid \eta(t)] \quad a.s.$$

Therefore

$$E[\hat{\varphi} \mid \mathcal{H}_t](\omega) = E[\varphi_1(g)\varphi_2 \mid \mathcal{H}_t](\omega)$$
$$= \varphi_1(g(\omega))E[\varphi_2 \mid \mathcal{H}_t](\omega)$$
$$= \varphi_1(g(\omega))E[\varphi_2 \mid \eta(t)](\omega).$$

On the other hand

$$h(x,\eta(t,\omega)) = E[\varphi_1(x)\varphi_2 \mid \eta(t)](\omega)$$
$$= \varphi_1(x)E[\varphi_2 \mid \eta(t)](\omega)$$

and then

$$h(g(\omega),\eta(t,\omega)) = \varphi_1(g(\omega))E[\varphi_2 \mid \eta(t)](\omega) \quad a.s.$$

which shows that (8.94) is true for the special considered case.

Further, let

$$\mathcal{M} = \{A \in \mathcal{B}(\mathbf{R}^n) \otimes \mathcal{R}_t \mid \chi_A \text{ satisfies } (8.94)\}$$
$$\mathcal{C} = \{U \times S \mid U \in \mathcal{B}(\mathbf{R}^n), S \in \mathcal{N}_t\}.$$

Since $\chi_{U \times S}(x,\omega) = \chi_U(x)\chi_S(\omega)$ it follows that $\mathcal{C} \subset \mathcal{M}$. One easily can verify that $\mathcal{C}$ is a π-system and $\mathcal{M}$ satisfy the conditions (i), (ii), (iii) of Theorem 1.1.1. Thus it results that $\mathcal{M}$ contains $\sigma[\mathcal{C}]$, $\sigma[\mathcal{C}]$ denoting the smallest σ-algebra containing $\mathcal{C}$, namely $\sigma[\mathcal{C}] = \mathcal{B}(\mathbf{R}^n) \otimes \mathcal{R}_t$. Thus we conclude that (8.94) is verified by any $A \in$

$\mathcal{B}(\mathbf{R}^n) \otimes \mathcal{R}_t$. Further let $0 \leq \varphi_k \leq \varphi_{k+1} \leq \varphi$, $\varphi_k(x,\omega)$ being a measurable function with respect to $\mathcal{B}(\mathbf{R}^n) \otimes \mathcal{R}_t$, and $\varphi_k(x,\omega) \to \varphi(x,\omega)$, for $k \to \infty \ \forall x, \omega$. Since (8.94) is true for φ_k, from Legesgue's theorem (see Theorem 1.2.2) one obtains that this relation is also true for a function φ verifying the assumptions in the statement and therefore the proof ends.   □

Consider now the system of stochastic nonlinear differential equations:

$$dx(t) = F_0(t,x(t),\eta(t))dt + \sum_{k=1}^{r} F_k(t,x(t),\eta(t))dw_k(t) \qquad (8.95)$$

where the functions $F_k : \mathbf{R}_+ \times \mathbf{R}^n \times \mathcal{D} \to \mathbf{R}^n$ have the following properties:

(j) $(t,x) \to F_k(t,x,i) : \mathbf{R}_+ \times \mathbf{R}^n \to \mathbf{R}^n$ are continuous functions and $F_k(t,0,i) = 0$, $t \geq 0, i \in \mathcal{D}, 0 \leq k \leq r$.
(jj) For any $\tau > 0$ it exists $v(\tau) > 0$ such that

$$|F_k(t,x_1,i) - F_k(t,x_2,i)| \leq v(\tau)|x_1 - x_2|, i \in \mathcal{D}, 0 \leq k \leq r,$$

$\forall x_1, x_2 \in \mathbf{R}^n, t \in [0,\tau]$.
(jjj) There exists $\delta > 0$ such that

$$|F_k(t,x,i)| \leq \delta|x|, \forall t \geq 0, x \in \mathbf{R}^n, i \in \mathcal{D}, 0 \leq k \leq r.$$

It is obvious that for any $\Delta \in \mathbf{\Delta}$ the perturbed system (8.92) satisfies the conditions (j), (jj), and (jjj). Applying Theorem 1.11.1 it follows that for any $t_0 \geq 0$ and $\xi \in \mathcal{X}_{t_0}$ the system (8.95) has a unique solution $x(t,t_0,\xi), t \geq 0$ such that $x(t_0,t_0,\xi) = \xi_0$.

Definition 8.3.4. The zero solution of the system (8.95) is ESMS-C if there exist $\alpha > 0, \beta > 0$ such that

$$E\left[|x(t,t_0,\xi)|^2 \mid \eta(t_0) = i\right] \leq \beta e^{-\alpha(t-t_0)}|\xi|^2,$$

$\forall t \geq t_0 \geq 0, \xi \in \mathbf{R}^n, i \in \mathcal{D}$.

The next result extends to the nonlinear case some results proved in Chap. 3 for the linear case.

Theorem 8.3.7. *The following assertions are equivalent:*

(i) The zero solution of the system (8.95) is ESMS-C.
(ii) There exists $c > 0$ such that

$$\int_t^\infty E\left[|x(s,t,\xi)|^2 \mid \eta(t) = i\right] ds \leq c|\xi|^2 \qquad (8.96)$$

$\forall\, t \geq 0,\, \xi \in \mathbf{R}^n$, *the constant c being independent of t and* ξ;

(iii) There exist $\alpha > 0$ *and* $\beta \geq 1$ *such that*

$$E\left[|x(t,t_0,\xi)|^2 \mid \eta(t_0) = i\right] \leq \beta e^{-\alpha(t-t_0)} E\left[|\xi|^2 \mid \eta(t_0) = i\right],$$

$\forall\, t \geq t_0 \geq 0,\, \xi \in \mathcal{X}_{t_0},\, i \in \mathcal{D}$.

Proof. (i) $\Rightarrow$ (ii) and (iii) $\Rightarrow$ (i) are obvious. We prove that (ii) $\Rightarrow$ (iii). Define

$$v(t,x,i) = \int_t^\infty h(s,t,x,i)\,ds$$

where

$$h(s,t,x,i) = E\left[|x(s,t,x,\cdot)|^2 \mid \eta(t) = i\right]$$

with $s \geq t \geq 0,\, x \in \mathbf{R}^n,\, i \in \mathcal{D}$. By virtue of Theorem 1.11.3 we can apply Lemma 8.3.6 for the function $\varphi(x,\omega) = |x(s,t,x,\omega)|^2,\ \forall\,(x,\omega) \in \mathbf{R}^n \times \Omega$ where $s \geq t$ are fixed and for the function $g(\omega) = x(t,t_0,\xi,\omega)$ with $t \geq t_0,\, \xi \in \mathcal{X}_{t_0}$ fixed. Therefore one obtains that:

$$h(s,t,x(t,t_0,\xi,\omega),\eta(t,\omega)) = E\left[|x(s,t,x(t,t_0,\xi,\omega),\omega)|^2 \mid \mathcal{H}_t\right]$$

$$= E\left[|x(s,t_0,\xi,\omega)|^2 \mid \mathcal{H}_t\right]. \qquad (8.97)$$

In the following we shall omit to write explicitly the argument ω. Define

$$v_i(t) = E\left[v(t,x(t,t_0,\xi),\eta(t)) \mid \eta(t_0) = i\right].$$

From (8.96) one deduces

$$v_i(t) \leq cE\left[|x(t,t_0,\xi)|^2 \mid \eta(t_0) = i\right]. \qquad (8.98)$$

Further, from (8.97) one obtains

$$v_i(t) = E\left[\int_t^\infty h(s,t,x(t,t_0,\xi),\eta(t))\,ds \mid \eta(t_0) = i\right]$$

$$= E\left[\int_t^\infty E\left[|x(s,t_0,\xi)|^2 \mid \mathcal{H}_t\right]\,ds \mid \eta(t_0) = i\right]$$

from which using the properties of conditional mean values, it immediately follows that

$$v_i(t) = \int_t^\infty E\left[|x(s,t_0,\xi)|^2 \mid \eta(t_0) = i\right]\,ds, \qquad (8.99)$$

$\forall\, t \geq t_0 \geq 0, \xi \in \mathcal{X}_{t_0},\ i \in \mathcal{D}$. From (8.99) it follows that the function $t \mapsto v_i(t)$ is absolutely continuous on $[t_0, \infty)$ and therefore it is derivable *a.e.* on $[t_0, \infty)$. Then from (8.99) one obtains that

$$\frac{d}{dt} v_i(t) = -E\left[|x(t,t_0,\xi)|^2 \mid \eta(t_0) = i \right].$$

Based on (8.98) it results that

$$\frac{d}{dt} v_i(t) \leq -\frac{1}{c} v_i(t) \quad a.e. \quad t \geq t_0. \tag{8.100}$$

Applying Theorem 1.10.2 to the function $|x|^2$ and to the system (8.95) one obtains:

$$E\left[|x(t,t_0,\xi)|^2 \mid \eta(t_0) = i \right] - E\left[|\xi|^2 \mid \eta(t_0) = i \right]$$

$$= E\left[\int_{t_0}^{t} \left\{ 2x^T(s,t_0,\xi) F_0(s,x(s,t_0,\xi),\eta(s)) \right. \right.$$

$$+ \sum_{k=1}^{r} |F_k(s,x(s,t_0,\xi),\eta(s))|^2 \tag{8.101}$$

$$\left. \left. + \sum_{j=1}^{d} q_{\eta(s),j} |x(s,t_0,\xi)|^2 \right\} ds \mid \eta(t_0) = i \right].$$

Taking into account (jjj) one obtains that

$$\left| 2x^T F_0(t,x,i) + \sum_{k=1}^{r} |F_k(t,x,i)|^2 \right| \leq \delta_0 |x|^2 \tag{8.102}$$

where $\delta_0 = \delta(2 + r\delta)$. Hence

$$2x^T F_0(t,x,i) + \sum_{k=1}^{r} |F_k(t,x,i)|^2 \geq -\delta_0 |x|^2.$$

Denoting

$$g_i(t) = E\left[|x(t,t_0,\xi)|^2 \mid \eta(t_0) = i \right],$$

from (8.101) it results that $g_i(\cdot)$ is an absolute continuous function on $[t_0, \infty)$ and

$$\frac{d}{dt} g_i(t) = E\left[2x^T(t,t_0,\xi) F_0(t,x(t,t_0,\xi),\eta(t)) + \sum_{j=1}^{d} q_{\eta(t),j} |x(t,t_0,\xi)|^2 \right.$$

$$\left. + \sum_{k=1}^{r} |F_k(t,x(t,t_0,\xi),\eta(t))|^2 \mid \eta(t_0) = i \right].$$

Using (8.102) one obtains that there exists $\delta_1 > 0$ such that

$$\frac{d}{dt} g_i(t) \geq -\delta_1 g_i(t) \quad a.e. \ , t \geq t_0,$$

which is equivalent with

$$\frac{d}{dt} \left[g_i(t) e^{\delta_1 t} \right] \geq 0,$$

which leads to

$$E \left[|x(t,t_0,\xi)|^2 \mid \eta(t_0) = i \right] \geq e^{-\delta_1 (t-t_0)} E \left[|\xi|^2 \mid \eta(t_0) = i \right]$$

$t \geq t_0 \geq 0, \xi \in \mathcal{X}_{t_0}, \ i \in \mathcal{D}$. From the last inequality one immediately obtains

$$h(s,t,x,i) \geq e^{-\delta_1 (s-t)} |x|^2$$

for all $s \geq t \geq 0, x \in \mathbf{R}^n, i \in \mathcal{D}$. Therefore $v(t,x,i) \geq \delta_1^{-1} |x|^2, t \geq 0, x \in \mathbf{R}^n, i \in \mathcal{D}$,

$$v_i(t) \geq \delta_1^{-1} E \left[|x(t,t_0,\xi)|^2 \mid \eta(t_0) = i \right].$$

From the above inequality together with (8.98) and (8.100) it follows that

$$E \left[|x(t,t_0,\xi)|^2 \mid \eta(t_0) = i \right] ds \leq \beta e^{-\alpha(t-t_0)} E \left[|\xi|^2 \mid \eta(t_0) = i \right]$$

with $\beta = \delta_1 c$ and $\alpha = 1/c$, and thus the proof is complete.   □

Before proving the main result of this subsection, let us notice that using the known constant matrices $A_k(i)$, $B_k(i)$, and $C(i)$ of the realization of the perturbed system (8.92) one can associate the following auxiliary system:

$$dx(t) = [A_0(\eta(t))x(t) + B_0(\eta(t))u(t)] dt$$

$$+ \sum_{k=1}^{r} [A_k(\eta(t))x(t) + B_k(\eta(t))u(t)] dw_k(t) \tag{8.103}$$

$$y(t) = C(\eta(t))x(t).$$

Then we have the following theorem.

Theorem 8.3.8. *Assume that the system $(A_0, \ldots A_r; Q)$ is stable. Then*

$$\rho_{NL} \{\mathbf{A}; Q \mid \mathbf{B}, C\} \geq \|\mathcal{T}\|^{-1}$$

where

$$\mathcal{T} : L^2_{\eta,w}([0,\infty), \mathbf{R}^m) \to L^2_{\eta,w}([0,\infty), \mathbf{R}^p)$$

*is the input–output operator associated with the auxiliary system (8.103) defined by
the matrices $A_k(i)$, $B_k(i)$, and $C(i)$, $0 \leq k \leq r$, $i \in \mathcal{D}$.*

Proof. We show that for every $\rho < \|\mathcal{T}\|^{-1}$ and for all $\Delta \in \Delta$ with $\|\Delta\| < \rho$, the zero
solution of the perturbed system (8.92) is ESMS-C. Denoting $\gamma = \rho^{-1}$ it follows
$\|\mathcal{T}\| < \gamma$ and $\|\Delta\| < \gamma^{-1}$, or

$$\sup\left\{\frac{|\Delta(t,y,i)|}{|y|}; t \geq 0, y \neq 0, i \in \mathcal{D}\right\} < \gamma^{-1}. \tag{8.104}$$

Using the implication (i) $\Rightarrow$ (iii) of Theorem 8.2.7 one deduces that the equation

$$A_0^T(i)X(i) + X(i)A_0(i) + \sum_{k=1}^r A_k^T(i)X(i)A_k(i) + \sum_{j=1}^d q_{ij}X(j)$$
$$+ \left[X(i)B_0(i) + \sum_{k=1}^r A_k^T(i)X(i)B_k(i)\right]\left[\gamma^2 I_m - \sum_{k=1}^r B_k^T(i)X(i)B_k(i)\right]^{-1}$$
$$\times \left[B_0^T(i)X(i) + \sum_{k=1}^r B_k^T(i)X(i)A_k(i)\right] + C^T(i)C(i) = 0 \tag{8.105}$$

has a stabilizing solution $\tilde{X} = (\tilde{X}(1), \ldots, \tilde{X}(d)) \geq 0$ such that

$$\gamma^2 I_m - \sum_{k=1}^r B_k^T(i)\tilde{X}(i)B_k(i) > 0 \tag{8.106}$$

for any $i \in \mathcal{D}$. Applying the Itô-type formula for the function $x^T \tilde{X}(i)x$ and for the
process $x(t) = x_\Delta(t, t_0, x_0)$ one obtains using (8.105) that:

$$E\left[\int_{t_0}^\tau \left\{|y(t)|^2 - \gamma^2 |\Delta(t, y(t), \eta(t))|^2\right\} dt \mid \eta(t_0) = i\right]$$
$$= x_0^T \tilde{X}(i)x_0 - E\left[x^T(\tau)\tilde{X}(\eta(\tau))x(\tau) \mid \eta(t_0) = i\right]$$
$$- E\left[\int_{t_0}^\tau \left(\Delta(t, y(t), \eta(t)) - \tilde{F}(\eta(t))x(t)\right)^T \right. \tag{8.107}$$
$$\times \left(\gamma^2 I_m - \sum_{k=1}^r B_k^T(\eta(t))\tilde{X}(\eta(t))B_k(\eta(t))\right)$$
$$\times \left.(\Delta(t, y(t), \eta(t)) - \tilde{F}(\eta(t))x(t)) dt \mid \eta(t_0) = i\right],$$

where $y(t) = C(\eta(t))x(t)$, $t \geq t_0$ and $\tilde{F}(i)$ denotes the stabilizing feedback
associated with the solution $\tilde{X}(i)$, $i \in \mathcal{D}$. Taking into account (8.106) it follows
that:

$$E\left[\int_{t_0}^\tau \left\{|y(t)|^2 - \gamma^2 |\Delta(t, y(t), \eta(t))|^2\right\} dt \mid \eta(t_0) = i\right]$$
$$\leq x_0^T \tilde{X}(i)x_0,$$

for any $\tau \geq t_0 \geq 0$, $x_0 \in \mathbf{R}^n$, $i \in \mathcal{D}$, which leads to

$$E\left[\int_{t_0}^{\infty} \left\{|y(t)|^2 - \gamma^2 |\Delta(t, y(t), \eta(t))|^2\right\} dt \mid \eta(t_0) = i\right] \quad (8.108)$$

$$\leq \tilde{\delta} |x_0|^2, \quad \forall\, t_0 \geq 0, x \in \mathbf{R}^n, i \in \mathcal{D}.$$

But

$$|\Delta(t, y(t), \eta(t))| \leq \|\Delta\| |y(t)|$$

$\forall\, t \geq 0, i \in \mathcal{D}, y \in \mathbf{R}^p$. On the other hand (8.104) gives $1 - \gamma^2 \|\Delta\|^2 > 0$ and then we deduce from (8.108) that

$$E\left[\int_{t_0}^{\infty} |y(t)|^2 dt \mid \eta(t_0) = i\right] \leq \frac{\tilde{\delta}}{(1 - \gamma^2) \|\Delta\|^2} |x_0|^2, \quad (8.109)$$

$\forall\, t_0 \geq 0$, $x_0 \in \mathbf{R}^n$, $i \in \mathcal{D}$. Finally, applying Theorem 3.6.1 and using (8.108) one obtains

$$E\left[\int_{t_0}^{\infty} |x_\Delta(t, t_0, x_0)|^2 dt \mid \eta(t_0) = i\right] \leq c |x_0|^2,$$

$\forall\, t_0 \geq 0$, $x_0 \in \mathbf{R}^n$, $i \in \mathcal{D}$, $c > 0$ being independent of t_0, x_0, i. Applying Theorem 8.3.7 we obtain that the zero solution of the perturbed system (8.92) is ESMS-C. Therefore $\rho_{NL}\{\mathbf{A}; Q \mid \mathbf{B}, C\} \geq \rho$. Since $\rho < \|\mathcal{T}\|^{-1}$ is arbitrary, it follows that $\rho_{NL}\{\mathbf{A}; Q \mid \mathbf{B}, C\} \geq \|\mathcal{T}\|^{-1}$ and thus the proof is complete. $\square$

At the end of this section we show that in a particular case of the system (8.92) we can obtain the exact value of the stability radius $\rho_{NL}\{\mathbf{A}; Q \mid \mathbf{B}, C\}$. To be more precise, consider the perturbed system:

$$\begin{aligned} dx(t) &= A_0 x(t) dt + \sum_{k=1}^{r_1-1} A_k x(t) dw_k(t) + \sum_{k=r_1}^{r} B_k \Delta(t, y(t)) dw_k(t) \\ y(t) &= Cx(t). \end{aligned} \quad (8.110)$$

The system (8.110) is a perturbation of the nominal system

$$dx(t) = A_0 x(t) dt + \sum_{k=1}^{r_1-1} A_k x(t) dw_k(t) \quad (8.111)$$

and it represents a particular case of the system (8.92), namely $\mathcal{D} = \{1\}, A_k = 0, r_1 \leq k \leq r, B_k = 0, 1 \leq k \leq r_1 - 1, q_{11} = 0$. In this particular case instead of $\rho_{NL}\{\mathbf{A}; Q \mid \mathbf{B}, C\}$ we shall denote the stability radius by $\rho_{NL}\{\mathbf{A} \mid \mathbf{B}, C\}$. Then the stability radius is given by the following result:

Theorem 8.3.9. *Assume that the zero solution of the nominal system (8.111) is ESMS-C. Then*

$$\rho_{NL}\{\mathbf{A} \mid \mathbf{B}, C\} = \tilde{\lambda}^{-\frac{1}{2}} \quad (8.112)$$

where $\tilde{\lambda}$ denotes the maximal eigenvalue of the matrix $\sum_{k=r_1}^{r} B_k^T \tilde{X} B_k$, $\tilde{X} \geq 0$ denoting the unique solution of the linear Lyapunov-type equation

$$A_0^T X + X A_0 + \sum_{k=1}^{r_1-1} A_k^T X A_k + C^T C = 0. \tag{8.113}$$

Proof. From Corollary 8.2.9 with $\mathcal{D} = \{1\}$ one obtains that $\tilde{\lambda}^{\frac{1}{2}} = \|\mathcal{T}\|$ where

$$\mathcal{T} : L_w^2([0,\infty), \mathbf{R}^m) \to L_w^2([0,\infty), \mathbf{R}^p)$$

is the input–output operator associated with the auxiliary system

$$\begin{aligned} dx(t) &= A_0 x(t) dt + \sum_{k=1}^{r_1-1} A_k x(t) dw_k(t) + \sum_{k=r_1}^{r} B_k u(t) dw_k(t) \\ y(t) &= C x(t). \end{aligned} \tag{8.114}$$

From Theorem 8.3.8 it follows that

$$\rho_{NL} \{\mathbf{A} \mid \mathbf{B}, C\} \geq \tilde{\lambda}^{-\frac{1}{2}}. \tag{8.115}$$

It order to prove (8.112) it is sufficient to show that for any $\varepsilon > 0$ there exists $\Delta_\varepsilon \in \mathbf{\Delta}$ with $\|\Delta_\varepsilon\| < \tilde{\lambda}^{-\frac{1}{2}} + \varepsilon$ for which the zero solution of (8.110) is not ESMS-C. Let $\lambda_\varepsilon \in \left(\tilde{\lambda}^{-\frac{1}{2}}, \tilde{\lambda}^{-\frac{1}{2}} + \varepsilon\right)$. Since $\lambda_\varepsilon^{-2} < \tilde{\lambda}$ it exists $u_\varepsilon \in \mathbf{R}^m$ with $|u_\varepsilon| = 1$ and

$$u_\varepsilon^T \left(I_m - \lambda_\varepsilon^2 \sum_{k=r_1}^{r} B_k^T \tilde{X} B_k \right) u_\varepsilon < 0. \tag{8.116}$$

Let

$$\Delta_\varepsilon(y) = \lambda_\varepsilon u_\varepsilon |y|. \tag{8.117}$$

Then it is obvious that $\Delta_\varepsilon \in \mathbf{\Delta}$ and $\|\Delta_\varepsilon\| = \lambda_\varepsilon$. We show that the zero solution of the system

$$\begin{aligned} dx(t) &= A_0 x(t) dt + \sum_{k=1}^{r_1-1} A_k x(t) dw_k(t) + \sum_{k=r_1}^{r} B_k \Delta_\varepsilon(t, y(t)) dw_k(t) \\ y(t) &= C x(t) \end{aligned} \tag{8.118}$$

is not ESMS. If the zero solution of (8.118) is ESMS, then there exists $\delta > 0$ such that

$$E \int_{t_0}^{\infty} |C x(t, t_0, x_0)|^2 dt \leq \delta |x_0|^2, \ \forall t_0 \geq 0, \ x_0 \in \mathbf{R}^n. \tag{8.119}$$

On the other hand applying the Itô-type formula to the function $x^T \tilde{X} x$ and to the system (8.118) and using (8.113) one obtains that

$$E \int_0^\tau \left(|y(t)|^2 - \sum_{k=r_1}^r \Delta_\varepsilon^T (y(t)) B_k^T \tilde{X} B_k \Delta_\varepsilon (y(t)) \right) dt \qquad (8.120)$$

$$= x_0^T \tilde{X} x_0 - E \left[x^T (\tau) \tilde{X} x(\tau) \right]$$

$\forall \, \tau > 0$, $x(t) = x(t, x_0)$ being the solution of (8.118) satisfying $x(0, x_0) = x_0$ and $y(t) = Cx(t, x_0)$. If the zero solution of (8.118) is ESMS, then (8.120) gives

$$E \int_0^\infty \left(|y(t)|^2 - \sum_{k=r_1}^r \Delta_\varepsilon^T (y(t)) B_k^T \tilde{X} B_k \Delta_\varepsilon (y(t)) \right) dt = x_0^T \tilde{X} x_0 \qquad (8.121)$$

$\forall \, x_0 \in \mathbf{R}^n$. Taking into account (8.117) one obtains that (8.121) becomes

$$u_\varepsilon^T \left(I_m - \lambda_\varepsilon^2 \sum_{k=r_1}^r B_k^T \tilde{X} B_k \right) u_\varepsilon E \int_0^\infty |y(t)|^2 dt = x_0^T \tilde{X} x_0, \ \forall \, x_0 \in \mathbf{R}^n$$

which contradicts (8.116), taking $x_0 \neq 0$ such that $x_0^T \tilde{X} x_0 > 0$ (since $\tilde{X} \geq 0$, (8.116) implies that there exists $x_0 \in R^n$ such that $x_0^T \tilde{X} x_0 > 0$). Thus the proof is complete.

$\square$

Notes and References

The theoretical developments presented in this chapter are new. They provide a unified approach of the stochastic version of the Bounded Real Lemma and stability radius for systems subject both to multiplicative white noise and to Markovian jumping. The stochastic version of the Bounded Real Lemma for systems with multiplicative white noise has been studied in [12,47,80,116,122] and for stochastic systems subject to Markov perturbations we cite [115]. For the case of stochastic systems subject both to multiplicative white noise and to Markovian jumping, a stochastic version of the Bounded Real Lemma was proved in [41]. A stochastic version of the Bounded Real Lemma for systems affected by Markov processes with an infinite countable number of states was proved in [140]. The stochastic counterpart of the Small Gain Theorem for systems with multiplicative white noise is given in [36] and [37, 38] for systems subject to Markov perturbations. As concerns the stability radius for systems with multiplicative white noise we cite [57, 81, 113, 114, 116] and for systems with Markovian jumping, see [115]. Some estimations for the stability radius in the case of stochastic systems with state multiplicative white noise and Markov jump perturbations are given in [41]. A different approach to estimate the stability radius for systems subject both to multiplicative white noise and to Markovian jumping can be found in [59].

Chapter 9
Robust Stabilization of Linear Stochastic Systems

In the present chapter we consider the robust stabilization problem of systems subject to both multiplicative white noise and to Markovian jumps with respect to some classes of parametric uncertainty. As it is already known, a wide variety of aspects of the robust stabilization problem can be embedded in a general disturbance attenuation problem (DAP) which extends the well-known H^∞ control problem in the case of deterministic invariant linear systems. A special attention will be paid in this chapter to the attenuation problem of exogenous perturbations with a specified level of attenuation. In the same time, some particular robust stabilization problems which solutions are derived using the results in the preceding chapter will be presented. The solution of the general attenuation problem will be given in terms of some linear matrix inequalities which provides necessary and sufficient solvability conditions. Throughout this chapter we assume that $\mathcal{D} = \{1, 2, \ldots, d\}$.

9.1 Formulation of the DAP

As it was shown in the preceding chapter, a measure of the robustness radius of stabilization with respect to a wide class of static or dynamic uncertainty can be characterized using the norm of the input–output operator associated with the nominal system. Based on this fact it follows that in order to achieve a certain level of robustness of stability one can design a stabilizing controller such that the norm of the input–output operator associated with the resulting system be less than the inverse of the imposed robustness radius.

The design problem of a stabilizing controller such that the norm of the input–output operator is less than a given level of attenuation is usually called in the literature the *DAP*. In this section the formulation of this problem will be given for the case of the stochastic linear systems considered in the present book.

V. Dragan et al., *Mathematical Methods in Robust Control of Linear Stochastic Systems*, DOI 10.1007/978-1-4614-8663-3_9, © Springer Science+Business Media New York 2013

Consider the following stochastic linear system:

$$dx(t) = [A_0(\eta(t))x(t) + G_0(\eta(t))v(t) + B_0(\eta(t))u(t)]\,dt$$

$$+ \sum_{k=1}^{r}[A_k(\eta(t))x(t) + G_k(\eta(t))v(t) + B_k(\eta(t))u(t)]\,dw_k(t)$$

$$z(t) = C_z(\eta(t))x(t) + D_{zv}(\eta(t))v(t) + D_{zu}(\eta(t))u(t) \tag{9.1}$$

$$y(t) = C_0(\eta(t))x(t) + D_0(\eta(t))v(t)$$

with two inputs namely $v(t) \in \mathbf{R}^{m_1}, u(t) \in \mathbf{R}^{m_2}$ and two outputs, $z(t) \in \mathbf{R}^{p_1}$, $y(t) \in \mathbf{R}^{p_2}$. The input variable $v(t)$ denotes exogenous signals, $u(t)$ includes the control variables, $z(t)$ is the regulated output, and $y(t)$ denotes the measured output. As usual, the state vector $x(t) \in \mathbf{R}^n$. The coefficients $A_k(i), G_k(i), B_k(i), 0 \le k \le r, C_z(i), D_{zv}(i), D_{zu}(i), C_0(i), D_0(i), i \in \mathcal{D}$ are known matrices with real coefficients with appropriate dimensions. The stochastic processes $\{\eta(t)\}_{t \ge 0}, \{w(t)\}_{t \ge 0}, w(t) = (w_1(t), \ldots, w_r(t))^T$ are defined as in the preceding chapters. The class of admissible controllers is described by the following equations:

$$dx_c(t) = [A_c(\eta(t))x_c(t) + B_c(\eta(t))y(t)]\,dt$$
$$u(t) \;\;\;= C_c(\eta(t))x_c(t) + D_c(\eta(t))y(t), \tag{9.2}$$

where $x_c \in \mathbf{R}^{n_c}$. In fact the controller (9.2) is characterized by the set of parameters $\{n_c, A_c(i), B_c(i), C_c(i), D_c(i), i \in \mathcal{D}\}$, where $n_c \ge 0$ is an integer number denoting the order of the controller and $A_c(i) \in \mathbf{R}^{n_c \times n_c}, B_c(i) \in \mathbf{R}^{n_c \times p_2}, C_c(i) \in \mathbf{R}^{m_2 \times n_c}, D_c(i) \in \mathbf{R}^{m_2 \times p_2}, i \in \mathcal{D}$. When coupling the controller (9.2) at the system (9.1) one obtains the following resulting system:

$$dx_{cl}(t) = [A_{0cl}(\eta(t))x_{cl}(t) + G_{0cl}(\eta(t))v(t)]\,dt$$
$$+ \sum_{k=1}^{r}[A_{kcl}(\eta(t))x_{cl}(t) + G_{kcl}(\eta(t))v(t)]\,dw_k(t) \tag{9.3}$$
$$z(t) \;\;\;= C_{cl}(\eta(t))x_{cl}(t) + D_{cl}(\eta(t))v(t),$$

where

$$A_{0cl}(i) = \begin{bmatrix} A_0(i) + B_0(i)D_c(i)C_0(i) & B_0(i)C_c(i) \\ B_c(i)C_0(i) & A_c(i) \end{bmatrix},$$

$$A_{kcl}(i) = \begin{bmatrix} A_k(i) + B_k(i)D_c(i)C_0(i) & B_k(i)C_c(i) \\ 0 & 0 \end{bmatrix}, 1 \le k \le r,$$

$$G_{0cl}(i) = \begin{bmatrix} G_0(i) + B_0(i)D_c(i)D_0(i) \\ B_c(i)D_0(i) \end{bmatrix}, \tag{9.4}$$

$$G_{kcl} = \begin{bmatrix} G_k(i) + B_k(i)D_c(i)D_0(i) \\ 0 \end{bmatrix}, 1 \le k \le r,$$

$$C_{cl}(i) = [C_z(i) + D_{zu}(i)D_c(i)C_0(i) \quad D_{zu}(i)C_c(i)],$$

$$D_{cl}(i) = D_{zv}(i) + D_{zu}(i)D_c(i)D_0(i), i \in \mathcal{D}.$$

Definition 9.1.1. A controller from the class (9.2) is a stabilizing controller of the system (9.1) if the zero solution of the system:

$$d\xi(t) = A_{0cl}(\eta(t))\xi(t)dt + \sum_{k=1}^{r} A_{kcl}(\eta(t))\xi(t)dw_k(t)$$

is ESMS.

For every stabilizing controller, define by:

$$\mathcal{T}_{cl} : L^2_{\eta,w}([0,\infty);\mathbf{R}^{m_1}) \to L^2_{\eta,w}([0,\infty);\mathbf{R}^{p_1})$$

the input–output operator defined by the closed-loop system (9.3), namely:

$$(\mathcal{T}_{cl}v)(t) = C_{cl}(\eta(t))x_{cl}(t,v) + D_{cl}(\eta(t))v(t), t \geq 0,$$

$\forall v \in L^2_{\eta,w}([0,\infty);\mathbf{R}^{m_1})$, where $x_{cl}(t,v)$ denotes the solution of the system (9.3) with the initial condition $x_{cl}(0,v) = 0$. As it was shown in Sect. 8.1 the input–output operator $\mathcal{T}_{cl}$ is a linear and bounded operator. We are now in position to formulate the DAP for the system (9.1) with an imposed level of attenuation $\gamma > 0$.

Problem formulation: Given $\gamma > 0$, find necessary and sufficient conditions for the existence of a stabilizing controllers for (9.1) such that $\|\mathcal{T}_{cl}\| < \gamma$. If such conditions are fulfilled give a procedure to determine a controller with the required properties.

Remark 9.1.1. Based on the definition of $\|\mathcal{T}_{cl}\|$ it follows that the γ-attenuation problem stated above is equivalent with

$$\sup_{v \in L^2_{\eta,w}([0,\infty);\mathbf{R}^{m_1})v \neq 0} \frac{\|z\|}{\|v\|} < \gamma.$$

9.2   Robust Stabilization of Linear Stochastic Systems. The Case of Full State Access

9.2.1   The Solution of DAP in the Case of Complete State Measurement

Consider the linear stochastic system described by:

$$\begin{aligned}
dx(t) &= [A_0(\eta(t))x(t) + G_0(\eta(t))v(t) + B_0(\eta(t))u(t)]dt \\
&\quad + \Sigma_{k=1}^{r}[A_k(\eta(t))x(t) + G_k(\eta(t))v(t) + B_k(\eta(t))u(t)]dw_k(t) \\
z(t) &= C_z(\eta(t))x(t) + D_{zv}(\eta(t))v(t) + D_{zu}(\eta(t))u(t) \\
y(t) &= x(t)
\end{aligned} \quad (9.5)$$

where $x(t) \in \mathbf{R}^n$, $v(t) \in \mathbf{R}^{m_1}$, $u(t) \in \mathbf{R}^{m_2}$, and $z(t) \in \mathbf{R}^{p_1}$ have the same meaning as in the system (9.1). Assume that the whole state vector is available for measurement. In fact the system (9.5) is a particular case of (9.1) with $p_2 = n$, $C_0(i) = I_n$, $D_0(i) = 0$, $i \in \mathcal{D}$. The class of admissible controllers is given by (9.2). We shall solve first the DAP in the case when zero-order controllers are used, namely $n_c = 0$. In this case (9.2) reduces to

$$u(t) = D_c(\eta(t)) x(t)$$

or, with a usual notation,

$$u(t) = F(\eta(t)) x(t),$$

where $F(i) \in \mathbf{R}^{m_2 \times n}$, $i \in \mathcal{D}$. The closed-loop system obtained with this controller is:

$$\begin{aligned}
dx(t) &= \{[A_0(\eta(t)) + B_0(\eta(t)) F(\eta(t))] x(t) + G_0(\eta(t)) v(t)\} dt \\
&\quad + \Sigma_{k=1}^r \{[A_k(\eta(t)) + B_k(\eta(t)) F(\eta(t))] x(t) \\
&\quad + G_k(\eta(t)) v(t)\} dw_k(t) \\
z(t) &= [C_z(\eta(t)) + D_{zu}(\eta(t)) F(\eta(t))] x(t) + D_{zv}(\eta(t)) v(t).
\end{aligned} \tag{9.6}$$

If $F = (F(1), \ldots, F(d))$ is a stabilizing state feedback for the system (9.5), we denote

$$\mathcal{T}_F : L_{\eta,w}^2([0,\infty); \mathbf{R}^{m_1}) \to L_{\eta,w}^2([0,\infty); \mathbf{R}^{p_1})$$

the input–output operator associated with (9.6). Therefore the control $u(t) = F(\eta(t)) x(t)$ solves the DAP with the level of attenuation γ if $\|\mathcal{T}_F\| < \gamma$. The following result provides necessary and sufficient conditions for the existence of such state feedback control.

Theorem 9.2.1. *For a given $\gamma > 0$ the following are equivalent*

(i) *There exists a control $u(t) = F(\eta(t)) x(t)$ that stabilizes the system (9.5) and $\|\mathcal{T}_F\| < \gamma$;*
(ii) *There exist $Y = (Y(1), \ldots, Y(d)) \in \mathcal{S}_n^d$ and $\Gamma = (\Gamma(1), \ldots, \Gamma(d)) \in \mathcal{M}_{m_2,n}^d$, $Y > 0$ satisfying the following system of LMI:*

$$\begin{bmatrix}
\mathcal{W}_{0,0}(Y,i) & \mathcal{W}_{0,1}(Y,i) & \cdots & \mathcal{W}_{0,r}(Y,i) & \mathcal{W}_{0,r+1}(Y,i) & \mathcal{W}_{0,r+2}(Y,i) \\
\mathcal{W}_{0,1}^T(Y,i) & \mathcal{W}_{1,1}(Y,i) & \cdots & \mathcal{W}_{1,r}(Y,i) & \mathcal{W}_{1,r+1}(Y,i) & \mathcal{W}_{1,r+2}(Y,i) \\
\vdots & \vdots & \ddots \vdots & \vdots & \vdots & \\
\mathcal{W}_{0,r}^T(Y,i) & \mathcal{W}_{1,r}^T(Y,i) & \cdots & \mathcal{W}_{r,r}(Y,i) & \mathcal{W}_{r,r+1}(Y,i) & \mathcal{W}_{r,r+2}(Y,i) \\
\mathcal{W}_{0,r+1}^T(Y,i) & \mathcal{W}_{1,r+1}^T(Y,i) & \cdots & \mathcal{W}_{r,r+1}^T(Y,i) & \mathcal{W}_{r+1,r+1}(Y,i) & \mathcal{W}_{r+1,r+2}(Y,i) \\
\mathcal{W}_{0,r+2}^T(Y,i) & \mathcal{W}_{1,r+2}^T(Y,i) & \cdots & \mathcal{W}_{r,r+2}^T(Y,i) & \mathcal{W}_{r+1,r+2}^T(Y,i) & \mathcal{W}_{r+2,r+2}(Y,i)
\end{bmatrix} < 0, \tag{9.7}$$

$i \in \mathcal{D}$, where

$$\mathcal{W}_{0,0}(Y,i) = A_0(i)Y(i) + Y(i)A_0^T(i) + q_{ii}Y(i) + B_0(i)\Gamma(i)$$
$$+ \Gamma^T(i)B_0^T(i) + G_0(i)G_0^T(i)$$

$$\mathcal{W}_{0,k}(Y,i) = Y(i)A_k^T(i) + \Gamma^T(i)B_k^T(i) + G_0(i)G_k^T(i), \ 1 \le k \le r$$

$$\mathcal{W}_{0,r+1}(i) = Y(i)C_z^T(i) + \Gamma^T(i)D_{zu}^T(i) + G_0(i)D_{zv}^T(i)$$

$$\mathcal{W}_{0,r+2} = \left[\sqrt{q_{i1}}Y(i) \ \cdots \ \sqrt{q_{i,i-1}}Y(i) \ \sqrt{q_{i,i+1}}Y(i) \ \cdots \ \sqrt{q_{id}}Y(i) \right]$$

$$\mathcal{W}_{l,k} = G_l(i)G_k^T(i), \ 1 \le l,k \le r, \ l \ne k$$

$$\mathcal{W}_{l,l} = G_l(i)G_l^T(i) - Y(i), \ 1 \le l \le r$$

$$\mathcal{W}_{l,r+1}(i) = G_l(i)D_{zv}^T(i), \ 1 \le l \le r$$

$$\mathcal{W}_{l,r+2}(i) = 0, \ 1 \le l \le r+1$$

$$\mathcal{W}_{r+1,r+1}(i) = D_{zv}(i)D_{zv}^T(i) - \gamma^2 I_{p_1}$$

$$\mathcal{W}_{r+2,r+2}(i) = diag\left(-Y(1) \ \cdots \ -Y(i-1) \ -Y(i+1) \ \cdots \ -Y(d)\right).$$

Moreover, if $(Y,\Gamma) \in \mathcal{S}_n^d \times \mathcal{M}_{m_2,n}^d$ is a solution of (9.7) with $Y > 0$, then the control $u(t) = F(\eta(t))x(t)$ with $F(i) = \Gamma(i)Y^{-1}(i)$ solves the γ-attenuation problem for the system (9.5).

Proof. The proof immediately follows applying Theorem 8.2.7 together with Proposition 8.2.11 to the system (9.6). $\qquad\square$

In the following we display the particular cases when the system (9.5) is subject only to Markovian jumping or to multiplicative white noise, respectively. Consider the linear stochastic system described by

$$\dot{x}(t) = A_0(\eta(t))x(t) + G_0(\eta(t))v(t) + B_0(\eta(t))u(t) \tag{9.8}$$
$$z(t) = C_z(\eta(t))x(t) + D_{zv}(\eta(t))v(t) + D_{zu}(\eta(t))u(t)$$
$$y(t) = x(t)$$

obtained from (9.5) with $A_k(i) = 0$, $G_k(i) = 0$, $B_k(i) = 0$, $1 \le k \le r$ and $i \in \mathcal{D}$. For the control $u(t) = F(\eta(t))x(t)$ one obtains the resulting system:

$$\dot{x}(t) = [A_0(\eta(t)) + B_0(\eta(t))F(\eta(t))]x(t) + G_0(\eta(t))v(t) \tag{9.9}$$
$$z(t) = [C_z(\eta(t)) + D_{zu}(\eta(t))F(\eta(t))]x(t) + D_{zv}(\eta(t))v(t).$$

Applying Corollary 8.2.12 for the system (9.9) we get the following corollary.

Corollary 9.2.2. *For a given $\gamma > 0$ the following are equivalent*

(i) *There exists a control $u(t) = F(\eta(t))x(t)$ stabilizing the system (9.8) such that the input–output operator $\mathcal{T}_F$ associated with (9.9) verifies $\|\mathcal{T}_F\| < \gamma$;*

(ii) *There exist $Y = (Y(1),\dots,Y(d)) \in \mathcal{S}_n^d$, $Y(i) > 0$ and $\Gamma = (\Gamma(1),\dots,\Gamma(d)) \in \mathcal{M}_{m_2,n}^d$ verifying the system of LMI:*

$$\begin{bmatrix} \mathcal{W}_{0,0}(Y,i) & \mathcal{W}_{0,r+1}(Y,i) & \mathcal{W}_{0,r+2}(Y,i) \\ \mathcal{W}_{0,r+1}^T(Y,i) & \mathcal{W}_{r+1,r+1}(Y,i) & \mathcal{W}_{r+1,r+2}(Y,i) \\ \mathcal{W}_{0,r+2}^T(Y,i) & \mathcal{W}_{r+1,r+2}^T(Y,i) & \mathcal{W}_{r+2,r+2}(Y,i) \end{bmatrix} < 0, \qquad (9.10)$$

where $\mathcal{W}_{ij}(Y,i)$ are the same as in (9.7). Moreover, if the pair $(Y,\Gamma) \in \mathcal{S}_n^d \times \mathcal{M}_{m_2,n}^d$ is a solution of (9.10) with $Y(i) > 0$, then the control $u(t) = F(\eta(t))x(t)$ with $F(i) = \Gamma(i)Y^{-1}(i)$ solves the γ-attenuation problem for the system (9.8).

In the case when $\mathcal{D} = \{1\}$ and $q_{11} = 0$ the system (9.5) becomes:

$$dx(t) = [A_0 x(t) + G_0 v(t) + B_0 u(t)] dt$$

$$+ \sum_{k=1}^r [A_k x(t) + G_k v(t) + B_k u(t)] dw_k(t) \qquad (9.11)$$

$$z(t) = C_z x(t) + D_{zv} v(t) + D_{zu} u(t)$$

$$y(t) = x(t).$$

Taking $u(t) = Fx(t)$ one obtains the closed-loop system:

$$dx(t) = [(A_0 + B_0 F)x(t) + G_0 v(t)] dt$$

$$+ \sum_{k=1}^r [(A_k + B_k F)x(t) + G_k v(t)] dw_k(t) \qquad (9.12)$$

$$z(t) = (C_z + D_{zu} F)x(t) + D_{zv} v(t).$$

Using Corollary 8.2.13 for the system (9.12) one obtains the following corollary.

Corollary 9.2.3. *For a given $\gamma > 0$ the following are equivalent:*

(i) *There exists F stabilizing the system (9.11) such that $\|\mathcal{T}_F\| < \gamma$ where $\mathcal{T}_F$ denotes the input–output operator associated with (9.12).*

(ii) *It exists $Y \in \mathcal{S}_n, Y > 0, \Gamma \in \mathbf{R}^{m_2 \times n}$ solving the following LMI:*

$$\begin{bmatrix} \mathcal{W}_{0,0}(Y) & \mathcal{W}_{0,1}(Y) & \cdots & \mathcal{W}_{0,r}(Y) & \mathcal{W}_{0,r+1}(Y) \\ \mathcal{W}_{0,1}^T(Y) & \mathcal{W}_{1,1}(Y) & \cdots & \mathcal{W}_{1,r}(Y) & \mathcal{W}_{1,r+1}(Y) \\ \vdots & \vdots & \ddots & \vdots & \vdots \\ \mathcal{W}_{0,r}^T(Y) & \mathcal{W}_{1,r}^T(Y) & \cdots & \mathcal{W}_{r,r}(Y) & \mathcal{W}_{r,r+1}(Y) \\ \mathcal{W}_{0,r+1}^T(Y) & \mathcal{W}_{1,r+1}^T(Y) & \cdots & \mathcal{W}_{r,r+1}^T(Y) & \mathcal{W}_{r+1,r+1}(Y) \end{bmatrix} < 0, \qquad (9.13)$$

where

$$W_{0,0}(Y) = A_0 Y + Y A_0^T + B_0 \Gamma + \Gamma^T B_0^T + G_0 G_0^T$$

$$W_{0,k}(Y) = Y A_k^T + \Gamma^T B_k^T + G_0 G_k^T, \ 1 \leq k \leq r$$

$$W_{0,r+1}(Y) = Y C_z^T + \Gamma^T D_{zu}^T + G_0 D_{zv}^T$$

$$W_{l,k}(Y) = G_l G_k^T, \ 1 \leq l, k \leq, \ l \neq k$$

$$W_{l,l}(Y) = G_l G_l^T - Y, \ 1 \leq l \leq r$$

$$W_{l,r+1}(Y) = G_l D_{zv}^T, \ 1 \leq l \leq r$$

$$W_{r+1,r+1}(Y) = D_{zv} D_{zv}^T - \gamma^2 I_{p_1}.$$

Moreover, if the pair $(Y, \Gamma) \in \mathcal{S}_n \times \mathbf{R}^{m_2 \times n}$, $Y > 0$ *is a solution of (9.13), then the control* $u(t) = \Gamma Y^{-1} x(t)$ *solves the γ-attenuation problem for the system (9.11).*

Consider now a controller in the set (9.2) defined by

$$(n_c, A_c(i), B_c(i), C_c(i), D_c(i); i \in \mathcal{D})$$

with $n_c > 0$, $A_c(i) \in \mathbf{R}^{n_c \times n_c}$, $B_c(i) \in \mathbf{R}^{n_c \times n}$, $C_c(i) \in \mathbf{R}^{m_2 \times n_c}$, $D_c(i) \in \mathbf{R}^{m_2 \times n}$, $i \in \mathcal{D}$. When coupling the controller to the system (9.5) one obtains a resulting system of form (9.3) with the matrix coefficients given by:

$$A_{0cl}(i) = \begin{bmatrix} A_0(i) + B_0(i) D_c(i) & B_0(i) C_c(i) \\ B_c(i) & A_c(i) \end{bmatrix},$$

$$A_{kcl}(i) = \begin{bmatrix} A_k(i) + B_k(i) D_c(i) & B_k(i) C_c(i) \\ 0 & 0 \end{bmatrix}, \ 1 \leq k \leq r,$$

$$G_{kcl}(i) = \begin{bmatrix} G_k(i) \\ 0 \end{bmatrix}, \ 0 \leq k \leq r, \tag{9.14}$$

$$C_{cl}(i) = [C_z(i) + D_{zu}(i) D_c(i) \quad D_{zu}(i) C_c(i)],$$

$$D_{cl}(i) = D_{zv}(i), \ i \in \mathcal{D}.$$

The next result shows that if the γ-attenuation problem can be solved with a dynamic controller (i.e., $n_c > 0$) then the same problem has also a solution expressed as a state feedback (i.e., $n_c = 0$).

Theorem 9.2.4. *For a $\gamma > 0$ the following are equivalent:*

(i) *There exists a dynamic controller (9.2) with $n_c > 0$ solving the DAP with the level of attenuation γ.*

(ii) *There exists a zero-order controller solving the DAP with the same level of attenuation γ.*

Proof. (i) $\Rightarrow$ (ii). Assume that there exists a dynamic controller of order $n_c > 0$ solving the γ-attenuation problem for the system (9.5). Therefore this controller stabilizes the system (9.5) and the input–output operator $\mathcal{T}_{cl}$ associated with the closed-loop system verifies the condition $\|\mathcal{T}_{cl}\| < \gamma$. Applying Theorem 8.2.7 and Proposition 8.2.11 for the system (9.3) with the coefficients (9.14) we deduce that there exists $\tilde{Y} = \left(\tilde{Y}(1),\ldots,\tilde{Y}(d)\right) \in \mathcal{S}_{n+n_c}^d, \tilde{Y}(i) > 0, i \in \mathcal{D}$ satisfying the following system of LMI:

$$\begin{bmatrix} \mathcal{W}_{0,0}\left(\tilde{Y},i\right) & \mathcal{W}_{0,1}\left(\tilde{Y},i\right) & \cdots & \mathcal{W}_{0,r}\left(\tilde{Y},i\right) & \mathcal{W}_{0,r+1}\left(\tilde{Y},i\right) & \mathcal{W}_{0,r+2}\left(\tilde{Y},i\right) \\ \mathcal{W}_{0,1}^T\left(\tilde{Y},i\right) & \mathcal{W}_{1,1}\left(\tilde{Y},i\right) & \cdots & \mathcal{W}_{1,r}\left(\tilde{Y},i\right) & \mathcal{W}_{1,r+1}\left(\tilde{Y},i\right) & \mathcal{W}_{1,r+2}\left(\tilde{Y},i\right) \\ \vdots & \vdots & \ddots \vdots & \vdots & \vdots \\ \mathcal{W}_{0,r}^T\left(\tilde{Y},i\right) & \mathcal{W}_{1,r}^T\left(\tilde{Y},i\right) & \cdots & \mathcal{W}_{r,r}\left(\tilde{Y},i\right) & \mathcal{W}_{r,r+1}\left(\tilde{Y},i\right) & \mathcal{W}_{r,r+2}\left(\tilde{Y},i\right) \\ \mathcal{W}_{0,r+1}^T\left(\tilde{Y},i\right) & \mathcal{W}_{1,r+1}^T\left(\tilde{Y},i\right) & \cdots & \mathcal{W}_{r,r+1}^T\left(\tilde{Y},i\right) & \mathcal{W}_{r+1,r+1}\left(\tilde{Y},i\right) & \mathcal{W}_{r+1,r+2}\left(\tilde{Y},i\right) \\ \mathcal{W}_{0,r+2}^T\left(\tilde{Y},i\right) & \mathcal{W}_{1,r+2}^T\left(\tilde{Y},i\right) & \cdots & \mathcal{W}_{r,r+2}^T\left(\tilde{Y},i\right) & \mathcal{W}_{r+1,r+2}^T\left(\tilde{Y},i\right) & \mathcal{W}_{r+2,r+2}\left(\tilde{Y},i\right) \end{bmatrix} < 0,$$

$$(9.15)$$

where

$$\mathcal{W}_{0,0}\left(\tilde{Y},i\right) = A_{0cl}(i)\tilde{Y}(i) + \tilde{Y}(i)A_{0cl}^T(i) + q_{ii}\tilde{Y}(i) + G_{0cl}(i)G_{0cl}^T(i),$$

$$\mathcal{W}_{0,k}\left(\tilde{Y},i\right) = \tilde{Y}(i)A_{kcl}^T(i) + G_{0cl}(i)G_{kcl}^T(i), \ 1 \le k \le r$$

$$\mathcal{W}_{0,r+1}\left(\tilde{Y},i\right) = \tilde{Y}(i)C_{cl}^T(i) + G_{0cl}(i)D_{cl}^T(i),$$

$$\mathcal{W}_{0,r+2}\left(\tilde{Y},i\right) = \left[\sqrt{q_{i1}}\tilde{Y}(i) \ \cdots \ \sqrt{q_{i,i-1}}\tilde{Y}(i) \ \sqrt{q_{i,i+1}}\tilde{Y}(i) \ \cdots \ \sqrt{q_{id}}\tilde{Y}(i)\right],$$

$$\mathcal{W}_{l,k}\left(\tilde{Y},i\right) = G_{lcl}(i)G_{kcl}^T(i), 1 \le l \ne k \le r,$$

$$\mathcal{W}_{l,l}\left(\tilde{Y},i\right) = G_{lcl}(i)G_{lcl}^T(i) - \tilde{Y}(i), 1 \le l \le r,$$

$$\mathcal{W}_{l,r+1}\left(\tilde{Y},i\right) = G_{lcl}(i)D_{cl}^T(i), 1 \le l \le r,$$

$$\mathcal{W}_{r+1,r+1}\left(\tilde{Y},i\right) = D_{cl}(i)D_{cl}^T(i) - \gamma^2 I_{p_1},$$

$$\mathcal{W}_{l,r+2}\left(\tilde{Y},i\right) = 0, 1 \le l \le r+1,$$

$$\mathcal{W}_{r+2,r+2}\left(\tilde{Y},i\right) = diag\left(-\tilde{Y}(1)\ldots-\tilde{Y}(i-1) \ -\tilde{Y}(i+1)\ldots-\tilde{Y}(d)\right).$$

Let

$$\tilde{Y}(i) = \begin{bmatrix} \tilde{Y}_{11}(i) & \tilde{Y}_{12}(i) \\ \tilde{Y}_{12}^T(i) & \tilde{Y}_{22}(i) \end{bmatrix}, i \in \mathcal{D},$$

be the partition of $\tilde{Y}(i)$ conformably with the partition of the matrix coefficients in (9.14), that is $\tilde{Y}_{11}(i) \in \mathcal{S}_n, \tilde{Y}_{22}(i) \in \mathcal{S}_{n_c}$. Define $\Psi \in \mathbf{R}^{\hat{n}\times\tilde{n}}, \hat{n} = (r+d) + p_1, \tilde{n} = (n+n_c)(r+d) + p_1$:

$$\Psi^T = diag\left(\underbrace{\Psi_0,\ldots,\Psi_0}_{r+1 \text{ times}}, I_{p_1}, \underbrace{\Psi_0,\ldots,\Psi_0}_{d-1 \text{ times}}\right)$$

where $\Psi_0 = [I_n \ \ 0_{n \times n_c}]$. By pre and post multiplication of (9.15) by Ψ^T and Ψ, respectively, one obtains the following system of LMI:

$$\begin{bmatrix} \mathcal{V}_{0,0}\left(\tilde{Y},i\right) & \mathcal{V}_{0,1}\left(\tilde{Y},i\right) & \cdots & \mathcal{V}_{0,r}\left(\tilde{Y},i\right) & \mathcal{V}_{0,r+1}\left(\tilde{Y},i\right) & \mathcal{V}_{0,r+2}\left(\tilde{Y},i\right) \\ \mathcal{V}_{0,1}^T\left(\tilde{Y},i\right) & \mathcal{V}_{1,1}\left(\tilde{Y},i\right) & \cdots & \mathcal{V}_{1,r}\left(\tilde{Y},i\right) & \mathcal{V}_{1,r+1}\left(\tilde{Y},i\right) & \mathcal{V}_{1,r+2}\left(\tilde{Y},i\right) \\ \vdots & \vdots & \ddots & \vdots & \vdots & \vdots \\ \mathcal{V}_{0,r}^T\left(\tilde{Y},i\right) & \mathcal{V}_{1,r}^T\left(\tilde{Y},i\right) & \cdots & \mathcal{V}_{r,r}\left(\tilde{Y},i\right) & \mathcal{V}_{r,r+1}\left(\tilde{Y},i\right) & \mathcal{V}_{r,r+2}\left(\tilde{Y},i\right) \\ \mathcal{V}_{0,r+1}^T\left(\tilde{Y},i\right) & \mathcal{V}_{1,r+1}^T\left(\tilde{Y},i\right) & \cdots & \mathcal{V}_{r,r+1}^T\left(\tilde{Y},i\right) & \mathcal{V}_{r+1,r+1}\left(\tilde{Y},i\right) & \mathcal{V}_{r+1,r+2}\left(\tilde{Y},i\right) \\ \mathcal{V}_{0,r+2}^T\left(\tilde{Y},i\right) & \mathcal{V}_{1,r+2}^T\left(\tilde{Y},i\right) & \cdots & \mathcal{V}_{r,r+2}^T\left(\tilde{Y},i\right) & \mathcal{V}_{r+1,r+2}^T\left(\tilde{Y},i\right) & \mathcal{V}_{r+2,r+2}\left(\tilde{Y},i\right) \end{bmatrix} < 0,$$

(9.16)

where

$$\mathcal{V}_{0,0}\left(\tilde{Y},i\right) = A_0\left(i\right)\tilde{Y}_{11}\left(i\right) + \tilde{Y}_{11}\left(i\right)A_0^T\left(i\right) + q_{ii}\tilde{Y}_{11}\left(i\right)$$

$$+ B_0\left(i\right)\left(D_c\left(i\right)\tilde{Y}_{11}\left(i\right) + C_c\left(i\right)\tilde{Y}_{12}^T\left(i\right)\right)$$

$$+ \left(D_c\left(i\right)\tilde{Y}_{11}\left(i\right) + C_c\left(i\right)Y_{12}^T\left(i\right)\right)^T B_0^T\left(i\right)$$

$$+ G_0\left(i\right)G_0^T\left(i\right),$$

$$\mathcal{V}_{0,k}\left(\tilde{Y},i\right) = \tilde{Y}_{11}\left(i\right)A_k^T\left(i\right) + \left(D_c\left(i\right)\tilde{Y}_{11}\left(i\right) + C_c\left(i\right)\tilde{Y}_{12}^T\left(i\right)\right)^T B_k^T\left(i\right)$$

$$+ G_0\left(i\right)G_k^T\left(i\right), 1 \le k \le r,$$

$$\mathcal{V}_{0,r+1}\left(\tilde{Y},i\right) = \tilde{Y}_{11}\left(i\right)C_z^T\left(i\right) + \left(D_c\left(i\right)\tilde{Y}_{11}\left(i\right) + C_c\left(i\right)\tilde{Y}_{12}^T\left(i\right)\right)^T D_{zu}^T\left(i\right)$$

$$+ G_0\left(i\right)D_{zv}^T\left(i\right),$$

$$\mathcal{V}_{0,r+2}\left(\tilde{Y},i\right) = \left[\sqrt{q_{i1}}\tilde{Y}_{11}\left(i\right) \ \cdots \ \sqrt{q_{i,i-1}}\tilde{Y}_{11}\left(i\right) \ \sqrt{q_{i,i+1}}\tilde{Y}_{11}\left(i\right) \ \cdots \ \sqrt{q_{id}}\tilde{Y}_{11}\left(i\right)\right],$$

$$\mathcal{V}_{l,k}\left(\tilde{Y},i\right) = G_l\left(i\right)G_k^T\left(i\right), 1 \le l \ne k \le r,$$

$$\mathcal{V}_{l,l}\left(\tilde{Y},i\right) = G_l\left(i\right)G_l^T\left(i\right) - \tilde{Y}_{11}\left(i\right), 1 \le l \le r,$$

$$\mathcal{V}_{l,r+1}\left(\tilde{Y},i\right) = G_l\left(i\right)D_{zv}^T\left(i\right), 1 \le l \le r,$$

$$\mathcal{V}_{r+1,r+1}\left(\tilde{Y},i\right) = D_{zv}\left(i\right)D_{zv}^T\left(i\right) - \gamma^2 I_{p_1},$$

$$\mathcal{V}_{l,r+2}\left(\tilde{Y},i\right) = 0, 1 \le l \le r+1,$$

$$\mathcal{V}_{r+2,r+2}\left(\tilde{Y},i\right) = diag\left(-\tilde{Y}_{11}\left(1\right)\ldots-\tilde{Y}_{11}\left(i-1\right) \ -\tilde{Y}_{11}\left(i+1\right)\ldots-\tilde{Y}_{11}\left(d\right)\right).$$

One can see that the LMI system (9.16) coincides with the LMI system (9.7) in Theorem 9.2.1 with Y replaced by $\tilde{Y}_{11}$ and with $\Gamma\left(i\right)$ replaced by $D_c\left(i\right)\tilde{Y}_{11}\left(i\right)$

$+C_c(i)\tilde{Y}_{12}^T(i)$, $i \in \mathcal{D}$. Applying Theorem 9.2.1 it follows that there exists a control $u(t) = F(\eta(t))x(t)$ solving the γ-attenuation problem for the system (9.5). More precisely,

$$F(i) = \left[D_c(i) + C_c(i)\tilde{Y}_{12}^T(i)\right]\tilde{Y}_{11}^{-1}(i), i \in \mathcal{D}.$$

Hence the first part of the proof is complete.

(ii) $\Rightarrow$ (i). Assume that there exists a stabilizing control state feedback $u(t) = F(\eta(t))x(t)$ solving the DAP with the level of attenuation γ for (9.5). Let $n_c > 0$ be a fixed integer and let $A_c(i) \in \mathbf{R}^{n_c \times n_c}$ be such that the zero solution of the system

$$\dot{x}_c(t) = A_c(\eta(t))x_c(t)$$

be ESMS. Then consider the controller $(n_c, A_c(i), 0_{n_c \times n}, 0_{m \times n_c}, F(i); i \in \mathcal{D})$. It is easy to check that this controller is stabilizing and the input–output operator associated with the closed loop system coincides with the input–output operator given by the state feedback control. Thus the proof ends. □

Remark 9.2.1. The smallest γ can be obtained by solving a semidefinite programming problem. Indeed, considering γ^2 as a new positive variable, the LMI (9.7) can be seen as a linear constraint in the minimization of γ^2.

9.2.2   Solution of Some Robust Stabilization Problems

Consider the system described by

$$\begin{aligned}
dx(t) = {} & \left\{\left[A_0(\eta(t)) + \hat{G}_0(\eta(t))\Delta_1(\eta(t))\hat{C}(\eta(t))\right]x(t)\right. \\
& + \left[B_0(\eta(t)) + \hat{B}_0(\eta(t))\Delta_2(\eta(t))\hat{D}(\eta(t))\right]u(t)\Big\}dt \qquad (9.17) \\
& + \sum_{k=1}^{r}\left\{\left[A_k(\eta(t)) + \hat{G}_k(\eta(t))\Delta_1(\eta(t))\hat{C}(\eta(t))\right]x(t)\right. \\
& + \left[B_k(\eta(t)) + \hat{B}_k(\eta(t))\Delta_2(\eta(t))\hat{D}(\eta(t))\right]u(t)\Big\}dw_k(t)
\end{aligned}$$

where $x(t) \in \mathbf{R}^n$ is the state, $u(t) \in \mathbf{R}^m$ is the control variable, $A_k(i) \in \mathbf{R}^{n \times n}, B_k(i) \in \mathbf{R}^{n \times m}, \hat{G}_k(i) \in \mathbf{R}^{n \times \hat{m}}, \hat{B}_k(i) \in \mathbf{R}^{n \times \tilde{m}}, \hat{C}(i) \in \mathbf{R}^{\hat{p} \times n}, \hat{D}(i) \in \mathbf{R}^{\tilde{p} \times m}, 0 \le k \le r, i \in \mathcal{D}$ are assumed known. The matrices $\Delta_1(i) \in \mathbf{R}^{\hat{m} \times \hat{p}}, \Delta_2(i) \in \mathbf{R}^{\tilde{m} \times \tilde{p}}$ are unknown and they describe the uncertainties of the system (9.17). It is assumed that the whole state vector is accessible for measurement. The robust stabilization problem considered here can be stated as follows: for a given $\rho > 0$ determine a control $u(t) = F(\eta(t))x(t)$ stabilizing (9.17) for any $\Delta_1 = (\Delta_1(1), \dots, \Delta_1(d))$ and $\Delta_2 = (\Delta_2(1), \dots, \Delta_2(d))$ such that

$$\max(|\Delta_1|, |\Delta_2|) < \rho$$

where

$$|\Delta_k| = \max_{i \in \mathcal{D}} \lambda_{max}^{\frac{1}{2}} \left(\Delta_k^T (i) \Delta_k (i) \right).$$

The closed loop system obtained with $u(t) = F(\eta(t))x(t)$ is given by:

$$
\begin{aligned}
dx(t) = &\left[A_0(\eta(t)) + B_0(\eta(t)) F(\eta(t)) + \hat{G}_0(\eta(t)) \Delta_1(\eta(t)) \hat{C}(\eta(t)) \right. \\
&+ \hat{B}_0(\eta(t)) \Delta_2(\eta(t)) \hat{D}(\eta(t)) F(\eta(t)) \big] x(t) dt \\
&+ \Sigma_{k=1}^r \left[A_k(\eta(t)) + B_k(\eta(t)) F(\eta(t)) + \hat{G}_k(\eta(t)) \Delta_1(\eta(t)) \hat{C}(\eta(t)) \right. \\
&+ \hat{B}_k(\eta(t)) \Delta_2(\eta(t)) \hat{D}(\eta(t)) F(\eta(t)) \big] x(t) dw_k(t).
\end{aligned}
\tag{9.18}
$$

Denoting by

$$G_k(i) = \begin{bmatrix} \hat{G}_k(i) & \hat{B}_k(i) \end{bmatrix}$$

$$C(i) = \begin{bmatrix} \hat{C}(i) \\ 0 \end{bmatrix}$$

$$D(i) = \begin{bmatrix} 0 \\ \hat{D}(i) \end{bmatrix}$$

$$\Delta(i) = \begin{bmatrix} \Delta_1(i) & 0 \\ 0 & \Delta_2(i) \end{bmatrix},$$

the system (9.18) can be rewritten as

$$
\begin{aligned}
dx(t) = &\{ A_0(\eta(t)) + B_0(\eta(t)) F(\eta(t)) + G_0(\eta(t)) \Delta(\eta(t)) \\
&\times [C(\eta(t)) + D(\eta(t)) F(\eta(t))] \} x(t) dt \\
&+ \Sigma_{k=1}^r \{ A_k(\eta(t)) + B_k(\eta(t)) F(\eta(t)) + G_k(\eta(t)) \Delta(\eta(t)) \\
&\times [C(\eta(t)) + D(\eta(t)) F(\eta(t))] \} x(t) dw_k(t).
\end{aligned}
\tag{9.19}
$$

Assume that $F(i)$ is such that the zero solution of the system

$$dx(t) = [A_0(\eta(t)) + B_0(\eta(t)) F(\eta(t))] x(t) dt$$

$$+ \sum_{k=1}^r [A_k(\eta(t)) + B_k(\eta(t)) F(\eta(t))] x(t) dw_k(t)$$

is ESMS. Then, applying Corollary 8.3.5 it follows that the zero solution of (9.19) is ESMS for all Δ with $|\Delta| < \rho$ if the input–output operator $\mathcal{T}_F$ associated with the system

$$
\begin{aligned}
dx(t) &= [(A_0(\eta(t)) + B_0(\eta(t)) F(\eta(t))) x(t) + G_0(\eta(t)) v(t)] dt \\
&\quad + \Sigma_{k=1}^r [(A_k(\eta(t)) + B_k(\eta(t)) F(\eta(t))) x(t) + G_k(\eta(t)) v(t)] dw_k(t) \\
z(t) &= (C(\eta(t)) + D(\eta(t)) F(\eta(t))) x(t)
\end{aligned}
$$

satisfies the condition $\|\mathcal{T}_F\| < 1/\rho$. Further notice that

$$|\Delta| = \max_{i \in \mathcal{D}} \lambda_{\max}^{\frac{1}{2}} \left(\Delta^T (i) \Delta (i) \right) = \max \left(|\Delta_1|, |\Delta_2| \right).$$

Therefore F is a robust stabilizing state feedback with the robustness radius ρ if it is a solution of the DAP with level of attenuation $\gamma = 1/\rho$ for the following system:

$$dx(t) = [A_0(\eta(t))x(t) + G_0(\eta(t))v(t) + B_0(\eta(t))u(t)] dt$$

$$+ \sum_{k=1}^{r} [A_k(\eta(t))x(t) + G_k(\eta(t))v(t) + B_k(\eta(t))u(t)] dw_k(t)$$

$$y(t) = x(t)$$

$$z(t) = C(\eta(t))x(t) + D(\eta(t))u(t)$$

with $G_k(i), C(i), D(i), i \in \mathcal{D}$ defined above.

Applying Theorem 9.2.1 we obtain the following theorem.

Theorem 9.2.5. *Suppose that there exist* $Y = (Y(1),\dots,Y(d)) \in \mathcal{S}_n^d, Y(i) > 0, \Gamma = (\Gamma(1),\dots,\Gamma(d)) \in \mathcal{M}_{m,n}^d$ *solving the following system of LMI:*

$$\begin{bmatrix}
\mathcal{W}_{0,0}(Y,i) & \mathcal{W}_{0,1}(Y,i) & \cdots & \mathcal{W}_{0,r}(Y,i) & \mathcal{W}_{0,r+1}(Y,i) & \mathcal{W}_{0,r+2}(Y,i) \\
\mathcal{W}_{0,1}^T(Y,i) & \mathcal{W}_{1,1}(Y,i) & \cdots & \mathcal{W}_{1,r}(Y,i) & \mathcal{W}_{1,r+1}(Y,i) & \mathcal{W}_{1,r+2}(Y,i) \\
\vdots & \vdots & \ddots & \vdots & \vdots & \vdots \\
\mathcal{W}_{0,r}^T(Y,i) & \mathcal{W}_{1,r}^T(Y,i) & \cdots & \mathcal{W}_{r,r}(Y,i) & \mathcal{W}_{r,r+1}(Y,i) & \mathcal{W}_{r,r+2}(Y,i) \\
\mathcal{W}_{0,r+1}^T(Y,i) & \mathcal{W}_{1,r+1}^T(Y,i) & \cdots & \mathcal{W}_{r,r+1}^T(Y,i) & \mathcal{W}_{r+1,r+1}(Y,i) & \mathcal{W}_{r+1,r+2}(Y,i) \\
\mathcal{W}_{0,r+2}^T(Y,i) & \mathcal{W}_{1,r+2}^T(Y,i) & \cdots & \mathcal{W}_{r,r+2}^T(Y,i) & \mathcal{W}_{r+1,r+2}^T(Y,i) & \mathcal{W}_{r+2,r+2}(Y,i)
\end{bmatrix} < 0,$$

$$(9.20)$$

where

$$
\begin{aligned}
\mathcal{W}_{0,0}(Y,i) &= A_0(i)Y(i) + Y(i)A_0^T(i) + q_{ii}Y(i) \\
&\quad + B_0(i)\Gamma(i) + \Gamma^*(i)B_0^T(i) + \hat{G}_0(i)\hat{G}_0^T(i) + \hat{B}_0(i)\hat{B}_0^T(i), \\
\mathcal{W}_{0,k}(Y,i) &= Y(i)A_k^T(i) + \Gamma^T(i)B_k^T(i) + \hat{G}_0(i)\hat{G}_k^*(i) + \hat{B}_0(i)\hat{B}_k^T(i), \ 1 \le k \le r, \\
\mathcal{W}_{0,r+1}(Y,i) &= \left[Y(i)\hat{C}^T(i) \quad \Gamma^T(i)\hat{D}^T(i) \right], \\
\mathcal{W}_{0,r+2}(Y,i) &= \left[\sqrt{q_{i1}}Y(i) \ \dots \ \sqrt{q_{i,i-1}}Y(i) \ \sqrt{q_{i,i+1}}Y(i) \ \dots \ \sqrt{q_{id}}Y(i) \right], \\
\mathcal{W}_{l,k}(Y,i) &= \hat{G}_l(i)\hat{G}_k^T(i) + \hat{B}_l(i)\hat{B}_k^T(i), \ 1 \le k \ne l \le r, \\
\mathcal{W}_{l,l}(Y,i) &= \hat{G}_l(i)\hat{G}_l^T(i) + \hat{B}_l(i)\hat{B}_l^T(i) - Y(i), \ 1 \le l \le r, \\
\mathcal{W}_{l,r+1}(Y,i) &= 0, \ 1 \le l \le r, \\
\mathcal{W}_{r+1,r+1}(Y,i) &= -\rho^{-2}I_{\hat{p}+\bar{p}}, \\
\mathcal{W}_{r+2,r+2}(Y,i) &= diag\left(-Y(1) \ \dots \ -Y(i-1) \ -Y(i+1) \ \dots \ -Y(d) \right).
\end{aligned}
$$

Then the state feedback gain $F(i) = \Gamma(i)Y^{-1}(i), i \in \mathcal{D}$ *is a solution of the robust stabilization problem.*

Consider now the system described by

$$
\begin{aligned}
dx(t) &= [A_0(\eta(t))x(t) + G_0(\eta(t))\Delta(\varphi(t),\eta(t)) + B_0(\eta(t))u(t)]dt \\
&\quad + \Sigma_{k=1}^r [A_k(\eta(t))x(t) + G_k(\eta(t))\Delta(\varphi(t),\eta(t)) \\
&\quad + B_k(\eta(t))u(t)]dw_k(t) \\
\varphi(t) &= C(\eta(t))x(t)
\end{aligned} \tag{9.21}
$$

where $x(t) \in \mathbf{R}^n$ is the state, $u(t) \in \mathbf{R}^m$ is the control variable and $A_k(i) \in \mathbf{R}^{n\times n}$, $B_k(i) \in \mathbf{R}^{n\times m}$, $G_k(i) \in \mathbf{R}^{n\times m_1}$, $0 \le k \le r$, $C(i) \in \mathbf{R}^{p_1\times n}$, $i \in \mathcal{D}$ are assumed to be known. The maps $y \to \Delta(y,i)$ are unknown functions including the uncertainties determined either by parameter variations or by truncation of nonlinear terms in the dynamic model. Denote by Δ the class of admissible uncertainty

$$
\Delta = (\Delta(y,1), \ldots, \Delta(y,d))
$$

where $y \to \Delta(y,i) : \mathbf{R}^{p_1} \to \mathbf{R}^{m_1}$ are Lipschitz continuous functions with $\Delta(0,i) = 0$, $i \in \mathcal{D}$. In the following it is assumed that in (9.21) the whole state is available for measurement. The robust stabilization problem considered can be stated as follows: for a given $\rho > 0$ find a control law $u(t) = F(\eta(t))x(t)$ stabilizing the system (9.21) for all $\Delta \in \Delta$ with $\|\Delta\| < \rho$. Recall that

$$
\|\Delta\| = \sup_{y\neq 0, y\in\mathbf{R}^{p_1}, i\in\mathcal{D}} \left\{ \frac{|\Delta(y,i)|}{|y|} \right\}.
$$

Let $u(t) = F(\eta(t))x(t)$ be such that the zero solution of the system

$$
\begin{aligned}
dx(t) &= [A_0(\eta(t)) + B_0(\eta(t))F(\eta(t))]x(t)\,dt \\
&\quad + \sum_{k=1}^r [A_k(\eta(t)) + B_k(\eta(t))F(\eta(t))]x(t)dw_k(t)
\end{aligned}
$$

is ESMS. When coupling this state feedback to (9.21) one obtains

$$
\begin{aligned}
dx(t) &= \{[A_0(\eta(t)) + B_0(\eta(t))F(\eta(t))]x(t) \\
&\quad + G_0(\eta(t))\Delta(\varphi(t),\eta(t))\}dt \\
&\quad + \Sigma_{k=1}^r \{[A_k(\eta(t)) + B_k(\eta(t))F(\eta(t))]x(t) \\
&\quad + G_k(\eta(t))\Delta(\varphi(t),\eta(t))\}dw_k(t) \\
\varphi(t) &= C(\eta(t))x(t).
\end{aligned} \tag{9.22}
$$

Applying Theorem 8.3.8 we deduce that the zero solution of (9.22) is ESMS for arbitrary $\Delta \in \Delta$ with $\|\Delta\| < \rho$ if the input–output operator $\mathcal{T}_F$ associated with the system

$$dx(t) = \{[A_0(\eta(t)) + B_0(\eta(t))F(\eta(t))]x(t) + G_0(\eta(t))v(t)\}dt$$
$$+ \sum_{k=1}^{r}\{[A_k(\eta(t))x(t) + B_k(\eta(t))F(\eta(t))]x(t) + G_k(\eta(t))v(t)\}dw_k(t)$$
$$z(t) = C(\eta(t))x(t)$$

satisfies the condition $\|\mathcal{T}_F\| < 1/\rho$. Therefore in order to obtain a robust state feedback control with a given robustness radius $\rho > 0$ it is sufficient to solve the DAP with the level of attenuation $\gamma = 1/\rho$ for the following auxiliary system

$$dx(t) = [A_0(\eta(t))x(t) + B_0(\eta(t))u(t) + G_0(\eta(t))v(t)]x(t)$$
$$+ \sum_{k=1}^{r}[A_k(\eta(t))x(t) + B_k(\eta(t))u(t) + G_k(\eta(t))v(t)]dw_k(t) \quad (9.23)$$
$$z(t) = C(\eta(t))x(t).$$

From Theorem 9.2.1 applied for the system (9.23) one obtains

Theorem 9.2.6. *Assume that there exist* $Y = (Y(1),\dots,Y(d)) \in \mathcal{S}_n^d$, $Y(i) > 0$, $\Gamma = (\Gamma(1),\dots,\Gamma(d)) \in \mathcal{M}_{m,n}^d$ *satisfying the following LMI:*

$$\begin{bmatrix}
\mathcal{W}_{0,0}(Y,i) & \mathcal{W}_{0,1}(Y,i) & \cdots & \mathcal{W}_{0,r}(Y,i) & \mathcal{W}_{0,r+1}(Y,i) & \mathcal{W}_{0,r+2}(Y,i) \\
\mathcal{W}_{0,1}^T(Y,i) & \mathcal{W}_{1,1}(Y,i) & \cdots & \mathcal{W}_{1,r}(Y,i) & \mathcal{W}_{1,r+1}(Y,i) & \mathcal{W}_{1,r+2}(Y,i) \\
\vdots & \vdots & \ddots & \vdots & \vdots & \vdots \\
\mathcal{W}_{0,r}^T(Y,i) & \mathcal{W}_{1,r}^T(Y,i) & \cdots & \mathcal{W}_{r,r}(Y,i) & \mathcal{W}_{r,r+1}(Y,i) & \mathcal{W}_{r,r+2}(Y,i) \\
\mathcal{W}_{0,r+1}^T(Y,i) & \mathcal{W}_{1,r+1}^T(Y,i) & \cdots & \mathcal{W}_{r,r+1}^T(Y,i) & \mathcal{W}_{r+1,r+1}(Y,i) & \mathcal{W}_{r+1,r+2}(Y,i) \\
\mathcal{W}_{0,r+2}^T(Y,i) & \mathcal{W}_{1,r+2}^T(Y,i) & \cdots & \mathcal{W}_{r,r+2}^T(Y,i) & \mathcal{W}_{r+1,r+2}^T(Y,i) & \mathcal{W}_{r+2,r+2}(Y,i)
\end{bmatrix} < 0,$$
$$(9.24)$$

$i \in \mathcal{D}$, where

$$\mathcal{W}_{0,0}(Y,i) = A_0(i)Y(i) + Y(i)A_0^T(i) + q_{ii}Y(i) + B_0(i)\Gamma(i)$$
$$+ \Gamma^T(i)B_0^T(i) + G_0(i)G_0^T(i)$$
$$\mathcal{W}_{0,k}(Y,i) = Y(i)A_k^T(i) + \Gamma^T(i)B_k^T(i) + G_0(i)G_k^T(i), \ 1 \le k \le r$$
$$\mathcal{W}_{0,r+1}(i) = Y(i)C^T(i)$$
$$\mathcal{W}_{0,r+2} = \left[\sqrt{q_{i1}}Y(i) \ \cdots \ \sqrt{q_{i,i-1}}Y(i) \ \sqrt{q_{i,i+1}}Y(i) \ \cdots \ \sqrt{q_{id}}Y(i)\right]$$
$$\mathcal{W}_{l,k} = G_l(i)G_k^T(i), \ 1 \le l,k \le r, \ l \ne k$$
$$\mathcal{W}_{l,l} = G_l(i)G_l^T(i) - Y(i), \ 1 \le l \le r$$
$$\mathcal{W}_{l,r+1}(i) = 0, \ 1 \le l \le r$$
$$\mathcal{W}_{l,r+2}(i) = 0, \ 1 \le l \le r+1$$
$$\mathcal{W}_{r+1,r+1}(i) = -\gamma^2 I_{p_1}$$
$$\mathcal{W}_{r+2,r+2}(i) = diag(-Y(1) \ \cdots \ -Y(i-1) \ -Y(i+1) \ \cdots \ -Y(d)).$$

Then the control $u(t) = F(\eta(t))x(t)$ *with* $F(i) = \Gamma(i)Y^{-1}(i)$, $i \in \mathcal{D}$ *provides a robust stabilizing feedback gain.*

Remark 9.2.2. In order to maximize the robustness radius one can use the idea presented in Remark 9.2.1 but with the constraint (9.24) instead of (9.7).

9.2.3   A Case Study

In order to illustrate the theoretical developments concerning the DAP in the case when the state is measurable we present in the following a case study for which some comparative aspects with the results provided by deterministic design approaches will be discussed.

Air launched unmanned air vehicles (UAV) are typically released with the wings folded in order to achieve a safe separation with respect to the launching aircraft. Its wings are deployed after several seconds when a glide slope maneuver is required. The wings deployment determines a "jump" of the aerodynamic coefficients leading to a transient of the angle of attack which must be minimized in order to prevent the loss of stability. The longitudinal short-period motion of the UAV has the following state-space equations:

$$\dot{x} = Ax + B\delta_{e_c} + Gv \qquad (9.25)$$

$$z = Cx + D\delta_{e_c}$$

where the state vector is

$$x = \begin{bmatrix} w \\ q \\ \delta_e \\ \xi \end{bmatrix},$$

with w denoting the vertical component of the true airspeed, q is the pitch rate, δ_e is the internal state of the actuator, and ξ denotes the state of the integral action $\dot{\xi} = a_z - a_{z_c}$ introduced in order to obtain zero steady-state tracking error of the normal acceleration a_z with respect to its commanded piecewise constant value a_{z_c}. The control variable is the elevon command δ_{e_c} and the input vector v includes the external reference a_{z_c} and disturbances, namely:

$$v = \begin{bmatrix} a_{z_c} \\ d_{\dot{w}} \\ d_{\dot{q}} \end{bmatrix},$$

$d_{\dot{w}}$ and $d_{\dot{q}}$ denoting the disturbances in $\dot{w}$ and $\dot{q}$, respectively. The quality output z has two components:

$$z = \begin{bmatrix} \beta\xi \\ \rho\delta_{e_c} \end{bmatrix}$$

where β and ρ are positive given weights. The matrix coefficients in (9.25) depend on the two flight conditions mentioned above, namely the situation when the UAV has the wings folded and the case when the wings are deployed, respectively. Therefore in this case the Markov chain has two states, that is $\mathcal{D} = \{1,2\}$. The numerical values corresponding to these two states are [134]:

$$A(1) = \begin{bmatrix} -0.1077 & 718.5340 & -31.3672 & 0 \\ -0.0219 & -0.7209 & -19.5316 & 0 \\ 0 & 0 & -30 & 0 \\ 0 & 2.8870 & 64.7283 & 0 \end{bmatrix},$$

$$A(2) = \begin{bmatrix} -0.4628 & 717.1890 & -16.7139 & 0 \\ -0.0333 & -0.7522 & -11.3638 & 0 \\ 0 & 0 & -30 & 0 \\ -0.2990 & 2.8210 & 39.1960 & 0 \end{bmatrix},$$

$$B(1) = B(2) = \begin{bmatrix} 0 \\ 0 \\ 30 \\ 0 \end{bmatrix},$$

$$G(1) = G(2) = \begin{bmatrix} 0 & 1 & 0 \\ 0 & 0 & 1 \\ 0 & 0 & 0 \\ -1 & 0 & 0 \end{bmatrix},$$

$$C(1) = C(2) = \begin{bmatrix} 0 & 0 & 0 & 20 \\ 0 & 0 & 0 & 0 \end{bmatrix},$$

$$D(1) = D(2) = \begin{bmatrix} 0 \\ 100 \end{bmatrix},$$

$\beta = 20, \rho = 100$. The transition rate matrix is

$$Q = \begin{bmatrix} -1 & 1 \\ 0.01 & -0.01 \end{bmatrix}.$$

The problem consists in determining a state feedback control $\delta_{e_c}(t) = F(\eta(t))x(t)$ such that the closed-loop system obtained when coupling it to (9.25), namely

$$\dot{x}(t) = [A(\eta(t)) + B(\eta(t))F(\eta(t))]x(t) + G(\eta(t))v(t)$$
$$z(t) = [C(\eta(t)) + D(\eta(t))F(\eta(t))]x(t)$$

is ESMS and its associated input–output operator has the norm less than a given $\gamma > 0$.

Applying Corollary 9.2.2 we obtained for $\gamma = 20$,

$$F(1) = [0.0290 \quad -2.7269 \quad -1.1120 \quad -1.5065], \tag{9.26}$$
$$F(2) = [0.0110 \quad -0.7722 \quad -0.4793 \quad -0.2112].$$

Table 9.1 Deterministic
comparison approach

$\|\mathcal{T}\|_\infty$	MJC	RC	SHD
$i=1$	18.3	32.9	18.1
$i=2$	15.7	22.5	12.9

Table 9.2 Stochastic
comparison approach

Method	MJC	RC	SHD
$\|\mathcal{T}\|$	20	32.4	76.7

In order to compare these results with the ones provided by other usual design methods we solved the same problem using two deterministic alternative approaches. The first one is the robust control (RC) design consisting in determining a unique "quadratically stabilizing" controller which stabilizes both systems corresponding to folded and unfolded wings situations. In this design we obtained using again an LMI based approach [12]:

$$F_{RC} = [10.57 \quad -425.6 \quad -180.7 \quad -305.7]$$

for the minimum closed-loop disturbance attenuation level $\gamma = 33.43$.

The second deterministic method consists in designing separate H^∞ state feedback zero-order controllers corresponding to each flight condition. This design will be abbreviated SDH and it gives for $\gamma = 18.1$ and for $\gamma = 12.9$, respectively, the following gains corresponding to the two flight conditions considered:

$$F_{SDH}(1) = [0.0040 \quad -0.0825 \quad -0.7510 \quad -0.4253],$$

$$F_{SDH}(2) = [0.1212 \quad 1.2540 \quad -1.7674 \quad -1.6579].$$

Two comparison approaches have been used: the first is completely deterministic and the second is entirely stochastic. In the first method, the H^∞ norm of the closed-loop system for $i = 1$ and $i = 2$ has determined for all three solutions obtained above. The results are presented in Table 9.1.

One can see that for MJC and SHD design the achieved H^∞ norms of the closed-loop system are very close and much lower than those of the RC-feedback gain.

In the second method we computed the levels of attenuation corresponding to the three solutions using the stochastic framework. To this end, we determined the closed-loop system with the corresponding feedback gains. Regarding these systems as stochastic systems with Markov jumps, we applied Theorem 8.2.7 to compute the corresponding level of attenuation. The obtained results are presented in Table 9.2.

The fact that in the stochastic design case (MJC) the level of attenuation is significantly lower as expected since the deterministic design (RC and SHD) do not take into consideration the parameter jumps.

The elements $P_{11}(t)$ and $P_{12}(t)$ of the transition probability matrix $P(t) = e^{Qt}$ as functions of time are illustrated in Fig. 9.1a. In Fig. 9.1b, c the time-responses of the angle of attack and of the elevon command to unit step acceleration are plotted. Inspecting these figures one can see that the angle of attack for all three methods is similar but the MJC uses considerably less control effort than both RC and SHD design.

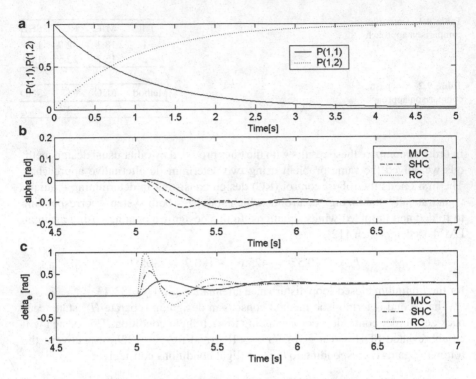

Fig. 9.1 (a) Transition probabilities (b) Angle of attack (c) Elevon angle

9.3 Solution of the DAP in the Case of Output Measurement

In this section we consider the DAP with an imposed level of attenuation $\gamma > 0$ in the case when the output is available for measurement. Our approach is based on an LMI technique and it extends to this framework the well-known results in the deterministic context. As in the deterministic case the necessary and sufficient conditions guaranteeing the existence of a γ-attenuating controller are obtained using the following result (see [12]):

Lemma 9.3.1 (Projection Lemma). *Let* $\mathcal{Z} \in \mathbf{R}^{\nu \times \nu}$, $\mathcal{Z} = \mathcal{Z}^T$, $\mathcal{U} \in \mathbf{R}^{\nu_1 \times \nu}$ *and* $\mathcal{V} \in \mathbf{R}^{\nu_2 \times \nu}$ *with* ν, ν_1, ν_2 *positive integers. Consider the following basic linear matrix inequality:*

$$\mathcal{Z} + \mathcal{U}^T \Theta \mathcal{V} + \mathcal{V}^T \Theta^T \mathcal{U} < 0 \qquad (9.27)$$

with the unknown variable $\Theta \in \mathbf{R}^{\nu_1 \times \nu_2}$. *Then the following are equivalent*

(i) There exists $\Theta \in \mathbf{R}^{\nu_1 \times \nu_2}$ *solving (9.27).*
(ii)

$$\mathcal{W}_{\mathcal{U}}^T \mathcal{Z} \mathcal{W}_{\mathcal{U}} < 0 \qquad (9.28)$$

and

$$W_{\mathcal{V}}^T \mathcal{Z} W_{\mathcal{V}} < 0 \qquad (9.29)$$

where $W_{\mathcal{U}}$ and $W_{\mathcal{V}}$ denote any bases of the null spaces $Ker\mathcal{U}$ and $Ker\mathcal{V}$, respectively.

Remark 9.3.1. It is known that if $\tilde{W}$ is a basis of *KerM* where M is a given matrix then any other basis of *KerM* can be expressed as $W = \tilde{W}\Gamma$ with $\det\Gamma \neq 0$. This shows that it is sufficient to check the conditions (9.28) and (9.29) for some suitable bases $W_{\mathcal{U}}$ and $W_{\mathcal{V}}$.

Lemma 9.3.2. *Let* $X, Y \in \mathcal{S}_n, N \in \mathbf{R}^{n \times n_c}$ *and* $S \in \mathcal{S}_{n_c}$ *with* $X > 0$ *and*

$$\begin{bmatrix} Y & N \\ N^T & S \end{bmatrix} > 0.$$

Then the following are equivalent

(i)

$$X = \left(Y - NS^{-1}N^T\right)^{-1};$$

(ii)

$$\text{rank} \begin{bmatrix} X\,I_n & 0 \\ I_n\,Y & N \\ 0 & N^T\,S \end{bmatrix} = n + n_c;$$

(iii)

$$\begin{bmatrix} Y & N \\ N^T & S \end{bmatrix}^{-1} = \begin{bmatrix} X & \star \\ \star & \star \end{bmatrix},$$

where $\star$ *denotes irrelevant entries.*

The next result provides necessary and sufficient conditions for the existence of a controller of type (9.2) solving the DAP for the system (9.1).

Theorem 9.3.3. *For a* $\gamma > 0$ *the following are equivalent*

(i) *There exists a controller of order* $n_c > 0$ *which solves the DAP with the level of attenuation* $\gamma > 0$ *for the system* (9.1);
(ii) *There exist* $X = (X(1),\ldots,X(d)) \in \mathcal{S}_n^d$, $X(i) > 0$, $i \in \mathcal{D}$, $Y = (Y(1),\ldots,Y(d))$ $\in \mathcal{S}_n^d$, $Y(i) > 0$, $S = (S(1),\ldots,S(d)) \in \mathcal{S}_n^d$, $S(i) > 0$, $N = (N(1),\ldots,N(d))$, N $\in \mathcal{M}_{n,n_c}^d$ *such that*

$$\begin{bmatrix} V_0^T(i) & V_1^T(i) \end{bmatrix} \mathcal{N}_i(X) \begin{bmatrix} V_0(i) \\ V_1(i) \end{bmatrix} < 0, \tag{9.30}$$

$$\begin{bmatrix} \Pi_{0,0}(i) & \Pi_{0,1}(i) & -U_1^T(i)N(i) & \cdots & -U_r^T(i)N(i) & \Pi_{0,r+1}(i) \\ \Pi_{0,1}^T(i) & -\gamma^2 I_{m_1} & 0 & \cdots & 0 & 0 \\ -N^T(i)U_1(i) & 0 & -S(i) & \cdots & 0 & 0 \\ \vdots & \vdots & \vdots & \ddots & \vdots & \vdots \\ -N^T(i)U_r(i) & 0 & 0 & \cdots & -S(i) & 0 \\ \Pi_{0,r+1}^*(i) & 0 & 0 & \cdots & 0 & \Pi_{r+1,r+1}(i) \end{bmatrix} < 0 \tag{9.31}$$

$$rank \begin{bmatrix} X(i) I_n & 0 \\ I_n & Y(i) & N(i) \\ 0 & N^*(i) S(i) \end{bmatrix} = n + n_c \tag{9.32}$$

where

$$\begin{bmatrix} V_0(i) \\ V_1(i) \end{bmatrix}$$

is a basis of $Ker\begin{bmatrix} C_0(i) & D_0(i) \end{bmatrix}$,

$$\begin{bmatrix} U_0(i) \\ \vdots \\ U_{r+1}(i) \end{bmatrix}$$

is a basis of $Ker\begin{bmatrix} B_0^T(i) & \cdots B_r^T(i) & D_{zu}^T(i) \end{bmatrix}$,

$$\mathcal{N}_i(X) = \begin{bmatrix} \mathcal{N}_{11}(X,i) & \mathcal{N}_{12}(X,i) \\ \mathcal{N}_{12}^T(X,i) & \mathcal{N}_{22}(X,i) \end{bmatrix},$$

$$\mathcal{N}_{11}(X,i) = A_0^T(i)X(i) + X(i)A_0(i) + \sum_{k=1}^r A_k^*(i)X(i)A_k(i)$$

$$+ \sum_{j=1}^d q_{ij}X(j) + C_z^T(i)C_z(i),$$

$$\mathcal{N}_{12}(X,i) = X(i)G_0(i) + \sum_{k=1}^r A_k^T(i)X(i)G_k(i) + C_z^T(i)D_{zv}(i),$$

$$\mathcal{N}_{22}(X,i) = -\gamma^2 I_{m_1} + D_{zv}^T(i)D_{zv}(i) + \sum_{k=1}^r G_k^T(i)X(i)G_k(i),$$

$$\Pi_{0,0}(i) = U_0^T(i)\left[A_0(i)Y(i) + Y(i)A_0^T(i) + q_{ii}Y(i)\right]U_0(i)$$

$$+ \sum_{k=1}^{r} U_0^T(i)Y(i)A_k^T(i)U_k(i) + U_0^T(i)Y(i)C_z^T(i)U_{r+1}(i)$$

$$+ U_{r+1}^T(i)C_z(i)Y(i)U_0(i) + \sum_{k=1}^{r} U_k^T(i)A_k(i)Y(i)U_0(i)$$

$$- \sum_{k=1}^{r} U_k^T(i)Y(i)U_k(i) - U_{r+1}^T(i)U_{r+1}(i),$$

$$\Pi_{0,1}(i) = \sum_{k=0}^{r} U_k^T(i)G_k(i) + U_{r+1}^T(i)D_{zv}(i)$$

$$\Pi_{0,r+1}(i) = U_0^T(i)[I_n \ 0]\left[\sqrt{q_{i1}}\tilde{Y}(i) \ \cdots \ \sqrt{q_{i,i-1}}\tilde{Y}(i) \ \sqrt{q_{i,i+1}}\tilde{Y}(i) \ \cdots \ \sqrt{q_{id}}\tilde{Y}(i)\right],$$

$$\Pi_{r+1,r+1}(i) = -diag\left(\tilde{Y}(1) \ \cdots \ \tilde{Y}(i-1), \ \tilde{Y}(i+1) \ \cdots \ \tilde{Y}(d)\right),$$

$$\tilde{Y}(i) = \begin{bmatrix} Y(i) & N(i) \\ N^*(i) & S(i) \end{bmatrix}, \ i \in \mathcal{D}.$$

Proof. The outline of the proof is similar to the one in the deterministic framework. The stochastic feature of the considered system does not appear explicitly in the following developments of the proof. This feature appears only in the specific formulae of the Bounded Real Lemma. Therefore the proof is also accessible for the readers who are not very familiar with stochastic systems.

(i) $\Rightarrow$ (ii). Assume that it exists a controller of the form (9.2) stabilizing the system (9.1) such that $\|\mathcal{T}_{cl}\| < \gamma$. Using the implication (i) $\Rightarrow$ (ii) of Theorem 8.2.7 (Bounded Real Lemma) for the closed-loop system we deduce that there exist

$$X_{cl} = (X_{cl}(1), \ldots, X_c(d)) \in \mathcal{S}_{n+n_c}^d, \ X_{cl}(i) > 0$$

such that

$$\mathcal{N}_i(X_{cl}, \gamma) < 0 \tag{9.33}$$

where

$$\mathcal{N}_i(X_{cl}, \gamma) = \begin{bmatrix} (\mathcal{L}_{cl}^* X_{cl})(i) + C_{cl}^T(i)C_{cl}(i) & \mathcal{P}_i^T(X_{cl}) \\ \mathcal{P}_i(X_{cl}) & \mathcal{R}_i(X_{cl}) \end{bmatrix},$$

$$(\mathcal{L}_{cl}^* X_{cl})(i) = A_{0cl}^T(i)X_{cl}(i) + X_{cl}(i)A_{cl}(i)$$

$$+ \sum_{k=1}^{r} A_{kcl}^{T}(i) X_{cl}(i) A_{kcl}(i) + \sum_{j=1}^{d} q_{ij} X_{cl}(j),$$

$$\mathcal{P}_i(X_{cl}) = G_{0cl}^{T}(i) X_{cl}(i) + \sum_{k=1}^{r} G_{kcl}^{T}(i) X_{cl} A_{kcl}(i)$$

$$+ D_{cl}^{T}(i) C_{cl}(i),$$

$$\mathcal{R}_i(X_{cl}) = -\gamma^2 I_{m_1} + \sum_{k=1}^{r} G_{kcl}^{T}(i) X_{cl}(i) G_{kcl}(i).$$

Based on Schur complements arguments it is easy to see that (9.33) is equivalent
with

$$\begin{bmatrix} (\mathcal{L}_0^* X_{cl})(i) & X_{cl}(i) G_{0cl}(i) & A_{1cl}^{T}(i) X_{cl}(i) & \cdots & A_{rcl}^{T}(i) X_{cl}(i) & C_{cl}^{T}(i) \\ G_{0cl}^{T}(i) X_{cl}(i) & -\gamma^2 I_{m_1} & G_{1cl}^{T}(i) X_{cl}(i) & \cdots & G_{rcl}^{T}(i) X_{cl}(i) & D_{cl}^{T}(i) \\ X_{cl}(i) A_{1cl}(i) & X_{cl}(i) G_{1cl}(i) & -X_{cl}(i) & \cdots & 0 & 0 \\ \vdots & \vdots & \vdots & \ddots & \vdots & \vdots \\ X_{cl}(i) A_{rcl}(i) & X_{cl}(i) G_{rcl}(i) & 0 & \cdots & -X_{cl}(i) & 0 \\ C_{cl}(i) & D_{cl}(i) & 0 & \cdots & 0 & -I_{p_1} \end{bmatrix} < 0 \quad (9.34)$$

where

$$(\mathcal{L}_0^* X_{cl})(i) = A_{0cl}^{T}(i) X_{cl}(i) + X_{cl}(i) A_{0cl}(i) + \sum_{j=1}^{d} q_{ij} X_{cl}(j).$$

Let us introduce the following notations:

$$\tilde{A}_k(i) = \begin{bmatrix} A_k(i) & 0 \\ 0 & 0 \end{bmatrix}, \tilde{G}_k(i) = \begin{bmatrix} G_k(i) \\ 0 \end{bmatrix}, 0 \leq k \leq r,$$

$$\tilde{B}_0(i) = \begin{bmatrix} 0 & B_0(i) \\ I_{n_c} & 0 \end{bmatrix}, \tilde{B}_k(i) = \begin{bmatrix} 0 & B_k(i) \\ 0 & 0 \end{bmatrix}, 1 \leq k \leq r,$$

$$\tilde{C}_0(i) = \begin{bmatrix} 0 & I_{n_c} \\ C_0(i) & 0 \end{bmatrix}, \tilde{C}_z(i) = [C_z(i) \quad 0],$$

$$\tilde{D}_{zu}(i) = [0 \quad D_{zu}(i)], \tilde{D}_0(i) = \begin{bmatrix} 0 \\ D_0(i) \end{bmatrix}, i \in \mathcal{D},$$

$$\Theta_c(i) = \begin{bmatrix} A_c(i) & B_c(i) \\ C_c(i) & D_c(i) \end{bmatrix}.$$

Using (9.4) one obtains

$$A_{kcl}(i) = \tilde{A}_k(i) + \tilde{B}_k(i)\Theta_c(i)\tilde{C}_0(i)$$

$$G_{kcl}(i) = \tilde{G}_k(i) + \tilde{B}_k(i)\Theta_c(i)\tilde{D}_0(i), \ 0 \le k \le r,$$

$$C_{cl}(i) = \tilde{C}_z(i) + \tilde{D}_{zu}(i)\Theta_c(i)\tilde{C}_0(i)$$

$$D_{cl}(i) = D_{zv}(i) + \tilde{D}_{zu}(i)\Theta_c(i)\tilde{D}_0(i), \ i \in \mathcal{D}.$$

With the above equations one can easily see that (9.34) can be written in the basic linear matrix inequality form

$$\mathcal{Z}(i) + \mathcal{U}^T(i)\Theta_c(i)\mathcal{V}(i) + \mathcal{V}^T(i)\Theta_c^T(i)\mathcal{U}(i) < 0, i \in \mathcal{D}, \tag{9.35}$$

where

$$\mathcal{Z}(i) = \begin{bmatrix} (\tilde{\mathcal{L}}_0^* X_{cl})(i) & X_{cl}(i)\tilde{G}_0(i) & \tilde{A}_1^T(i)X_{cl}(i) & \cdots & \tilde{A}_r^T(i)X_{cl}(i) & \tilde{C}_z^T(i) \\ \tilde{G}_0^T(i)X_{cl}(i) & -\gamma^2 I_{m_1} & \tilde{G}_1^T(i)X_{cl}(i) & \cdots & \tilde{G}_r^T(i)X_{cl}(i) & D_{zv}^T(i) \\ X_{cl}(i)\tilde{A}_1(i) & X_{cl}(i)G_1(i) & -X_{cl}(i) & \cdots & 0 & 0 \\ \vdots & \vdots & \vdots & \ddots & \vdots & \vdots \\ X_{cl}(i)\tilde{A}_r(i) & X_{cl}(i)\tilde{G}_r(i) & 0 & \cdots & -X_{cl}(i) & 0 \\ \tilde{C}_z(i) & D_{zv}(i) & 0 & \cdots & 0 & -I_{p_1} \end{bmatrix},$$

$$\mathcal{U}(i) = \begin{bmatrix} \tilde{B}_0^T(i)X_{cl}(i) & 0_{(m_2+n_c)\times m_1} & \tilde{B}_1^T(i)X_{cl}(i) & \cdots & \tilde{B}_r^T(i)X_{cl}(i) & \tilde{D}_{zu}^T(i) \end{bmatrix},$$

$$\mathcal{V}(i) = \begin{bmatrix} \tilde{C}_0(i) & \tilde{D}_0(i) & 0_{(p_2+n_c)\times[p_1+r(n+n_c)]} \end{bmatrix}, \ i \in \mathcal{D}. \tag{9.36}$$

with

$$(\tilde{\mathcal{L}}_0^T X_{cl})(i) = \tilde{A}_{0cl}^T(i)X_{cl}(i) + X_{cl}(i)\tilde{A}_{0cl}(i) + \sum_{j=1}^d q_{ij}X_{cl}(j)$$

Therefore the existence of a stabilizing γ-attenuation controller for (9.1) is equivalent with the solvability of (9.35). Based on Lemma 9.3.1, (9.35) is solvable if and only if there exist:

$$\mathcal{W}_{\mathcal{U}(i)}^T \mathcal{Z}(i)\mathcal{W}_{\mathcal{U}(i)} < 0 \tag{9.37}$$

$$\mathcal{W}_{\mathcal{V}(i)}^* \mathcal{Z}(i)\mathcal{W}_{\mathcal{V}(i)} < 0, \ i \in \mathcal{D}, \tag{9.38}$$

where $\mathcal{W}_{\mathcal{U}(i)}, \mathcal{W}_{\mathcal{V}(i)}$ denote bases of the null spaces of $\mathcal{U}(i)$ and $\mathcal{V}(i)$, respectively. It is easy to see that a basis of the null space of $\mathcal{U}(i)$ is

$$\mathcal{W}_{\mathcal{U}(i)} = \mathcal{X}^{-1}(i)\mathcal{W}_{\tilde{\mathcal{U}}(i)} \tag{9.39}$$

where

$$\mathcal{X}(i) = diag\left(X_{cl} \; I_{m_1} \; X_{cl}(i) \; \cdots \; X_{cl}(i) \; I_{p_1}\right)$$

and $\mathcal{W}_{\tilde{\mathcal{U}}(i)}$ is a basis of the null subspace of the matrix:

$$\tilde{\mathcal{U}}(i) = \left[\tilde{B}_0^T(i) \; 0_{(m_2+n_c) \times m_1} \; \tilde{B}_1^T(i) \; \cdots \; \tilde{B}_r^T(i) \; \tilde{D}_{zu}^T(i) \right].$$

A basis of the null subspace of $\tilde{\mathcal{U}}(i)$ is:

$$\mathcal{W}_{\tilde{\mathcal{U}}(i)} = \begin{bmatrix} T_0(i) & 0 & 0 & \cdots & 0 \\ 0 & I_{m_1} & 0 & \cdots & 0 \\ T_1(i) & 0 & L & \cdots & 0 \\ \vdots & \vdots & \vdots & \ddots & \vdots \\ T_r(i) & 0 & 0 & \cdots & L \\ U_{r+1}(i) & 0 & 0 & \cdots & 0 \end{bmatrix} \tag{9.40}$$

where

$$T_k(i) = \begin{bmatrix} U_k(i) \\ 0 \end{bmatrix}, 0 \le k \le r, \; L = \begin{bmatrix} 0 \\ I_{n_c} \end{bmatrix}$$

and

$$\begin{bmatrix} U_0(i) \\ \vdots \\ U_{r+1}(i) \end{bmatrix}$$

is a basis of the null subspace of the matrix:

$$\left[B_0^T(i) \; B_1^T(i) \; \cdots \; B_r^T(i) \; D_{zu}^T(i) \right].$$

A suitable choice for $\mathcal{W}_{\mathcal{V}(i)}$ is the following:

$$\mathcal{W}_{\mathcal{V}(i)} = \begin{bmatrix} V_0(i) & 0 \\ 0 & 0 \\ V_1(i) & 0 \\ 0 & I_{p_1 + r(n+n_c)} \end{bmatrix} \tag{9.41}$$

where $\begin{bmatrix} V_0(i) \\ V_1(i) \end{bmatrix}$ is a basis of the null subspace of the matrix $[C_0(i) \; D_0(i)]$.

Consider the partition of $X_{cl}(i)$:

$$X_{cl}(i) = \begin{bmatrix} X(i) & M(i) \\ M^T(i) & \tilde{X}(i) \end{bmatrix}$$

with $X(i) \in \mathbf{R}^{n \times n}$. Then by direct computations one obtains

$$\mathcal{W}_{\mathcal{V}(i)}^T \mathcal{Z}(i) \mathcal{W}_{\mathcal{V}(i)} \tag{9.42}$$

$$= \begin{bmatrix} \Psi_{0,0}(i) & \Psi_{0,1}(i) & \cdots & \Psi_{0,r}(i) & \Psi_{0,r+1}(i) \\ \Psi_{0,1}^T(i) & -X_{cl}(i) & \cdots & 0 & 0 \\ \vdots & \vdots & \ddots & \vdots & \vdots \\ \Psi_{0,r}^T(i) & 0 & \cdots & -X_{cl}(i) & 0 \\ \Psi_{0,r+1}^T(i) & 0 & \cdots & 0 & -I_{p_1} \end{bmatrix}$$

where we denoted

$$\Psi_{0,0}(i) = V_0^T(i) \left[A_0^T(i) X(i) + X(i) A_0(i) + \sum_{j=1}^{d} q_{ij} X(j) \right] V_0(i)$$

$$+ V_0^T(i) X(i) G_0(i) V_1(i) + V_1^T(i) G_0^T(i) X(i) V_0(i)$$

$$- \gamma^2 V_1^T(i) V_1(i),$$

$$\Psi_{0,k}(i) = \left(\begin{bmatrix} V_0^T(i) & 0 \end{bmatrix} \tilde{A}_k^T(i) + V_1^T(i) \tilde{G}_k^T(i) \right) X_{cl}, \quad 1 \le k \le r,$$

$$\Psi_{0,r+1}(i) = V_0^T(i) C_z^T(i) + V_1^T(i) D_{zv}^T(i).$$

Using again Schur complement arguments it follows that condition (9.38) together with (9.42) is equivalent with

$$\Psi_{0,0}(i) + \sum_{k=1}^{r} \Psi_{0,k}(i) X_{cl}^{-1}(i) \Psi_{0,k}^T(i) + \Psi_{0,r+1}(i) \Psi_{0,r+1}^T(i) < 0.$$

Detailing the coefficients in the above inequality, (9.37) directly follows.
In order to explicit the condition (9.37) one first computes

$$\mathcal{X}^{-1}(i) \mathcal{Z}(i) \mathcal{X}^{-1}(i) \tag{9.43}$$

$$= \begin{bmatrix} (\mathcal{L}_0^* \tilde{Y})(i) & \tilde{G}_0(i) & \tilde{Y}(i)\tilde{A}_1^T(i) & \cdots & \tilde{Y}(i)\tilde{A}_r^T(i) & \tilde{Y}(i)\tilde{C}_z^T(i) \\ \tilde{G}_0^T(i) & -\gamma^2 I_{m_1} & \tilde{G}_1^T(i) & \cdots & \tilde{G}_r^T(i) & D_{zv}^*(i) \\ \tilde{A}_1(i)\tilde{Y}(i) & \tilde{G}_1^T(i) & -\tilde{Y}(i) & \cdots & 0 & 0 \\ \vdots & \vdots & \vdots & \ddots & \vdots & \vdots \\ \tilde{A}_r(i)\tilde{Y}(i) & \tilde{G}_r(i) & 0 & \cdots & -\tilde{Y}(i) & 0 \\ \tilde{C}_z(i)\tilde{Y}(i) & D_{zv}(i) & 0 & \cdots & 0 & -I_{p_1} \end{bmatrix}$$

where

$$\left(\mathcal{L}_0^*\tilde{Y}\right)(i) = \tilde{A}_0(i)\,\tilde{Y}(i) + \tilde{Y}(i)\,\tilde{A}_0^T(i) \tag{9.44}$$

$$+ \sum_{j=1}^{d} q_{ij}\tilde{Y}(i)\,\tilde{Y}^{-1}(j)\,\tilde{Y}(i)$$

$$\tilde{Y}(i) = X_{cl}^{-1}(i). \tag{9.45}$$

We also introduce the notation

$$\tilde{Y}(i) = \begin{bmatrix} Y(i) & N(i) \\ N^T(i) & S(i) \end{bmatrix}, \ Y(i) \in \mathbf{R}^{n \times n}.$$

Using (9.40), (9.43), (9.44), and (9.39) one obtains that (9.37) becomes

$$\begin{bmatrix} \tilde{\Pi}_{0,0}(i) & \Pi_{0,1}(i) & -U_1^T(i)N(i) & \cdots & -U_r^T(i)N(i) \\ \Pi_{0,1}^T(i) & -\gamma^2 I_{m_1} & 0 & \cdots & 0 \\ -N^T(i)U_1(i) & 0 & -S(i) & \cdots & 0 \\ \vdots & \vdots & \vdots & \ddots & \vdots \\ -N^T(i)U_r(i) & 0 & 0 & \cdots & -S(i) \end{bmatrix} < 0 \tag{9.46}$$

where

$$\tilde{\Pi}_{0,0}(i) = U_0^T(i)\left\{A_0(i)Y(i) + Y(i)A_0^T(i)\right.$$

$$+ \sum_{j=1}^{d} q_{ij}[Y(i) \ \ N(i)]\tilde{Y}^{-1}(j)\begin{bmatrix} Y(i) \\ N^T(i) \end{bmatrix}\bigg\}U_0(i)$$

$$+ \sum_{k=1}^{r} U_0^T(i)Y(i)A_k^T(i)U_k(i) + U_0^T(i)Y(i)C_z^T(i)U_{r+1}(i)$$

$$+ U_{r+1}^T(i)C_z(i)Y(i)U_0(i) + \sum_{k=1}^{r} U_k^T(i)A_k(i)Y(i)U_0(i)$$

$$- \sum_{k=1}^{r} U_k^T(i)Y(i)U_k(i) - U_{r+1}^T(i)U_{r+1}(i),$$

$$\Pi_{0,1}(i) = \sum_{k=0}^{r} U_k^T(i)G_k(i) + U_{r+1}^T(i)D_{zv}(i).$$

By Schur complement arguments one can see that (9.46) is equivalent with an extended LMI which coincides with (8.79). Taking into account that

$$
rank \begin{bmatrix} X(i) & I & 0 \\ I & Y(i) & N(i) \\ 0 & N^T(i) & S(i) \end{bmatrix}
$$

$$
= rank \begin{bmatrix} X(i) - \left(Y(i) - N(i)S^{-1}(i)N^T(i) \right)^{-1} & 0 & 0 \\ 0 & Y(i) - N(i)S^{-1}(i)N^T(i) & 0 \\ 0 & 0 & S(i) \end{bmatrix}
$$

and $S(i) > 0$, $Y(i) - N(i)S^{-1}(i)N^T(i) > 0$ it follows that (9.45) gives

$$
X(i) = \left(Y(i) - N(i)S^{-1}(i)N^T(i) \right)^{-1}
$$

from which (9.32) directly follows.

(ii) $\Rightarrow$ (i). Assume that there exist $X(i), Y(i), N(i)$ and $S(i)$ verifying (9.30)–(9.32). From (9.31) it follows that $\Pi_{r+1,r+1}(i) < 0$ and therefore

$$
\tilde{Y}(i) = \begin{bmatrix} Y(i) & N(i) \\ N^T(i) & S(i) \end{bmatrix} > 0.
$$

Hence $\tilde{Y}(i)$ is invertible. From Lemma 9.3.2 it follows that $\tilde{Y}^{-1}(i)$ has the structure

$$
\begin{bmatrix} X(i) & \star \\ \star & \star \end{bmatrix}.
$$

where by $\star$ we denoted the irrelevant entries. From the developments performed to prove the implication (i) $\Rightarrow$ (ii) it follows that (9.37) and (9.38) are verified by

$$
X_{cl}(i) = \tilde{Y}^{-1}(i)
$$

and hence (9.35) has a solution which fact guarantees the existence of a stabilizing and γ-attenuating controller. Thus the proof ends. □

Remark 9.3.2. In the case of the static output feedback ($n_c = 0$), in the above theorem we have to remove all variables $n_c, N(i)$ and $S(i)$, $i \in \mathcal{D}$.

Remark 9.3.3. According to the proof of the above result, the algorithm to determine a solution of the DAP is the following:

Step 1. Solve the system of LMI (9.30) and (9.31) with the constraint (9.32);
Step 2. Compute $\mathcal{Z}(i), \mathcal{U}(i)$ and $\mathcal{V}(i), i \in \mathcal{D}$ according to (9.36);
Step 3. Solve the basic LMI (9.35) with respect to Θ_c. Then the solution of the DAP is given by the partition

$$
\Theta_c(i) = \begin{bmatrix} A_c(i) & B_c(i) \\ C_c(i) & D_c(i) \end{bmatrix}.
$$

Obviously, if $n_c = 0$ then $\Theta_c(i) = D_c(i)$.

In the following we shall emphasize the important particular cases when the system (9.1) is subject only to Markovian jumping or to multiplicative white noise.

In the situation when $A_k(i) = 0$, $B_k(i) = 0$, $G_k(i) = 0$, $1 \leq k \leq r$, $i \in \mathcal{D}$ the system (9.1) becomes

$$\dot{x}(t) = A_0(\eta(t))x(t) + G_0(\eta(t))v(t) + B_0(\eta(t))u(t)$$
$$z(t) = C_z(\eta(t))x(t) + D_{zv}(\eta(t))v(t) + D_{zu}(\eta(t))u(t) \qquad (9.47)$$
$$y(t) = C_0(\eta(t))x(t) + D_0(\eta(t))v(t).$$

The closed loop system obtained by coupling a controller of the form (9.2) to the system (9.47) has the following state-space realization:

$$\dot{x}_{cl}(t) = A_{0cl}(\eta(t))x_{cl}(t) + G_{0cl}(\eta(t))v(t)$$
$$z(t) = C_{cl}(\eta(t))x_{cl}(t) + D_{cl}(\eta(t))v(t) \qquad (9.48)$$

where the matrix coefficients are defined as in (9.4). The results in the previous theorem leads for the particular system (9.47) to the following theorem.

Theorem 9.3.4. *For a $\gamma > 0$ the following are equivalent*

(i) *There exists a controller of order $n_c \geq 0$ of type (9.2) which stabilizes the system (9.47) such that the input–output operator associated with the system (9.48) verifies $\|\mathcal{T}_{cl}\| < \gamma$;*
(ii) *There exist $X = (X(1),\ldots,X(d)) \in \mathcal{S}_n^d$, $Y = (Y(1),\ldots,Y(d)) \in \mathcal{S}_n^d$, $S = (S(1),\ldots,S(d)) \in \mathcal{S}_n^d$, $N \in (N(1),\ldots,N(d))$, $N \in \mathcal{M}_{n,n_c}^d$ such that:*

$$\begin{bmatrix} V_0(i) \\ V_1(i) \end{bmatrix}^T \begin{bmatrix} A_0^T(i)X(i) + X(i)A_0(i) \\ + \sum_{j=1}^d q_{ij}X(j) + C_z^T(i)C_z(i) & X(i)G_0(i) + C_z^T(i)D_{zv}(i) \\ G_0^T(i)X(i) + D_{zv}^T(i)C_z(i) & -\gamma^2 I_{m_1} + D_{zv}^T(i)D_{zv}(i) \end{bmatrix}$$
$$\times [V_0(i) \quad V_1(i)] < 0$$

$$(9.49)$$

$$\begin{bmatrix} \Pi_{0,0}(\gamma,i) & \Pi_{0,r+1}(\gamma,i) \\ \Pi_{0,r+1}^T(\gamma,i) & \Pi_{r+1,r+1}(\gamma,i) \end{bmatrix} < 0 \qquad (9.50)$$

$$rank \begin{bmatrix} X(i) & I & 0 \\ I & Y(i) & N(i) \\ 0 & N^T(i) & S(i) \end{bmatrix} = n + n_c, \quad i \in \mathcal{D}, \qquad (9.51)$$

where

$$\Pi_{0,0}(\gamma, i) = \begin{bmatrix} U_0(i) \\ U_{r+1}(i) \end{bmatrix}^T$$

$$\times \begin{bmatrix} A_0(i)Y(i) + Y(i)A_0^T(i) & Y(i)C_z^T(i) + \gamma^{-2}G_0(i)D_{zv}^T(i) \\ +q_{ii}X(i) + \gamma^{-2}G_0(i)G_0^T(i) & \\ C_z(i)Y(i) + \gamma^{-2}D_{zv}(i)G_0^T(i) & -I_{p_1} + \gamma^{-2}D_{zv}(i)D_{zv}^T(i) \end{bmatrix}$$

$$\times [U_0(i) \quad U_{r+1}(i)],$$

$\Pi_{0,r+1}(\gamma, i)$ and $\Pi_{r+1,r+1}(\gamma, i)$ are as in Theorem (9.3.3),

$$\begin{bmatrix} U_0(i) \\ U_{r+1}(i) \end{bmatrix}$$

is a basis of the null subspace of $\begin{bmatrix} B_0^*(i) & D_{zu}^T(i) \end{bmatrix}$ and

$$\begin{bmatrix} V_0(i) \\ V_1(i) \end{bmatrix}$$

is a basis of the null subspace of $[C_0(i) \quad D_0(i)]$.

Remark 9.3.4. From the above theorem one can see that the necessary and sufficient conditions guaranteeing the solvability of the DAP involve the same unknown variables, namely $X(i), Y(i), S(i), N(i), i \in \mathcal{D}$ as in the general case of the system (9.1). It seems that this is the price paid to obtain a controller of order $n_c < n$. In the particular case when a full order controller (n_c) is required the rank condition (9.32) in the statement of Theorem 9.3.3 is removed (see Theorem 9.4.1 from below).

Consider now the case when $\mathcal{D} = \{1\}$. Then the system (9.1) becomes:

$$dx(t) = [A_0 x(t) + G_0 v(t) + B_0 u(t)] dt$$

$$+ \sum_{k=1}^r [A_k x(t) + G_k v(t) + B_k u(t)] dw_k(t)$$

$$z(t) = C_z x(t) + D_{zv} v(t) + D_{zu} u(t) \tag{9.52}$$

$$y(t) = C_0 x(t) + D_0 v(t)$$

where the matrices $A_k, B_k, G_k, 0 \le k \le r, C_z, D_{zu}, D_{zv}, C_0, D_0$ are given matrices of appropriate dimensions. The class of admissible controllers consists in deterministic controllers of the form:

$$\dot{x}_c(t) = A_c x_c(t) + B_c y(t) \tag{9.53}$$

$$u(t) = C_c x_c(t) + D_c y(t).$$

The closed-loop system obtained when coupling (9.53) to (9.52) is:

$$dx_c(t) = [A_{0cl}x_{cl}(t) + G_{0cl}v(t)]dt \tag{9.54}$$

$$+ \sum_{k=1}^{r}[A_{kcl}x_{cl}(t) + G_{kcl}v(t)]dw_k(t)$$

$$z(t) = C_{cl}x_{cl}(t) + D_{cl}v(t)$$

where the matrix coefficients are as in (9.4) with $d = 1$.

The next result provides the version of Theorem 9.3.3 for the particular case of the system (9.52).

Theorem 9.3.5. *For a given* $\gamma > 0$ *the following are equivalent:*

(i) *There exist an* $n_c \geq 0$ *order controller stabilizing (9.52) such that the input–output operator associated with the system (9.54) verifies* $\|\mathcal{T}_{cl}\| < \gamma$.

(ii) *There exist* $X, Y \in \mathcal{S}_n, S \in \mathcal{S}_{n_c}, N \in \mathbf{R}^{n \times n_c}$ *satisfying* $X > 0, Y > 0, S > 0$ *such that*

$$\begin{bmatrix} V_0 \\ V_1 \end{bmatrix}^T \begin{bmatrix} A_0^T X + X A_0 & X G_0 + \sum_{k=1}^{r} A_k^T X G_k \\ + \sum_{k=1}^{r} A_k^T X A_k + C_z^T C_z & + C_z^T D_{zv} \\ G_0^T X + \sum_{k=1}^{r} G_k^T X A_k & -\gamma^2 I_{m_1} + D_{zv}^T D_{zv} \\ + D_{zv}^T C_z & + \sum_{k=1}^{r} G_k^T X G_k \end{bmatrix} \tag{9.55}$$

$$\times \begin{bmatrix} V_0 \\ V_1 \end{bmatrix} < 0$$

$$\begin{bmatrix} \Pi_{0,0}(Y) & \Pi_{0,1} & -U_1^T N & \cdots & -U_r^T N \\ \Pi_{0,1}^T & -\gamma^2 I_{m_1} & 0 & \cdots & 0 \\ -N^T U_1 & 0 & -S & \cdots & 0 \\ \vdots & \vdots & \vdots & \ddots & \vdots \\ -N^T U_r & 0 & 0 & \cdots & -S \end{bmatrix} < 0 \tag{9.56}$$

$$\text{rank} \begin{bmatrix} X & I & 0 \\ I & Y & N \\ 0 & N^T & S \end{bmatrix} = n + n_c \tag{9.57}$$

where $\begin{bmatrix} V_0 \\ V_1 \end{bmatrix}$ *and* $\begin{bmatrix} U_0 \\ \vdots \\ U_{r+1} \end{bmatrix}$ *are bases of the null subspaces of* $[C_0 \ D_0]$ *and*

$[B_0^T \ B_1^T \ \cdots \ B_r^T \ D_{zu}^T]$, *respectively, and*

$$\Pi_{0,0}(Y) = U_0^T \left[A_0 Y + Y A_0^T\right] U_0 + \sum_{k=1}^{r} U_k^T A_k Y U_0$$

$$+ \sum_{k=1}^{r} U_k^T A_k Y U_0 + U_0^T Y C_z^T U_{r+1} + U_{r+1}^T C_z Y U_0$$

$$- \sum_{k=1}^{r} U_k^T Y U_k - U_{r+1}^T U_{r+1}$$

$$\Pi_{0,1} = \sum_{k=0}^{r} U_k^T G_k + U_{r+1}^T D_{zv}.$$

The next result is well known in the deterministic case; however, for the sake of completeness we shall briefly present it in the following lemma.

Lemma 9.3.6. *Let $X_{cl} \in \mathbf{R}^{n \times n_c}$ be partitioned as*

$$X_{cl} = \begin{bmatrix} X & M \\ M^T & \tilde{X} \end{bmatrix}, X \in \mathcal{S}_n, \tilde{X} \in \mathcal{S}_{n_c},$$

where $n_c \geq 1$. Assume that $X_{cl} > 0$ and consider the following partition of X_{cl}^{-1}:

$$X_{cl}^{-1} = \begin{bmatrix} Y & N \\ N^T & S \end{bmatrix}, Y \in \mathcal{S}_n, S \in \mathcal{S}_{n_c}.$$

Then we have

$$X \geq Y^{-1} > 0 \tag{9.58}$$

$$rank\left(X - Y^{-1}\right) \leq n_c. \tag{9.59}$$

Conversely, if there exist $X \in \mathcal{S}_n$, $Y \in \mathcal{S}_n$ verifying conditions (9.58) and (9.59) then there exist $M \in \mathbf{R}^{n \times n_c}$, $\tilde{X} \in \mathcal{S}_{n_c}$, $N \in \mathbf{R}^{n \times n_c}$, $S \in \mathcal{S}_{n_c}$ such that

$$\begin{bmatrix} X & M \\ M^T & \tilde{X} \end{bmatrix} > 0$$

and

$$\begin{bmatrix} X & M \\ M^T & \tilde{X} \end{bmatrix} \begin{bmatrix} Y & N \\ N^T & S \end{bmatrix} = \begin{bmatrix} I & 0 \\ 0 & I \end{bmatrix}. \tag{9.60}$$

Proof. From $X_{cl} > 0$ it follows that $X > 0, \tilde{X} > 0, S > 0$. From the condition $X_{cl} X_{cl}^{-1} = I$ one obtains that

$$X - Y^{-1} = Y^{-1} N \tilde{X} N^T Y^{-1}$$

and therefore (9.58) immediately follows. The above conditions also leads to

$$rank\left(X - Y^{-1}\right) = rank\left(N\right) \le n_c$$

and hence (9.59) results.

Conversely, let $X, Y \in \mathcal{S}_n$ satisfying (9.58) and (9.59). Define $M \in \mathbf{R}^{n \times n_c}$ as the Cholesky factor

$$X - Y^{-1} = MM^T$$

and

$$N = -YM$$
$$\tilde{X} = I_{n_c}$$
$$S = I_{n_c} + M^T YM.$$

Then it follows that:

$$X - M\tilde{X}^{-1}M^T = Y^{-1} > 0$$
$$S - N^T Y^{-1} N = I_{n_c} > 0.$$

Then (9.60) follows by direct computations and thus the proof ends.   □

The next result shows that it is possible to remove the unknown variables N and S but in this case the condition (9.56) in Theorem 9.3.5 becomes nonlinear.

Theorem 9.3.7. *For a given $\gamma > 0$ the following are equivalent*

(i) There exists a stabilizing controller with order $n_c > 0$ of the form (9.53) solving the DAP for the system (9.52).

(ii) There exist $X, Y \in \mathcal{S}_n$, $X > 0$, $Y > 0$ satisfying the following conditions

$$\begin{bmatrix} X & I \\ I & Y \end{bmatrix} \ge 0 \tag{9.61}$$

$$rank \begin{bmatrix} X & I \\ I & Y \end{bmatrix} \le n + n_c \tag{9.62}$$

$$\begin{bmatrix} V_0 \\ V_1 \end{bmatrix}^T \begin{bmatrix} A_0^T X + X A_0 & X G_0 + \sum_{k=1}^r A_k^T X G_k \\ + \sum_{k=1}^r A_k^T X A_k + C_z^T C_z & + C_z^T D_{zv} \\ G_0^T X + \sum_{k=1}^r G_k^T X A_k & -\gamma^2 I_{m_1} + D_{zv}^T D_{zv} \\ + D_{zv}^T C_z & + \sum_{k=1}^r G_k^T X G_k \end{bmatrix}$$
$$\times [V_0 \quad V_1] < 0 \tag{9.63}$$

$$U^T \Lambda(Y, \gamma) U < 0 \tag{9.64}$$

where $\begin{bmatrix} V_0 \\ V_1 \end{bmatrix}$ is a basis of the null subspace of $[C_0 \ D_0]$,

$$U = \begin{bmatrix} U_0 \\ U_1 \\ \vdots \\ U_{r+1} \end{bmatrix}$$

is a basis of the null subspace of $[B_0^T \ B_1^T \ \cdots \ B_r^T \ D_{zu}^T]$ and

$$\Lambda = \begin{bmatrix} \Lambda_{0,0} & \cdots & \Lambda_{0,r} & \Lambda_{0,r+1} \\ \vdots & \ddots & \vdots & \vdots \\ \Lambda_{0,r}^T & \cdots & \Lambda_{r,r} & \Lambda_{r+1,r} \\ \Lambda_{0,r+1}^T & \cdots & \Lambda_{r+1,r}^T & \Lambda_{r+1,r+1} \end{bmatrix},$$

$$\Lambda_{0,0} = A_0 Y + Y A_0^T + \gamma^{-2} G_0 G_0^T,$$
$$\Lambda_{0,k} = Y A_k^T + \gamma^{-2} G_0 G_k^T, \ 1 \le k \le r,$$
$$\Lambda_{0,r+1} = Y C_z^T + \gamma^{-2} G_0 D_{zv}^T,$$
$$\Lambda_{l,k} = \gamma^{-2} G_l G_k^T, \ 1 \le l \ne k \le r,$$
$$\Lambda_{l,l} = \gamma^{-2} G_l G_l^T - X^{-1},$$
$$\Lambda_{l,r+1} = \gamma^{-2} G_l D_{zv}^T, \ 1 \le l \le r,$$
$$\Lambda_{r+1,r+1} = -I_{p_1} + \gamma^{-2} D_{zv} D_{zv}^T.$$

Proof. (i) $\Rightarrow$ (ii). If (i) in the statement is fulfilled, then using the implication (i) $\Rightarrow$ (ii) of Theorem 9.3.5 we deduce that there exist $X, Y \in \mathcal{S}_n$, $S \in \mathcal{S}_{n_c}$, $N \in \mathbf{R}^{n \times n_c}$ such that (9.55)–(9.57) are satisfied. One can see that (9.55) is just (9.63). On the other hand (9.57) leads to:

$$X = (Y - NS^{-1}N^T)^{-1}. \tag{9.65}$$

This means that X is the $(1,1)$ block of the matrix

$$\begin{bmatrix} Y & N \\ N^T & S \end{bmatrix}^{-1}.$$

Applying Lemma 9.3.6 for

$$X_{cl} = \begin{bmatrix} Y & N \\ N^T & S \end{bmatrix}^{-1}$$

it follows that

$$X - Y^{-1} \geq 0 \qquad\qquad (9.66)$$

and

$$rank\left(X - Y^{-1}\right) \leq n_c. \qquad\qquad (9.67)$$

It is obvious that (9.66) is equivalent with (9.61) and (9.67) is equivalent with (9.62). But (9.56) leads to

$$
\begin{aligned}
& U_0^T\left(A_0 Y + Y A_0^T\right) U_0 + \sum_{k=0}^r \sum_{l=0}^r \gamma^{-2} U_k^T G_k G_l^T U_k \\
& + \sum_{k=1}^r \left(U_0^T Y A_k^T U_k + U_k^T A_k Y U_0\right) + U_0^T Y C_z^T U_{r+1} \\
& + U_{r+1} C_z Y U_0 + \sum_{k=0}^r \gamma^{-2}\left(U_k^T G_k D_{zv}^T U_{r+1} + U_{r+1}^T D_{zv} G_k^T U_k\right) \\
& - U_{r+1}^T\left(I_{p_1} - \gamma^{-2} D_{zv} D_{zv}^T\right) U_{r+1} - \sum_{k=1}^r U_k^T\left(Y - N S^{-1} N^T\right) U_k < 0.
\end{aligned}
\qquad (9.68)
$$

Using (9.65), (9.68) becomes (9.64). Therefore there exist $X, Y \in S_n, X > 0, Y > 0$ verifying (9.61)–(9.64). Suppose now that (ii) holds. From (9.61) one deduces that $X \geq Y^{-1} > 0$ and $rank\left(X - Y^{-1}\right) \leq n_c$. Then according to Lemma 9.3.6 there exist $N \in \mathbf{R}^{n \times n_c}, M \in \mathbf{R}^{n \times n_c}, \tilde{X} \in \mathbf{R}^{n_c \times n_c}, S \in \mathbf{R}^{n_c \times n_c}$ such that

$$
\begin{bmatrix} X & M \\ M^T & \tilde{X} \end{bmatrix} = \begin{bmatrix} Y & N \\ N^T & S \end{bmatrix}^{-1} > 0. \qquad (9.69)
$$

and therefore

$$X^{-1} = Y - N S^{-1} N^T. \qquad\qquad (9.70)$$

Thus (9.64) becomes (9.68) and therefore (9.56) holds. Moreover (9.69) and (9.70) implies (9.57). Taking into account that (9.63) is just (9.55) we conclude that if (ii) in the statement holds then the condition (ii) in Theorem 9.3.5 are also verified. Then the implication (ii) $\Rightarrow$ (i) in Theorem 9.3.5 shows that (i) in the statement is fulfilled and hence the proof ends.   □

Remark 9.3.5. In order to solve the system (9.61)–(9.64) one can suggest the following algorithm:

Step 1.   Solve (9.63) with respect to X.
Step 2.   Introduce X determined at Step 1 in (9.61), (9.62) and (9.64) and solve the obtained LMI system with respect to Y.

Consider now the particular case when in (9.52), $B_k = 0, k = 1, \ldots, r$. In this situation the base U becomes

$$
U = \begin{bmatrix} \tilde{U}_0 & 0 \\ 0 & I_{nr} \\ \tilde{U}_{r+1} & 0 \end{bmatrix}
$$

where

$$\begin{bmatrix} \tilde{U}_0 \\ \tilde{U}_{r+1} \end{bmatrix}$$

is a basis of the null subspace of the matrix $\begin{bmatrix} B_0^T & D_{zu}^T \end{bmatrix}$. Then condition (9.64) becomes

$$\begin{bmatrix} \tilde{\Pi}_{0,0} & \tilde{\Pi}_{0,1} & \cdots & \tilde{\Pi}_{0,r} \\ \tilde{\Pi}_{0,1}^T & \tilde{\Pi}_{1,1} & \cdots & \tilde{\Pi}_{1,r} \\ \vdots & \vdots & \ddots & \vdots \\ \tilde{\Pi}_{0,r}^T & \tilde{\Pi}_{1,1}^T & \cdots & \tilde{\Pi}_{r,r} \end{bmatrix} < 0 \qquad (9.71)$$

where

$$\tilde{\Pi}_{0,0} = \begin{bmatrix} \tilde{U}_0 \\ \tilde{U}_{r+1} \end{bmatrix}^T$$

$$\times \left(\begin{bmatrix} A_0 Y + Y A_0^T & Y C_z^T \\ C_z Y & -I_{p_1} \end{bmatrix} + \gamma^{-2} \begin{bmatrix} G_0 \\ D_{zv} \end{bmatrix} \begin{bmatrix} G_0^T & D_{zv}^T \end{bmatrix} \right)$$

$$\times \begin{bmatrix} \tilde{U}_0 \\ \tilde{U}_{r+1} \end{bmatrix}$$

$$\tilde{\Pi}_{0,l} = \tilde{U}_0^T Y A_l^T + \gamma^{-2} \begin{bmatrix} \tilde{U}_0 \\ \tilde{U}_{r+1} \end{bmatrix}^T \begin{bmatrix} G_0 \\ D_{zv} \end{bmatrix} G_l^T, \ 1 \leq l \leq r$$

$$\tilde{\Pi}_{0,k} = \gamma^{-2} G_l G_k^T, \ 1 \leq l \neq k \leq r$$

$$\tilde{\Pi}_{l,l} = \gamma^{-2} G_l G_l^T - X^{-1}, \ 1 \leq l \leq r.$$

By Schur complement arguments it results that (9.71) is equivalent with the extended inequality

$$\begin{bmatrix} \Lambda_{0,0}(Y,\gamma) & \tilde{U}_0^T Y A_1^T & \cdots & \tilde{U}_0^T Y A_r^T & \tilde{U}_0^T G_0 + \tilde{U}_{r+1}^T D_{zv} \\ A_1 Y \tilde{U}_0 & -X^{-1} & \cdots & 0 & G_1 \\ \vdots & \vdots & \ddots & \vdots & \vdots \\ A_r Y \tilde{U}_0 & 0 & \cdots & X^{-1} & G_r \\ G_0^T \tilde{U}_0 + D_{zv}^T \tilde{U}_{r+1} & G_1^T & \cdots & G_r^T & -\gamma^2 I_{m_1} \end{bmatrix} < 0$$

where

$$\Lambda_{0,0}(Y,\gamma) = \begin{bmatrix} \tilde{U}_0 \\ \tilde{U}_{r+1} \end{bmatrix}^T \begin{bmatrix} A_0 Y + Y A_0^T & Y C_z^T \\ C_z Y & -I_{p_1} \end{bmatrix} \begin{bmatrix} \tilde{U}_0 \\ \tilde{U}_{r+1} \end{bmatrix}.$$

Taking the Schur complement of $diag\left(-X^{-1},\dots,-X^{-1}\right)$ in the above inequality, one obtains:

$$\begin{bmatrix} \Lambda_{0,0}\,(Y,\gamma) & \tilde{U}_0^T G_0 + \tilde{U}_{r+1}^T D_{zv} \\ +\sum_{k=1}^r \tilde{U}_0 Y A_k^T X A_k Y \tilde{U}_0 & +\sum_{k=1}^r \tilde{U}_0^T Y A_k^T X G_k \\ G_0^T \tilde{U}_0 + D_{zv}^T \tilde{U}_{r+1} & -\gamma^2 I_{m_1} + \sum_{k=1}^r G_k^T X G_k \\ +\sum_{k=1}^r G_k^T X A_k Y \tilde{U}_0 & \end{bmatrix} < 0.$$

The above inequality together with (9.61)–(9.63) are the necessary and sufficient conditions derived in [81].

In the final part of this section we shall discuss two problems of robust stabilization with respect to parametric uncertainty.

Consider the system described by

$$\begin{aligned}
dx(t) &= \Big\{\left[A_0\left(\eta\left(t\right)\right)+\hat{G}_0\left(\eta\left(t\right)\right)\Delta_1\left(\eta\left(t\right)\right)\hat{C}\left(\eta\left(t\right)\right)\right]x(t) \\
&\quad + \left[B_0\left(\eta\left(t\right)\right)+\hat{B}_0\left(\eta\left(t\right)\right)\Delta_2\left(\eta\left(t\right)\right)\hat{D}\left(\eta\left(t\right)\right)\right]u(t)\Big\}dt \\
&\quad + \sum_{k=1}^r \Big\{\left[A_k\left(\eta\left(t\right)\right)+\hat{G}_k\left(\eta\left(t\right)\right)\Delta_1\left(\eta\left(t\right)\right)\hat{C}\left(\eta\left(t\right)\right)\right]x(t) \qquad (9.72) \\
&\quad + \left[B_k\left(\eta\left(t\right)\right)+\hat{B}_k\left(\eta\left(t\right)\right)\Delta_2\left(\eta\left(t\right)\right)\hat{D}\left(\eta\left(t\right)\right)\right]u(t)\Big\}dw_k(t) \\
y(t) &= C_0\left(\eta\left(t\right)\right)x(t)
\end{aligned}$$

where $x(t)\in\mathbf{R}^n$ denotes the state, $u(t)\in\mathbf{R}^m$ is the control variable, and $y\in\mathbf{R}^{p_2}$ is the measured output. The matrices $A_k(i)\in\mathbf{R}^{n\times n}, B_k(i)\in\mathbf{R}^{n\times m}, \hat{G}_k(i)\in\mathbf{R}^{n\times\hat{m}_1}, \hat{B}_k(i)\in\mathbf{R}^{n\times\tilde{m}_1}, 0\le k\le r, \hat{C}(i)\in\mathbf{R}^{\hat{p}_1\times n}, \hat{D}(i)\in\mathbf{R}^{\tilde{p}_1\times m}, C_0(i)\in\mathbf{R}^{p_2\times n}$ are known matrices and $\Delta_1\in\mathbf{R}^{\hat{m}_1\times\hat{p}_1}, \Delta_2\in\mathbf{R}^{\tilde{m}_1\times\tilde{p}_1}$ are unknown matrices describing the parametric uncertainty.

The robust stabilization problem which we address has the following statement: find a stabilizing controller of form (9.2) for the system (9.72) for arbitrary Δ_1,Δ_2 with $\max\left(\left|\Delta_1\right|,\left|\Delta_2\right|\right)<\rho$ for a prescribed $\rho>0$, where $\left|\Delta_l\right|=\max_{i\in\mathcal{D}}\left|\Delta_l\left(i\right)\right|, l=1,2$. The closed-loop system obtained when coupling the controller (9.2)–(9.72) has the following state-space representation

$$\begin{aligned}
dx(t) &= \Big\{\left[A_0\left(\eta\left(t\right)\right)+B_0\left(\eta\left(t\right)\right)D_c\left(\eta\left(t\right)\right)C_0\left(\eta\left(t\right)\right)\right]x(t) \\
&\quad + B_0\left(\eta\left(t\right)\right)C_c\left(\eta\left(t\right)\right)x_c(t)+\left[\hat{G}_0\left(\eta\left(t\right)\right)\Delta_1\left(\eta\left(t\right)\right)\hat{C}\left(\eta\left(t\right)\right)\right. \\
&\quad \left. +\hat{B}_0\left(\eta\left(t\right)\right)\Delta_2\left(\eta\left(t\right)\right)\hat{D}\left(\eta\left(t\right)\right)D_c\left(\eta\left(t\right)\right)C_0\left(\eta\left(t\right)\right)\right]x(t) \\
&\quad + \hat{B}_0\left(\eta\left(t\right)\right)\Delta_2\left(\eta\left(t\right)\right)\hat{D}\left(\eta\left(t\right)\right)C_c\left(\eta\left(t\right)\right)x_c(t)\Big\}dt \\
&\quad + \sum_{k=1}^r\Big\{\left[A_k\left(\eta\left(t\right)\right)+B_k\left(\eta\left(t\right)\right)D_c\left(\eta\left(t\right)\right)C_0\left(\eta\left(t\right)\right)\right]x(t) \qquad (9.73) \\
&\quad + B_k\left(\eta\left(t\right)\right)C_c\left(\eta\left(t\right)\right)x_c(t)+\left[\hat{G}_k\left(\eta\left(t\right)\right)\Delta_1\left(\eta\left(t\right)\right)\hat{C}\left(\eta\left(t\right)\right)\right. \\
&\quad \left. +\hat{B}_k\left(\eta\left(t\right)\right)\Delta_2\left(\eta\left(t\right)\right)\hat{D}\left(\eta\left(t\right)\right)D_c\left(\eta\left(t\right)\right)C_0\left(\eta\left(t\right)\right)\right]x(t) \\
&\quad + \hat{B}_k\left(\eta\left(t\right)\right)\Delta_2\left(\eta\left(t\right)\right)\hat{D}\left(\eta\left(t\right)\right)C_c\left(\eta\left(t\right)\right)x_c(t)\Big\}dw_k(t) \\
dx_c(t) &= \left[B_c\left(\eta\left(t\right)\right)C_0\left(\eta\left(t\right)\right)x(t)+A_c\left(\eta\left(t\right)\right)x_c(t)\right]dt.
\end{aligned}$$

Denoting

$$G_k(i) = [\hat{G}_k(i) \quad \hat{B}_k(i)] \in \mathbf{R}^{n \times (\hat{m}_1 + \tilde{m}_1)}, 0 \le k \le r$$

$$C_z(i) = \begin{bmatrix} \hat{C}(i) \\ 0 \end{bmatrix} \in \mathbf{R}^{(\hat{p}_1 + \tilde{p}_1) \times n} \tag{9.74}$$

$$D_{zu}(i) = \begin{bmatrix} 0 \\ \hat{D}(i) \end{bmatrix} \in \mathbf{R}^{(\hat{p}_1 + \tilde{p}_1) \times m}$$

$$\Delta = \begin{bmatrix} \Delta_1(i) & 0 \\ 0 & \Delta_2(i) \end{bmatrix},$$

the system (9.73) can be rewritten in a compact form as follows:

$$\begin{aligned}
d\xi(t) &= [A_{0cl}(\eta(t)) + G_{0cl}(\eta(t)) \Delta(\eta(t)) C_{cl}(\eta(t))] \xi(t) dt \\
&\quad + \sum_{k=1}^{r} [A_{kcl}(\eta(t)) + G_{kcl}(\eta(t)) \Delta(\eta(t)) C_{cl}(\eta(t))] \xi(t) dw_k(t)
\end{aligned} \tag{9.75}$$

where $A_{kcl}(i)$, are defined as in (9.4) and

$$G_{kcl}(i) = \begin{bmatrix} G_k(i) \\ 0 \end{bmatrix},$$

$$\begin{aligned}
C_{cl}(i) &= [C_z(i) + D_{zu}(i) D_c(i) C_0(i) \quad D_{zu}(i) C_c(i)] \\
&= \begin{bmatrix} \hat{C}(i) & 0 \\ \hat{D}(i) D_c(i) C_0(i) & \hat{D}(i) C_c(i) \end{bmatrix}, i \in \mathcal{D}.
\end{aligned}$$

Therefore the closed-loop system can be viewed as a perturbation of the system

$$d\xi(t) = A_{0cl}(\eta(t)) \xi(t) dt + \sum_{k=1}^{r} A_{kcl}(\eta(t)) \xi(t) dw_k(t)$$

obtained by coupling the controller (9.2) to the nominal system (9.72) obtained with $\Delta_1 = 0, \Delta_2 = 0$. Applying Corollary 8.3.5 to the system (9.75) it follows that a controller of type (9.2) stabilizes (9.72) for any Δ_1, Δ_2 with $\max(|\Delta_1|, |\Delta_2|) < \rho$ if the input–output operator $\mathcal{T}_{cl}$ associated with the fictious system:

$$\begin{aligned}
d\xi_{cl}(t) &= [A_{0cl}(\eta(t)) \xi(t) + G_{0cl}(\eta(t)) v(t)] dt \\
&\quad + \sum_{k=1}^{r} [A_{kcl}(\eta(t)) \xi(t) + G_{kcl}(\eta(t)) v(t)] dw_k(t) \\
z(t) &= C_{cl}(\eta(t)) \xi(t)
\end{aligned}$$

verifies the condition $\|\mathcal{T}_{cl}\| < 1/\rho$. Then a stabilizing controller (9.2) providing the robustness radius ρ can be obtained as a solution of the DAP with $\gamma = \rho^{-1}$ corresponding to the two-input, two-output generalized system

$$dx(t) = [A_0(\eta(t))x(t) + G_0(\eta(t))v(t) + B_0(\eta(t))u(t)]dt$$
$$+ \Sigma_{k=1}^{r}[A_k(\eta(t))x(t) + G_k(\eta(t))v(t) + B_k(\eta(t))u(t)]dw_k(t) \quad (9.76)$$
$$z(t) = C_z(\eta(t))x(t) + D_{zu}(\eta(t))u(t)$$
$$y(t) = C_0(\eta(t))x(t)$$

where $G_k(i), 0 \leq k \leq r, C_z(i), D_{zu}(i), i \in \mathcal{D}$ are defined as in (9.74). Then a robust stabilizing controller with the robustness radius ρ may be obtained applying Theorem 9.3.3 to the system (9.76) for $\gamma = \rho^{-1}$.

The second robust stabilization problem with respect to parametric uncertainty considered in the final part of this section is the following: find a stabilizing controller of type (9.2) for the system

$$dx(t) = [A_0(\eta(t))x(t) + G_0(\eta(t))\Delta(\varphi(t),\eta(t)) + B_0(\eta(t))u(t)]dt$$
$$+ \Sigma_{k=1}^{r}[A_k(\eta(t))x(t) + G_k(\eta(t))\Delta(\varphi(t),\eta(t)) \quad (9.77)$$
$$+ B_k(\eta(t))u(t)]dw_k(t)$$
$$y(t) = C_0(\eta(t))x(t) + D_0(\eta(t))v(t)$$

where $\varphi(t) = C(\eta(t))x(t)$ and Δ are unknown Lipschitz functions with $\Delta(0,i) = 0$ and

$$\sup_{i \in \mathcal{D}, z \in \mathbf{R}^{p_1}, z \neq 0} \frac{|\Delta(z,i)|}{|z|} < \rho. \quad (9.78)$$

When coupling a controller of type (9.2) to the system (9.77), the closed-loop system has the following state-space equation:

$$dx_{cl}(t) = [A_{0cl}(\eta(t))x_{cl}(t) + G_{0cl}(\eta(t))\Delta(\varphi(t),\eta(t))]dt$$
$$+ \Sigma_{k=1}^{r}[A_{kcl}(\eta(t))x_{cl}(t) + G_{kcl}(\eta(t))\Delta(\varphi(t),\eta(t))]dw_k(t) \quad (9.79)$$

where $A_{kcl}(i), G_{kcl}(i)$ are defined as in (9.4), $0 \leq k \leq r$. Invoking Theorem 8.3.8 for the system (9.79), it follows that a controller (9.28) stabilizes (9.77) for any nonlinear perturbation Δ satisfying (9.78) if $\rho < 1/\|\mathcal{T}_{cl}\|$ where $\mathcal{T}_{cl}$ is the input–output operator of the system:

$$d\xi(t) = [A_{0cl}(\eta(t))\xi(t) + G_{0cl}(\eta(t))v(t)]dt$$
$$+ \Sigma_{k=1}^{r}[A_{kcl}(\eta(t))\xi(t) + G_{kcl}(\eta(t))v(t)]dw_k(t) \quad (9.80)$$
$$z(t) = [C(\eta(t)) \quad 0]\xi(t).$$

Hence a robust stabilizing controller for (9.77) can be obtained by solving the DAP for $\gamma = 1/\rho$ for the system

$$dx(t) = [A_0(\eta(t))x(t) + G_0(\eta(t))v(t) + B_0(\eta(t))u(t)]dt$$
$$+ \Sigma_{k=1}^{r}[A_k(\eta(t))x(t) + G_k(\eta(t))v(t)$$
$$+ B_k(\eta(t))u(t)]dw_k(t) \quad (9.81)$$
$$z(t) = C(\eta(t))x(t)$$
$$y(t) = C_0(\eta(t))x(t) + D_0(\eta(t))v(t).$$

Solvability conditions for this DAP are provided by Theorem 9.3.3.

9.4 DAP for Linear Stochastic Systems with Markovian Jumping

In this section we shall investigate the γ-attenuation problem for linear stochastic systems of form (9.47) looking for strictly proper n-order controllers with $D_c(i) = 0$, $i \in \mathcal{D}$. More precisely, the class of considered controllers is given by

$$
\begin{aligned}
\dot{x}_c(t) &= A_c(\eta(t))x_c(t) + B_c(\eta(t))y(t) \\
u(t) &= C_c(\eta(t))x_c(t)
\end{aligned}
\tag{9.82}
$$

where $A_c(i) \in \mathbf{R}^{n \times n}, B_c(i) \in \mathbf{R}^{n \times p_2}, C_c(i) \in \mathbf{R}^{m_2 \times n}, i \in \mathcal{D}$. When coupling the controller (9.82) to the system (9.47) one obtains:

$$
\begin{aligned}
\dot{x}_{cl}(t) &= A_{cl}(\eta(t))x_{cl}(t) + G_{cl}(\eta(t))v(t) \\
z(t) &= C_{cl}(\eta(t))x_{cl}(t) + D_{cl}(\eta(t))v(t),
\end{aligned}
$$

where

$$
\begin{aligned}
A_{cl}(i) &= \begin{bmatrix} A_0(i) & B_0(i)C_c(i) \\ B_c(i)C_0(i) & A_c(i) \end{bmatrix} \\
G_{cl}(i) &= \begin{bmatrix} G_0(i) \\ B_c(i)D_0(i) \end{bmatrix} \\
C_{cl}(i) &= \begin{bmatrix} C_z(i) & D_{zu}(i)C_c(i) \end{bmatrix} \\
D_{cl}(i) &= D_{zv}(i).
\end{aligned}
\tag{9.83}
$$

The following result provides necessary and sufficient conditions which guarantee the existence of a solution of form (9.82) of the DAP.

Theorem 9.4.1. *For $\gamma > 0$ the following are equivalent*

(i) *There exists a controller of form (9.82) stabilizing (9.47) and solving the DAP with the level of attenuation γ;*
(ii) *There exist $X = (X(1), \ldots, X(d)) \in \mathcal{S}_n^d, Y = (Y(1), \ldots, Y(d)) \in \mathcal{S}_n^d, F = (F(1), \ldots, F(d)) \in \mathcal{M}_{m_2 n}^d, K = (K(1), \ldots, K(d)) \in \mathcal{M}_{n p_2}^d$ which verify*

$$
X(i) > 0,
$$

$$
V(i) = \begin{bmatrix} V_{11}(i) & V_{12}(i) \\ V_{12}^*(i) & V_{22}(i) \end{bmatrix} < 0
\tag{9.84}
$$

$$
W(i) = \begin{bmatrix} W_{11}(i) & W_{12}(i) & W_{13}(i) \\ W_{12}^*(i) & W_{22}(i) & 0 \\ W_{13}^*(i) & 0 & W_{33}(i) \end{bmatrix} < 0
\tag{9.85}
$$

$$\begin{bmatrix} Y(i) \, I_n \\ I_n \quad X(i) \end{bmatrix} > 0, \tag{9.86}$$

where

$$V_{11}(i) = A_0^T(i) X(i) + X(i) A_0(i) + K(i) C_0(i) + C_0^T(i) K^T(i)$$

$$+ \sum_{j=1}^{d} q_{ij} X(j) + C_z^T(i) C_z(i)$$

$$V_{12}(i) = X(i) G_0(i) + K(i) D_0(i) + C_z^T(i) D_{zv}(i)$$

$$V_{22}(i) = -\gamma^2 I_{m_1} + D_{zv}^*(i) D_{zv}(i)$$

and

$$W_{11}(i) = A_0(i) Y(i) + Y(i) A_0^T(i) + B_0(i) F(i) + F^T(i) B_0^T(i)$$

$$+ q_{ii} Y(i) + \gamma^{-2} G_0(i) G_0^T(i)$$

$$W_{12}(i) = Y(i) C_z^T(i) + F^T(i) D_{zu}^T(i) + \gamma^{-2} G_0(i) D_{zv}^T(i)$$

$$W_{13}(i) = \left[\sqrt{q_{i,1}} Y(i) \dots \sqrt{q_{i,i-1}} Y(i) \ \sqrt{q_{i,i+1}} Y(i) \dots \sqrt{q_{i,d}} Y(i) \right]$$

$$W_{22}(i) = -I_{p_1} + \gamma^2 D_{zv}(i) D_{zv}^T(i)$$

$$W_{33}(i) = -diag\left(Y(1) \dots Y(i-1) \ Y(i+1) \dots Y(d) \right),$$

Moreover, if (9.84)–(9.86) are feasible, then a controller of the form (9.82) is given by

$$A_c(i) = \left[X(i) - Y^{-1}(i) \right]^{-1} \left\{ A_0^T(i) + X(i) A_0(i) Y(i) + X(i) B_0(i) F(i) \right.$$

$$+ K(i) C_0(i) Y(i) + C_z^T(i) \left[C_z(i) Y(i) + D_{zu}(i) F(i) \right]$$

$$+ \left[X(i) G_0(i) + K(i) D_0(i) + C_z^T(i) D_{zv}(i) \right] \left[\gamma^2 I_{m_1} - D_{zv}^T(i) D_{zv}(i) \right]^{-1}$$

$$\times \left[G_0^T(i) + D_{zv}^T(i) C_z(i) Y(i) + D_{zv}^T(i) D_{zu}(i) F(i) \right]$$

$$+ \left. \sum_{j=1}^{d} q_{ij} Y(i) Y^{-1}(j) \right\} Y^{-1}(i) \tag{9.87}$$

$$B_c(i) = \left[Y^{-1}(i) - X(i) \right]^{-1} K(i)$$

$$C_c(i) = F(i) Y^{-1}(i).$$

Proof. (i) ⇒ (ii) Assume that there exists a controller of the form (9.82) such that the zero solution of the system (9.83) for $v(t) = 0$ is ESMS and $\|\mathcal{T}_{cl}\| < \gamma$ where $\mathcal{T}_{cl}$

denotes the input–output operator associated with (9.83). Applying Corollary 8.2.12 for the system (9.83) we deduce that there exists $X_{cl} = (X_{cl}(1), \ldots, X_{cl}(d)) \in \mathcal{S}_{2n}^d$, $X_{cl}(i) > 0$, $i \in \mathcal{D}$ such that:

$$\Pi_i(X_{cl}) = \begin{bmatrix} \Pi_{i,11}(X_{cl}) & \Pi_{i,12}(X_{cl}) \\ \Pi_{i,12}^T(X_{cl}) & \Pi_{i,22}(X_{cl}) \end{bmatrix} < 0 \tag{9.88}$$

where we denoted:

$$\Pi_{i,11}(X_{cl}) = A_{cl}^T(i) X_{cl}(i) + X_{cl}(i) A_{cl}(i) + \sum_{j=1}^d q_{ij} X_{cl}(j)$$

$$+ C_{cl}^T(i) C_{cl}(i)$$

$$\Pi_{i,12}(X_{cl}) = X_{cl}(i) G_{cl}(i) + C_{cl}^T(i) D_{cl}(i)$$

$$\Pi_{i,22}(X_{cl}) = -\gamma^2 I_{m_1} + D_{cl}^T(i) D_{cl}(i).$$

By a Schur complement reasoning, (9.88) leads to the following two conditions:

$$\begin{aligned} & A_{cl}^T(i) X_{cl}(i) + X_{cl}(i) A_{cl}(i) + \sum_{j=1}^d q_{ij} X_{cl}(j) + C_{cl}^T(i) C_{cl}(i) \\ & + \left[X_{cl}(i) G_{cl}(i) + C_{cl}^T(i) D_{cl}(i) \right] \left[\gamma^2 I_{m_1} - D_{cl}^T(i) D_{cl}(i) \right]^{-1} \\ & \times \left[G_{cl}^T(i) X_{cl}(i) + D_{cl}^T(i) C_{cl}(i) \right] < 0 \end{aligned} \tag{9.89}$$

$$\gamma^2 I_{m_1} - D_{cl}^T(i) D_{cl}(i) > 0. \tag{9.90}$$

Consider the following partition of $X_{cl}(i)$:

$$X_{cl}(i) = \begin{bmatrix} X(i) & M(i) \\ M^T(i) & \tilde{X}(i) \end{bmatrix}$$

and

$$X_{cl}^{-1}(i) = \begin{bmatrix} Y(i) & N(i) \\ N^T(i) & S(i) \end{bmatrix}$$

where $X(i), Y(i) \in \mathcal{S}_n^d$, and $M(i), N(i) \in \mathbf{R}^{n \times n}$. Without losing the generality one can assume that $M(i)$ is invertible for every $i \in \mathcal{D}$. Indeed, if $M(i)$ is not invertible for some $i \in \mathcal{D}$, then one can replace X_{cl} by

$$X_\varepsilon = X_{cl} + \begin{bmatrix} 0 & \varepsilon I_n \\ \varepsilon I_n & 0 \end{bmatrix} \text{ with some } \varepsilon > 0$$

such that $X_\varepsilon > 0$, $\Pi_i(X_\varepsilon) < 0$ for all $i \in \mathcal{D}$ and in addition $M_\varepsilon(i) = M(i) + \varepsilon I_n$ is invertible for every $i \in \mathcal{D}$. Since $X(i)N(i) + M(i)S(i) = 0$ it follows that $N(i) = -X^{-1}(i)M(i)S(i)$ and then $N(i)$ is invertible, too. Let us define

$$T(i) = \begin{bmatrix} Y(i) & I_n \\ N^T(i) & 0 \end{bmatrix}.$$

It is obvious that $T(i)$ is invertible and

$$T^{-1}(i) = \begin{bmatrix} 0 & \left(N^{-1}(i)\right)^T \\ I_n & -Y(i)\left(N^{-1}(i)\right)^T \end{bmatrix}.$$

Then we have:

$$T^T(i)X_{cl}(i) = \begin{bmatrix} I_n & 0 \\ X(i) & M(i) \end{bmatrix} \tag{9.91}$$

and

$$T^T(i)X_{cl}(i)T(i) = \begin{bmatrix} Y(i) & I_n \\ I_n & X(i) \end{bmatrix}. \tag{9.92}$$

From (9.91) together with $X_{cl}(i) > 0$ give (9.86). By pre and post multiplication of (9.89) by $T^T(i)$ and $T(i)$, respectively, one obtains:

$$T^T(i)\hat{\Pi}_i(X_{cl})T(i) < 0. \tag{9.93}$$

where $\hat{\Pi}_i(X_{cl})$ is the left-hand side of the inequality (9.89). Let

$$\Lambda(i) = T^T(i)\hat{\Pi}_i(X_{cl})T(i) = \begin{bmatrix} \Lambda_{11}(i) & \Lambda_{21}^T(i) \\ \Lambda_{21}(i) & \Lambda_{22}(i) \end{bmatrix},$$

where by direct computations, based on (9.89)–(9.92), we have

$$\Lambda_{11}(i) = A_0(i)Y(i) + B_0(i)C_c(i)N^T(i) + Y(i)A_0^T(i) + N(i)C_c^T(i)B_0^T(i)$$
$$+ \left[G_0(i) + \left(Y(i)C_z^T(i) + N(i)C_c^T(i)D_{zu}^T(i)\right)D_{zv}(i)\right]$$
$$\times \left[\gamma^2 I_{m_1} - D_{zv}^T(i)D_{zv}(i)\right]^{-1}$$
$$\times \left[G_0^T(i) + D_{zv}^T(i)\left(C_z(i)Y(i) + D_{zu}(i)C_c(i)N^T(i)\right)\right]$$
$$+ \left[Y(i)C_z^T(i) + N(i)C_c^T(i)D_{zu}^T(i)\right]\left[C_z(i)Y(i) + D_{zu}(i)C_c(i)N^T(i)\right]$$
$$+ \Sigma_{j=1}^d q_{ij}\left[Y(i)X(j)Y(i) + N(i)M^T(j)Y(i) + Y(i)M(j)N^T(i)\right.$$
$$\left. + N(i)\tilde{X}(j)N^T(i)\right]$$
$$\Lambda_{21}(i) = A_0^T(i) + X(i)A_0(i)Y(i) + X(i)B_0(i)C_c(i)N^T(i) + M(i)B_c(i)C_0(i)Y(i)$$
$$+ M(i)A_c(i)N^T(i) + \left[X(i)G_0(i) + M(i)B_c(i)D_0^T(i) + C_z^T(i)D_{zv}(i)\right]$$
$$\times \left[\gamma^2 I_{m_1} - D_{zv}^T(i)D_{zv}(i)\right]^{-1}$$
$$\times \left[G_0^T(i) + D_{zv}^T(i)C_z(i)Y(i) + D_{zv}^T(i)D_{zu}(i)C_c(i)N^*(i)\right]$$
$$+ C_z^T(i)\left[C_z(i)Y(i) + D_{zu}(i)C_c(i)N^T(i)\right]$$
$$+ \Sigma_{j=1}^d q_{ij}\left[Y(i)X(j) + N(i)M^T(j)\right]$$
$$\Lambda_{22}(i) = A_0^T(i)X(i) + X(i)A_0(i) + M(i)B_c(i)C_0(i) + C_0^T(i)B_c^T(i)M(i)$$
$$+ \left[X(i)G_0(i) + M(i)B_c(i)D_0^T(i) + C_z^T(i)D_{zv}(i)\right]$$
$$\times \left[\gamma^2 I_{m_1} - D_{zv}^T(i)D_{zv}(i)\right]^{-1}$$
$$\times \left[G_0^T(i)X(i) + D_0(i)B_c^T(i)M^T(i) + D_{zv}^T(i)C_z(i)\right]$$
$$+ C_z^T(i)C_z(i) + \Sigma_{j=1}^d q_{ij}X(j).$$

Let us introduce the following notations

$$K(i) = M(i)B_c(i)$$
$$F(i) = C_c(i)N^T(i).$$

Thus one obtains

$$\Lambda_{11}(i) = A_0(i)Y(i) + Y(i)A_0^T(i) + B_0(i)F(i) + F^T(i)B_0^T(i)$$
$$+ \left[G_0(i) + \left(Y(i)C_z^T(i) + F^T(i)D_{zu}^T(i)\right)D_{zv}(i)\right]$$
$$\times \left[\gamma^2 I_{m_1} - D_{zv}^T(i)D_{zv}(i)\right]^{-1} \qquad\qquad (9.94)$$
$$\times \left[G_0^T(i) + D_{zv}^T(i)\left(C_z(i)Y(i) + D_{zu}(i)F(i)\right)\right]$$
$$+ \left[Y(i)C_z^T(i) + F^T(i)D_{zu}^T(i)\right]\left[C_z(i)Y(i) + D_{zu}(i)F(i)\right]$$
$$+ \Sigma_{j=1}^d q_{ij}\left[Y(i)X(j)Y(i) + N(i)M^T(j)Y(i) + Y(i)M(j)N^T(i)\right.$$
$$\left. + N(i)\tilde{X}(j)N^T(i)\right]$$

$$\Lambda_{21}(i) = A_0^T(i) + X(i)A_0(i)Y(i) + X(i)B_0(i)F(i) + K(i)C_0(i)Y(i)$$
$$+ M(i)A_c(i)N^T(i) + \left[X(i)G_0(i) + K(i)D_0^T(i) + C_z^T(i)D_{zv}(i)\right]$$
$$\times \left[\gamma^2 I_{m_1} - D_{zv}^T(i)D_{zv}(i)\right]^{-1} \qquad\qquad (9.95)$$
$$\times \left[G_0^T(i) + D_{zv}^T(i)C_z(i)Y(i) + D_{zv}^T(i)D_{zu}(i)F(i)\right]$$
$$+ C_z^T(i)\left[C_z(i)Y(i) + D_{zu}(i)F(i)\right]$$
$$+ \Sigma_{j=1}^d q_{ij}\left[Y(i)X(j) + N(i)M^T(j)\right]$$

$$
\begin{aligned}
\Lambda_{22}(i) = {}& A_0^T(i)X(i) + X(i)A_0(i) + K(i)C_0(i) + C_0^T(i)K^T(i) \\
& + \left[X(i)G_0(i) + K(i)D_0^T(i) + C_z^T(i)D_{zv}(i) \right] \\
& \times \left[\gamma^2 I_{m_1} - D_{zv}^T(i)D_{zv}(i) \right]^{-1} \\
& \times \left[G_0^T(i)X(i) + D_0(i)K^T(i) + D_{zv}^T(i)C_z(i) \right] \\
& + \sum_{j=1}^d q_{ij}X(j) + C_z^T(i)C_z(i).
\end{aligned}
\tag{9.96}
$$

The condition (9.93) leads to

$$
\Lambda_{11}(i) < 0 \tag{9.97}
$$

$$
\Lambda_{22}(i) < 0. \tag{9.98}
$$

Using (9.96) and (9.98), by a Schur complement argument (9.84) directly follows. On the other hand we may write

$$
\begin{aligned}
& Y(i)X(j)Y(i) + N(i)M^T(j)Y(i) + Y(i)M(j)N^T(i) + N(i)\tilde{X}(j)N^T(i) \\
& = Y(i)\left[X(j) - M(j)\tilde{X}^{-1}(j)M^T(j) \right]Y(i) + Y(i)M(j)\tilde{X}^{-1}(j)M^T(j)Y(i) \\
& \quad + N(i)M^T(j)Y(i) + Y(i)M(j)N^T(i) + N(i)\tilde{X}(j)N^T(i) \\
& = Y(i)Y^{-1}(j)Y(i) + \left[Y(i)M(j) + N(i)\tilde{X}(j) \right]\tilde{X}^{-1}(j) \\
& \quad \times \left[M^T(j)Y(i) + \tilde{X}(j)N^T(i) \right].
\end{aligned}
$$

Then (9.97) and (9.94) leads to

$$
\begin{aligned}
& A_0(i)Y(i) + Y(i)A_0^T(i) + B_0(i)F(i) + F_0^T(i)B_0^T(i) \\
& + \left[G_0(i) + \left(Y(i)C_z^T(i) + F^T(i)D_{zu}^T(i) \right)D_{zv}(i) \right] \\
& \times \left[\gamma^2 I_{m_1} - D_{zv}^T(i)D_{zv}(i) \right]^{-1} \\
& \times \left[G_0^T(i) + D_{zv}^T(i)\left(C_z(i)Y(i) + D_{zu}(i)F(i) \right) \right] \\
& + \left[Y(i)C_z^T(i) + F^T(i)D_{zu}^T(i) \right]\left[C_z(i)Y(i) + D_{zu}(i)F(i) \right] \\
& + \sum_{j=1}^d q_{ij}Y(i)Y^{-1}(j)Y(i) < 0.
\end{aligned}
\tag{9.99}
$$

Using again Schur complement arguments one can easily see that the above inequality together with (9.90) implies (9.85) in the statement. Thus the implication (i) $\Rightarrow$ (ii) is proved.

(ii) $\Rightarrow$ (i) Assume that there exist $X(i) > 0$, $Y(i) > 0$, $F(i)$, $K(i)$, $i \in \mathcal{D}$ verifying (9.84)–(9.86). From (9.86) we obtain that $X(i) - Y^{-1}(i) > 0$. Consider

$$
X_{cl}(i) = \begin{bmatrix} X(i) & Y^{-1}(i) - X(i) \\ Y^{-1}(i) - X(i) & X(i) - Y^{-1}(i) \end{bmatrix}.
$$

Then we have

$$X(i) - \left(Y^{-1}(i) - X(i)\right)\left(X(i) - Y^{-1}(i)\right)^{-1}\left(Y^{-1}(i) - X(i)\right)$$
$$= X(i) + Y^{-1}(i) - X(i) = Y^{-1}(i) > 0.$$

Therefore $X_{cl}(i) > 0$. Using (9.87) one obtains the closed-loop system

$$\dot{x}_{cl}(t) = \tilde{A}_{cl}(\eta(t))x_{cl}(t) + \tilde{G}_{cl}(\eta(t))v(t)$$
$$z(t) = \tilde{C}_{cl}(\eta(t))x_{cl}(t) + \tilde{D}_{cl}(\eta(t))v(t)$$

with the coefficients defined as in (9.83). Let

$$\tilde{\Pi}_i(X_{cl}) = \begin{bmatrix} \tilde{\Pi}_{i,11}(X_{cl}) & \tilde{\Pi}_{i,12}(X_{cl}) \\ \tilde{\Pi}_{i,12}^*(X_{cl}) & \tilde{\Pi}_{i,22}(X_{cl}) \end{bmatrix}$$

where

$$\tilde{\Pi}_{i,11}(X_{cl}) = \tilde{A}_{cl}^T(i)X_{cl}(i) + X_{cl}(i)\tilde{A}_{cl}(i) + \sum_{j=1}^{d} q_{ij}X_{cl}(j)$$
$$+ \tilde{C}_{cl}^T(i)\tilde{C}_{cl}(i)$$
$$\tilde{\Pi}_{i,12}(X_{cl}) = X_{cl}(i)\tilde{G}_{cl}(i) + \tilde{C}_{cl}^T(i)\tilde{D}_{cl}(i)$$
$$\tilde{\Pi}_{i,22}(X_{cl}) = -\gamma^2 I_{m_1} + \tilde{D}_{cl}^T(i)\tilde{D}_{cl}(i).$$

Then for

$$\tilde{T}(i) = \begin{bmatrix} Y(i) & I_n \\ Y(i) & 0 \end{bmatrix}$$

direct computations give:

$$\tilde{T}^*(i)\overline{\Pi}_i(X_{cl})\tilde{T}(i) = \begin{bmatrix} \tilde{\Lambda}_{11}(i) & 0 \\ 0 & \tilde{\Lambda}_{22}(i) \end{bmatrix}$$

where

$$\overline{\Pi}_i(X_{cl}) = \tilde{\Pi}_{1,11}(X_{cl}) + \tilde{\Pi}_{1,12}(X_{cl})\left(\gamma^2 I_{m_1} - \tilde{D}_{cl}^T(i)\tilde{D}_{cl}(i)\right)^{-1}$$
$$\times \tilde{\Pi}_{1,11}(X_{cl}),$$
$$\tilde{\Lambda}_{11}(i) = A_0(i)Y(i) + Y(i)A_0^T(i) + B_0(i)F(i) + F_0^*(i)B_0^*(i)$$
$$+ \left[G_0(i) + \left(Y(i)C_z^T(i) + F^T(i)D_{zu}^T(i)\right)D_{zv}(i)\right]$$
$$\times \left[\gamma^2 I_{m_1} - D_{zv}^T(i)D_{zv}(i)\right]^{-1}$$
$$\times \left[G_0^T(i) + D_{zv}^T(i)\left(C_z(i)Y(i) + D_{zu}(i)F(i)\right)\right]$$
$$+ \left[Y(i)C_z^T(i) + F^*(i)D_{zu}^T(i)\right]\left[C_z(i)Y(i) + D_{zu}(i)F(i)\right]$$
$$+ \Sigma_{j=1}^{d} q_{ij}Y(i)Y^{-1}(j)Y(i)$$

and $\tilde{\Lambda}_{22}(i) = \Lambda_{22}(i)$ defined in (9.96). From (9.84) and (9.85) by Schur complement arguments it follows that

$$\tilde{\Lambda}_{11}(i) < 0$$

$$\tilde{\Lambda}_{22}(i) < 0,$$

respectively, and therefore $\overline{\Pi}_i(X_{cl}) < 0$. Moreover, from (9.84) it results that $\gamma^2 I_{m_1} - D_{zv}^T(i) D_{zv}(i) > 0$ which coincides with the condition $\gamma^2 I_{m_1} - D_{cl}^T(i) D_{cl}(i) > 0$. This last condition together with $\overline{\Pi}_i(X_{cl}) < 0$ leads to an inequality of the form (9.88) for $\tilde{\Pi}_i(X_{cl})$ which shows that the controller (9.87) is a solution of the DAP and thus the proof ends.   □

9.5   An H_∞-Type Filtering Problem for Signals Corrupted with Multiplicative White Noise

In this section we consider a particular filtering problem in which the measured output is subject to multiplicative white noise. Its solution is derived via an H_∞-type method based on the Bounded Real Lemma version proved in Corollary 8.2.13.

Consider the following linear stable system

$$dx(t) = [Ax(t) + Bu(t)] dt \tag{9.100}$$

$$dy_1(t) = C_1 x(t) (dt + \sigma dw(t))$$

$$y_2(t) = C_2 x(t)$$

where $x(t) \in \mathbf{R}^n$ denotes the state, $u(t) \in \mathbf{R}^m$ is an input variable, $y_1 \in \mathbf{R}^{p_1 \times n}$ denotes the measured output, $y_2 \in \mathbf{R}^{p_2 \times n}$ is a quality output, $\sigma \in \mathbf{R}$ and $w(t)$ is a scalar standard Wiener process. Given $\gamma > 0$, the problem consists in determining an n_f-order deterministic filter where $n_f > 0$ is given, with the input y_1 and the output $y_f \in \mathbf{R}^{p_2}$, having the state-space equations:

$$\dot{x}_f(t) = A_f x_f(t) + B_f y_1(t) \tag{9.101}$$

$$y_f(t) = C_f x_f(t)$$

such that the resulting system obtained by coupling it to (9.100) is ESMS and the input–output operator:

$$\mathcal{T} : L_w^2([0, \infty), \mathbf{R}^m) \to L_w^2([0, \infty), \mathbf{R}^{p_2})$$

by $u \mapsto z$, where $z(t) = y_2(t) - y_f(t)$, has the norm less than γ.

The solution of this problem is provided by the following theorem.

Theorem 9.5.1. *The filtering problem has a solution if and only if there exist the matrices* $P, X \in \mathcal{S}_n$, $\tilde{X} \in \mathcal{S}_{n_f}$, $P > 0, X > 0, \tilde{X} > 0$ *and* $\tilde{M} \in \mathbf{R}^{n \times n_f}$ *such that*

$$\begin{bmatrix} A^T P + PA + \sigma^2 C_1^T U^T \tilde{X} U C_1 & PB \\ B^T P & -\gamma^2 I \end{bmatrix} < 0 \qquad (9.102)$$

$$\begin{bmatrix} A^T X + XA + MUC_1 + C_1^T U^T M^T \\ + \sigma^2 C_1^T U^T \tilde{X} U C_1 + C_2^T C_2 & XB \\ B^T X & -\gamma^2 I \end{bmatrix} < 0 \qquad (9.103)$$

$$\begin{bmatrix} X & M \\ M^T & \tilde{X} \end{bmatrix} > 0 \qquad (9.104)$$

$$rank\left(\begin{bmatrix} P - X & M \\ M^T & -\tilde{X} \end{bmatrix} \right) = n_f, \qquad (9.105)$$

where

$$U = \begin{cases} \begin{bmatrix} I_{p_1} \\ 0_{(n_f - p_1) \times p_1} \end{bmatrix} & \text{if } n_f \geq p_1 \text{ and} \\ \begin{bmatrix} I_{n_f} & 0_{n_f \times (p_1 - n_f)} \end{bmatrix} & \text{if } n_f < p_1. \end{cases} \qquad (9.106)$$

Proof. When coupling the filter (9.101) to the system (9.100) one obtains the resulting system:

$$dx(t) = [Ax(t) + Bu(t)] dt$$
$$dx_f(t) = [A_f x_f(t) + C_1 x(t)] dt + \sigma B_f C_1 x(t) dw(t)$$
$$z(t) = C_2 x(t) - C_f x_f(t)$$

or equivalently

$$d \begin{bmatrix} x(t) \\ x_f(t) \end{bmatrix} = \left(\begin{bmatrix} A & 0 \\ B_f C_1 & A_f \end{bmatrix} \begin{bmatrix} x(t) \\ x_f(t) \end{bmatrix} + \begin{bmatrix} B \\ 0 \end{bmatrix} u(t) \right) dt$$

$$+ \begin{bmatrix} 0 & 0 \\ \sigma B_f C_1 & 0 \end{bmatrix} \begin{bmatrix} x(t) \\ x_f(t) \end{bmatrix} dw(t) \qquad (9.107)$$

$$z(t) = \begin{bmatrix} C_2 & -C_f \end{bmatrix} \begin{bmatrix} x(t) \\ x_f(t) \end{bmatrix}.$$

Let us introduce the following notations:

$$\mathcal{A}_0 = \begin{bmatrix} A & 0 \\ B_f C_1 & A_f \end{bmatrix}; \mathcal{A}_1 = \begin{bmatrix} 0 & 0 \\ \sigma B_f C_1 & 0 \end{bmatrix}; \tag{9.108}$$

$$\mathcal{B}_0 = \begin{bmatrix} B_0 \\ 0 \end{bmatrix}; \mathcal{C} = \begin{bmatrix} C_2 & -C_f \end{bmatrix}.$$

Applying Theorem 8.2.7 of Chap. 8 for the resulting system (9.107), it follows that it is ESMS and its associated input–output operator has the norm less than γ if and only if there exists $\mathcal{X} > 0$ such that

$$\begin{bmatrix} \mathcal{A}_0^T \mathcal{X} + \mathcal{X} \mathcal{A}_0 + \mathcal{A}_1^T \mathcal{X} \mathcal{A}_1 + \mathcal{C}^T \mathcal{C} & \mathcal{X} \mathcal{B}_0 \\ \mathcal{B}_0^T \mathcal{X} & -\gamma^2 I \end{bmatrix} < 0. \tag{9.109}$$

Further consider the partition of $\mathcal{X}$

$$\mathcal{X} = \begin{bmatrix} X & M \\ M^T & \tilde{X} \end{bmatrix},$$

where $X \in \mathbf{R}^{n \times n}, \tilde{X} \in \mathbf{R}^{n_f \times n_f}$ and $M \in \mathbf{R}^{n \times n_f}$. Then using (9.108), the condition (9.109) becomes:

$$\mathcal{N}(X,M,\tilde{X},A_f,B_f,C_f) = \begin{bmatrix} \mathcal{N}_{11} & \mathcal{N}_{12} & \mathcal{N}_{13} & 0 \\ \mathcal{N}_{12}^T & \mathcal{N}_{22} & \mathcal{N}_{23} & \mathcal{N}_{24} \\ \mathcal{N}_{13}^T & \mathcal{N}_{23}^T & -\gamma^2 I_m & 0 \\ 0 & \mathcal{N}_{24}^T & 0 & -I_{p_2} \end{bmatrix} < 0 \tag{9.110}$$

where

$$\mathcal{N}_{11} = A^T X + X A + M B_f C_1 + C_1^T B_f^T M^T$$
$$+ \sigma^2 C_1^T B_f^T \tilde{X} B_f C_1 + C_2^T C_2,$$

$$\mathcal{N}_{12} = A^T M + C_1^T B_f^T \tilde{X} + M A_f - C_2^T C_f, \tag{9.111}$$

$$\mathcal{N}_{13} = X B,$$

$$\mathcal{N}_{22} = A_f^T \tilde{X} + \tilde{X} A_f,$$

$$\mathcal{N}_{23} = M^T B,$$

$$\mathcal{N}_{24} = -C_f^T.$$

Assume that B_f is full rank. This is not a restrictive assumption since in the case when the filtering problem state above has a solution with B_f non-full rank, then one can always find a small enough perturbation of B_f such that the perturbed matrix $\tilde{B}_f$ be full rank and verifying (9.110). Then, it exists a nonsingular transformation T such that

$$
T\tilde{B}_f = \begin{cases} \begin{bmatrix} \Sigma \\ 0_{(n_f-p_1)\times p_1} \end{bmatrix} & \text{if } n_f \geq p_1 \text{ or,} \\[2em] \begin{bmatrix} \Psi & 0_{n_f\times(p_1-n_f)} \end{bmatrix} & \text{if } n_f < p_1. \end{cases}
$$

where Σ and Ψ are nonsingular. It follows that applying to $\tilde{B}_f$ the nonsingular transformation

$$
\begin{bmatrix} \Sigma^{-1} & 0 \\ 0 & I \end{bmatrix} T \text{ if } n_f \geq p_1 \text{ or,}
$$

$$
\Psi^{-1} T \text{ if } n_f < p_1,
$$

one obtains that $\tilde{B}_f = U$ with U given by (9.106). Therefore, without loosing the generality one can choose $B_f = U$.

The condition (9.110) can be expressed as

$$
\mathcal{Z} + \mathcal{P}^T \Omega \mathcal{Q} + \mathcal{Q}^T \Omega^T \mathcal{P} < 0 \tag{9.112}
$$

where we denoted

$$
\mathcal{Z} = \begin{bmatrix} \mathcal{N}_{11} & A^T M + C_1^T B_f^T \tilde{X} & \mathcal{N}_{13} & 0 \\ M^T A + \tilde{X} B_f C_1 & 0 & \mathcal{N}_{23} & 0 \\ \mathcal{N}_{11} & \mathcal{N}_{11} & -\gamma^2 I_m & 0 \\ 0 & 0 & 0 & -I_{p_2} \end{bmatrix},
$$

$$
\mathcal{P} = \begin{bmatrix} M^T & \tilde{X} & 0 & 0 \\ -C_2 & 0 & 0 & -I_{p_2} \end{bmatrix}, \quad \mathcal{Q} = \begin{bmatrix} 0 & I_{n_f} & 0 & 0 \end{bmatrix}, \tag{9.113}
$$

$$
\Omega = \begin{bmatrix} A_f \\ C_f \end{bmatrix}.
$$

Using the Projection Lemma (Lemma 9.3.1), it follows that (9.112) has a solution Ω if and only if

$$
W_{\mathcal{P}}^T Z W_{\mathcal{P}} < 0 \tag{9.114}
$$

$$
W_{\mathcal{Q}}^* Z W_{\mathcal{Q}} < 0 \tag{9.115}
$$

where $W_{\mathcal{P}}$ and $W_{\mathcal{Q}}$ denote bases of the null subspaces of $\mathcal{P}$ and $\mathcal{Q}$, respectively. Further, perform the partition of $\mathcal{X}^{-1}$ according to the partition of $\mathcal{X}$:

$$
\mathcal{X}^{-1} = \begin{bmatrix} Y & N \\ N^T & \tilde{Y} \end{bmatrix}.
$$

With these notations, one obtains that

$$
W_{\mathcal{P}}^T = \begin{bmatrix} Y\,N\,0 & -YC_2^T \\ 0\,0\,-I_{n_f} & 0 \end{bmatrix},
$$

$$
W_{\mathcal{Q}} = \begin{bmatrix} I_n & 0 & 0 \\ 0 & 0 & 0 \\ 0 & I_m & 0 \\ 0 & 0 & I_{p_2} \end{bmatrix}.
$$

Direct algebraic computations using that $Y^{-1} - X = MN^T Y^{-1}$ show that (9.114) is equivalent with (9.102), where $P = Y^{-1}$ and (9.115) is equivalent with (9.103). The rank condition (9.105) directly follows from the relationship between $\mathcal{X}$ and $\mathcal{X}^{-1}$ and it shows that $Y^{-1} = X - M\tilde{X}^{-1}M^T$. Thus the proof ends.   □

If the necessary and sufficient conditions in Theorem 9.5.1 are fulfilled, then a solution of the filtering problem can be easily obtained by solving the basic LMI (9.112) with respect to Ω.

In the following we present a numerical example illustrating the above result. The *instrumental landing system* (ILS) is a radioelectronic equipment providing at the board of the aircrafts online information concerning the aircraft position relative to some glideslope references in the landing phase of the flight. The glideslope signal is expressed as

$$
i_{gs} = Ki_0, \tag{9.116}
$$

where the multiplicative factor K depends on the glideslope sensitivity and i_o denotes the nominal signal. The offset in the glideslope sensitivity depends on the performance category of the ILS. If σ denotes the mean square deviation of K, then $P(|K(t) - K_0| < 3\sigma) \geq 0.997$, where K_0 denotes the nominal value of the multiplicative factor. This probability increases when $\sigma \to 0$. Then, taking $\sigma = 0.06$ for which $3\sigma = 0.18$ one can obtain a maximum deviation from the glideslope sensitivity of 18 %, conformably with the international standards (Category II of ILS). Therefore the multiplication factor K in (9.116) can be replaced by

$$
K = K_0 + \sigma\xi \tag{9.117}
$$

where ξ is a white noise with unitary covariance. If the altitude dynamics is approximated by $\dot{x} = Ax + Bu$ with $i_0 = Cx$, then according to (9.116) and (9.117) the glideslope measured signal is $i_{gs} = (K_0 + \sigma\xi)Cx$. Thus one obtains a stochastic system of form (9.100) with the output subject to multiplicative white noise, for which a deterministic filter is designed. For $A = -1/30, B = 50/30, C_1 = C_2 = 1$ and $K_0 = 1$, using the result stated in Theorem 9.5.1, we obtained for the level of attenuation $\gamma = 5$, the following solution of the system of inequalities

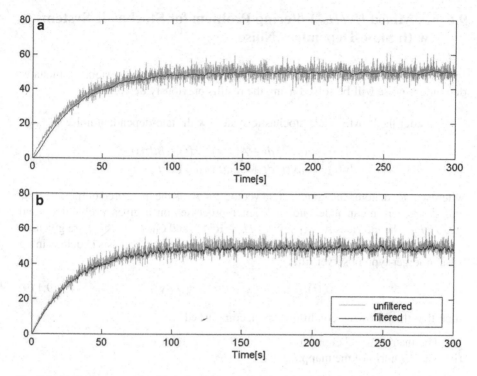

Fig. 9.2 Time responses of unfiltered and filtered signals: (**a**) H_∞ filtering (**b**) Kalman filtering

(9.102–9.105): $X = 1.9457$; $M = -0.6692$; $\tilde{X} = 0.3132$; $P = 0.5161$. Solving the LMI (9.112) it results $\Omega = \begin{bmatrix} -0.4073 \\ 0.4450 \end{bmatrix}$ and therefore the solution of the filtering problem is given by:

$$\dot{x}_f = -0.4073x_f + y_1$$

$$y_f = 0.4045x_f.$$

In Fig. 9.2a the unfiltered and the filtered signals are plotted. For comparison reasons we further determined a Kalman filter for the attitude dynamics by tuning the covariance matrices Q_0 and R_0 corresponding to the control and to the output additive white noise perturbations. For $Q_0 = 100$ and $R_0 = 0.1$ the resulting Kalman filter provides the results shown in Fig. 9.2b where the filtered and unfiltered signals are represented.

Analyzing the numerical results illustrated in the above figures, one concludes, as it is expected, that a filter designed using the specific multiplicative feature of the stochastic perturbation provides better results with respect to the ones given by Kalman filters which are suitable in the case of additive stochastic perturbations.

9.6 A Mixed H_2/H_∞ Filtering Problem for Stochastic Systems with State-Dependent Noise

In this section a mixed H_2/H_∞ filtering problem for stochastic systems with state-dependent noise will be solved using the results previously derived.

Consider the ESMS linear stochastic system with state-dependent noise

$$
\begin{aligned}
dx(t) &= A_0 x(t)\,dt + A_1 x(t)\,dv(t) + B\,d\eta(t) \\
dy(t) &= C_0 x(t)\,dt + C_1 x(t)\,d\mu(t) + d\xi(t)
\end{aligned}
\tag{9.118}
$$

where $x \in \mathbf{R}^n$ denotes the system state vector, $y \in \mathbf{R}^p$ is the measured output, v, η, μ and ξ are zero-mean independent Wiener processes on a given probability field $(\Omega, \mathcal{F}, P)$. The matrices $A_0, A_1 \in \mathbf{R}^{n \times n}$, $B \in \mathbf{R}^{n \times m}$, and $C_0, C_1 \in \mathbf{R}^{p \times n}$ are given.

The mixed H_2/H_∞ filtering problem treated in this section consists in determining a Luenberger-type observer filter

$$
d\hat{x}(t) = (A_0 - LC_0)\hat{x}(t)\,dt + L\,dy(t)
\tag{9.119}
$$

such that the following conditions are accomplished

(a) The matrix $A_0 - LC_0$ is Hurwitz;
(b) The H_2 norm of the mapping

$$
\begin{bmatrix} \eta(t) \\ \xi(t) \end{bmatrix} \mapsto e(t) := x(t) - \hat{x}(t)
$$

is minimized under the H_∞ type constraint

$$
E \int_0^\infty \left[|e(t)|^2 - \gamma^2 \left(|u(t)|^2 + |v(t)|^2 \right) \right] dt < 0
$$

for any $u \in L^{2,m}$, $v \in L^{2,p}$, where in (9.118) $d\eta(t)$ and $d\xi(t)$ are replaced by $u(t)\,dt$ and $v(t)\,dt$, respectively, and $\gamma > 0$ is a given level of attenuation.

The solution of the mixed Kalman/H_∞ estimation problem formulated in the previous section is given by the following result.

Theorem 9.6.1. *The optimal gain L of the filter (9.119) is given by*

$$
L = YC_0^T K^{-1}
\tag{9.120}
$$

where Y is the stabilizing solution of the filtering type Riccati equation

$$
\begin{aligned}
& A_0 Y + Y A_0^T + Y \left(\gamma^{-2} I - C_0^T K^{-1} C_0 \right) Y \\
& + BB^T + A_1 P_c A_1^T = 0,
\end{aligned}
\tag{9.121}
$$

$P_c \geq 0$ *denoting the unique solution of the Lyapunov equation*

$$A_0 P_c + P_c A_0^T + A_1 P_c A_1^T + BB^T = 0 \qquad (9.122)$$

and where, by notation

$$K = I + C_1 P_c C_1^T. \qquad (9.123)$$

Proof. When coupling the observer (9.119) to the system (9.118) one obtains the resulting system

$$
\begin{aligned}
dx &= A_0 x dt + A_1 x dv + B d\eta \\
d\hat{x} &= (A_0 - LC_0)\hat{x} dt + LC_0 x dt + LC_1 x d\mu + L d\xi \\
e &= x - \hat{x}.
\end{aligned}
$$

Subtracting the first two equations above one obtains the following equivalent system

$$
\begin{aligned}
\begin{bmatrix} de \\ dx \end{bmatrix} &= \begin{bmatrix} A_0 - LC_0 & 0 \\ 0 & A_0 \end{bmatrix}\begin{bmatrix} e \\ x \end{bmatrix} dt + \begin{bmatrix} 0 & A_1 \\ 0 & A_1 \end{bmatrix}\begin{bmatrix} e \\ x \end{bmatrix} dv \\
&+ \begin{bmatrix} 0 & -LC_1 \\ 0 & 0 \end{bmatrix}\begin{bmatrix} e \\ x \end{bmatrix} d\mu + \begin{bmatrix} B & -L \\ B & 0 \end{bmatrix}\begin{bmatrix} d\eta \\ d\xi \end{bmatrix} \qquad (9.124) \\
z &:= e = \begin{bmatrix} I & 0 \end{bmatrix}\begin{bmatrix} e \\ x \end{bmatrix}.
\end{aligned}
$$

Denote

$$
\begin{aligned}
\mathcal{A}_0 &= \begin{bmatrix} A_0 - LC_0 & 0 \\ 0 & A_0 \end{bmatrix}, \mathcal{A}_1 = \begin{bmatrix} 0 & A_1 \\ 0 & A_1 \end{bmatrix}, \mathcal{A}_2 = \begin{bmatrix} 0 & -LC_1 \\ 0 & 0 \end{bmatrix}, \\
\mathcal{B} &= \begin{bmatrix} B & -L \\ B & 0 \end{bmatrix}, \mathcal{C} = \begin{bmatrix} I & 0 \end{bmatrix}. \qquad (9.125)
\end{aligned}
$$

Then the minimization of the H_2 norm of the stochastic system (9.125) with the H_∞-type constraint is equivalent with the optimization problem

$$\inf_{\mathcal{Y}} Tr\left(\mathcal{C} \mathcal{Y} \mathcal{C}^T\right) \qquad (9.126)$$

with the constraints $\mathcal{Y} > 0$ and

$$\mathcal{A}_0 \mathcal{Y} + \mathcal{Y} \mathcal{A}_0^T + \mathcal{A}_1 \mathcal{Y} \mathcal{A}_1^T + \mathcal{A}_2 \mathcal{Y} \mathcal{A}_2^T + \gamma^{-2} \mathcal{Y} \mathcal{C}^T \mathcal{C} \mathcal{Y} + \mathcal{B} \mathcal{B}^T < 0. \qquad (9.127)$$

Performing the partition

$$\mathcal{Y} = \begin{bmatrix} Y & U \\ U^T & Z \end{bmatrix}$$

where $Y, Z \in \mathbf{R}^{n \times n}$ are symmetric, the blocks $(1,1)$ and $(2,2)$, of (9.127) give

$$
A_0 Y + Y A_0^T + A_1 Z A_1^T + BB^T + Y \left(\gamma^{-2} I - C_0^T K^{-1} C_0 \right) Y \\
+ \left(L - Y C_0^T K^{-1} \right) K \left(L - Y C_0^T K^{-1} \right)^T < 0
\tag{9.128}
$$

and

$$
A_0 Z + Z A_0^T + A_1 Z A_1^T + BB^T < 0,
\tag{9.129}
$$

respectively, where K is defined in the statement of the theorem. Using the fact that the stabilizing solution is increasing with respect to the free term (see Theorem 5.3.6 and [2]) it follows that the solution of (9.126) coincides with the minimal solution of (9.128). Similarly, the controllability Gramian $P_c > 0$ solving the Lyapunov equation (9.122) (see Theorem 2.7.7 (ii)) has the property that $P_c \leq Z$ for any solution Z of (9.129). Thus it follows that the minimal solution of (9.128) is obtained when its free term $A_1 Z A_1^T + \left(L - Y C_0^T K^{-1} \right) K \left(L - Y C_0^T K^{-1} \right)^T$ is minimal that is for $Z = P_c$ and for L given by (9.120), respectively. □

In order to illustrate the above developments, the following numerical example is considered

$$
A_0 = \begin{bmatrix} 0 & 1 \\ -1 & -0.4 \end{bmatrix}, A_1 = \begin{bmatrix} 0.5 & 0.3 \\ 0.3 & -0.12 \end{bmatrix}, B = \begin{bmatrix} -1 \\ 1 \end{bmatrix}, \\
C_0 = \begin{bmatrix} -0.5 & 1 \end{bmatrix}, C_1 = \begin{bmatrix} 1 & 2 \end{bmatrix}.
$$

The minimal value of γ for which the Riccati equation (9.121) has a stabilizing solution is $\gamma^* = 3.06$. For $\gamma = 3.1$ one obtains the mixed H_2/H_∞ gain

$$
L_{KH} = \begin{bmatrix} -0.7664 \\ 0.8233 \end{bmatrix}
$$

for which the eigenvalues of $A_0 - L_{KH} C_0$ are $\{-0.8032 \pm 0.9289 j\}$.

In Fig. 9.3, the following time responses are plotted: in (a), the true states of the stochastic system (9.118), in (b) the measured output y and in (c), the estimated states.

Making $\gamma \to \infty$ and applying Theorem 9.6.1 one obtains the optimal gain L_K corresponding to the Kalman-type filter for the stochastic system (9.118). For the numerical example considered in this section,

$$
L_K = \begin{bmatrix} -0.2974 \\ 0.3636 \end{bmatrix}
$$

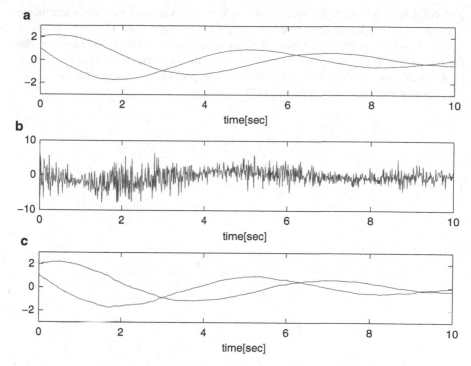

Fig. 9.3 Time responses of (**a**) true states, (**b**) measured output and (**c**) estimated states

one determines the minimal value $\gamma = 3.7$ for which the Riccati equation

$$A_0 Y + Y A_0^T + A_1 Z A_1^T + BB^T + Y \left(\gamma^{-2} I - C_0^T K^{-1} C_0 \right) Y$$
$$+ \left(L_K - Y C_0^T K^{-1} \right) K \left(L_K - Y C_0^T K^{-1} \right)^T = 0$$

has a stabilizing solution. This shows that the optimal Kalman filter provides a γ attenuation level with about 20 % greater than the smallest value for which the mixed Kalman/H_∞ filtering problem is feasible.

Notes and References

Most of the results derived in this chapter are presented for the first time in the first edition of the book. State feedback H_∞ control for linear systems with multiplicative white noise has been studied in several works. Among them we cite [12, 58, 122] and the references therein. For the time-varying case corresponding results can be found in [35]. In the Markovian systems situation, the problem has been addressed in [37, 39, 134] for the time-varying case. The design problem of a stabilizing γ-attenuating controller for systems with state-dependent white noise is given in [80]. The result derived in Sect. 9.4 is inspired from [30]. The H_∞-type filtering problem presented in Sect. 9.5 has been considered in [135] based on the formulation in [70] where deterministic filters with the same order as the generator systems are derived. The mixed H_2/H_∞ filtering problem treated at the end of this chapter has also been presented in [137].

Bibliography

1. H. Abou-Kandil, G. Freiling, G. Jank, Solution and asymptotic behavior of coupled Riccati equations in jump linear systems. IEEE Trans. Automat. Contr. **39**, 1631–1636 (1994)
2. H. Abou-Kaudil, G. Freiling, V. Ionescu, G. Jank, *Matrix Riccati Equations in Control Systems Theory* (Birhäuser, Basel, 2003)
3. M. Ait-Rami, L. El-Ghaoui, Robust stabilization of jump linear systems using linear matrix inequalities, in *Proceedings of IFAC Symposium on Robust Control Design Rio de Janeiro*, RJ (1994), pp. 148–151
4. M. Ait-Rami, X.Y. Zhou, Linear matrix inequalities, Riccati equations and indefinite stochastic linear quadratic controls. IEEE Trans. Automat. Contr. **45**(6), 1131–1143 (2000)
5. W. Arendt, Resolvent positive operators. Proc. Lond. Math. Soc. **54**(3), 321–349 (1987)
6. L. Arnold, *Stochastic Differential Equations; Theory and Applications* (Wiley, New York, 1974)
7. L. Arnold, *Random Dynamical Systems* (Springer, Berlin, 1988)
8. J. Baczynski, M. Fragoso, Maximal versus strong solution to algebraic Riccati equations arising in infinite Markov jump linear systems. Syst. Contr. Lett. **57**, 246–254 (2008)
9. R. Belmann, *Introduction to Matrix Analysis* (McGraw-Hill, New York, 1960)
10. A. Berman, M. Neumann, R.J. Stern, *Nonnegative Matrices in Dynamic Systems* (Wiley, New York, 1989)
11. J.M. Bismut, Linear quadratic optimal control with random coefficients. SIAM J. Contr. Optim. **14**, 419–444 (1976)
12. S. Boyd, L. El Ghaoui, E. Feron, V. Balkrishan, *Linear Matrix Inequalities in Systems and Control Theory* (SIAM, Philadelphia, 1994)
13. R.W. Brockett, Parametrically stochastic linear differential equations. Math. Program. Study **5**, 8–21 (1979)
14. H. Bunke, *Gewöhnliche Differential-gleichungen mit zufäligen Parametern* (Academie, Berlin, 1972)
15. H.F. Chen, On stochastic observability. Sci. Sin. **20**, 305–323 (1977)
16. H.F. Chen, On the stochastic observability and controllability, in *Proceedings of 7th IFAC World Congress*, Helsinki (1978), pp. 1115–1162
17. S. Chen, X. Li, Y. Zhou, Stochastic linear quadratic regulators with indefinite control weight costs. SIAM J. Contr. Optim. **36**(5), 1685–1702 (1998)
18. G. Ciucu, C. Tudor, Teoria Probalitatilor si Aplicatii, Edit.Stiintifica si Enciclopedica, Bucuresti, 1983
19. O.L.V. Costa, J.B.R. do Val, J.C. Geromel, Continuous-time state-feedback H_2 control of Markovian jump linear systems via convex analysis. Automatica **35**, 259–268 (1999)
20. E.F. Costa, J.B.R. do Val, An algorithm for solving a perturbed algebraic Riccati equation. Eur. J. Contr. **10**, 576–580 (2004)

21. K.L. Chung, *Markov Chains with Stationary Transition Probabilities* (Springer, Berlin, 1967)
22. R.F. Curtain, *Stability of Stochastic Dynamical Systems*. Lecture Notes in Mathematics, vol. 294 (Springer, Berlin, 1972)
23. G. Da Prato, A. Ichikawa, Stability and quadratic control for linear stochastic equation with unbounded coefficients. Boll. Un Mat. **6**, 989–1001 (1985)
24. G. Da Prato, A. Ichikawa, Quadratic control for linear periodic systems. Appl. Math. Optim. **18**, 39–66 (1988)
25. G. Da Prato, A. Ichikawa, Quadratic control for linear time-varying stochastic systems. SIAM J. Contr. Optim. **28**, 359–381 (1990)
26. G. Da Prato, J. Zabczyk, *Stochastic Equations in Infinite Dimension* (Cambridge University Press, Cambridge, 1992)
27. T. Damm, *Rational Matrix Equations in Stochastic Control*. Lecture Notes in Control and Information Sciences, vol. 297 (Springer, Berlin, 2004)
28. T. Damm, D. Hinrichsen, Newton's method for a rational matrix equation occurring in stochastic control. Linear Algebra Appl. **332/334**, 81–109 (2001)
29. T. Damm, D. Hinrichsen, Newton's method for concave operators with resolvent derivatives in ordered Banach spaces. Linear Algebra Appl. **363**, 43–64 (2003)
30. D.P. de Farias, J.C. Geromel, J.B.R. do Val, O.L.V. Costa, Output feedback control of Markov jump linear systems in continuous-time. IEEE Trans. Automat. Contr. **45**(2), 944–948 (2000)
31. Ju.L. Daleckii, M.G. Krein, *Stability of Solutions of Differential Equations in Banach Space*. Translations of Mathematical Monographs, vol. 43 (American Mathematical Society, Providence, 1974)
32. J.L. Doob, *Stochastic Processes* (Wiley, New York, 1967)
33. J.L. Doob, *Measure Theory*. Graduate Text in Mathematics (Springer, Berlin, 1994)
34. V. Dragan, A. Halanay, T. Morozan, Optimal stabilizing compensator for linear systems with state-dependent noise. Stoch. Anal. Appl. **10**(5), 557–572 (1992)
35. V. Dragan, T. Morozan, Global solutions to a game-theoretic Riccati equation of stochastic control. J. Differ. Equat. **138**(2), 328–350 (1997)
36. V. Dragan, A. Halanay, A. Stoica, A small gain theorem for linear stochastic systems. Syst. Contr. Lett. **30**, 243–251 (1997)
37. V. Dragan, A. Halanay, A. Stoica, A small gain and robustness for linear systems with jump Markov perturbations, in *Proceedings of Mathematical Theory of Networks and Systems (MTNS)*, Padova, Italy, 6–10 July 1998, ThE11, pp. 779–782
38. V. Dragan, A. Halanay, A. Stoica, The γ-attenuation problem for systems with state-dependent noise. Stoch. Anal. Appl. **17**(3), 395–404 (1999)
39. V. Dragan, T. Morozan, Game-theoretic coupled Riccati equations associated to controlled linear differential systems with jump Markov perturbations. Stoch. Anal. Appl. **19**(5), 715–751 (2001)
40. V. Dragan, T. Morozan, Exponential stability for a class of linear time varying singularly perturbed stochastic systems. Dyn. Cont. Discrete Impulsive Syst. Ser. B: Appl. Algorithms **9**, 213–231 (2002)
41. V. Dragan, T. Morozan, Stability and robust stabilization to linear stochastic systems described by differential equations with Markov jumping and multiplicative white noise. Stoch. Anal. Appl. **20**(1), 33–92 (2002)
42. V. Dragan, T. Morozan, A. Stoica, Iterative procedure for stabilizing solutions of differential Riccati type equations arising in stochastic control, in *Analysis and Optimization of Differential Systems*, ed. by V. Barbu, I. Lasiecka, C. Varsan (Kluwer, Boston, 2003), pp. 133–144
43. V. Dragan, T. Morozan, A. Stoica, Optimal H^2 state-feedback control for linear systems with Markovian jumps. Rev. Roum. Sci. Tech. Tome **48**, 523–537 (2003)
44. V. Dragan, T. Morozan, Systems of matrix rational differential equations arising in connection with linear stochastic systems with Markovian jumping. J. Differ. Equat. **194**(1), 1–38 (2003)
45. V. Dragan, T. Morozan, The linear quadratic optimization problem and tracking problem for a class of linear stochastic systems with multiplicative white noise and Markovian jumping. IEEE-T-AC **49**, 665–676 (2004)

46. V. Dragan, T. Morozan, Stochastic observability and applications. IMA J. Math. Contr. Inf. **21**, 323–344 (2004)
47. V. Dragan, T. Morozan, A. Stoica, Stochastic version of bounded real lemma and applications. Math. Rep. **6**(56), 389–430 (2004)
48. V. Dragan, T. Morozan, A. Stoica, H^2 optimal control for linear stochastic systems. Automatica **40**, 1103–1113 (2004)
49. V. Dragan, G. Freiling, A. Hochhaus, T. Morozan, A class of nonlinear differential equations on the space of symmetric matrices. Electron. J. Differ. Equat. **2004**(96), 1–48 (2004)
50. V. Dragan, T. Damm, G. Freiling, T. Morozan, Differential equations with positive evolutions and some applications. Results Math. **48**, 206–236 (2005)
51. V. Dragan, T. Morozan, A.M. Stoica, *Mathematical Methods in Robust Control of Discrete-Time Linear Stochastic Systems* (Springer, Berlin, 2010)
52. V. Dragan, T. Morozan, Criteria for exponential stability of linear differential equations with positive evolution on ordered Banach spaces. IMA J. Math. Contr. Inf. **27**(3), 267–307 (2010)
53. V. Dragan, Stabilizing composite control for a class of linear systems modeled by singularly perturbed Ito differential equations. Automatica **46**(1), 122–126 (2011)
54. V. Dragan, H. Mukaidani, P. Shi, The linear quadratic regulator problem for a class of controlled systems modeled by singularly perturbed ITO differential equations. SIAM J. Contr. Optim. **50**(1), 448–470 (2012)
55. N. Dunford, J. Schwartz, *Linear Operators: Part I* (Interscience Publishers, New York, 1958)
56. M. Ehrhardt, W. Kliemann, Controllability of linear stochastic systems. Report no. 50, Institut of Dynamische Systems, Bremen, August 1981
57. A. El Bouhtouri, A.J. Protchard, Stability radii of linear systems with respect to stochastic perturbations. Syst. Contr. Lett. **19**, 29–33 (1992)
58. L. El-Ghaoui, State-feedback control of systems with multiplicative noise via linear matrix inequalities. Syst. Contr. Lett. **24**, 223–228 (1995)
59. K. El Hadri, Robustesse de la stabilite et stabilization d'une classe de systemes a sauts Markoviens assujettis a des perturbations structurees incertaines, Ph.D. thesis, 2001
60. L. Elsner, Quasimonotonie und Ungleichungen in halbgeordneten Räumen. Linear Algebra Appl. **8**, 249–261 (1974)
61. X. Feng, K.A. Loparo, Y. Ji, H.J. Chizeck, Stochastic stability properties of jump linear systems. IEEE Trans. Automat. Contr. **37**(1), 38–53 (1992)
62. A. Fischer, D. Hinrichsen, N.K. Son, Stability radii of Metzler operators. Vietnam J. Math. **26**, 147–163 (1998)
63. M.D. Fragoso, J. Baczynski, Optimal control for continuous time LQ-problems with infinite Markov jump parameters. SIAM J. Contr. Optim. **40**(1), 270–297 (2001)
64. M.D. Fragoso, O.L.V. Costa, A unified approach for mean square stability of continuous-time linear systems with Markov jumping parameters and additive disturbances. LNCC Internal Report no. 11/1999 (1999)
65. M.D. Fragoso, O.L.V. Costa, C.E. de Souza, A new approach to linearly perturbed Riccati equations arising in stochastic control. Appl. Math. Optim. **37**, 99–126 (1998)
66. M.D. Fragoso, O.L.V. Costa, A unified approach for stochastic and mean square stability of continuous time linear systems with Markovian jumping parameters and additive disturbances. SIAM J. Contr. Optim. **44**(4), 1165–1191 (2005)
67. G. Freiling, A. Hochhaus, On a class of rational matrix differential equations arising in stochastic control. Linear Algebra Appl. **379**, 43–68 (2004)
68. A. Friedman, *Stochastic Differential Equations and Applications*, vol. I (Academic, New York, 1975)
69. Z. Gajic, I. Borno, Liapunov iterations for optimal control of jump linear systems at steady state. IEEE Trans. Automat. Contr. **40**, 1971–1975 (1995)
70. E. Gershon, D.J.N. Limebeer, U. Shaked, I. Yaesh, H_∞ filtering of continuous-time linear systems with multiplicative noise. *Proceedings of the 3rd IFAC Symposium on Robust Control Design (ROCOND)*, Prague, Czech Republic, 2000.

71. I. Ghihman, A. Skorohod, *Introduction to the Theory of Random Processes* (W.B. Saunders Co., Philadelphia, 1969) (translated from Russian)
72. I. Ghihman, A. Skorohod, *Stochastic Differential Equations* (Springer, Berlin, 1972)
73. C.H. Guo, Iterative solution of a matrix Riccati equation arising in stochastic control. Oper. Theor.: Adv. Appl. **130**, 209–221 (2001)
74. A. Halanay, *Differential Equations, Stability, Oscillations, Time Lag* (Academic, New York, 1966)
75. P. Halmos, *Measure Theory* (Van Nostrand, New York, 1951)
76. A. Haurie, A. Leizarowitz, Overtracking optimal regulation and tracking of piecewise diffusion linear systems. SIAM J. Contr. Optim. **30**(4), 816–832 (1992)
77. U.G. Hausmann, Optimal stationary control with state and control dependent noise. SIAM J. Contr. Optim. **9**, 184–198 (1971)
78. U.G. Haussmann, Stability of linear systems with control dependent noise. SIAM J. Contr. **11**(2), 382–394 (1973)
79. E. Hille, R. Phillips, *Functional Analysis and Semigroups*, vol. 31 (American Mathematical Society Colloquium Publications, Providence, 1957)
80. D. Hinrichsen, A.J. Pritchard, Stochastic H^∞. SIAM J. Contr. Optim. **36**, 1504–1538 (1998)
81. D. Hinrichsen, A.J. Pritchard, Stability radii of systems with stochastic uncertainty and their optimization by output feedback. SIAM J. Contr. Optim. **34**, 1972–1998 (1996)
82. A. Ichikawa, Optimal control of linear stochastic evolution equation with state and control dependent noise, in *Proceedings of IMA Conference on Recent Theoretical Developments in Control*, Leicester (Academic, New York, 1978), pp. 383–401
83. A. Ichikawa, Dynamic programming approach to stochastic evolution equation. SIAM J. Contr. Optim. **17**, 162–174 (1979)
84. A. Ichikawa, Filtering and control of stochastic differential equations with unbounded coefficients. Stoch. Anal. Appl. **4**, 187–212 (1986)
85. N. Ikeda, S. Watanabe, *Stochastic Differential Equations* (North Holland, Amsterdam, 1981)
86. Y. Ji, H.J. Chizeck, Controllability, stabilizability and continuous-time Markovian jump linear quadratic control. IEEE Trans. Automat. Contr. **35**(7), 777–788 (1990)
87. Y. Ji, H.J. Chizeck, Jump linear quadratic Gaussian control: steady-state solution and testable conditions. Contr. Theor. Adv. Technol. **7**(2), 247–270 (1991)
88. T. Kaczorek, *Positive 1D and 2D Systems* (Springer, Berlin, 2002)
89. R.E. Kalman, Contributions to the theory of optimal control. Bull. Soc. Math. Mex. **5**, 102–119 (1960)
90. L.V. Kantorovich, G.B. Akilov, *Functional Analysis* (Russian) (Nauka, Moskow, 1977)
91. I. Kats, N.N. Krasovskii, On stability of systems with random parameters (in Russian). Prikl. Mat. Mekh. **24**, 809–823 (1960)
92. R.Z. Khasminskii, *Stochastic Stability of Differential Equations* (Sÿthoff and Noordhoff, Alpen aan den Ryn, 1980)
93. J. Klamka, L. Socha, Some remarks about stochastic controllability. IEEE Trans. Automat. Contr. **27**, 880–881 (1977)
94. J. Klamka, L. Socha, Stochastic controllability of dynamical systems (in Polish). Podstawy Sterowonia **8**, 191–200 (1978)
95. M.A. Krasnosel'skij, *Positive Solutions of Operator Equations* (Russian) (Fizmatgiz, Moskow, 1962)
96. M.A. Krasnosel'skij, Je.A. Lifshits, A.V. Sobolev, *Positive Linear Systems—The Method of Positive Operators*. Sigma Series in Applied Mathematics, vol. 5 (Heldermann, Berlin, 1989)
97. M.G. Krein, R. Rutman, Linear operators leaving invariant a cone in a Banach space. Am. Math. Soc. Transl. Ser. 1 **10**, 199–325 (1962) (originally Uspehi Mat. Nauk (N.S.) 1 (23), 1–95 (1948))
98. H. Kushner, *Stochastic Stability and Control* (Academic, New York, 1967)
99. G.S. Ladde, V. Lakshmikantham, *Random Differential Inequalities* (Academic, New York, 1980)

100. M.V. Levit, V.A. Yakubovich, Algebraic criteria for stochastic stability of linear systems with parametric excitation of white noise type. Prikl. Mat. **1**, 142–147 (1972)

101. M. Lewin, On the boundedness measure and stability of solutions of an Itô equation perturbed by a Markov chain. Stoch. Anal. Appl. **4**(4), 431–487 (1986)

102. A. Lipser, A.N. Shiryaev, *Statistics of Random Processes* (Nauka, Moscow, 1974; Springer, Berlin, 1977)

103. K.A. Loparo, Stochastic stability of coupled linear systems: a survey of method and results. Stoch. Anal. Appl. **2**, 193–228 (1984)

104. X. Mao, Stability of stochastic differential equations with Markov switching. Stoch. Process. Appl. **79**, 45–67 (1999)

105. X. Mao, C. Yuan, *Stochastic Differential Equations with Markovian Switching* (Imperial College Press, London, 2006)

106. M. Mariton, *Jump Linear Systems in Automatic Control* (Dekker, New York, 1990)

107. J.L. Massera, J.J. Schaffer, *Linear Differential Equations and Function Spaces* (Academic, New York, 1966)

108. M. Mariton, Almost sure and moments stability of jump linear systems. Syst. Contr. Lett. **11**, 393–397 (1988)

109. T. Morozan, Optimal stationary control for dynamic systems with Markov perturbations. Stoch. Anal. Appl. **1**(3), 299–325 (1983)

110. T. Morozan, Stabilization of some control differential systems with Markov perturbations (in Romanian). Stud. Cerc. Mat. **37**(3), 282–284 (1985)

111. T. Morozan, Stochastic uniform observability and Riccati equations of stochastic control. Rev. Roum. Math. Pures Appl. **38**(9), 771–781 (1993)

112. T. Morozan, Stability and control for linear systems with jump Markov perturbations. Stoch. Anal. Appl. **13**(1), 91–110 (1995)

113. T. Morozan, Stability radii of some stochastic differential equations. Stoch. Stoch. Rep. **54**, 281–291 (1995)

114. T. Morozan, Stability radii of some time-varying linear stochastic differential systems. Stoch. Anal. Appl. **15**(3), 387–397 (1997)

115. T. Morozan, Parametrized Riccati equations for controlled linear differential systems with jump Markov perturbations. Stoch. Anal. Appl. **14**(4), 661–682 (1998)

116. T. Morozan, Parametrized Riccati equation and input–output operators for time-varying stochastic differential equations with state-dependent noise. Studii si Cercetari matematice **1**, 99–115 (1999)

117. T. Morozan, Linear quadratic and tracking problems for time-invariant stochastic differential systems perturbed by Markov chain. Rev. Roum. Math. Pures Appl. **25**(6), 783–804 (2001)

118. I.P. Natanson, *Theory of Functions of a Real Variable* (in Russian) (Gost, Moscow, 1950)

119. J. Neveu, *Bases mathematiques du Calcul des Probabilités* (Masson, Paris, 1970)

120. J. Neveu, Integrales stochastiques et applications, Notes de cours multigraphiées, 1972

121. B. Oksendal, *Stochastic Differential Equations* (Springer, Berlin, 1998)

122. I.R. Petersen, V.A. Ugrinovskii, A.V. Savkin, *Robust Control Design Using H^∞ Methods* (Springer, Berlin, 2000)

123. A.J. Pritchard, J. Zabczyk, Stability and stabilizability of infinite dimensional systems. SIAM Rev. **23**, 25–52 (1981)

124. M.A. Rami, X.Y. Zhou, Linear matrix inequalities, Riccati equations, and indefinite stochastic linear quadratic controls. IEEE Trans. Automat. Contr. **45**, 1131–1143 (2000)

125. W. Rudin, *Functional Analysis* (McGraw-Hill, New York, 1973)

126. T. Sasagawa, LP-stabilization problem for linear-stochastic control systems with multiplicative noise. J. Optim. Theor. Appl. **61**(3), 451–471 (1989)

127. T. Sasagawa, Stabilization of linear stochastic systems with delay by using a certain class of controls, in *Transactions of the 8th Prague Conference on Information Theory, Statistical Decision Functions* 28 August-1 September 1978, pp. 295–300

128. H. Schneider, M. Vidyasagar, Cross-positive matrices. SIAM J. Numer. Anal. **7**(4), 508–519 (1970)

129. L. Schwartz, *Analyse Mathématique*, vols. I, II (Hermann, Paris, 1967)
130. A.N. Shiryayev, *Probability* (Springer, New York, 1984)
131. Y. Shunahara, S. Aihara, K. Kishino, On the stochastic observability and controllability of nonlinear systems. IEEE Trans. Automat. Contr. **19**, 49–54 (1974)
132. Y. Shunahara, S. Aihara, K. Kishino, On the stochastic observability and controllability of nonlinear systems. Int. J. Contr. **22**, 65–82 (1975)
133. A. Skorohod, *Random Processes with Independent Increments* (in Russian) (Nauka, Moscow, 1964)
134. A. Stoica, I. Yaesh, Robust H_∞ control of wings deployment loops for an uninhabited air vehicle—the jump Markov model approach. J. Guid. Contr. Dyn. **25**, 407–411 (2002)
135. A. Stoica, H^∞ filtering of signals subject to multiplicative white noise, in *Proceedings to IFAC Conference*, Barcelona, 21–26 July 2002
136. A.-M. Stoica, V. Dragan, I. Yaesh, Kalman type filtering for stochastic systems with state-dependent noise and Markovian jumps, in *15th IFAC Symposium on System Identification, SYSID 2009*, St. Malo, 2009
137. A.-M. Stoica, A mixed H_2/H_∞ estimation problem for linear systems with state dependent noise, in *Proceedings of IECON2011—the 37th Annual Conference of the IEEE Industrial Electronics Society*, Melbourne, 7–10 November 2011
138. J.C. Taylor, *An Introduction to Measure and Probability* (Springer, Berlin, 1996)
139. G. Tessitore, Some remarks on the Riccati equation arising in an optimal control problem with state and control dependent noise. SIAM J. Contr. Optim. **30**(3), 714–744 (1992)
140. M.G. Todorov, M.D. Fragoso, Infinite Markov jump-bounded real lemma. Syst. Contr. Lett. **57**, 64–70 (2008)
141. C. Tudor, *Procesos estochasticas* (Sociedad Matematica Mexicana, Mexico City, 1994)
142. V.M. Ungureanu, Stochastic uniform observability of linear differential equations with multiplicative noise. J. Math. Anal. Appl. **343**(1), 446–463 (2008)
143. V.M. Ungureanu, Stochastic uniform observability of general linear differential equations with multiplicative noise. Dyn. Syst. **23**(3), 333–350 (2008)
144. V.M. Ungureanu, V. Dragan, Nonlinear differential equations of Riccati type on ordered Banach spaces. Electron. J. Qual. Theor. Differ. Equat. Proc. 9th Coll. QTDE **17**, 1–22 (2012)
145. V.M. Ungureanu, V. Dragan, Stability of discrete-time positive evolution operators on ordered Banach spaces and applications. J. Diff. Equat. Appl. doi:10.1080/10236198.2012.7043694
146. V.M. Ungureanu, V. Dragan, T. Morozan, Global solutions of a class of discrete-time backward nonlinear equations on ordered Banach spaces with applications to Riccati equations of stochastic control. Optim. Contr. Appl. Meth. (2012). doi:10.1002/oca.2015
147. J.L. Willems, Mean square stability criteria for linear white noise stochastic systems. Probl. Contr. Inf. Theor. **2**(3–4), 199–217 (1973)
148. J.L. Willems, in *Stochastic Problems in Dynamics*, ed. by B.L. Clarkson (Pitman, London, 1977)
149. J.L. Willems, J.C. Willems, Feedback stabilization for stochastic systems with state and control dependent noise. Automatica **12**, 277–283 (1976)
150. J.L. Willems, J.C. Willems, Robust stabilization of uncertain systems. SIAM J. Contr. **21**, 395–409 (1983)
151. W.M. Wonham, Random differential equations in control theory, in *Probabilistic Methods in Applied Mathematics*, vol. 2 (Academic, New York, 1970), pp. 131–212
152. G.G. Yin, Q. Zhang, *Continuous-Time Markov Chains and Applications. A Singular Perturbation Approach* (Springer, Berlin, 1997)
153. G. Yin, C. Zhu, *Hybrid Switching Diffusions: Properties and Applications* (Springer, New York, 2010)
154. J. Zabczyk, An introduction to probability theory, in *Control Theory and Topics in Functional Analysis*, vol. I (IAEA, Vienna, 1976)
155. J. Zabczyk, Controllability of stochastic linear systems. Syst. Contr. Lett. **1**(1), 25–31 (1981)

Printed in the United States
By Bookmasters